BIOPHYSICAL
PLANT PHYSIOLOGY
AND ECOLOGY

BIOPHYSICAL PLANT PHYSIOLOGY AND ECOLOGY

Park S. Nobel
University of California, Los Angeles

W. H. FREEMAN AND COMPANY
New York

Project Editor: Patricia Brewer
Manuscript Editor: James Liljenwall
Designer: Robert Ishi
Production Coordinator: William Murdock
Illustration Coordinator: Richard Quiñones
Artists: John and Judy Waller
Compositor: Syntax International
Printer and Binder: The Maple-Vail Book Manufacturing Group

Library of Congress Cataloging in Publication Data

Nobel, Park S.
  Biophysical plant physiology and ecology.

  Based on the author's Introduction to biophysical
plant physiology.
  Includes bibliographies and index.
  1. Plant physiology. 2. Botany—Ecology. 3. Plant
cells and tissues. 4. Cell physiology. I. Title.
QK711.2.N59 1983   581.1   82-20974
ISBN 0-7167-1447-7

Printed in the United States of America

2 3 4 5 6 7 8 9 0 MP 1 0 8 9 8 7 6 5 4

Contents

Preface

Plant physiology is traditionally divided into such areas as water relations, mineral nutrition, metabolism, photosynthesis, and growth and development. Many of these areas overlap with cell physiology, where the differences between plants and animals tend to break down. One goal of a physiologist is to understand such fields in formal physical and chemical terms. Accurate models can then be constructed and organismal responses to the environment predicted. Such physicochemical explanations are so far available for only certain areas of plant physiology, for example, ion and water movements across cellular membranes, the photosynthetic conversion of radiant energy from the sun into other forms of energy, and gaseous fluxes for leaves. These are some of the topics discussed in this book.

Chapters 1 through 3 present the physical description of water relations and ion transport involving plant cells. In Chapter 1, after discussing the concept of diffusion, we consider the physical barriers to diffusion caused by cellular and organelle membranes. The other physical barrier associated with plant cells is the massive cell wall, which limits the size of the cells. In the treatment of the movement of water through cells in response to specific forces in Chapter 2, we make use of the thermodynamic argument of chemical potential gradients. Chapter 3 considers the topic of solute movement into and out of plant cells, leading to an explanation of electrical potential differences across membranes and the formal criteria for distinguishing diffusion from active transport. Using concepts from irreversible thermo-dynamics, an important parameter called the reflection coefficient is derived, which permits a precise evaluation of the influence of osmotic potentials across membranes. The thermodynamic arguments used to describe ion and water movements are equally applicable to animal cells.

The next three chapters deal primarily with the interconversion of various forms of energy. In Chapter 4 we consider the properties of light and its absorption. After light is absorbed by nonliving material, its radiant energy is usually rapidly converted to heat. The arrangement of photosynthetic pigments and their special molecular structures allow some radiant energy from the sun to be converted by plants into other forms of energy before its eventual degradation to thermal energy. In Chapter 5 we discuss the particular features of chlorophyll and the accessory pigments for photo-synthesis that allow this energy conversion. Light energy absorbed by chloroplasts leads to the formation of ATP and NADPH. These compounds represent currencies for carrying chemical and electrical (redox potential) energy, respectively. How much energy they actually carry is discussed in Chapter 6.

In the last three chapters we consider the various forms in which energy and matter enter and leave a plant. Such relationships between the environment, fluxes, and plants are a part of physiological or biophysical plant ecology. The resistances (or their reciprocals, conductances) affecting the movement of both water vapor accompanying transpiration and carbon dioxide during photosynthesis are discussed in detail. The physical quantities involved in an energy budget analysis are presented so that the relative importance of the various factors affecting the temperature of leaves or other plant parts can be quantitatively evaluated. The movement of water from the soil through the plant to the atmosphere is also discussed quanti-tatively. Since these and other topics depend on material introduced else-where in the book, the text is extensively cross-referenced so that particular aspects can be selected for consideration.

This text has evolved from *Introduction to Biophysical Plant Physiology* (W. H. Freeman and Company, 1974), which in turn evolved from *Plant Cell Physiology: A Physicochemical Approach* (W. H. Freeman and Company, 1970). Some of the changes involved updating based on the ever-increasing quantity and quality of plant research. A major change for the present book was a switch to the units and conventions of the "Inter-national System of Units." Such SI (Système International) units are some-times peculiar, occasionally bizarre, but their international acceptance warrants the extra effort to master them (other units are often included parenthetically in the text, and an extensive set of conversion factors is given in an appendix). Numerous other changes have emanated from the best laboratory for any book, the classroom.

This book emphasizes physical concepts that are important for a broad biological understanding of processes, particularly in vivo. It is not intended to be a compendium of facts, but rather an attempt to integrate at an introductory level certain relevant parts of the physical sciences with plant biology. Numerous references provide further details, although in some cases the simple principles carry the reader to the forefront of current research. Calculations are used to point out the physiological or ecological consequences of the equations developed (problems at the end of chapters provide further such exercises). Although some knowledge of elementary calculus, chemistry, and physics is useful (an appendix on calculus is included), the intent is to encourage rigorous development without placing undue demands on the student's background or memory.

The author gratefully acknowledges the many colleagues who offered suggestions on improvement of the earlier books as well as the researchers and publishers who granted permission to use their figures. The text was substantially improved by the valuable comments of Peter Jordan, Gary Geller, Virtue Ishihara, Debbie Orbach, and Wayne Barcikowski. Marjorie Macdonald did an outstanding job on the typing.

December 24, 1982   *Park S. Nobel*

Ye rigid Ploughmen! Bear in mind
Your labour is for future hours.
Advance! Spare not! Nor look behind!
Plough deep and straight with all your powers!

from *The Plough*, R. H. Horne, 1848

CHAPTER 1

Cells and Diffusion

CELL STRUCTURE

Before formal consideration of diffusion and other biophysical topics, we will outline the structure of certain plant cells and tissues, thus introducing many of the anatomical terms needed throughout the rest of the book.

Generalized Plant Cell

Figure 1.1 depicts a representative leaf cell from a higher plant and illustrates the larger subcellular structures found in a plant cell. Surrounding the *protoplast* is the *cell wall*, composed mainly of cellulose and other polysaccharides; these polymers help provide rigidity to individual cells as well as to the whole plant. Since the cell wall contains numerous relatively large interstices, it does not serve as the main permeability barrier to water or solute movement into plant cells. The main barrier, a cell membrane known as the plasma membrane or *plasmalemma*, occurs inside the cell wall and surrounds the *cytoplasm*. The permeability of this membrane varies with the particular solute so the plasmalemma can regulate what enters and leaves a plant cell. The cytoplasm contains organelles like *chloroplasts* and *mitochondria*, distinct membrane-surrounded compartments in which energy can be converted from one form to another. Also in the cytoplasm are microbodies such as peroxisomes, numerous ribosomes, proteins, as well as many

1

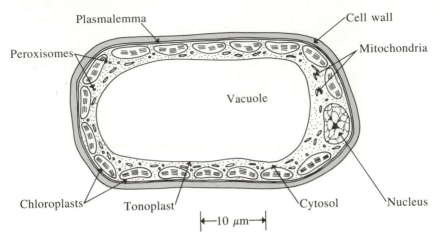

Figure 1.1
Schematic representation of a mature mesophyll cell from
the leaf of a higher plant, suggesting some of the complexity
resulting from the presence of many membrane-surrounded
subcellular bodies.

other macromolecules and structures that influence the thermodynamic
properties of water. The term *cytoplasm* includes the organelles (but often
not the nucleus), while the increasingly used term *cytosol* refers to the
cytoplasmic solution delimited by the plasmalemma and the tonoplast (to
be discussed next) but exterior to all the organelles (see Tolbert).

In mature cells of higher plants as well as a number of lower (less advanced)
plants, there is a large central aqueous phase referred to as the *vacuole*. This
vacuole is surrounded by a membrane known as the *tonoplast*. The tonoplast
is usually quite large in area, since the central vacuole can occupy up to 90%
of the volume of a mature cell. The relatively simple aqueous solution in the
central vacuole contains mainly inorganic ions or organic acids as solutes,
although considerable amounts of sugars and amino acids may be present
in some species. Water uptake into this vacuole occurs during cell growth.

One immediate impression of plant cells is the great prevalence of membranes. In addition to surrounding the cytoplasm, membranes also separate
various compartments in the cytoplasm. Diffusion of substances across these
membranes is much more difficult than diffusion within the compartments.
Thus organelle and vacuolar membranes can effectively control the contents
and consequently the reactions occurring in the particular compartments
which they surround. Diffusion can also impose limitations on the overall size
of a cell, since the time for diffusion increases with the square of the distance,
as we will quantitatively consider in the next section.

Although many plant cells share most of the features indicated in Figure 1.1, they are remarkably diverse in size. The nearly spherical cells of the green alga *Chlorella* are approximately 4×10^{-6} metre (4 μm) in diameter (see App. I for a complete list of abbreviations). On the other hand, some species of the intertidal green alga *Valonia* have multinucleated cells as large as 20 mm in diameter. The genera *Chara* and *Nitella* include fresh and brackish water green algae having large internodal cells (see Fig. 3.9) that may be 100 mm long and 1 mm in diameter. Such giant algal cells have proved extremely useful for studying ion fluxes, as we will consider in Chapter 3.

Leaf Cells

A transverse section of a leaf as diagrammed in Figure 1.2 further illustrates some of the heterogeneity of cell types encountered in plants. It also introduces certain anatomical features important for photosynthesis and transpiration. Leaves are generally 4 to 10 cells thick, which corresponds to a few hundred μm (Fig. 1.2). An *epidermis* occurs on both the upper and the

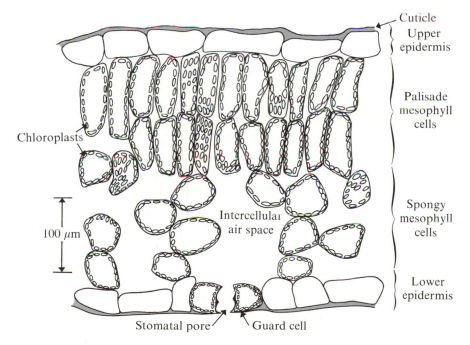

Figure 1.2
Schematic transverse section through a leaf, indicating the arrangement of various cell types. There are often about 30 to 40 mesophyll cells per stoma.

lower sides of a leaf and is usually one layer of cells thick. The cytoplasm of epidermal cells generally appears colorless since very few if any chloroplasts are present there (depending on the plant species). Epidermal cells have a relatively thick waterproof *cuticle* on the atmospheric side (Fig. 1.2). The cuticle contains cutin, a diverse group of complex polymers composed principally of esters of 16- and 18-carbon monocarboxylic acids that have two or three hydroxyl groups. Since cutin is relatively inert and also resists enzymatic degradation by microorganisms, it is often well preserved in fossil material. We will be mainly concerned with its role in preventing water loss from a leaf.

Between the two epidermal layers is the leaf *mesophyll*, which is usually differentiated into chloroplast-containing "palisade" and "spongy" cells. The palisade cells are usually elongated at right angles to the upper epidermis and are found immediately beneath it (Fig. 1.2). The spongy mesophyll cells, between the lower epidermis and the palisade mesophyll cells, are loosely packed with conspicuous intercellular air spaces. In fact, most of the surface area of both spongy and palisade mesophyll cells is exposed to the air in the intercellular spaces. A spongy mesophyll cell from a pea leaf is often rather spherical, is about 40 μm in diameter, and contains approximately 40 chloroplasts. (As Fig. 1.2 illustrates, the cells are by no means geometrically regular in shape, so that dimensions here indicate only approximate size.) A neighboring palisade cell is generally more oblong; it can be 80 μm long, can contain 60 chloroplasts, and might be represented by a cylinder 30 μm in diameter with hemispherical ends. In many leaves about 70% of the chloroplasts are in the palisade cells, which generally outnumber the spongy mesophyll cells nearly 2 to 1.

The pathway of least resistance for gases to cross an epidermis—and thus to enter or to exit from a leaf—is through the adjustable space between a pair of guard cells (Fig. 1.2). This pore is called a *stoma* or *stomate* (plural: stomata and stomates, respectively), and it can open and close. The stomatal pores allow for the entry into the leaf of CO_2 and for the release of the photosynthetically produced O_2. The inevitable loss of water vapor by transpiration also occurs mainly through the stomatal pores, as we will discuss in Chapter 7. The stomata thus serve as a control, which helps strike a balance between freely admitting the CO_2 needed for photosynthesis and at the same time preventing excessive loss from the plant of the water vapor which evaporates from the walls of cells within the leaf and then diffuses out into the surrounding atmosphere. Air pollutants such as ozone (O_3) and sulfur dioxide (SO_2) also enter plants primarily through the open stomata.

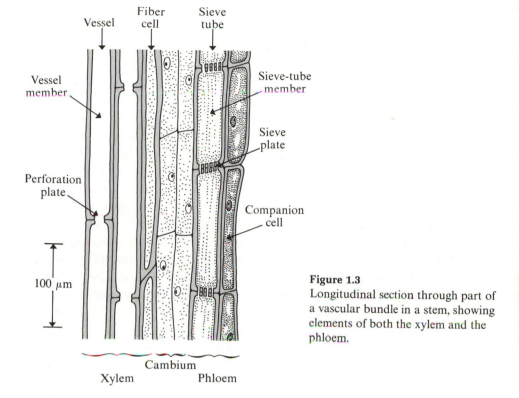

Figure 1.3
Longitudinal section through part of a vascular bundle in a stem, showing elements of both the xylem and the phloem.

Cells of Vascular Tissue

The *xylem* and the *phloem* make up the vascular systems found in the roots, stems, and leaves of plants (Fig. 1.3). In a tree trunk the phloem constitutes a layer of the bark while the xylem constitutes almost all of the wood. The xylem commonly provides structural support for plants. Conduction in the xylem of a tree often occurs only in the outermost annual ring*, which lies just inside the vascular *cambium* (region of meristematic activity from which xylem and phloem cells differentiate). Outside the functioning phloem are other phloem cells that can be shed as pieces of bark slough off. Phloem external to the xylem, as in a tree, is the general pattern for the stems of plants. As we follow the vascular tissue from the stem along a petiole and

---

* The rings in trees are not always annual. In many desert species a ring forms when large and, later small xylem cells are produced following a suitable rainy period, and this can occur more than once or sometimes not at all in a given year.

into a leaf, we observe that the xylem and phloem often form a vein, which sometimes conspicuously protrudes from the lower surface of the leaf. Reflecting the orientation in the stem or the trunk, the phloem is found abaxial to the xylem in the vascular tissue of a leaf; i.e., the phloem is located on the side of the lower epidermis. The vascular system branches and rebranches as it crosses a dicotyledonous leaf, becoming smaller (in cross section) at each step. In contrast to the net or reticulate venation in dicots, monocotyledons characteristically have parallel-veined leaves. Individual mesophyll cells in the leaf are never further than a few cells from the vascular tissue.

The upward movement of water and nutrients from the soil to the upper portions of a plant occurs primarily in the xylem. The xylem sap generally contains about 10 mol m^{-3} (10 mM)* inorganic nutrients plus organic forms of nitrogen that are metabolically produced in the root. The xylem is a tissue of various cell types which we will consider in more detail in the final chapter (p. 493) when water movement in plants is discussed in terms of water potential. The conducting cells in the xylem are the narrow, elongated *tracheids* and the *vessel members*, which tend to be shorter and wider than the tracheids. Vessel members are joined end-to-end in long linear files, their adjoining end walls or *perforation plates* having from one large to many small holes. The conducting cells lose their protoplasts, and the remaining cell walls thus form a low resistance channel for the passage of solutions. Water moves from the root, up the stem, through the petiole, and then to the leaves in these hollow xylem "cells," with motion occurring in the direction of decreasing hydrostatic pressure (see pp. 495–498). Ions and other solutes accompany such movement of water to the leaf. Some solutes leave the xylem along the stem on the way to a leaf, while others diffuse or are actively transported across the plasmalemmas of various leaf cells adjacent to the conducting cells of the xylem.

The movement of most organic compounds throughout the plant takes place in the other vascular tissue, the phloem. A portion of the photosynthetic products made in the mesophyll cells of the leaf diffuses or is actively transported across cellular membranes until it reaches the conducting cells of the leaf phloem. By means of the phloem, the photosynthetic products— which are then mainly in the form of sucrose—are distributed throughout the plant. The carbohydrates produced by photosynthesis and certain other substances generally move in the phloem toward regions of lower con-

---

* Molarity (moles of solute per litre of solution, symbolized by M) is a useful unit for concentration, but it is not recommended by the international unit convention, Le Système International d'Unités (SI). Nevertheless, we will use molarity in addition to the SI unit of mol m^{-3}. We also note that SI recommends the spellings "litre" and "metre." See Incoll et al. or National Bureau of Standards.

centration, although diffusion is not the mechanism for the movement, as we will indicate later (pp. 501, 505). The phloem is a tissue consisting of several types of cells. In contrast with the xylem, however, the conducting cells of the phloem contain cytoplasm. They are called *sieve cells* and *sieve-tube members* and are joined end-to-end, thus forming a transport system through-out the plant. Although these phloem cells often contain no nuclei at maturity, they remain metabolically active.

Root Cells

Roots anchor plants in the ground as well as absorb water and nutrients from the soil and then conduct these substances upward to the stem. To help understand uptake into a plant, we will examine the functional zones that occur along the length of a root.

At the extreme tip of a root is the root cap (Fig. 1.4a). It consists of relatively undifferentiated cells (often starch-containing), which are scraped off as the root grows into new regions of the soil. Cell walls in the root cap are often

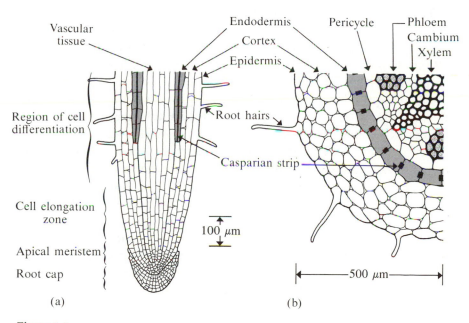

Figure 1.4
Schematic diagrams of a root: (a) longitudinal section, indicating the zones that can occur near the root tip, and (b) cross-sectional view approximately 10 mm back from the tip, indicating the arrangement of the various cell types.

mucilaginous, which might serve to reduce friction with soil particles. Behind the root cap is a meristematic region where the cells rapidly divide. Cells in this *apical meristem* tend to be isodiametric and have thin cell walls. Next to this is a region of *cell elongation* in the direction of the root axis. Such elongation mechanically pushes the root tip through the soil, causing cells of the root cap to slough off by abrasion with soil particles. Sometimes the region of dividing cells is not spatially distinct from the elongation zone. Also, cell size and the extent of the zones vary with both plant species and physiological status (see Esau for details).

The next region indicated in Figure 1.4a is that of *cell differentiation*, where the cells begin to assume more highly specialized functions. The cell walls become thicker and elongation is greatly diminished. The epidermal cells develop fine projections, radially outward from the root, called *root hairs*. These root hairs greatly increase the surface area across which water and nutrients can enter a plant. In fact, most of the water usually enters the root system in the region where the root hairs are located (generally 2 to 50 mm from the tip). As we follow the root upward, its surface becomes less permeable to water and its interior more involved with conducting water toward the stem. Since water movement into the root is discussed in Chapter 9, the discussion here will be restricted to some morphological features.

The region of the root where water absorption occurs has little or no waxy cuticle, as is consistent with its function of taking up water from the soil. Figure 1.4b shows a cross section of a root at the level where root hairs are found. Starting from the outside, we observe first the root epidermis and then a number of layers of cells known as the *cortex*. There are abundant intercellular air spaces in the cortex, facilitating the diffusion of O_2 and CO_2 within this tissue (such air spaces are generally lacking in vascular tissue). Inside the cortex is a single layer of cells known as the *endodermis*. The radial and transverse walls of the endodermal cells are impregnated with waxy material, forming a band around the cells known as the *casparian strip*, which prevents passage of water and solutes across that part of the cell wall. Since there are no air spaces between endodermal cells, and the radial walls are blocked by the waterproof casparian strip, water must pass through the lateral walls and enter the cytoplasm of endodermal cells to continue across the root. The endodermal cells can represent the only place in the entire pathway for water movement from the soil, through the plant, to the air where it is mandatory that the water enter a cell's cytoplasm.* In the rest of the pathway, water

---

* In the roots of many species a subepidermal layer or layers of *hypodermis* occurs. Radial walls of hypodermal cells can also be blocked with a waxy material analogous to the casparian strip in the endodermis.

can move in cell walls or in the hollow lumens of xylem vessels, a region referred to as the *apoplast*.

Immediately inside the endodermis is the *pericycle*, which is typically one cell thick in angiosperms. The cells of the pericycle can divide and form a meristematic region, which can produce lateral or branch roots in the region just above the root hairs. Radially inside the pericycle is the vascular tissue. The phloem generally occurs in two to eight or more strands located around the root axis. The xylem often occurs along the root axis with sections radiating out between the phloem strands. Hence water does not have to cross the phloem to reach the root xylem. As in stems, the tissue between the xylem and the phloem is the cambium, which through cell division and differentiation produces xylem to the inside and phloem to the outside.

The endodermis surrounds the pericycle and vascular tissue for only a short length of the root. It does not begin until after the zone of cell elongation (Fig. 1.4a) and until a few phloem cells have differentiated. It ceases to function just above the region of root hairs, where the root begins to widen appreciably as secondary xylem and secondary phloem are formed by the radial divisions of cambial layer cells. The main region for water entry into most roots is the zone of primary vascular tissue where there is an intact sleeve of endodermal cells. Toward the stem from this region the vascular tissue becomes larger in cross section, the xylem is more devoted to conduction of water upward, and the phloem supplies all the branch roots that have converged.

Our rather elementary discussion of leaves, vascular tissues, and roots leads to the following oversimplified but useful picture. The roots take up water from the soil along with nutrients required for growth. These are then conducted in the xylem to the leaves. Leaves of a photosynthesizing plant lose the water to the atmosphere along with a release of O_2 and an uptake of CO_2. Carbon from the latter ends up in photosynthate translocated in the phloem back to the root. Thus the xylem and the phloem serve as the "plumbing" which connects the two types of plant organs that functionally interact with the environment. To understand the details of such physiological processes we must turn to fields like calculus, thermodynamics, and photochemistry for the analytical methods. Our next step is to bring the abstract ideas of these fields into the realm of cells and plants, which means we need to make calculations using appropriate assumptions and approximations.

We will begin by describing diffusion (Ch. 1). To discuss water (Ch. 2) and solutes (Ch. 3), we will introduce the thermodynamic concept of chemical potential. This leads to a quantitative description of fluxes, electrical potentials across membranes, and the energy requirements for active transport of solutes. Some important energy conversion processes take place in the organelles. For instance, light energy is absorbed (Ch. 4) by photosynthetic pigments located in the internal membranes of chloroplasts (Ch. 5) and then converted into other forms of energy useful to a plant (Ch. 6) or dissipated as heat (Ch. 7). Leaves (Ch. 8) as well as groups of plants (Ch. 9) also interact with the environment through exchanges of water vapor and CO_2. In our problem-solving approach to these topics we will pay particular attention to dimensions and ranges for the parameters as well as to the insights which can be gained only by developing the relevant formulae and then making calculations.

DIFFUSION

Diffusion is a spontaneous process leading to the net movement of a substance from some region to adjacent ones where that substance has a lower concentration. It takes place in both the liquid and the gas phases associated with plants. Diffusion results from the random thermal motions of the molecules either of the solute(s) and the solvent in the case of a solution or of gases in the case of air. The net movement caused by diffusion is a statistical phenomenon: there is a greater probability of molecules moving from the concentrated to the dilute region than vice versa. In other words, more molecules per unit volume are present in the concentrated region than in the dilute one, and thus more are available for diffusing toward the dilute region than are available for movement in the opposite direction. If isolated from external influences, diffusion of a neutral species will tend to even out any macroscopic concentration differences originally present in adjoining regions of a liquid or a gas. In fact, the randomizing tendency of such molecular Brownian movement is a good example of the increase in entropy, or decrease in order, that accompanies all spontaneous processes.

Diffusion is involved in many processes for plants, such as gas exchange and the movement of nutrients toward root surfaces. For instance, diffusion is the mechanism for most, if not all, steps by which CO_2 from the air reaches the sites of photosynthesis in chloroplasts. CO_2 diffuses from the atmosphere up to the leaf surface and then diffuses through the stomatal pores. After entering the leaf, diffusion of CO_2 occurs within intercellular air spaces, such as those indicated in Figure 1.2. Next, CO_2 diffuses across the wet cell wall, crosses the plasmalemma of a leaf mesophyll cell, and then diffuses through the cytosol to reach the chloroplasts. Finally, CO_2 enters a chloroplast and diffuses up to the enzymes that are involved in carbohydrate formation. If the enzymes were to fix all the CO_2 in their vicinity, and no other CO_2 were to *diffuse* in from the atmosphere surrounding the plant, photosynthetic processes would stop.* In this chapter we will develop the mathematical formulation necessary for understanding both diffusion across a membrane and diffusion in free solution.

Fick's First Law

Fick in 1855 was one of the first to examine diffusion quantitatively. For such an analysis we need to consider the concentration (c_j) of some species j in a solution. (The subscript j calls attention to the fact that we are considering only one species out of the many that could be present.) We will assume that the concentration of species j in some region is less than in a neighboring one. A net migration of molecules then occurs by diffusion from the concentrated to the dilute region. Such a molecular flow down a concentration gradient is analogous to the flow of heat down a temperature gradient from a warmer region to a cooler one. The analogy is actually rather good (especially for gases), since both processes depend on the random thermal motion of the molecules. In fact, the differential equations and their solutions which are used to describe diffusion are those which had previously been developed to describe heat flow. (See Crank for a discussion of the mathematics of such general transport relations.)

To express diffusion quantitatively, we will consider a diffusive flux or flow of species j. We will restrict our attention to diffusion involving planar fronts of uniform concentration. (Fortunately, as we shall see, this has widespread application to situations of interest in biology.) We will let J_j be the amount of some species j crossing a certain area per unit time, e.g., moles of particles per m^2 in a second, which is termed the *flux density*.[†] Reasoning by analogy with heat flow, Fick deduced that the "force," or causative agent, leading to the net molecular movement depends on the concentration gradient. A gradient is simply a measure of how a certain parameter changes with distance; e.g., the gradient in concentration of species j in the x-direction is $\partial c_j / \partial x$. (See App. V for the definition and properties of a partial derivative.) In general, *the flux density of some substance is proportional to the appropriate force*, a relation which we will use repeatedly in this text. In the present instance the driving force is the negative of the concentration gradient of

---

* Note that in solution "CO_2" can occur in the form of bicarbonate, and also that the crossing of membranes does not have to be by diffusion, refinements that we will return to on pp. 423–427.

† Although the SI convention recommends the term flux density, most of the diffusion literature refers to J_j as a flux.

species j, which we will represent by $-\partial c_j/\partial x$ for diffusion in one dimension. (To appreciate why a minus sign occurs, we should recall that the direction of net diffusion is toward regions of lower concentration.) We can now write down the following relation showing the dependence of the flux density on the driving force:

$$J_j = -D_j \frac{\partial c_j}{\partial x} \tag{1.1}$$

Equation 1.1 is commonly known as Fick's first law of diffusion, where D_j is the *diffusion coefficient* of species j. The partial derivative is used in Equation 1.1 to indicate the change in concentration in the x-direction of Cartesian coordinates at some moment in time (constant t) and for specified values of y and z. (For the cases that we will consider, the flux density in the x-direction actually has the same magnitude at any value of y and z.) For J_j in mol m^{-2} s^{-1} and c_j in mol m^{-3} (hence, $\partial c_j/\partial x$ in mol m^{-4}), D_j has units of m^2 s^{-1}. Since D_j varies with concentration and temperature, it is properly called a *coefficient* in the general case. In certain applications, however, we can obtain sufficient accuracy by treating D_j as if it were a constant. By convention, a net flow in the direction of increasing x is positive (from left to right in Fig. 1.5). Since a net flow occurs toward regions of lower concentration, the minus sign is indeed needed in Equation 1.1. Fick's first law—which has been amply demonstrated experimentally—is the starting point for our discussion of diffusion.

Continuity Equation and Fick's Second Law

As we noted earlier, diffusion in a solution is an important process in the movement of solutes across plant cells and tissues. How rapid are such processes? For example, if we release a certain amount of material in one location, how long will it take before we can detect that substance at various distances? To discuss such phenomena adequately, we must determine the dependence of the concentration on both time and distance. We can readily derive such a time-distance relationship if we first consider the concept of conservation of mass, which is necessary if we are to transform Equation 1.1 into an expression convenient for describing the actual solute distributions caused by diffusion. In particular, we want to eliminate J_j from Equation 1.1 so that we can see how c_j depends on x and t.

The amount of species j per unit time crossing a given area, here considered

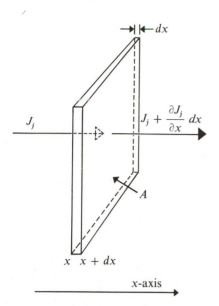

Figure 1.5
Diagram showing the dimensions and flux densities that form the geometrical basis for the continuity equation.

to be a planar area perpendicular to the x-axis (Fig. 1.5), can change with position along the x-axis. Let us imagine a volume element of thickness dx in the direction of flow and of cross-sectional area A (Fig. 1.5). At x, we will let the flux density across the surface of area A be J_j. At $x + dx$, the flux density has changed to $J_j + (\partial J_j/\partial x)\, dx$, where $\partial J_j/\partial x$ is the gradient of the flux density of j in the x-direction; i.e., the rate of change of J_j with position, $\partial J_j/\partial x$, times the distance, dx, gives the overall change in the flux density, $(\partial J_j/\partial x)\, dx$. The change in the amount of species j in the volume $A\, dx$ in unit time for this one-dimensional problem is simply the amount flowing into the volume element per unit time, $J_j A$, minus that flowing out, $[J_j + (\partial J_j/\partial x)\, dx]A$. The change in the amount of species j in the volume element in unit time can also be expressed as the change in the concentration of species j with time, $\partial c_j/\partial t$, multiplied by the volume in which the change in concentration occurs, $A\, dx$. Equating these two different expressions that describe the change in the amount of species j in the volume $A\, dx$, we obtain the following relation:

$$J_j A - \left(J_j + \frac{\partial J_j}{\partial x}\, dx \right) A = \frac{\partial c_j}{\partial t}\, A\, dx \qquad (1.2)$$

The two $J_j A$ terms on the left side of Equation 1.2 cancel each other. After division through by $A\, dx$, Equation 1.2 leads to the very useful expression known as the *continuity equation*:

$$-\frac{\partial J_j}{\partial x} = \frac{\partial c_j}{\partial t} \tag{1.3}$$

The continuity equation is a mathematical way of stating that matter cannot be created or destroyed under ordinary conditions. Thus, if the flux density of some species decreases as we move in the x-direction ($\partial J_j/\partial x < 0$), Equation 1.3 indicates that its concentration must consequently be increasing with time, as the material is then accumulating locally. When we substitute Fick's first law (Eq. 1.1) into the continuity equation (Eq. 1.3), we obtain *Fick's second law*. For the important special case of constant D_j, this general equation for diffusion becomes

$$\frac{\partial c_j}{\partial t} = -\frac{\partial}{\partial x}\left(-D_j\frac{\partial c_j}{\partial x}\right) = D_j\frac{\partial^2 c_j}{\partial x^2} \tag{1.4}$$

Solution of Equation 1.4 describes how the concentration of some solute changes with position and time as a result of diffusion. To determine the particular function which satisfies this important differential equation, we need to know the specific conditions for the situation under consideration. Nevertheless, a representative solution useful for the consideration of diffusion under simple initial and boundary conditions will be sufficient for the present purpose of describing the characteristics of solute diffusion in general terms. For example, we will assume that there are no obstructions in the x-direction and that species j is initially placed in a plane at the origin ($x = 0$). In this case, the following expression for the concentration of species j satisfies the differential form of Fick's second law when D_j is constant, and also satisfies our rather simple conditions:*

$$c_j = \frac{M_j}{2(\pi D_j t)^{1/2}} e^{-x^2/4D_j t} \tag{1.5}$$

---

* To show that Equation 1.5 is indeed a possible solution of Fick's second law, it can be substituted into Equation 1.4 and the differentiations performed (M_j and D_j are constant; $\partial ax^n/\partial x = anx^{n-1}$, $\partial e^{ax^n}/\partial x = anx^{n-1}e^{ax^n}$, and $\partial uv/\partial x = u\,\partial v/\partial x + v\,\partial u/\partial x$, as indicated in App. V). The solution of Equation 1.4 becomes progressively more difficult when more complex initial and boundary conditions or molecular interactions (which cause variations in D_j) are considered. (See Crank or Jacobs for analytical solutions in such cases.)

In Equation 1.5, M_j is the total amount of solute j per unit area initially ($t = 0$) placed in a plane located at the origin of the x-direction (i.e., at $x = 0$, while y and z can have any value, which defines the plane considered here) and c_j is its concentration at position x at any later time t. For M_j to have this useful meaning, the factor $1/[2(\pi D_j)^{1/2}]$ is necessary in Equation 1.5.* Moreover, the solute in this case is allowed to diffuse for an unlimited distance in either the plus or minus x-direction and no additional solute is added at times $t > 0$. Often this idealized situation can be achieved by inserting a radioactive tracer in a plane at the origin of the x-direction. Equation 1.5 is only one of the possible solutions to the second order partial differential equation representing Fick's second law. The form is relatively simple compared with other solutions, and, more importantly, the initial condition of having a finite amount of material released at a particular location is actually realistic for certain applications to biological problems.

Time-Distance Relation for Diffusion

Although the functional form of c_j given by Equation 1.5 is only one particular solution to Fick's second law (Eq. 1.4) and is restricted to the case of constant D_j, it nevertheless proves to be an extremely useful expression for understanding diffusion. It can be used to relate the distance a substance diffuses to the time necessary to reach that distance. The expression involves the diffusion coefficient of species j, D_j, which can be determined experimentally. In fact, Equation 1.5 itself is often employed to determine a particular D_j.

In the plane at the origin of the x-direction ($x = 0$), Equation 1.5 indicates that the concentration is $M_j/[2(\pi D_j t)^{1/2}]$, which becomes infinitely large as t is turned back to 0, the starting time. This infinite value for c_j at $x = 0$ corresponds to having all the solute initially placed in a plane at the origin. For t greater than 0, the material begins to diffuse away from the origin. The distribution of molecules along the x-axis at two successive times is indicated in Figures 1.6a and 1.6b, while 1.6c explicitly shows the movement of the concentration profiles along the time axis. Since the total amount of species j does not change (it remains at M_j per unit area of the y-z plane, i.e., in a volume element parallel to the x-axis and extending from x values of $-\infty$ to $+\infty$), the area under each of the concentration profiles is the same.

---

* Note that $\int_{-\infty}^{\infty} c_j(x, t)\, dx = M_j$, where $c_j(x, t)$ is the concentration function which depends on position and time as given by Equation 1.5 and the probability integral, $\int_{-\infty}^{\infty} e^{-a^2 u^2}\, du$, equals $\sqrt{\pi}/a$.

Comparing Figure 1.6a with 1.6b, we see that the average distance of the diffusing molecules from the origin increases with time. Also, Figure 1.6c shows how the concentration profiles flatten out as time increases, since the diffusing species is then distributed over a greater region of space. In estimating how far molecules diffuse in time t, a useful parameter is the distance at which the concentration has dropped to $1/e$ or 37% of its value in the plane at the origin. Although somewhat arbitrary, this parameter describes the shift of the statistical distribution of the molecules with time. From Equation 1.5, the concentration at the origin is $M_j/[2(\pi D_j t)^{1/2}]$. (See also Fig. 1.6.) The concentration drops to 37% of the value at the origin when the exponent of e in Equation 1.5 is -1. From Equation 1.5, this distance, x_e, is therefore given by

$$x_e^2 = 4D_j t_e \tag{1.6}$$

The distance x_e along the x-axis is also indicated in Figures 1.6a and 1.6b.

Equation 1.6 is an extremely important relationship that indicates a fundamental characteristic of diffusion processes: the distance a population of molecules of a particular species diffuses—for the one-dimensional case where the molecules are released in a plane at the origin—is proportional to the square root of both the diffusion coefficient of the species and the time for diffusion. In other words, the time for diffusion increases with the square of the distance to be traversed. An individual molecule may diffuse a greater or lesser distance in time t_e than is indicated by Equation 1.6 (see Fig. 1.6), since the latter refers to the time required for the concentration of species j at position x_e to become $1/e$ of the value at the origin; i.e., we are dealing with the characteristics of a whole population of molecules, not the details of an individual molecule. Furthermore, the factor 4 is rather arbitrary, since some criterion other than $1/e$ would cause this numerical factor to be somewhat different, although the basic form of Equation 1.6 would be preserved. For example, the numerical factor is 2.8 if the criterion is to drop to half of the value at the origin.

Table 1.1 lists the magnitudes of diffusion coefficients for various solutes in water at 25°C.* For ions and other small molecules, D_j's in aqueous solutions are approximately 10^{-9} m^2 s^{-1}. Since proteins have higher

---

* The symbol °C in this text represents degrees on the Celsius temperature scale. By the SI system the Celsius degree as well as the kelvin degree or unit (abbreviated K) is $1/273.16$ of the thermodynamic temperature of the triple point of water (0.01000°C) and absolute zero is at -273.15°C. Although it is still in use, the term "centigrade" is no longer recommended.

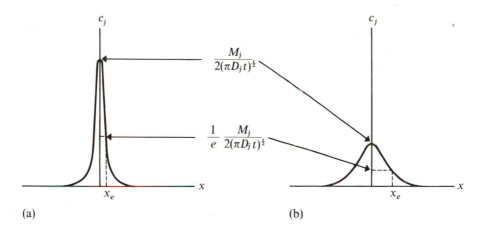

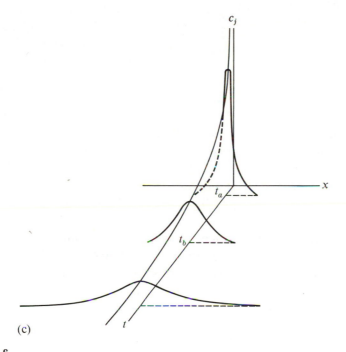

Figure 1.6

Concentration of species j, c_j, as a function of position x for molecules diffusing according to Fick's second law. The molecules were initially placed in a plane at the origin of the x-direction, i.e., at $x = 0$. For a given value of x, c_j is the same throughout a plane in the y- and z-directions. (a) Distribution of concentrations along the x-axis occurring at a time t_a, (b) distribution occurring at a subsequent time t_b, and (c) portrayal of the concentration profiles as a function of time. Note that x_e is the location where the concentration of species j has dropped to $1/e$ of its value at the origin.

Table 1.1
Diffusion coefficients in aqueous solutions and air. Values are for dilute solutions at 25°C or air under standard atmospheric pressure at 20°C. See Sober or Weast.

Small solutes in water		Globular proteins in water	
Substance	D_j ($m^2 s^{-1}$)	Molar mass ($kg\ mol^{-1}$)	D_j ($m^2 s^{-1}$)
Alanine	0.92×10^{-9}	15	10^{-10}
Citrate	0.66×10^{-9}	1 000	10^{-11}
Glucose	0.67×10^{-9}	Gases in air	
Glycine	1.1×10^{-9}		
Sucrose	0.52×10^{-9}		
Ca^{2+} (with Cl^-)	1.2×10^{-9}	Gas	D_j ($m^2 s^{-1}$)
K^+ (with Cl^-)	1.9×10^{-9}		
Na^+ (with Cl^-)	1.5×10^{-9}	CO_2	1.5×10^{-5}
CO_2	1.7×10^{-9}	H_2O	2.4×10^{-5}
		O_2	1.9×10^{-5}

relative molecular masses (i.e., higher molecular weights[*]) than the small solutes, their diffusion coefficients are lower (Table 1.1). Also, because of the greater frictional interaction between water molecules and fibrous proteins than with the more compact globular ones, fibrous proteins often have approximately twofold lower diffusion coefficients than do globular proteins of the same molecular weight. (See Sober or Tanford for data on various D_j's of proteins.)

To illustrate the time-distance consequences of Equation 1.6, we will quantitatively consider the diffusion of small molecules in an aqueous solution. How long, on the average, would it take a small solute with a D_j of $10^{-9}\ m^2\ s^{-1}$ to diffuse 50 μm, the distance across a typical leaf cell? From Equation 1.6, the time required for 37% of the population of molecules to shift this distance is

$$t_e = \frac{(50 \times 10^{-6}\ m)^2}{(4)(10^{-9}\ m^2\ s^{-1})} = 0.6\ s$$

---

[*] Molecular weight is a dimensionless number indicating the molecular mass of a substance relative to that of a carbon atom taken as 12. Molar mass is the mass per mole; e.g., sucrose has a molar mass of 0.342 kg mol^{-1} and a molecular weight (relative molecular mass) of 342.

Thus, diffusion is a fairly rapid process over subcellular distances. Next, let us consider the diffusion of the same substance over a distance of 1 m. The time needed is

$$t_e = \frac{(1 \text{ m})^2}{(4)(10^{-9} \text{ m}^2 \text{ s}^{-1})} = 2.5 \times 10^8 \text{ s} \cong 8 \text{ years}$$

Diffusion is indeed not a rapid process over long distances. Thus, inorganic nutrients in xylary sap would not ascend a tree by diffusion at a rate necessary to sustain life. On the other hand, diffusion is often sufficient for the movement of solutes within leaf cells and especially inside organelles such as chloroplasts and mitochondria. In summary, diffusion in a solution is fairly fast over short distances (less than about 100 μm) but extremely slow for very long distances.

In the cytoplasm of living cells there is a mechanical mixing due to *cytoplasmic streaming*, which can lead to even more rapid movement than by diffusion. This cytoplasmic streaming, whose cessation is often a good indicator that cellular damage has occurred, requires energy. Energy is usually supplied in the form of adenosine triphosphate (ATP), and movement may involve a similar mechanism to that in muscle (see Kamiya).

Diffusion of gases in the air surrounding and within leaves is necessary for both photosynthesis and transpiration. For instance, water vapor evaporating from the cell walls of leaf cells diffuses across the intercellular air spaces (Fig. 1.2) to reach the stomata and from there diffuses across an air boundary layer into the atmosphere (considered in detail in Ch. 8). CO_2 diffuses from the atmosphere through the open stomata to the surfaces of leaf cells while the photosynthetically evolved O_2 traverses the same pathway in the reverse direction, also primarily by diffusion. The experimentally determined diffusion coefficients of these three gases in air at sea level (standard atmospheric pressure) and 20°C are about 10^{-5} m^2 s^{-1} (see Table 1.1). Such diffusion coefficients in air are approximately 10^4 times greater than the D_j describing the diffusion of a small solute in a liquid, indicating that diffusion coefficients depend markedly on the medium. In particular, many more intermolecular collisions occur per unit time in a liquid phase than in a less dense gas phase. Thus, a molecule can move further in air than in an aqueous solution before being influenced by other molecules. Most cells in animals are bathed by fluids, so for larger animals a circulatory system is necessary to transport O_2 to their cells and CO_2 away. Plants, on the other hand, often have conspicuous intercellular air spaces where the large values of D_{O_2} and D_{CO_2} in a gas phase facilitate diffusion.

Our above examples have shown that larger molecules tend to have lower diffusion coefficients in both liquids and gases, but the relation with molecular weight can be fairly complicated (see Bull, Stein, or Paganelli et al.). Diffusion coefficients depend inversely on the viscosity of the medium (discussed on p. 120, where the temperature dependence of D_j is also considered). Since diffusion coefficients of gases in air are inversely proportional to ambient pressure (Equation 8.9), they become larger at higher altitudes.

Because of a leaf's anatomy, the pathways for the movement of gas molecules in the intercellular air spaces can be tortuous. Nevertheless, calculations using Equation 1.6 can give useful estimates of the diffusion times for the many processes involving gaseous movements in a leaf. An upper limit for the diffusion pathway in such intercellular air spaces of a leaf might be 1 000 μm. Using Equation 1.6 and the diffusion coefficients given in Table 1.1, we can calculate that the times needed for water vapor, CO_2, and O_2 to diffuse 1 000 μm in air are from 10 to 16 ms. Thus diffusion of molecules in a gas is a fairly rapid process. However, when illumination and temperature are not limiting, the rate of photosynthesis in plants is generally limited by the amount of CO_2 diffusing to the chloroplasts, while the rate of transpiration is determined by diffusion of water vapor from the cell walls within the leaf to the outside air. In the latter case, the limitation posed by diffusion helps prevent excessive water loss from the plant and therefore is quite useful, physiologically speaking.

MEMBRANE STRUCTURE

The plasmalemma presents a major barrier to the diffusion of solutes into and out of plant cells, while the organelle membranes play an analogous role for the various subcellular compartments, and the tonoplast for the vacuole. For instance, although H_2O and CO_2 readily penetrate the plasmalemma, ATP and metabolic intermediates usually do not diffuse across it at an appreciable rate. Before we mathematically describe the penetration of membranes by solutes, we will briefly review certain features of the structure of membranes.

Membrane Models

Gorter and Grendel in 1925 estimated that the area covered by the lipids extracted from erythrocytes (red blood cells), when spread as a monomolecular layer on water, was about twice the surface area of the cells. The amount of lipid present was apparently sufficient to form a double layer in the membrane. Moreover, the penetration of a series of species across membranes often appeared to depend primarily on the relative lipid solubility of the molecules (see Davson and Danielli, Diamond and Wright, or Stein). This circumstantial evidence led to the concept of a biological membrane composed primarily of a lipid bilayer.

To help understand the properties of lipid bilayers, we must consider the charge distribution within lipid molecules. The arrangement of atoms in the hydrocarbon (containing only C and H) region of lipid molecules leads to bonding in which no local charge imbalance develops. This hydrocarbon part of the molecule is thus nonpolar and tends to avoid water, and is therefore called *hydrophobic*. Most lipid molecules in membranes also have a phosphate or an amine group or both. This region of the molecule becomes charged in aqueous solutions. Such charged regions interact electrostatically with the polar parts of other molecules. Since they interact attractively with water, the polar regions are termed *hydrophilic*.

Many of the phospholipids in membranes are glycerol derivatives that have two esterified fatty acids plus a charged side chain joined by a phosphate ester linkage. A typical example is phosphatidyl choline (also known by the older name *lecithin*), a major component of most membranes:

$$
\begin{array}{l}
\quad\quad\quad\quad O \\
\quad\quad\quad\quad \| \\
CH_2{-}O{-}C{-}R \\
\quad\quad\quad\quad O \\
\quad\quad\quad\quad \| \\
CH{-}O{-}C{-}R' \\
\quad\quad\quad\quad O \\
\quad\quad\quad\quad \| \\
CH_2{-}O{-}P{-}O{-}CH_2CH_2\overset{+}{N}(CH_3)_3 \\
\quad\quad\quad\quad | \\
\quad\quad\quad\quad O^-
\end{array}
$$

where R and R' are the hydrocarbon parts of the fatty acids. Various fatty acids commonly esterified to this phospholipid include palmitic (16C and no double bonds, represented as 16:0), palmitoleic (16:1), stearic (18:0), oleic (18:1), linoleic (18.2), and linolenic (18:3). (For example, the major fatty acid combined in various ways in the membranes of higher plant chloroplasts is linolenic.) The hydrocarbon side chains from the esterified fatty acids affect the packing of the lipid molecules in a membrane. As the

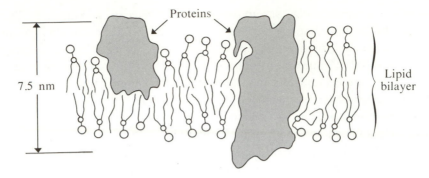

Figure 1.7
Membrane model where globular proteins are interspersed within a
lipid bilayer. The ionic "head" of phospholipids is represented by ○—○
and the fatty acid side chains leading to the nonpolar "tail" are indicated
by the two wavy lines emanating from the head.

number of double bonds in the fatty acid side chain increases, the side
chain tends to be less straight, so that the area per lipid molecule in a mono-
layer becomes greater. This change in intermolecular distances affects the
permeability of such lipid layers and, presumably, that of biological
membranes.

To form a bilayer, the lipid molecules have their nonpolar portions
adjacent to each other (Fig. 1.7), facilitating hydrophobic interactions. The
polar or hydrophilic regions are then on the outside. These outer charged
regions of the lipids can attract water molecules and the charged parts
of proteins, which are both present in membranes. For instance, up to 70%
of the dry weight of mitochondrial membranes can be protein, whereas
chloroplast membranes are generally less than 50% protein. Membranes
also contain a small amount of oligosaccharides, often bound to proteins.
Membranes can be about half water by weight. When the water content
of membranes is reduced below about 20%, as can occur during prolonged
desiccation, the lipid bilayer configuration and membrane integrity are
lost (see Simon). Thus, a minimal water content is necessary to create the
hydrophilic environment that stabilizes the lipid bilayer.

Both thermodynamic predictions and many experimental results suggest
that membrane proteins occur not as monolayers adsorbed on either side
of a central lipid bilayer but as globular forms embedded within the membranes (see Singer and Nicolson, or Harrison and Lunt). For instance, the
α-helical content of proteins in membranes is consistent with that of globular
proteins, but far greater than that expected for monolayers. If globular

proteins were attached to the outer surfaces of a lipid bilayer, the thickness would be greater than that observed for membranes. Proteins may thus occur as globular subunits embedded within a lipid matrix, as in Figure 1.7, and indeed some proteins may extend all the way across the membrane. Globular proteins apparently have their hydrophobic portions buried within the membrane, while a hydrophilic (polar or charged) portion sticks out into the aqueous solution next to the membrane. Amino acids whose side chains dissociate—aspartate, glutamate, arginine, lysine—tend to be exposed to water, while amino acids with hydrophobic side chains—leucine, isoleucine, valine—tend to the interior of the membrane where they interact with the fatty acid side chains of the phospholipids, which are also hydrophobic. In this model the lipid occurs as a bilayer, and the hydrophilic portions of the lipids interact directly with water on either side of the membrane (Fig. 1.7).

Organelle Membranes

Both mitochondria and chloroplasts have extensive internal membrane systems, and both are highly involved with cellular metabolism. When membranes of such organelles are carefully investigated with electron microscopy, a globular substructure is indeed found (see Park and Sane, or Lee et al.). For example, particles with major axes near 10 nm and 17 nm have been observed within the internal membranes of chloroplasts using the freeze-etch technique (the frozen specimens are fractured, water is removed by sublimation, and replicas of the specimen are examined using an electron microscope). These subunits appear to be embedded in the membrane, not adsorbed as monolayers on the surface.

Proteins implementing electron transfer in respiration and in photosynthesis are found in the interior membranes of mitochondria and chloroplasts, respectively. Such membrane subunits are able to move or vibrate thermally because of their own kinetic energy while remaining in the membrane. In fact, diffusion of globular proteins *within* the plane of the membrane has been observed. Diffusion coefficients for such movements are generally somewhat less than 10^{-14} m^2 s^{-1}, compared with nearly 10^{-10} m^2 s^{-1} for the same proteins in solution, suggesting that membranes have rather high viscosities (see Singer and Nicolson). Such diffusion of membrane proteins could allow successive interactions of some bound substrate with various enzymes located in the membrane. Furthermore, the side-by-side location of various components involved with electron transfer in a semi-

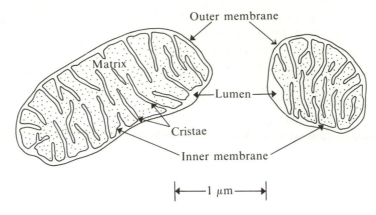

Figure 1.8
Representative mitochondria as seen in an electron micrograph
of a 50-nm-thick section of plant tissue.

solid part of the membrane system could ensure an orderly, rapid, directed
passage of electrons from enzyme to enzyme.

The proteins involved with electron transfer vary in size and shape, and
thus the internal membranes of chloroplasts and mitochondria would not
be expected to be uniform and regular. Proteins taking part in ion transport
and possibly cell wall synthesis may be embedded in the plasmalemma,
while other proteins involved with transport also presumably occur in the
tonoplast. On the other hand, many globular proteins in membranes could
simply serve a structural role. Between the globular proteins there may be
a lipid bilayer arrangement (as illustrated in Fig. 1.7), which would account
for the correlation between lipid solubility and membrane permeation
noted above.

Of the two mitochondrial membranes (Fig. 1.8), the outer one appears
to be much more permeable than the inner one to sucrose, various small
anions and cations (e.g., H^+), adenine nucleotides, and many other compounds. The inner membrane apparently invaginates to from the mitochondrial *cristae*, in which are embedded the enzymes responsible for
electron transfer and the accompanying ATP formation. For instance, the
inner membrane system has various dehydrogenases, an ATPase, and
cytochromes a, a_3, b, and c associated with it. (Cytochromes are discussed
in Chs. 5 and 6.) These proteins with enzymatic activity occur in a globular
form, and they can be an integral part of the membrane (Fig. 1.7) or loosely
bound to its periphery (as appears to be the case for cytochrome c). Electron
microscopy has revealed small particles attached by stalks to the cristae;
these particles are apparently proteins involved in the phosphorylation
accompanying respiration.

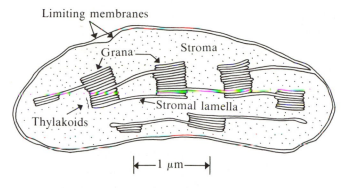

Figure 1.9
Generalized chloroplast from a leaf mesophyll cell.

Inside the inner membrane of a mitochondrion is a viscous region known as the *matrix* (Fig. 1.8). The citric acid cycle (Krebs cycle) enzymes, as well as others, are located there. For substrates to be catabolized via the citric acid cycle, they must cross two membranes to pass from the cytosol to the inside of a mitochondrion. Often the slowest or rate-limiting step in the oxidation of such substrates is their entry into the mitochondrial matrix. Since the inner mitochondrial membrane is highly impermeable to most molecules, transport across the membrane using a "carrier" or "transporter" (see p. 154) is generally invoked to explain how various substances get into the matrix. These carriers, situated in the inner membrane, might shuttle important substrates from the lumen between the outer and inner mito-chondrial membranes to the matrix. Because of the inner membrane, impor-tant ions and substrates in the mitochondrial matrix do not leak out. Such permeability barriers between various subcellular compartments can im-prove the overall efficiency of a cell.

Chloroplasts (Fig. 1.9) are also surrounded by two *limiting* membranes, the outer one of which is more permeable than the inner to small solutes. These limiting membranes are relatively high in galactolipid and low in protein (see Douce and Joyard, or Heber and Heldt). The internal *lamellar* membranes of chloroplasts are about half lipid and half protein by dry weight. The chlorophyll and most other photosynthetic pigments are bound to the proteins. These proteins, as well as other components involved with photosynthetic electron transfer (see Chs. 5 and 6), are anchored in the lamellar membranes, apparently by hydrophobic forces (see Fig. 1.7). Each lamella consists of a pair of closely apposed membranes 6 to 8 nm thick. In many regions of a chloroplast the lamellae form flattened sacs that are referred to as *thylakoids* or *discs* (Fig. 1.9). When seen in an electron micrograph, a transverse section of a thylakoid shows a pair of apposed membranes joined at the ends.

The actual organization of lamellar membranes within the chloroplast varies greatly with the species. The chloroplasts of red algae appear to have the simplest internal structure, since the lamellae are in the form of single large thylakoids separated by appreciable distances from each other. For most higher plant chloroplasts, the characteristic feature is stacks of about 10 or more thylakoids known as *grana* (Fig. 1.9), which are typically 0.4 to 0.5 μm in diameter, 10 to 50 of these grana occurring in a single chloroplast. The lamellar extensions between grana are called intergranal or stromal lamellae. The remainder of the chloroplast volume is known as the *stroma*, which contains the enzymes involved with the fixation of CO_2 into the various products of photosynthesis. Blue-green algae and photosynthetic bacteria do not contain chloroplasts, but their photosynthetic pigments are also generally located in membranes, often in lamellae immediately underlying the cell membrane. In some photosynthetic bacteria the lamellae appear to have pinched off and formed discrete subcellular bodies sometimes referred to as *chromatophores*.

"Microbodies" are also quite numerous in plant cells. They are generally divided into two classes, *glyoxysomes* and *peroxisomes*, the latter being about three times as prevalent as mitochondria in many leaf cells. Microbodies are usually spherical and from 0.4 to 1.5 μm in diameter. They differ from mitochondria and chloroplasts in that they are surrounded by a single membrane. Since both glyoxysomes and peroxisomes carry out only a portion of an overall metabolic pathway, these subcellular compartments depend on reactions in the cytosol and in other organelles. In Chapter 7 we will briefly consider the role of peroxisomes in photorespiration, where O_2 is consumed and CO_2 is released in the light in mesophyll cells. Glyoxysomes contain the enzymes necessary for the breakdown of fatty acids and are prevalent in fatty tissues of germinating seeds (see Tolbert).

MEMBRANE PERMEABILITY

With this knowledge of the general structure of membranes, let us now turn to a quantitative analysis of the interactions between membranes and diffusing solutes. In Chapter 3 we will discuss active transport, which is very important for moving specific solutes across membranes and thus overcoming limitations posed by diffusion.

The rate-limiting step for the movement of many molecules into and out of plant cells is diffusion through the plasmalemma. Because of the close packing and the interactions between the component molecules of a membrane, such diffusion is greatly restricted compared with the relatively free movement in an aqueous phase like the cytosol. In other words, the solute has great difficulty in threading its way between the molecules composing the membrane phase. The average diffusion coefficient of a small solute in a membrane is often about 10^6 times lower than in the adjacent aqueous solutions. Also, a membrane represents a different type of molecular environment than does an aqueous solution, so the relative solubility of a species in the two phases must be taken into account to describe membrane permeability. As we will show, Fick's first law can be modified to describe the diffusion of molecules across a membrane. Once past the barrier presented by the membrane, the molecules may be distributed throughout the cell relatively rapidly by diffusion, as well as by cytoplasmic streaming.

Concentration Difference Across a Membrane

The driving force for diffusion of uncharged molecules into or out of plant cells can be regarded as the negative of the concentration gradient of that species across the plasmalemma. Since the actual concentration gradient of some species j is not known in the plasmalemma (or in any other membrane for that matter), the driving force is generally approximated by the negative of the average gradient of that species across the membrane:

$$-\frac{\partial c_j}{\partial x} \cong -\frac{\Delta c_j}{\Delta x} = -\frac{c_j^i - c_j^o}{\Delta x} = \frac{c_j^o - c_j^i}{\Delta x} \qquad (1.7)$$

where c_j^o is the concentration of species j outside the cell, c_j^i its concentration in the cytosol, and Δx the thickness of the plasmalemma, which acts as the barrier restricting the penetration of the molecules into the cell (see Fig. 1.10). The concentrations c_j^o and c_j^i in Equation 1.7 can also represent values in the phases separated by any other membrane, such as the tonoplast. If c_j^i is less than c_j^o as we move in the positive x-direction (see Fig. 1.10), $\partial c_j/\partial x$ is then negative and so $-\partial c_j/\partial x$ is positive; the net flux density would then occur in the positive x-direction (see Eq. 1.1, $J_j = -D_j \partial c_j/\partial x$). Alternatively, if we designate a net flow into a cell as positive—which is the commonly used convention—then the minus sign in $-\partial c_j/\partial x$ is incorporated

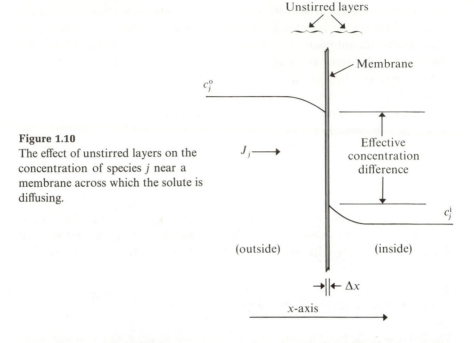

Figure 1.10
The effect of unstirred layers on the concentration of species j near a membrane across which the solute is diffusing.

into the concentration difference $(c_j^o - c_j^i)$ that we must use to describe the flux density, namely, $J_j = D_j(c_j^o - c_j^i)/\Delta x$.

The magnitude of what may be called the "effective" concentration difference across a membrane is made somewhat uncertain by the existence of unstirred boundary layers (Fig. 1.10; see Dainty 1963). Boundary layers in which there is no turbulent mixing occur at the interface between any fluid (liquid or gas) and a solid, such as the air boundary layers on either side of a leaf (discussed in Ch. 7). A substance moves across these relatively unstirred layers next to the membrane or other solid not by mechanical mixing but by diffusion, indicating that a concentration gradient must also exist in the boundary layers (see Fig. 1.10). When the mixing in the solutions adjacent to a membrane is increased—for example, either by the turbulence resulting from cytoplasmic streaming on the inside of the cell or by rapid stirring on the outside—the thicknesses of the unstirred layers are reduced. But they are not eliminated. Under actual experimental conditions with vigorous stirring the external unstirred layer may be from 10 to 100 μm thick, which is much thicker than the membrane itself (see Green and Otori). Because of cytoplasmic streaming in plant cells, the internal unstirred layer is generally thinner than the external one. Although diffusion is more rapid in aqueous solutions than in membranes, the unstirred layers do represent an appreciable distance for molecules to diffuse across. In fact, diffusion through the unstirred layer in some cases can become the rate-limiting factor for the entry into cells or organelles of those molecules which rapidly penetrate the membrane itself. For convenience, the difference in concentration across a membrane will be represented by $c_j^i - c_j^o$, but this is an overestimate of the effective concentration difference, as is indicated in Figure 1.10.

The difference in concentration actually determining the diffusion of molecules across a membrane is the concentration just inside one side of the membrane minus that just within the other side. In Equation 1.7 the concentrations are those in the aqueous phases outside on either side of the membrane. Since membranes are quite different as solvents compared with aqueous solutions, the concentrations just inside the membrane can differ appreciably from those just outside in the aqueous solution. A correction factor must therefore be applied to Equation 1.7 to give the actual concentration gradient existing in the membrane. This factor is known as the *partition coefficient*, K_j. It is defined as the ratio of the concentration of a solute in the material of the membrane to that in equilibrium outside it in the aqueous phase (c_j^o or c_j^i), and so it is dimensionless. As a guide to the situation in the membrane, the partition coefficient is generally determined by measuring the ratio of the equilibrium concentration of some solute in a lipid phase—such as olive oil, which mimics the membrane lipids—to the concentration in an adjacent and immiscible aqueous phase (which mimics the solutions on either side of the membrane). This rather simple convention for obtaining partition coefficients is based on the high lipid content of membranes and the numerous experimental results which show that the relative ease of penetration of a membrane depends on the lipid solubility of the molecules (see Davson and Danielli, as well as Diamond and Wright). Partition coefficients have a wide range of values, most being between 10^{-6} and 10 (see Wright and Diamond, where the effect on K_j of various lipids as solvents is described). For example, an inorganic ion might have a K_j near 10^{-5} for membranes, while a nonpolar hydrophobic substance might have a value near 1. Usually, K_j is assumed to be the same coming from either side of the membrane. These considerations lead us to write $K_j(c_j^o - c_j^i)$ for the concentration difference leading to diffusion across a membrane, where K_j is characteristic of a particular substance.

Permeability Coefficient

We will next use Fick's first law ($J_j = -D_j \partial c_j/\partial x$, Eq. 1.1) to obtain an expression convenient for describing the movement of a substance across a membrane. The negative concentration gradient will be replaced by ($c_j^o -$

$c_j^i)/\Delta x$ (Eq. 1.7). Moreover, since D_j is the diffusion coefficient of species j within the substance of the membrane, we must use the actual concentration drop within the membrane, $K_j(c_j^o - c_j^i)$. These various considerations lead us to the following expression describing the diffusion of species j:

$$J_j = D_j \frac{K_j(c_j^o - c_j^i)}{\Delta x}.$$

$$= P_j(c_j^o - c_j^i) \tag{1.8}$$

where P_j is called the *permeability coefficient* of species j.

Since permeability coefficients have so many applications in physiology, we will define P_j explicitly:

$$P_j = \frac{D_j K_j}{\Delta x} \tag{1.9}$$

which follows directly from Equation 1.8. The permeability coefficient conveniently replaces three quantities which describe the diffusion of some solute across a membrane or other barrier. In this regard, the partition coefficients of membranes are generally determined by using a lipid phase such as olive oil or ether, not the actual lipids occurring in membranes, and thus K_j will have some uncertainty. Also, measurements of Δx and D_j for membranes tend to be rather indirect. On the other hand, P_j is a single, readily measured quantity characterizing the diffusion of some solute across a membrane or other barrier. The units of P_j are length per time, e.g., m s^{-1}.

A typical permeability coefficient for a small nonelectrolyte or uncharged molecule (e.g., isopropanol or phenol) is 10^{-6} m s^{-1} for the plasmalemma (see Stein, or Atman and Dittmer), whereas P_j for a small ion (e.g., K$^+$ or Na$^+$) might be about 10^{-9} m s^{-1}. The lower permeability coefficients for charged particles are mainly due to the much lower partition coefficients of these solutes compared with nonelectrolytes, which tend to be hydrophobic (recall that $P_j = D_j K_j/\Delta x$, Eq. 1.9). In other words, because of its charge, the electrolyte tends to be much less soluble in a membrane than is a neutral molecule, so that the effective concentration gradient driving the charged species across the membrane is generally smaller for given concentrations in the aqueous phases on either side of the barrier. By comparison, water can rapidly enter or leave cells and can have a permeability coefficient of up to about 10^{-4} m s^{-1} for Characean algae, although values from 10^{-6} to 10^{-5} m s^{-1} are more usual for plant cells (see Stadelmann). (Actually, P_w is hard to measure because of the relative importance of the unstirred layers; see Dainty 1963.) Although adequate measurements on plant membranes have so far not been made for O_2 and CO_2, their P_j's for the plasmalemma should be quite high. For instance, O_2, which is lipophilic and has a fairly high partition coefficient of 4.4, has a very high permeability coefficient of about 0.3 m s^{-1} for erythrocyte membranes at 25°C (see Fischkoff and Vanderkooi).

Diffusion and Cellular Concentration

Instead of calculating the flux density of some species diffusing into a cell— the amount entering per unit area per unit time—we often focus our attention on the total amount of that species diffusing in over a certain time interval. Let s_j be the amount of species j inside the cell, where s_j can be expressed in moles. If that substance is not involved in any other reactions, ds_j/dt represents the rate of entry of species j into the cell. The flux density J_j is the rate of entry of the substance per unit area or $(1/A)(ds_j/dt)$, where A is the area of the cellular membrane across which the substance is diffusing. Using this expression for J_j, we can express Equation 1.8 as follows:

$$ds_j/dt = J_j A = P_j A(c_j^o - c_j^i) \qquad (1.10)$$

When the external concentration of species j (c_j^o) is greater than the internal one (c_j^i), species j will enter the cell and ds_j/dt will then be positive, as Equation 1.10 indicates.

A question that often arises in cellular physiology is how the internal concentration of a penetrating solute changes with time. How long, for example, would it take the internal concentration of an initially absent species to build up to 70% of the external concentration? To solve the general case, we must first express s_j in Equation 1.10 in a suitable fashion and then integrate. It is useful to introduce the approximation—particularly appropriate to plant cells—that the cell volume does not change during the time interval of interest. In other words, because of the rigid cell wall, we will assume that the plant cell has a nearly constant volume (V) during the entry or exit of the solute being considered.

The average internal concentration of species j (c_j^i) is equal to the amount of that particular solute inside the cell (s_j) divided by the cellular volume (V), or s_j/V. Therefore, s_j can conveniently be replaced by Vc_j^i, and thus ds_j/dt in Equation 1.10 equals Vdc_j^i/dt when we can assume that the cell volume does not change with time. We are also assuming that species j is not produced or consumed by any reaction within the cell, so that the diffusion of species j into or out of the cell is the only process affecting c_j^i. For simplicity, we are treating the concentration of species j as if it were uniform within the cell. (In a more complex case we might need a relation of the form of Equation 1.10 for each membrane-surrounded compartment.) Moreover, we are presupposing either mechanical mixing or rapid diffusion inside the cell, so that the drop in concentration $(c_j^o - c_j^i)$ occurs essentially only across the membrane. Finally, let us assume that P_j is independent of concentration, a condition that is often satisfactorily met over a limited range of concentration. We can now proceed with the integration of Equation 1.10. Upon replacement of ds_j/dt by $V\,dc_j^i/dt$, rearrangement to separate variables (see App. V), and the insertion of integral signs, Equation 1.10 becomes

$$\int_{c_j^i(0)}^{c_j^i(t)} \frac{dc_j^i}{c_j^o - c_j^i} = \frac{P_j A}{V} \int_0^t dt \tag{1.11}$$

where $c_j^i(0)$ is the initial internal concentration of species j, i.e., when $t = 0$, and $c_j^i(t)$ is that at a later time t.

The volume outside certain unicellular algae and other membrane-surrounded entities can be quite large compared with V. In such cases the external concentration (c_j^o) does not change appreciably. Such an approximation can also be appropriate for experiments of short duration or where special arrangements are made to maintain c_j^o at some fixed value. For those cases where c_j^o is constant, Equation 1.11 can be integrated to give the following expression:[*]

$$\frac{P_j A t}{V} = \ln \frac{c_j^o - c_j^i(0)}{c_j^o - c_j^i(t)} \tag{1.12}$$

Starting with a known c_j^o outside a cell and determining the internal concentration both initially, i.e., for $t = 0$, and at some subsequent time t, we can calculate P_j from Equation 1.12 if A/V is known. Even when A/V is not known, the relative permeability coefficients for two substances can be determined from the time dependencies of the respective c_j's. The above derivation can easily be extended to the case where c_j^o is zero. In that case,

---

[*] Note that $\int dx/(a - x) = -\int d(a - x)/(a - x) = -\ln(a - x) + b$, where ln is the natural logarithm to the base e. See Appendix IV for certain logarithmic identities and Appendix V for a detailed consideration of the integration of Equation 1.11.

P_jAt/V equals $\ln[c_j^i(0)/c_j^i(t)]$, an expression that can be used to describe the diffusion of a photosynthetic product out of a chloroplast or some substance from a cell into a large external solution initially devoid of that species. Such studies can be greatly simplified by using radioactive tracers.

When diffusion occurs from all directions across a membrane surrounding a cell (or an organelle), the time to reach a given internal concentration can be directly proportional to the dimensions of the cell. For convenience in illustrating this point, let us consider a spherical cell of radius r. The volume V then equals $(4/3)\pi r^3$ and the surface area A is $4\pi r^2$. Hence, V/A is $r/3$. Equation 1.12 then becomes

$$t^{\text{sphere}} = \frac{r}{3P_j} \ln \frac{c_j^o - c_j^i(0)}{c_j^o - c_j^i(t)} \tag{1.13}$$

For the same P_j, a given level inside a small cell is reached sooner than it would inside a large cell. But when diffusion across the membrane into a spherical cell occurs from all directions, the time to reach a given concentration inside is *linearly* proportional to the radius (assuming a uniform c_j^i). On the other hand, the time it takes for a planar front to diffuse in one direction is proportional to the square of the distance traveled ($x_e^2 = 4D_j t_e$, Eq. 1.6).

Using Equation 1.13, we can calculate the time required for a substance initially absent from a cell [$c_j^i(0) = 0$] to achieve an internal concentration equal to half the external concentration [$c_j^i(t) = \frac{1}{2}c_j^o$]. The conditions on the concentrations mean that $[c_j^o - c_j^i(0)]/[c_j^o - c_j^i(t)]$ equals $(c_j^o - 0)/(c_j^o - \frac{1}{2}c_j^o)$, which is 2. Substituting these relations into Equation 1.13 indicates that the time needed is $(r/3P_j) \ln 2$, where $\ln 2$ equals 0.693. For a cell 50 μm in diameter—which is a reasonable dimension for spongy mesophyll cells in certain leaves (see Fig. 1.2)—the time for an initially absent nonelectrolyte with a P_j of 10^{-6} m s^{-1} to reach half of the external concentration therefore is

$$t_{1/2}^{\text{sphere}} = \frac{(25 \times 10^{-6} \text{ m})}{(3)(10^{-6} \text{ m s}^{-1})} (0.693) = 6 \text{ s}$$

and the time is 1000-fold greater (2 hours) for an electrolyte with a P_j of 10^{-9} m s^{-1}. Hence, the external membrane is an extremely good barrier for electrolytes, markedly hindering their entry into or egress from the cell. On the other hand, small nonelectrolytes can fairly readily diffuse in and out of plant cells, depending to a large extent on their relative lipid solubility.

CELL WALLS

Cell walls play many roles in plants. Their rigidity helps determine the size and shape of a cell and ultimately that of a plant. This supportive role is performed in conjunction with the internal hydrostatic pressure, which causes a distension of the cell walls. The cell wall is also intrinsically involved in many aspects of the ion and water relations of a plant. Since it surrounds the plasmalemma of each cell, all fluxes of water and solutes into or out of a plant cell protoplast must cross the cell wall, usually by diffusion. The cell wall generally has a large negative charge, and so it can interact differently with cations than with anions. Water evaporating from a plant during transpiration comes directly from cell walls. (Some aspects of this process will be considered quantitatively in Chs. 2 and 8.) The space surrounded by the cell wall in certain specialized cells can act as a channel through which solutions move. In the xylem, for example, the conducting cells have lost their protoplasts, with the result that the pathway for the conduction of solutions in the xylem consists essentially of hollow pipes, or conduits, made exclusively of cell walls.

Cell walls vary from about 0.1 to 10 μm in thickness, and they are generally divided into three regions: primary wall, secondary wall, and middle lamella. The *primary cell wall* surrounds dividing meristematic cells as well as the elongating cells during the period of cell enlargement. The cell wall often becomes thickened by the elaboration of a *secondary cell wall* inside the primary one (Fig. 1.11), which makes the cell mechanically much less flexible. Hence, cells whose walls have undergone secondary thickening, such as in xylem vessel members, are generally incapable of subsequent elongation. The cell wall also includes an amorphous region between contiguous cells called the *middle lamella*, the extent of which is not universally agreed upon. Although cellulose also occurs there, the middle lamella may be composed mainly of the calcium salts of pectic acids, the presence of which causes adjacent cells to adhere together.

Chemistry and Morphology

Cellulose is the most abundant organic component in living organisms. It is the characteristic substance of the plant cell wall, constituting from 25% to 50% of the cell wall organic material. Cellulose is a linear (unbranched) polysaccharide consisting of 1,4-linked β-D-glucopyranose units:

The polymer is about 0.8 nm in its maximum width, 0.33 nm^2 in cross-sectional area, and can contain about 10 000 glucose residues with their rings in the same plane. In the cell wall these polymers are organized into *microfibrils* that can be 5 nm by 9 nm in cross section. These microfibrils apparently consist of an inner core of about 50 parallel chains of cellulose arranged in a crystalline array surrounded by a similar number of cellulose and other polymers in a paracrystalline array (see Preston). Microfibrils are the basic unit of the cell wall and are readily observed in electron micrographs. Although great variation exists, they tend to be interwoven in the the primary cell wall and parallel to each other in the secondary wall (see Fig. 1.11).

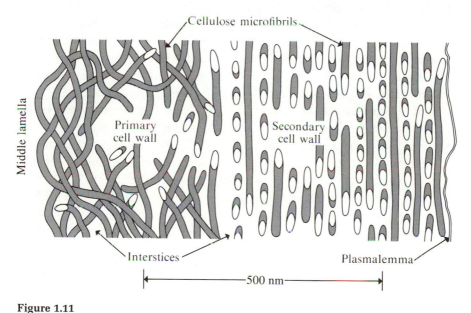

Figure 1.11
Hypothetical thin section through a cell wall, indicating the cellulose microfibrils in the primary and secondary cell walls. The interstices are filled with noncellulosic material including appreciable water, the solvent for solute diffusion across the cell wall.

Interstices between the cellulose microfibrils are usually 5 to 30 nm across. In these interstices is a matrix of amorphous components that can occupy a larger volume of the cell wall than do the microfibrils themselves. In fact, by weight, the main constituent of the cell wall is actually water, some consequences of which we will consider here and in Chapter 2.

The cell wall matrix contains noncellulosic polysaccharides such as pectin, lignins (secondary wall only), a small amount of protein, bound and free water, appreciable calcium, other cations, and sometimes silicates. Lignins are complex polymers based on phenylpropanoid subunits (a 6-carbon ring, to which is attached a 3-carbon chain as in phenylalanine, a lignin precursor) plus certain other residues. Lignins constitute the second most abundant class of organic molecules in living organisms and are about half as prevalent as cellulose. They are quite resistant to enzymatic degradation, and consequently lignins are important in peat and coal formation. They also make the cell wall more rigid. Since the plant cannot break down the lignin polymers, cells are unable to expand after extensive lignification of their cell walls, as occurs in the secondary walls of xylem vessel members. (See Sarkanen and Ludwig.) Pectin consists primarily of 1,4-linked α-D-galacturonic acid residues, the carboxyl groups of which are normally dissociated and have a negative charge (galacturonic acid differs from galactose by having —COOH instead of —CH_2OH in the 6-carbon position). The negative charge of the dissociated carboxyl groups leads to the tremendous cation-binding capacity of cell walls. In particular, much of the divalent cation calcium (Ca^{2+}) is bound, which may help link the various polymers together. Polymers based on 1,4-linked β-D-xylopyranose units (xylans), as well as on many other residues, can also be extracted from cell walls. They are loosely referred to as hemicelluloses, e.g., xylans, mannans, galactans, and glucans. In general, hemicelluloses tend to have a low molecular weight (in the ten thousands) compared with pectin (about five to ten times larger) or cellulose. (See McNeil et al, Northcote, Preston, or Tolbert for further details on cell wall chemistry.) The presence of the negatively charged pectins hinders the entry of anions into plant cells. As we will consider next, however, ions and other solutes generally pass through the cell wall much more easily than through the plasmalemma.

Diffusion Across Cell Walls

How would we calculate the ease with which molecules might diffuse across a cell wall? A good place to start is Equation 1.9, which indicates that the permeability coefficient of species j equals $D_jK_j/\Delta x$. We must first consider what we mean by D_j and K_j in a cell wall. The diffusion of solutes across a cell wall occurs mainly in the water located in the numerous interstices, which are often about 10 nm in diameter. Thus, the movement from the external solution up to the plasmalemma can be in aqueous channels through the cell wall. By the definition of a partition coefficient (solubility in barrier/ solubility in water), we recognize that K_j would be unity for the water-filled interstices of the cell wall, it would be very low for the solid phases (cellulose, lignin, and the other polymers present in the wall), and it would have some intermediate value for the cell wall as a whole. Since diffusion takes place primarily in the cell wall water, we will let K_j be unity. If we use this con- venient definition for K_j in cell walls, we must be careful how we define D_j. In particular, diffusion coefficients will have to be expressed on the basis of the whole cell wall, and not on the much smaller area presented by the interstices. The value of D_j in the water of the interstices could be very similar to that in a free solution, whereas its value averaged over the whole cell wall would be considerably less, since the interstices are not straight channels through the cell wall and since they do not occupy the entire cell wall volume. Such effective diffusion coefficients of small solutes are at least three times lower when averaged over the cell wall than are the D_j's of the same species in an extended aqueous solution.

Next, let us calculate the permeability coefficient for a solute that has a diffusion coefficient of 2×10^{-10} m^2 s^{-1} for a cell wall. We will assume a representative value of 1 μm for the cell wall thickness. Using Equation 1.9 $(P_j = D_j K_j / \Delta x)$, we can thus estimate that P_j for the cell wall would be as follows:

$$P_j = \frac{(2 \times 10^{-10} \text{ m}^2 \text{ s}^{-1})}{(10^{-6} \text{ m})} = 2 \times 10^{-4} \text{ m s}^{-1}$$

Most of the permeability coefficients for small solutes crossing the plasma- lemma range from 10^{-10} to 10^{-6} m s^{-1}. Hence, a cell wall generally has a higher permeability coefficient than does a membrane, which means that the wall is usually more permeable for small solutes than is the plasmalemma. For comparison, let us consider a permeability coefficient that might be appropriate for an unstirred liquid layer adjacent to a cell wall or membrane. Specifically, D_j for a small solute may be 10^{-9} m^2 s^{-1} in water, K_j is unity in the aqueous solution, and let us assume Δx is 30 μm for the unstirred layer. In such a case we have

$$P_j = \frac{(10^{-9} \text{ m}^2 \text{ s}^{-1})(1)}{(30 \times 10^{-6} \text{ m})} = 3 \times 10^{-5} \text{ m s}^{-1}$$

which is less than the above value for a cell wall.

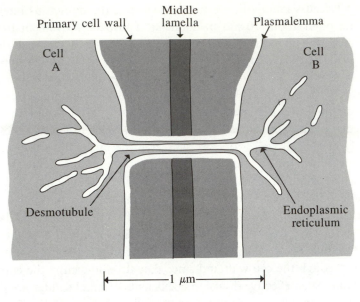

Figure 1.12
Longitudinal section through a plasmodesma, showing that the primary cell walls are locally thinner (which is generally the case) and that the plasmalemma of the two adjacent cells are continuous.

Molecules diffuse less readily across a given distance in a plasmalemma than in a cell wall or the adjacent unstirred layer. For the above numerical values, $D_j K_j$ is 10^{-9} m^2 s^{-1} in the aqueous solution and 2×10^{-10} m^2 s^{-1} in the cell wall, yet for a plasmalemma 10 nm thick, $D_j K_j$ is only 10^{-18} to 10^{-14} m^2 s^{-1}. Membranes do indeed provide very effective barriers for the diffusion of solutes.

Secondary cell walls are often interrupted by localized pits. A pit in the wall of a given cell usually occurs opposite a complementary pit in an adjacent cell, the cytoplasm of two adjacent cells being brought into close proximity at such a pit-pair. The local absence of extensive cell wall substance facilitates the diffusion of molecules from one cell into the other. An easier way for molecules to move between plant cells is by means of the *plasmo-desmata* (singular: *plasmodesma*). These are fine, membrane-flanked, cyto-plasmic threads that pass from a protoplast, through a pore in the cell wall, directly into the protoplast of a second cell (see Fig. 1.12). The pores generally occur in locally thin regions of the primary cell wall, referred to as primary pit-fields, which can contain many plasmodesmata. (If a secondary wall is deposited, pits in that wall occur over the primary pit-fields.) Plasmodesmata can be 50 nm in diameter (range, 20 to 200 nm) and typically have a frequency of 2 to 10 per μm^2 of cell surface (see Robards). They usually occupy about 0.1% to 0.5% of the surface area of a cell. The continuum of communicating cytoplasm created by such intercellular connections is commonly referred to as the *symplasm*.* The plasmodesmata are not simple aqueous channels between cells, since they contain desmotubules and have other ultrastructural characteristics including special constrictions and associations of the desmotubules with the membranes of the endoplasmic reticulum (see Gunning and Robards). Nevertheless, the presence of the plasmodesmata can provide a particularly effective pathway for solute movement between adjacent cells where permeability barriers in the form of either cell walls or membranes can be avoided (see also Tyree), although the details of the movement and the function of the desmotubules are not yet fully understood.

Since movement from cell to cell through the symplasm may be quite important physiologically, let us make a rather oversimplified calculation so that we can compare it with the concomitant diffusion across the plasma-lemma. We will use Fick's first law as presented in Equation 1.8, namely, $J_j = D_j K_j \Delta c_j / \Delta x = P_j \Delta c_j$. Let us consider two adjacent cells in which the glucose concentration differs by 1 mol m^{-3} (1 mM). Glucose is rather insol-uble in membrane lipids and might have a permeability coefficient of about 10^{-9} m s^{-1} for a plasmalemma. For movement from cell to cell across the two plasmalemmas in series, the effective P_j would be 0.5×10^{-9} m s^{-1} (inclusion of the cell wall as another series barrier for diffusion would only slightly reduce the effective P_j in this case; see Problem 1.4 for handling barriers in series). Thus, the flux density of glucose across the plasmalemmas toward the cell with the lower concentration would be

$$J_{glucose}^{plasmalemma} = (0.5 \times 10^{-9} \text{ m s}^{-1})(1 \text{ mol m}^{-3}) = 5 \times 10^{-10} \text{ mol m}^{-2} \text{ s}^{-1}$$

Let us next consider the flux for the symplasm. In the aqueous part of the plasmodesmata, $D_{glucose}$ should be similar to its value in water, 0.7×10^{-9} m^2 s^{-1} (see p. 18), and $K_{glucose}$ would be 1.0. We will let the plasmodesmata be 1 μm long. The flux density in the pores would thus be

$$J_{glucose}^{pores} = \frac{(0.7 \times 10^{-9} \text{ m}^2 \text{ s}^{-1})(1.0)}{(10^{-6} \text{ m})} (1 \text{ mol m}^{-3}) = 7 \times 10^{-4} \text{ mol m}^{-2} \text{ s}^{-1}$$

---

* The term *symplast* is defined as the continuum of proto*plasts* together with the plasmodesmata which interconnect them, and hence means the same as symplasm (compare *apoplast*, p. 9).

If the aqueous channels or pores occupy 10% of the area of the plasmodes-mata, which in turn occupy 0.2% of the surface area of the cells, then the rate of glucose diffusion through the plasmodesmata per unit area of the *cells* would be $(0.1)(0.002) \, J^{pores}_{glucose}$, 14×10^{-8} mol m^{-2} s^{-1}. This is more than a hundredfold greater than the simultaneously occurring diffusion across the plasmalemmas.

The permeability coefficients of the plasmalemmas for phosphorylated (charged) sugars like ribose-5-phosphate or ribulose-1,5-diphosphate would be less than that for glucose, while the diffusion coefficients in the plasmo-desmata would be about the same as for glucose. Thus, the discrepancy between flux densities of charged species through the plasmodesmata and those across the plasmalemmas would be even greater than the difference calculated here for glucose. The joining of cytoplasms into a symplasm indeed facilitates the diffusion of solutes from one cell to another.

Stress-Strain Relations of Cell Wall

Cell walls of mature plant cells are generally quite resistant to mechanical stretching, especially when there is appreciable thickening of the secondary cell walls. Nevertheless, cell walls will stretch when a stress is applied. Reversible elastic properties are generally described by a measure of elas-ticity known as *Young's modulus*, which is the ratio of applied stress (force per unit area) to the resulting strain (fractional change in length):

$$\text{Young's modulus} = \frac{\text{stress}}{\text{strain}} = \frac{\text{force/area}}{\Delta l/l} \qquad (1.14)$$

Since $\Delta l/l$ is dimensionless, Young's modulus has the dimensions of force per unit area, or pressure. A high value of this modulus of elasticity means that a large stress must be applied to produce a noticeable extension.

Young's modulus for dry cotton fibers, which are nearly pure cellulose, is quite large—about 10^{10} N m^{-2}, or 10 000 MPa, which is 5% of that for steel (see Preston).* Because of both the complicated three-dimensional array of microfibrils in the cell wall and the presence of many other com-ponents, Young's modulus for a cell wall is considerably less than for pure

---

* The newton, abbreviated N, and the pascal, abbreviated Pa, are the SI units for force and pressure, respectively (1 N m^{-2} = 1 Pa). Pressures in plant studies are often expressed in bars, where 1 bar equals 10^6 dynes (dyn) cm^{-2}, 0.987 atmosphere, or 0.1 MPa. See App. III for further conversion factors.

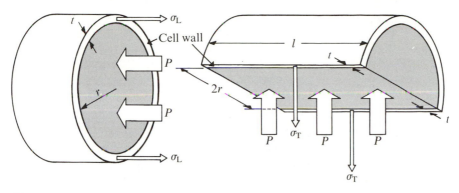

Figure 1.13
Schematic sections of a hypothetical cylindrical cell resembling the internodal cells of *Nitella* or *Chara*, illustrating various dimensions and the stresses (σ_L and σ_T) existing in the cell wall. One way to calculate the stresses is to imagine that the cellular contents are removed, leaving only the cell wall, which has a uniform hydrostatic pressure P acting perpendicular to its inside surface. The projection of this P over the appropriate area gives the force acting in a certain direction, while the reaction to this force is an equal force in the opposite direction in the cell wall. By dividing the force in the cell wall by the area over which it occurs, we can determine the cell wall stress.

cellulose. For example, the modulus of elasticity for the cell wall of *Nitella* is about 700 MPa (see Kamiya et al.). We will use this value when we indicate the fractional stretching that can occur for plant cells.

The hydrostatic (turgor) pressure (force per unit area in a liquid), P, acts uniformly in all directions in a cell. This internal pressure pushes against the plasmalemma, which itself is closely appressed to the cell wall. This tends to cause the cell to expand and also leads to tensions (stresses) in the cell wall. The magnitude of these stresses varies with the physiological state of the plant as well as with direction, an aspect we will consider next.

A cylinder, which closely approximates the shape of the large internodal cells of *Chara* or *Nitella* (Fig. 3.9), is a useful geometrical model for evaluating the various cell wall stresses. We will let the radius of the cylinder be r (Fig. 1.13). The force on an end wall is the pressure, P, times the area of the end wall, πr^2, and so it equals $P\pi r^2$. This force is balanced by a force arising from the *longitudinal stress*, σ_L, that occurs in the cell wall. The area over which σ_L acts is shown by the cut portion of the cell wall in the left part of Figure 1.13. The longitudinal stress occurs in an annulus of circumference $2\pi r$ (approximately) and a width equal to the thickness of the cell wall, t. The force in the cell wall is thus $(\sigma_L)(2\pi r)(t)$. This force is an equal and opposite reaction to $P\pi r^2$, and so $\sigma_L 2\pi rt$ equals $P\pi r^2$, or

$$\sigma_L = \frac{rP}{2t} \tag{1.15}$$

The longitudinal stress acts parallel to the major axis of the cylinder and resists the lengthwise deformation of the cell.

A *tangential stress*, σ_T, also exists in the cell wall in response to the internal pressure; it limits the radial expansion of the cell. To determine the magnitude of this stress, we will consider a cell split in half along its axis (right side of Fig. 1.13). The split part of the cell has an area of $2rl$ in the plane of the cut; this area is acted on by the pressure P, leading to a force of $P2rl$. This force is resisted by the tangential stress in the cell wall. As shown in Figure 1.13, σ_T acts along two cell wall surfaces, each of width t and length l. The total force in the cut part of the cell wall is thus the area ($2tl$) times the tangential stress. Equating this force ($\sigma_T 2tl$) with that due to the internal pressure in the cell ($P2rl$), we obtain the following relationship for the tangential stress:

$$\sigma_T = \frac{rP}{t} \tag{1.16}$$

The tangential stress in the cell wall given by Equation 1.16 is twice as large as the longitudinal stress (Eq. 1.15). Thus, this simple cylindrical model illustrates the important point that the stresses in a cell wall can vary with direction (see Lockhart). Young's modulus also varies with direction, reflecting the anisotropic orientation of the microfibrils in the cell wall.

To estimate the magnitudes of the stresses and the resulting strains in the cell wall, let us consider a *Nitella* or *Chara* cell 1 mm in diameter with a cell wall 5 μm thick. In this case, r/t is as follows:

$$\frac{r}{t} = \frac{(0.5 \times 10^{-3} \text{ m})}{(5 \times 10^{-6} \text{ m})} = 100$$

A reasonable estimate of P is 0.5 MPa (5 bars). Using Equation 1.15, we can calculate the longitudinal stress in the cell wall of such a cell:

$$\sigma_L = \frac{(100)(0.5 \text{ MPa})}{(2)} = 25 \text{ MPa}$$

which is an appreciable tension. By Equation 1.16, the tangential stress would be twice as great, or 50 MPa. Ignoring changes in the radial direction, we can calculate the strain produced by the longitudinal stress using the definition of Young's modulus (Eq. 1.14) and its particular value in the longitudinal direction for the cell wall of *Nitella*, 700 MPa. From Equation 1.14, the fractional change in length is then

$$\frac{\Delta l}{l} = \frac{(25 \text{ MPa})}{(700 \text{ MPa})} = 0.036$$

i.e., only about 4%. (If changes in the radial direction were included, the length change would be about 3% in the present case; see Vinters et al.) Hence, even with an internal pressure of 0.5 MPa, the cell wall (and consequently the whole cell) is not extended very much. The cell wall is indeed rigid and therefore well suited both for delimiting individual cells and for providing the structural support of a plant.

Elastic properties of plant cell walls are crucial for understanding the cell expansion accompanying growth as well as other features of water movement. For this, we are interested in what volume change (ΔV) is caused by a given pressure change (ΔP); this relation can be quantified using the volumetric elastic modulus (ε):

$$\varepsilon = \frac{\Delta P}{\Delta V/V} \tag{1.17}$$

Values of ε, determined experimentally, generally range from 1 to 50 MPa (see Dainty 1976 or Zimmermann), meaning that cells change in volume by 0.2% to 10% for each 0.1 MPa change in pressure, but only a few plant species have been examined so far. Also, ε depends both on P (it is smaller at lower P), V, and the developmental stage of the cells. We will return to a consideration of the elastic modulus when discussing the water relations of cells in Chapter 2.

Before concluding this discussion of cell walls, it should be noted that the case of elasticity or reversible deformability is only one extreme of stress-strain behavior. At the opposite extreme is plastic (irreversible) extension. If the amount of strain is directly proportional to the time that a certain stress is applied, and if the strain persists when the stress is removed, we have an example of viscous flow. The cell wall actually exhibits intermediate properties and is said to be *viscoelastic*. When a given stress is applied to a viscoelastic material, the resulting strain is approximately proportional to the logarithm of time. Such extension is partly elastic (reversible) and partly plastic (irreversible). Underlying the viscoelastic behavior of the cell wall are the crosslinks between the various polymers. For example, if a bond from one cellulose polymer to another is broken while the cell wall is under tension, the opportunity exists for a new bond to form in a less strained configuration, leading to an irreversible or plastic extension of the cell wall. The quantity responsible for the tension in the cell wall—which in turn leads to such viscoelastic extension—is the hydrostatic pressure within the cell (see Cleland or Preston).

Problems

1.1. A thin layer of some solution is inserted into a long column of water. One hour later, the concentration of the solute is 100 mol m^{-3} (0.1 M) at the plane of insertion and 37 mol m^{-3} (0.037 M) at a distance 3 mm away. (a) What is its diffusion coefficient? (b) When the concentration 90 mm away is 37% of the value at the plane of insertion, how much time has elapsed? (c) How many moles of solute per unit area were initially inserted into the column of water? (d) Suppose that a trace amount of a substance having a diffusion coefficient 100 times smaller than that of the main solute was also initially introduced. For the time in (b), where would its concentration drop to $1/e$ of the value at the plane of insertion?

1.2. Let us suppose that mitochondria with a volume of 0.30 μm^3 each and a density of 1 110 kg m^{-3} (1.10 g cm^{-3}) diffuse like a chemical species. (a) What is the "molecular weight" of mitochondria? (b) Suppose that a chemically similar species of molecular weight 200 has a diffusion coefficient of 0.5×10^{-9} m^2 s^{-1}. If diffusion coefficients are inversely proportional to the cube root of molecular weights for this series of similar species, what is $D_{mitochondria}$? (c) If we assume that Equation 1.6 can adequately describe such motion, how long would it take on the average for a mitochondrion to diffuse 0.2 μm (a distance just discernible using a light microscope)? How long would it take for the mitochondrion to diffuse 50 μm (the distance across a typical leaf cell)? (d) If D_{ATP} is 0.3×10^{-9} m^2 s^{-1}, how long would it take ATP to diffuse 50 μm? Relate your answer to cellular use of the ATP produced by respiration—see part (c) of this problem.

1.3. Suppose that an unstirred air layer 1 mm wide is adjacent to a guard cell with a cell wall 2 μm thick. (a) Assume that an (infinitely) thin layer of $^{14}CO_2$ is introduced at the surface of the guard cell. If D_{CO_2} is 10^6 times larger in air than in the cell wall, what are the relative times for $^{14}CO_2$ to diffuse across the two barriers? (b) If it takes $^{14}CO_2$ just as long to cross an 8 nm plasmalemma as it does to cross the cell wall, what are the relative sizes of the two diffusion coefficients (assume that the $^{14}CO_2$ was introduced in a plane between the two barriers)? (c) Assuming that the partition coefficient for CO_2 is 100 times greater in the cell wall than in the plasmalemma, in which barrier is the permeability coefficient larger, and by how much?

1.4. Without correcting for the effect of an unstirred layer 20 μm thick outside a membrane 7.5 nm in thickness, the apparent (total) permeability coefficients were measured to be 1.0×10^{-4} m s^{-1} for D_2O, 2.0×10^{-5} m s^{-1} for methanol, and 3.0×10^{-8} m s^{-1} for L-leucine. For barriers in series, the overall permeability coefficient for species j (P_j^{total}) is related to those of the individual barriers (P_j^i) as follows: $1/P_j^{total} = \sum_j 1/P_j^i$. For purposes of calculation, we will assume that in the present case the unstirred layer on the inner side of the membrane is negligibly thin. (a) What is P_j for the external unstirred layer for each of the compounds? Assume that D_{D_2O} is 2.6×10^{-9} m^2 s^{-1}, $D_{methanol}$ is 0.80×10^{-9} m^2 s^{-1}, and $D_{leucine}$ is 0.20×10^{-9} m^2 s^{-1} in water at 25°C. (b) What are the permeability coefficients of the three compounds for the membrane? (c) From the results in (a) and (b), what are the main barriers for the diffusion of the three compounds in this case? (d) What are the highest possible values at 25°C of P_j^{total} for each of the three compounds moving across an unstirred layer of 20 μm and an extremely permeable membrane in series?

1.5. Consider a solute having a permeability coefficient of 10^{-6} m s^{-1} for the plasmalemma of a cylindrical *Chara* cell that is 100 mm long and 1 mm in diameter. Assume that its concentration remains essentially uniform within the cell. (a) How much time would it take for 90% of the solute to diffuse out into a large external solution initially devoid of that substance? (b) How much time would it take if diffusion occurred only at the two ends of the cell? (c) How would the times calculated in (a) and (b) change for 99% of the solute to diffuse out? (d) How would the times change if P_j were 10^{-8} m s^{-1}?

1.6. A cylindrical *Nitella* cell is 100 mm long and 1 mm in diameter, a spherical *Valonia* cell 10 mm in diameter, and a sperical *Chlorella* cell 4 μm in diameter. (a) What is the area/volume in each case? (b) Which cell has the largest amount of surface area per unit volume? (c) If it takes 1 s for the internal concentration of ethanol, which is initially absent from the cells, to reach half the external concentration for *Chlorella*, how long would it take for *Nitella* and *Valonia*? Assume that $P_{ethanol}$ is the same for all the cells. (d) Assume that the cell walls are equal in thickness. For a given internal pressure, which cell would have the highest cell wall stress (consider only the lateral wall for *Nitella*)?

References

Altman, P. L., and D. S. Dittmer. 1966. *Environmental Biology*. Federation of American Societies for Experimental Biology, Bethesda, Maryland.

Bull, H. B. 1964. *An Introduction to Physical Biochemistry*, F. A. Davis, Philadelphia.

Cleland, R. 1971. Cell wall extension. *Annual Review of Plant Physiology* 22: 197–222.

Crank, J. 1975. *The Mathematics of Diffusion*, 2nd ed. Clarendon Press, Oxford.

Dainty, J. 1963. Water relations of plant cells. *Advances in Botanical Research 1*: 279–326.

Dainty, J. 1976. Water Relations of plant cells. In *Transport in Plants II, Part A, Cells,* U. Lüttge and M. G. Pitman, eds. *Encyclopedia of Plant Physiology, New Series,* Vol. 2. Springer-Verlag, Berlin. Pp. 12–35.

Davson, H., and J. F. Danielli. 1952. *The Permeability of Natural Membranes,* 2nd ed. Cambridge University Press, Cambridge.

Diamond, J. M., and E. M. Wright. 1969. Biological membranes: The physical basis of ion and nonelectrolyte selectivity. *Annual Review of Physiology 31*:581–646.

Douce, R., and J. Joyard. 1979. Structure and function of the plastid envelope. *Advances in Botanical Research 7*:1–117.

Esau, K. 1977. *Anatomy of Seed Plants,* 2nd ed. Wiley, New York.

Fischkoff, S., and J. M. Vanderkooi. 1975. Oxygen diffusion in biological and artificial membranes determined by the fluorochrome pyrene. *Journal of General Physiology 65*:663–676.

Green, K., and T. Otori. 1970. Direct measurements of membrane unstirred layers. *Journal of Physiology 207*:93–102.

Gunning, B. E. S., and A. W. Robards, eds. 1976. *Intercellular Communication in Plants: Studies on Plasmodesmata.* Springer-Verlag, Berlin.

Harrison, R., and G. G. Lunt. 1980. *Biological Membranes: Their Structure and Function,* 2nd ed. Halsted Press, Wiley, New York.

Heber, U., and H. W. Heldt. 1981. The chloroplast envelope: structure, function, and role in leaf metabolism. *Annual Review of Plant Physiology 32*:139–168.

Incoll, L. D., S. P. Long, and M. R. Ashmore. 1977. SI units in publications in plant science. *Current Advances in Plant Science—Commentaries in Plant Science 28*:331–343.

Jacobs, M. H. 1967. *Diffusion Processes.* Springer-Verlag, New York.

Kamiya, N., M. Tazawa, and T. Takata. 1963. The relation of turgor pressure to cell volume in *Nitella* with special reference to mechanical properties of the cell wall. *Protoplasma 57*:501–521.

Kamiya, N. 1981. Physical and chemical basis of cytoplasmic streaming. *Annual Review of Plant Physiology 32*:205–236.

Lee, C. P., G. Schatz, and G. Daller. 1981. *Mitochondria and Microsomes.* Addison-Wesley, Reading, Massachusetts.

Lockhart, J. A. 1965. Cell extension. In *Plant Biochemistry,* J. Bonner and J. E. Varner, eds. Academic Press, New York. Pp. 826–849.

McNeil, M., A. G. Darvill, and P. Albersheim. 1979. The structural polymers of the primary cell wall of dicots. *Progress in the Chemistry of Organic Natural Products 37*:191–249.

National Bureau of Standards. 1977. *The International System of Units (SI),* 3rd ed. NBS Special Publication 330, U.S. Government Printing Office, Washington, D.C.

Northcote, D. H. 1972. Chemistry of the plant cell wall. *Annual Review of Plant Physiology 23*:113–132.

Paganelli, C. V., A. Ar, H. Rahn, and O. D. Wangensteen. 1975. Diffusion in the gas phase: the effects of ambient pressure and gas composition. *Respiration Physiology 25*:247–258.

Park, R. B., and P. V. Sane. 1971. Distribution of function and structure in chloroplast lamellae. *Annual Review of Plant Physiology 22*:395–430.

Preston, R. B. 1974. *The Physical Biology of Plant Cell Walls*. Chapman and Hall, London.

Robards, A. W. 1975. Plasmodesmata. *Annual Review of Plant Physiology 26*:13–29.

Sarkanen, K. V., and C. H. Ludwig, eds. 1971. *Lignins: Occurrence, Formation, Structure and Reactions*. Interscience, Wiley, New York.

Simon, E. W. 1974. Phospholipids and plant membrane permeability. *New Phytologist 73*:377–420.

Singer, S. J., and G. L. Nicolson. 1972. The fluid mosaic model of the structure of cell membranes. *Science 175*:720–731.

Sober, H. A., ed. 1970. *Handbook of Biochemistry with Selected Data for Molecular Biology*, 2nd ed. Chemical Rubber, Cleveland.

Stadelmann, E. J. 1977. Passive transport parameters of plant cell membranes. In *Regulation of Cell Membrane Activities in Plants*, E. Marré and O. Cifferi, eds. North Holland Publishers, Amsterdam. Pp. 3–18.

Stein, W. D. 1967. *The Movement of Molecules Across Cell Membranes*. Academic Press, New York.

Tanford, C. 1961. *Physical Chemistry of Macromolecules*. Wiley, New York.

Tolbert, N. E., ed. 1980. *The Plant Cell*. In *The Biochemistry of Plants: A Comprehensive Treatise*, P. K. Stumpf and E. E. Conn, eds., Vol. 1. Academic Press, New York.

Tyree, M. T. 1970. The symplast concept. *Journal of Theoretical Biology 26*:181–214.

Vinters, H., J. Dainty, and M. T. Tyree. 1977. Cell wall elastic properties of *Chara corallina*. *Canadian Journal of Botany 55*:1933–1939.

Weast, R. C., ed. 1981. *Handbook of Chemistry and Physics*, 62nd ed. Chemical Rubber, Cleveland.

Wright, E. M., and J. M. Diamond. 1969. Patterns of non-electrolyte permeability. *Proceedings of the Royal Society (Series B) 172*:227–271.

Zimmerman, U. 1978. Physics of turgor- and osmoregulation. *Annual Review of Plant Physiology 29*:121–148.

CHAPTER 2

Water

Water is the main constituent of most plant cells, as our discussion of both vacuoles and cell walls in the previous chapter has already suggested. The actual cellular water content varies with cell type and physiological condition. For example, the carrot root is about 85% water by weight; the young inner leaves of lettuce contain up to 95% water. Water makes up only 5% of certain dry seeds and spores, but when these become metabolically active, an increase in water content is essential for the transformation.

The physical and chemical properties of water make it suitable for a variety of purposes in plants. It is the medium in which diffusion of solutes takes place in plant cells. It is a solvent and often a participant in biochemical reactions. Its incompressibility means that water uptake can lead to cell expansion and that intracellular hydrostatic pressures important for plant support can develop. It is well suited to temperature regulation, since it has a high heat of vaporization, a high specific heat on a mass basis, and a high thermal conductivity for a liquid. Water is also an extremely good general solvent, in part owing to the small size of its molecules. Its polar character makes it suitable for dissolving other polar substances. Its high dielectric constant makes it a particularly appropriate solvent for ions. This latter property has far-reaching consequences for life, since most biologically important solutes are electrically charged. The mineral nutrients needed for growth and the organic products of photosynthesis are transported throughout the plant in aqueous solutions. In actively growing land plants, water continuity exists from the soil, through the plant, to the evaporation sites in the leaves. Water is quite transparent to visible irradiation, enabling sunlight to reach chloroplasts within the cells in leaves and to reach plants submerged at appreciable depths in lakes and oceans. Water is also intrinsically involved with metabolism. It is the source of the O_2 evolved in photosynthesis and the hydrogen used for CO_2 reduction. The generation of the important energy currency, ATP, involves the extraction of the components of water from ADP plus phosphate; in other words, such a phosphorylation is a *dehydration* process, taking place in an *aqueous* solution under biological conditions. At another level, the facts that ice floats and that water is densest near 4°C are important for organisms living in streams or rivers subject to freezing. Thus, an understanding of the physical and chemical properties of water is crucial for an overall understanding of physiology and ecology.

A number of isotopically different forms of water can be prepared, greatly facilitating experimental studies. If deuterium (^2H) replaces both of the usual hydrogens, the result is "heavy water," or deuterium oxide, with a molecular weight of 20. The entry of water into chemical reactions in the plant can then be studied by analyzing the deuterium content of the various substances involved as reactants or products. Tritium (^3H), a radioactive isotope with a half-life of 12.4 years, can also be incorporated into water. Such tritiated water has been used to measure the rate of water diffusion in plant tissue. Another alternative for tracing the pathway of water is to replace the usual ^{16}O isotope with ^{18}O. This "labeling" of water with ^{18}O was used in determining that the O_2 evolved in photosynthesis does in fact come from water and not from CO_2 (Ch. 5).

PHYSICAL PROPERTIES

Water differs markedly from substances having closely related electronic structures, a point we can illustrate by considering the series CH_4, NH_3, H_2O, HF, and Ne (Fig. 2.1). Each member contains 10 protons and 10 electrons, but the number of hydrogen atoms decreases along the series from methane to neon. The relatively high melting and boiling points for water, compared with substances having a similar electronic structure, are consequences of its strong intermolecular forces. In other words, thermal agitation does not easily disrupt the water-water bonding. This strong attraction between molecules is responsible for many characteristic properties of water.

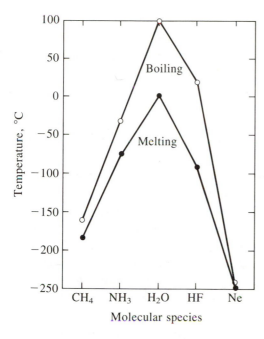

Figure 2.1
Boiling and melting points for molecules with 10 protons and 10 electrons, showing the strikingly high values for water. Nearly all biological processes take place from 0°C to 50°C, where water can be in the liquid state but the other substances cannot.

Hydrogen Bonding

The strong intermolecular forces in water result from the structure of the H_2O molecule (Fig. 2.2). The internuclear distance between the oxygen and each of the two hydrogens is approximately 0.099 nm; the H—O—H bond

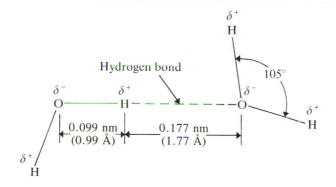

Figure 2.2
Schematic structure of water molecules, indicating the hydrogen bonding resulting from the electrostatic attraction between the net positive charge on a hydrogen (δ^+) in one molecule and the net negative charge on an oxygen (δ^-) in a neighboring water molecule.

angle is about 105°. The oxygen atom is strongly electronegative and tends to draw electrons away from the hydrogen atoms. The oxygen thus has a partial negative charge (δ^- in Fig. 2.2), and the two hydrogens each have a partial positive charge (δ^+). These positively charged hydrogens are electrostatically attracted to the negatively charged oxygens of two neighboring water molecules. This leads to *hydrogen bonding* between water molecules, with an energy of about 20 kilojoules (kJ) mol^{-1} of hydrogen bonds. Such bonding of water molecules to each other leads to increased order in aqueous solutions. In fact, liquid water becomes nearly crystalline in local regions, an extremely important condition for determining the molecular interactions and orientations that occur in aqueous solutions.

Ice is a coordinated crystalline structure in which essentially all the water molecules are joined by hydrogen bonds. When enough heat is added to melt the ice, some of these intermolecular hydrogen bonds are broken. The heat of fusion of ice at 0°C is 6.03 kJ mol^{-1}. Total rupture of the intermolecular hydrogen bonds involving both of its hydrogens would require 40 kJ mol^{-1} of water.* The heat of fusion thus indicates that (6.03 kJ mol^{-1})/(40 kJ mol^{-1}), or at most 0.15 (15%), of the hydrogen bonds are broken when ice melts. (Some energy is needed to overcome van der Waals attractions,† so less than 15% of the hydrogen bonds are actually broken upon melting.) Conversely, over 85% of the hydrogen bonds remain intact for liquid water at 0°C. Since 75.4 J is needed to heat 1 mol of water 1°C (the specific heat of water on a mole basis), it takes (0.0754 kJ mol^{-1} °C^{-1})(25°C), or 1.9 kJ mol^{-1}, to heat water from 0°C to 25°C. Even if all this energy were used to break hydrogen bonds, over 80% of the bonds would still remain intact at 25°C. Such bonding leads to the semicrystalline order found in aqueous solutions. The extensive intermolecular hydrogen bonding present in the liquid state contributes to the unique and biologically important properties of water that we will discuss throughout this chapter.

The energy required to separate molecules from a liquid and move them

---

* The actual magnitude of the hydrogen bond energy assigned to ice depends somewhat on the particular operational definition used in the measurement of the various bonding energies. Therefore, the values in the literature vary somewhat. See Eisenberg and Kauzmann, Pauling, or Schuster et al.

† Van der Waals forces are the electrostatic attractions between electrons in one molecule and the nucleus of an adjacent molecule minus the molecules' interelectronic and internuclear repulsive forces. In 1929 London showed these forces to be caused by the attraction between an electric dipole in some molecule and the electric dipole induced in an adjacent one. Therefore, van der Waals forces result from random fluctuations of charge and are important only for molecules that are very close together—in particular, for neighboring molecules.

into an adjacent vapor phase without a change of temperature is called the *heat of vaporization*. For water, the heat of vaporization at 100°C is 2.257 MJ kg^{-1}, or 40.7 kJ mol^{-1}. Per unit mass, this is the highest heat of vaporization of any known liquid, and it reflects the large amount of energy required to disrupt the extensive hydrogen bonding in aqueous solutions. More pertinent is the heat of vaporization of water at temperatures encountered by plants. At 25°C each mole of water evaporated requires 44.0 kJ (see App. II), meaning that substantial heat loss by the plant accompanies the evaporation of water in transpiration. Most of this vaporization energy is needed to break hydrogen bonds so that the water molecules can become separated in the gaseous phase. For example, if 80% of the hydrogen bonds remained at 25°C, then (0.80)(40 kJ mol^{-1}), or 32 kJ mol^{-1}, would be needed just to rupture them. Other energy is needed to overcome the attractive van der Waals forces and for the expansion involved in going from a liquid to a gas. The heat loss accompanying the evaporation of water is one of the primary means of temperature stabilization in land plants. It dissipates much of the energy gained from the absorption of solar irradiation. (Other important heat losses occur through conduction and convection; see Ch. 7.)

Surface Tension

Water has an extremely high surface tension, evident at an interface between water and air. Surface tension can be defined as the force per unit length pulling perpendicularly to a line in the plane of the surface. Because of its high surface tension, water can support a steel needle placed carefully on its surface. The surface tension at such an air-water interface is 0.0728 N m^{-1} (72.8 dyn cm^{-1}) at 20°C (see App. II for values at other temperatures). Another way of understanding surface tension is to regard it as the amount of energy required to expand a surface by unit area. (We note that surface tension has the dimensions not only of force per unit length, but also of energy per unit area.)

To see why energy is required to expand the water surface, let us consider water molecules brought from the interior of an aqueous phase to an air-water interface. If this involves only an expansion of the surface area—i.e., if there is no accompanying movement of other water molecules from the surface to the interior—then a loss in the water-water attraction of some of the intermolecular hydrogen bonds occurs with no compensating air-water attraction. Energy is thus needed to break the hydrogen bonds lost in moving water molecules from the interior of the solution to the air-water interface (0.0728 J to increase the area by 1 m^2). In fact, the term "surface free energy" is more appropriate from a thermodynamic point of view than the conventional term "surface tension."

The surface tension of an aqueous solution usually is only slightly influenced by the composition of the adjacent gas phase, but it can be markedly affected by certain solutes. Molecules are relatively far apart in a gas—the density of dry air at 0°C and one standard atmosphere (0.1013 MPa, 1.013 bar, or 760 mm Hg) is 1.293 kg m^{-3}—so the frequency and intensity of interactions between molecules in the gas phase and those in the liquid phase are correspondingly low. Certain solutes, such as sucrose or KCl, do not preferentially collect at the air-liquid interface and consequently have little effect on the surface tension of an aqueous solution. But fatty acids and certain lipids may become concentrated at interfaces and greatly reduce surface tensions. For example, 10 mol m^{-3} (10 mM) caproic acid (a 6-carbon fatty acid) lowers the surface tension about 0.015 N m^{-1} below the value for pure water, while only 0.05 mol m^{-3} capric acid (a 10-C fatty acid) lowers it by about 0.025 N m^{-1} (see Bull). Substances such as soaps (salts of fatty acids) or denatured proteins with large hydrophobic side chains may collect nearly exclusively at the interface and reduce the surface tension of aqueous solutions to as low as 0.020 N m^{-1}. A common feature of such "surfactant" molecules is that they have both polar and nonpolar regions.

Capillary Rise

The intermolecular attraction occurring between like molecules in the liquid state, such as the water-water attraction based on hydrogen bonds, is called *cohesion*. The attractive interaction between a liquid and a solid phase, such as between water and the walls of a small capillary (a cylindrical tube with a small internal diameter), is called *adhesion*. When the water-wall attraction is appreciable compared with the water-water cohesion, the walls are said to be *wettable*, and water then rises in such a capillary. At the opposite extreme, where the intermolecular cohesive forces within the liquid are substantially greater than the adhesion between the liquid and the wall material, the upper level of the liquid in a capillary is lower than the surface of the free solution. Such capillary depression occurs for liquid mercury in glass capillaries. For water in glass capillaries or in xylem vessels, the attraction between the water molecules and the walls is great, and so the liquid rises. Since capillary rise has important implications in plant physiology, we will discuss its characteristics quantitatively.

As an example appropriate to the evaluation of water rise in a plant, let us consider a capillary of inner radius r with wettable walls dipping into

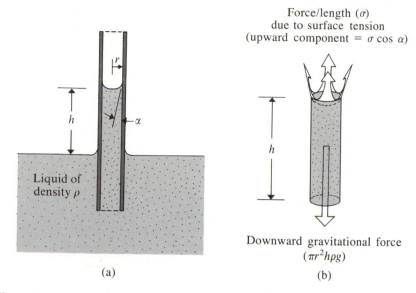

Figure 2.3

Capillary rise of a liquid: (a) parameters involved, and (b) force diagram indicating that surface tension projected in the upward direction is balanced by gravity acting downward.

some aqueous solution (Fig. 2.3a). The strong adhesion of water molecules to the wettable wall tends to cause the fluid to rise up the walls of the capillary. Since there is also a strong water-water cohesion in the bulk solution, water is concomitantly pulled up into the lumen of the capillary as water rises along the walls. Viewed another way, the air-water surface greatly resists being stretched, a property reflected in the high surface tension of water at air-water interfaces. This resistance tends to minimize the area of the air-water interface, a condition achieved if water also moves up in the lumen as well as along the sides of the capillary. The tendency for water to rise along the walls of the capillary is thus transmitted to a volume of fluid. We will call the height that the liquid rises in the capillary h and the contact angle that the liquid makes with the capillary wall α (Fig. 2.3). The extent of the rise depends on α, so the properties of the contact angle will now be examined a little more closely.

The size of the contact angle depends on the magnitude of the liquid-solid adhesive force compared with that of the liquid-liquid cohesive force. Specifically, Young's equation indicates that

$$\text{adhesion} = \frac{1 + \cos \alpha}{2} \text{ cohesion} \qquad (2.1)$$

(Equation 2.1 is sometimes called the Young and Dupré equation; see Adamson, Bikerman, or Davies and Rideal for an authoritative treatment.) When the adhesive force equals (or exceeds) the cohesive force in the liquid, $\cos \alpha$ must be unity; therefore the contact angle α then equals zero ($\cos 0° = 1$). This is the case for water in capillaries made of clear, smooth glass or having walls with polar groups on the exposed surface. When the adhesive force equals half the cohesive force, $\cos \alpha$ is zero and the contact angle is then $90°$ ($\cos 90° = 0$) by Equation 2.1. In this case, the level of the fluid in the capillary is the same as that in the bulk of the solution. This latter condition is closely approached in water-polyethylene adhesion, where α equals $94°$. As the liquid-solid adhesive force becomes relatively less compared with the intermolecular cohesion in the liquid phase, the contact angle (indicated in Fig. 2.3a) increases toward $180°$, and capillary depression occurs. For instance, water has an α of about $110°$ for paraffin, while the contact angle for mercury interacting with a glass surface is near $150°$. In such cases, the level of the liquid is lower in the capillary than in the bulk solution (see Fig. 2.3a, but imagine that $\alpha > 90°$, so the liquid curves downward where it intersects the surface of the capillary).

We can calculate the extent of capillary rise by considering the balance of two forces: (1) gravity acting downward and (2) surface tension, which leads to an upward force in the case of wettable walls in a vertical tube (see Fig. 2.3b). The force pulling upward acts along the inside perimeter of the capillary, a distance of $2\pi r$, with a force per unit length equal to σ, the surface tension. The component of this force acting vertically upward is $2\pi r \sigma \cos \alpha$, where α is the contact angle illustrated in Figure 2.3a. This upward force is balanced by the gravitational force acting on the liquid of density ρ and volume approximately $\pi r^2 h$.* The gravitational force is simply the mass involved times the gravitational acceleration g ($F = ma$, Newton's second law of motion). In the present case, gravity acts on a mass of fluid of approximately $\pi r^2 h \rho$, so the gravitational force is $\pi r^2 h \rho g$, acting downward. Since this force is balanced by $2\pi r \sigma \cos \alpha$ pulling in the opposite direction, the extent of rise, h, is given by equating the two forces ($\pi r^2 h \rho g = 2\pi r \sigma \cos \alpha$), which leads to

$$h = \frac{2\sigma \cos \alpha}{r \rho g} \tag{2.2a}$$

---

* Some fluid is also held in the "rim" of the meniscus, as indicated in Figure 2.3a. For narrow capillaries, the fluid in the meniscus increases the effective height of the column by about $r/3$.

Equation 2.2a indicates the readily demonstrated property that the extent of liquid rise in a capillary is inversely proportional to the radius of the tube. For water in glass capillaries as well as in many of the fine channels encountered in plants where the cell walls have a large number of exposed polar groups, the contact angle can be near zero, in which case cos α in Equation 2.2a can be set equal to 1. The density of water at 20°C is 998 kg m^{-3} (actually, ρ in Eq. 2.2a is the difference between the liquid density and the density of the displaced air, the latter being about 1 kg m^{-3}), and the acceleration due to gravity, g, is about 9.80 m s^{-2} (see App. II). Using a surface tension for water at 20°C of 0.0728 N m^{-1} (App. II), we obtain the following relation between the height of the rise and the capillary radius when the contact angle is zero:

$$h_{(m)} = \frac{1.49 \times 10^{-5} \ m^2}{r_{(m)}} \tag{2.2b}$$

where the subscript (m) in Equation 2.2b means that the dimensions involved are expressed in meters.

Capillary Rise in the Xylem

Although Equation 2.2 refers to the height of capillary rise only in a static sense, it still has important implications concerning the movement of water in plants. To be specific, let us consider a xylem vessel having a lumen radius of 20 μm. From Equation 2.2b, we calculate that water will rise in it to the following height:

$$h_{(m)} = \frac{(1.49 \times 10^{-5} \ m^2)}{(20 \times 10^{-6} \ m)} = 0.75 \ m$$

Such a capillary rise would be sufficient to account for the extent of the upward movement of water in small plants, although it says nothing about the rate of such movement (we will consider the rate in Ch. 9). For water to reach the top of a 30 m tree by capillary action, however, the vessel would have to be 0.5 μm in radius. This is much smaller than that observed for xylem vessels, indicating that capillary rise in channels of the size of the xylem cells cannot account for the extent of the water rise in tall trees. Furthermore, the lumens of the xylem vessels are not open to the air at the upper end, and thus they are not really analogous to the capillary depicted in Figure 2.3.

The numerous interstices in the cell wall of xylem vessels form a mesh-work of many small, tortuous capillaries, which can lead to an extensive capillary rise of water in a tree. A representative "radius" for these channels in the cell wall might be 5 nm. According to Equation 2.2b, a capillary of 5 nm radius could support a water column of 3 km—far in excess of the needs of any plant. In other words, the cell wall could act as a very effective wick for water rise in its numerous small interstices, although the actual rate of such movement up a tree would generally be far from sufficient to replace the water lost by transpiration (see Ch. 9).

Because of the appreciable water-wall attraction which can develop both at the top of a xylem vessel and in the numerous interstices of its cell walls, water already present in the lumen of a xylem vessel can be sustained or supported at great heights. The upward force, transmitted to the rest of the solution in the xylem vessel by water-water cohesion, overcomes the gravitational pull downward. Thus, the key to holding up water already present in the xylem vessel against the pull of gravity is the very potent attractive interaction (adhesion) between water and the cell wall surfaces at the top of the vessel. Two aspects of this warrant further consideration. First, what happens if the lumen of the xylem vessel becomes filled with air? Will water then refill it? The capillary rise of water is not sufficient to refill most air-filled xylem vessels greater than about 1 m in length, so that, in general, air-filled vessels are lost for conduction, as occurs for the inner annual rings of most trees. The second aspect is more subtle, and we will consider it more extensively.

Tensile Strength

The pulling on water columns that occurs in capillary rise and in the sustaining of water in the xylem requires that water be put under tension, or negative pressure. In other words, water must have a definite tensile strength—the maximum tension (force per unit area) that it can withstand before breaking. The intermolecular hydrogen bonds lead to this tensile strength by resisting the pulling apart of water in a column. According to experiment, water withstands negative pressures (tensions) of up to approximately 30 MPa (300 bars) at 20°C without breaking. (Tensile strengths are observed to depend on the wall material, the diameter of the vessel in which the determinations are made, and any solutes present in the water; see Hayward or Oertli.) This tensile strength for water is nearly 10% of that for copper or aluminum, and is sufficiently high to meet the demands encountered for water movement in plants. The great cohesive forces between water molecules thus allow an appreciable tension to exist in an uninterrupted water column in a wettable capillary or tube such as a xylem vessel. This property is important for the continuous movement of water from the root through the plant to the surrounding atmosphere during transpiration.

In contrast to metals, which are in a stable state, water under tension is actually in a metastable state. If gas bubbles form in the water under tension in the xylem vessels, the water column can be ruptured. Thus, the introduction of another phase can disrupt the metastable state describing water under tension. Minute air bubbles sometimes spontaneously form in the xylem as the external temperature rises. These usually adhere to the walls of the xylem vessel (or the pores of the perforation plate), and the gas in them slowly redissolves. If, however, they grow large enough or a number of small bubbles coalesce, that xylem vessel could cease functioning. Moreover, freezing of the solution in xylem vessels can lead to bubbles in the ice (the solubility of air is quite low in ice), and these bubbles could then interrupt the water columns upon thawing. Some plants are in fact damaged by freezing and thawing of xylem sap and the consequent loss of water continuity in the xylem vessels (see Zimmermann and Brown).

Electrical Properties

Another important physical characteristic of water is its extremely high dielectric constant, which is also a consequence of its molecular structure. A high dielectric constant for a solvent lowers the electrical forces between charged solutes. To quantify the magnitude of electrical effects occurring in a fluid, let us consider two ions having charges Q_1 and Q_2 and separated from each other by a distance r. The electrical force exerted by one ion on the other is expressed in Coulomb's law,

$$\text{electrical force} = \frac{Q_1 Q_2}{4\pi\varepsilon_0 D r^2} \tag{2.3}$$

where ε_0 is a proportionality constant known as the permittivity of a vacuum and D is a dimensionless quantity called the dielectric constant. D equals unity in a vacuum, and it is 1.00058 in air at 0°C at a pressure of one standard atmosphere.

Any substance composed of highly polar molecules, such as water, generally has a high dielectric constant. Specifically, D for water is 80.2 at

20°C and 78.4 at 25°C—extremely high values for liquids. By contrast, the dielectric constant of the nonpolar liquid hexane is only 1.87, a low value typical of many organic solvents. Based on these two vastly different dielectric constants and using Equation 2.3, we can calculate that the attractive electrical force between ions such as Na^+ and Cl^- is (80.2/1.87), or 43 times greater in hexane than in water. The much stronger attraction between Na^+ and Cl^- in hexane reduces the amount of NaCl that will dissociate, compared with the dissociation of this salt in aqueous solutions. Stated another way, the much weaker electrical forces between ions in aqueous solutions, compared with those in organic solvents, permits a larger concentration of ions to be in solution. Water is thus a good solvent for charged particles.

The electrostatic interaction between ions and water partially cancels or screens out the local electrical fields of the ions (Fig. 2.4). Cations attract the

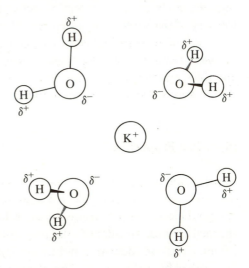

Figure 2.4
Orientation of water molecules around a potassium ion, indicating the charge arrangement that leads to screening of the local electrical field.

negatively charged oxygens of the water molecule; anions attract its positively charged hydrogen atoms. The water molecules oriented around the charged particles produce local electrical fields opposing the fields of the ions. The resultant screening diminishes the electrical interaction between the ions and allows more of them to remain in solution, which is the molecular basis for water's high dielectric constant (see Fig. 2.4).

The energy of attraction between water and nonpolar molecules is usually less than the energy required to break water-water hydrogen bonds. Nonpolar compounds are therefore not very soluble in water. Certain substances, such as detergents, phospholipids, and proteins, can have both a nonpolar and also a polar region in the same molecule. In aqueous solutions these compounds can form aggregates, called *micelles*, in which the nonpolar groups of the molecules are in the center, while the charged or polar groups are external and interact with water. The lack of appreciable attractive electrostatic interaction between polar or charged species and the nonpolar (hydrophobic) regions of membrane lipids underlies the ability of biological membranes to limit the passage of such solutes into and out of cells and organelles.

CHEMICAL POTENTIAL

The chemical potential of species j thermodynamically indicates the free energy associated with it and available for performing work. Since there are many concepts to be mastered before understanding such a statement, we will first briefly explore the concept of free energy. (Further details are presented later in this chapter, in Ch. 6, and in App. VI.) We can then introduce specific terms in the chemical potential of species j and, finally, consider the extremely important case of water.

Free Energy and Chemical Potential

Plants and animals require a continual input of free energy. If we were to remove the sources of free energy, organisms would drift toward equilibrium and consequent cessation of life. The ultimate source of free energy is the sun. Photosynthesis converts its plentiful radiant energy into free energy, which is stored first in intermediate energy "currencies" like ATP and then in the altered chemical bonds that result when CO_2 and H_2O react to form a carbohydrate and O_2. When the overall photosynthetic reactions are reversed during respiration, the available free energy is reconverted into suitable energy currencies like ATP. In turn, such free energy is used to perform biological work, such as transporting amino acids into cells, pumping blood, powering the electrical machinery of the brain, or lifting weights. For every such process on both a molecular and a global scale, the free energy decreases. In fact, the structure of cells, as well as that of ecosystems, is governed by the initial supply of free energy and by the inexorable laws of thermodynamics which describe the expenditure of that free energy.

Instead of viewing the whole universe, a thermodynamicist focuses on some small part of it referred to as a "system." Such a system might be K^+ dissolved in water, a plant cell, a leaf, or even a whole plant. Something happens to the system; the K^+ concentration is increased, some protein is made in a cell, a leaf abscises, or water moves up a tree. We then say that the system changes from state A to state B. (The change may of course be brought about in many different ways.) The minimum amount of work needed to cause the change from A to B can be defined as the free energy increase associated with it. Alternatively, the free energy decrease in going from state B back to state A represents the maximum amount of work that can be derived from the transition. (If there were no frictional or other irreversible losses, we could do without the words "minimum" and "maximum.") Thus, the limits to the work done on or by a system when it undergoes a transition from one state to another concern changes in free energy. Knowledge of the free energy under one condition compared with another allows us to predict the direction of spontaneous change or movement: a spontaneous change in a system at constant temperature and pressure always proceeds in the direction of decreasing free energy.

Most systems of interest in biology are subject to a constant pressure (atmospheric) and remain at a constant temperature, at least for short periods. In discussing the energetics of processes for systems at constant temperature and pressure, the appropriate quantity is known as the *Gibbs free energy*. The Gibbs free energy has a very useful property: it always decreases for a spontaneous process at constant temperature and pressure. Under such conditions, the decrease in Gibbs free energy equals the maximum amount of energy available for work, while if it increases for some transition, the change in Gibbs free energy represents the minimum amount of work required. (We will write down an equation for the Gibbs free energy in Ch. 6; App. VI deals with a number of formal matters associated with its use.) For present purposes, we will simply note that the Gibbs free energy of an entire system is made up of additive contributions from each of the species present. We will therefore shift our emphasis to the individual components of the system and the manner in which their free energy can be described.

To every chemical component in a system we can assign a free energy per mole of that species. This quantity is called the *chemical potential* of species j and is given the symbol μ_j. We can view μ_j as a property of species j, telling how that species will react to a given change, e.g., a transition of the system from state A to state B. During the transition, the chemical potential of species j changes from μ_j^A to μ_j^B. If μ_j^B is less than μ_j^A, then the free energy per mole of species j decreases. Such a process can take place spontaneously as far as species j is concerned—water flowing downhill is an example of such a spontaneous process. (We are restricting our attention to changes that occur when the overall system is at a constant temperature and subjected to a constant pressure.) In principle, we can harness the spontaneous change to do work. The maximum amount of work that can be done per mole of species j is $\mu_j^A - \mu_j^B$ (Fig 2.5). Suppose now that μ_j^B equals μ_j^A. As far as species j is concerned, no work is involved in the transition from state A to state B; species j is then at equilibrium (see Fig. 2.5). We will show later not only that living systems are as a whole far from equilibrium, but also that many chemical species are not in equilibrium across cellular membranes. Finally, let us consider the case where μ_j^B is greater than μ_j^A. A change increasing the chemical potential of a substance can only occur if some other change in the system supplies the free energy required. The pumping of blood along arteries, using ATP to cause contraction of the heart muscles, is an instance where the increase in chemical potential of water in the blood is accompanied by an even larger decrease in the free energy associated with ATP (we will consider ATP in detail in Ch. 6). The minimum amount of work that must be done per mole of species j to cause the energetically uphill transition is $\mu_j^B - \mu_j^A$ (Fig. 2.5).

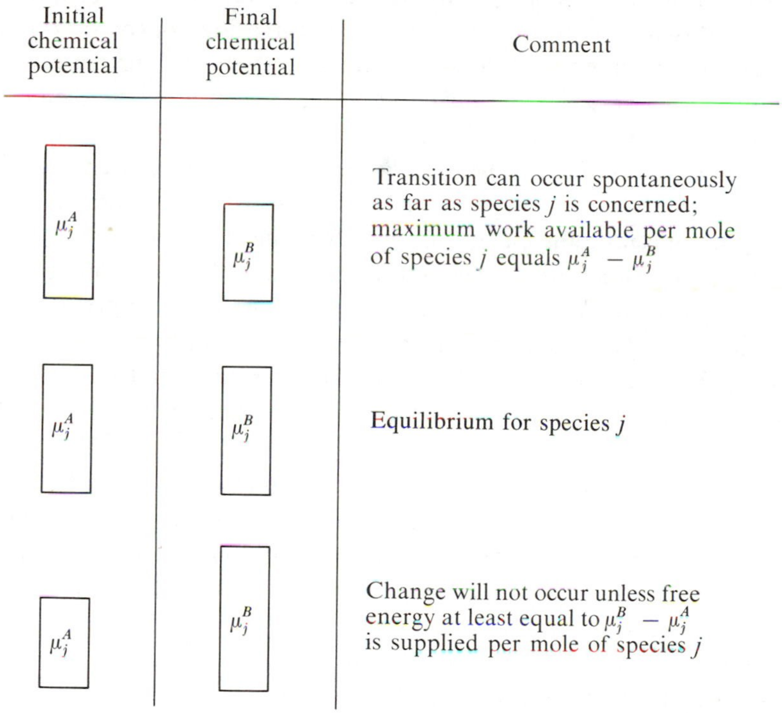

Initial chemical potential	Final chemical potential	Comment
μ_j^A	μ_j^B	Transition can occur spontaneously as far as species j is concerned; maximum work available per mole of species j equals $\mu_j^A - \mu_j^B$
μ_j^A	μ_j^B	Equilibrium for species j
μ_j^A	μ_j^B	Change will not occur unless free energy at least equal to $\mu_j^B - \mu_j^A$ is supplied per mole of species j

Figure 2.5
Possible changes in the chemical potential of species j that can accompany a transition from initial state A to final state B—as might occur in a chemical reaction or in crossing a membrane. The heights of the bars representing μ_j correspond to the relative values of the chemical potential.

The condition for a spontaneous change—a decrease in chemical potential—has important implications for discussing fluxes from one region to another. In particular, we can use the chemical potential difference between two locations as a measure of the "driving force" for the movement of that component. Thus, the bigger the chemical potential difference, $\mu_j^A - \mu_j^B$, the faster the spontaneous change takes place—in this case, the larger the flux density of species j from region A to region B.

The chemical potential of a substance clearly depends on its chemical composition, but it is also influenced by other factors. Here we shall discuss in an intuitive way the environmental factors that might influence μ_j. To begin with, we note that chemical potential depends on the "randomness" (entropy) of the system, and that concentration is a quantitative description of this randomness. For example, the passive process of diffusion discussed in Chapter 1 describes neutral molecules spontaneously moving from some region to another where their concentration is lower. In more formal terms, the net diffusion of species j proceeds in the direction of decreasing μ_j, which in this case is the same as that of decreasing concentration. The concept of chemical potential was thus implicit in the development of Fick's equations in Chapter 1. The importance of concentration gradients—or differences—to the movement of a substance was also established in the discussion of diffusion. Here we will present the effect of concentration on chemical potential in a somewhat more sophisticated manner, using that part of the concentration which is thermodynamically active.

Chemical potential also depends on pressure,* which for situations of interest in plant physiology and ecology means hydrostatic pressure. The existence of pressure gradients can cause movements of fluids—the flow of crude oil in long-distance pipelines, blood in arteries, and sap in xylem vessels—which is another way of saying that pressure differences can affect the direction for spontaneous changes.

Since electrical potential also affects the chemical potential of charged particles, it must be considered when predicting the direction of their movement. It is a matter of definition that work must be done to move a positively charged particle to a higher electrical potential. Accordingly, if an electrical potential difference is imposed across an electrolyte solution containing a uniform concentration of ions throughout, then cations will spontaneously move in one direction (toward the cathode, i.e., to regions of lower electrical

---

* Although local pressure effects can be readily incorporated into μ_j, the system as a whole must experience a constant external pressure for the Gibbs free energy to have the useful properties discussed above.

potential), while anions will move in the opposite direction (toward the anode). It is clear, then, that to describe the chemical potential of a charged species we must include an electrical term in μ_j.

Another contributor to chemical potential is gravity. We can readily appreciate that position in a gravitational field affects μ_j, inasmuch as work must be done to move the substance upward. Although the gravitational term can be neglected for ion and water movements across plant cells and membranes, it is important for water movement in a tall tree.

In summary, the chemical potential of a species should depend on its concentration, the pressure, the electrical potential, and gravity. We can compare the chemical potentials of a species on two sides of a barrier to decide whether or not the species is in equilibrium. If μ_j is the same on the two sides, we would not expect a net movement of species j to occur spontaneously across the barrier. We will use a comparison of the relative values of the chemical potential of species j at various locations to predict the directions for spontaneous movement of that chemical substance (toward lower μ_j), just as we can use a comparison of temperatures to predict the direction for heat flow (toward lower T). Finally, we will find that $\Delta\mu_j$ from one region to another gives a convenient measure of the driving force on species j.

Analysis of Chemical Potential

For convenience—and also because it has proved experimentally valid—we will represent the chemical potential of any species j by the following sum of the various components into which it can be analyzed:

$$\mu_j = \mu_j^* + RT \ln a_j + \bar{V}_j P + z_j FE + m_j gh \qquad (2.4)$$

One measure of the elegance of a mathematical expression is the amount of information it contains. Based on this criterion, Equation 2.4 represents one of the most elegant relations in all biology. After defining and describing the various contributions to μ_j indicated in Equation 2.4, we will consider the various terms in greater detail for the important special case where species j is water.

Chemical potential, like electrical or gravitational potential, is a relative quantity. By this we mean that it must be expressed relative to some arbitrary level. Thus, an unknown additive constant, or reference level, μ_j^*, is included in Equation 2.4. Since it contains an unknown constant, the actual value of the chemical potential is not determinable. But for most applications of chemical potential, we are interested in the difference in the chemical potential between two particular locations, so only relative values of the chemical potential are important anyway. The arbitrariness of μ_j^* is of little consequence because it is added to each of the chemical potentials under comparison. Therefore, μ_j^* cancels out when the chemical potential in one location is subtracted from that in another to obtain the chemical potential difference between the two locations. Figure 2.5 further illustrates the usefulness of considering differences in chemical potential instead of actual values. Note that the units of μ_j and of μ_j^* are energy per mole of the substance, e.g., J mol^{-1}.

In certain cases we can adequately describe the chemical properties of species j by using the concentration of that solute, c_j. Owing to molecular interactions, however, this usually requires that the total solute concentration be low. Molecules of solute species j interact with each other as well as with other solutes in the solution, and this influences the behavior of species j. Such intermolecular interactions increase as the solution becomes more concentrated. The use of concentrations for describing the thermodynamic properties of some solute thus indicates an approximation, except in the limiting case of infinite dilution where interactions between solute molecules are negligible. Where exactness is required, *activities*—which may be regarded as "corrected" concentrations—are employed. Consequently, for general thermodynamic considerations, as in Equation 2.4, the influence of the amount of a particular species j on its chemical potential is handled not by its concentration but by the activity of that species, a_j. The activity of solute j is related to its concentration by means of an activity coefficient, γ_j:

$$a_j = \gamma_j c_j \tag{2.5}$$

The activity coefficient is usually less than unity, since the thermodynamically *effective* concentration of a species—its *activity*—is generally less than is its actual concentration.*

For an ideal solute, γ_j is unity, and the activity of species j equals its concentration. This condition can be approached for real solutes in certain dilute aqueous solutions, especially for neutral species. Activity coefficients for charged species can be appreciably less than unity because of the importance of electrical interactions (see pp. 110–112).

The activity of a solvent is defined differently from that of a solute (the solvent is the species having the highest mole fraction in a solution; we will

---

* See Bull or Edsall and Wyman for the very special cases where γ_j can be greater than unity for high concentrations of certain ions.

define mole fraction on p. 73). For a solvent, a_j is $\gamma_j N_j$, where N_j is its mole fraction. An ideal solvent has $\gamma_{solvent}$ equal to 1, meaning that the interactions of solvent molecules with the surrounding molecules are indistinguishable from their interactions in the pure solvent. An ideal solution has all activity coefficients equal to unity.

In the expression for chemical potential, Equation 2.4, the term involving the activity is $RT \ln a_j$. Therefore, the greater the activity of species j—or, loosely speaking, the higher its concentration—the larger will be its chemical potential. The logarithmic form can be "justified" in a number of different ways, all based on agreement with empirical observations. (The particular justifications invoked in this text are given on pp. 121 and 575. It should be noted that the appearance of a logarithmic term in energy expressions is not without precedent. For example, the expression for work done in compressing a gas also leads to a logarithmic term.) The factor RT multiplying $\ln a_j$ in Equation 2.4, where R is the gas constant (see App. II), gives to the activity term the units of energy per mole.

The term $\overline{V}_j P$ in Equation 2.4 represents the effect of pressure on chemical potential. Since essentially all measurements in plant physiology and ecology are made on systems subjected to atmospheric pressure, it is convenient to define P as the pressure in excess of this, and we will adopt such a convention here. $\overline{V}_j$ is the differential increase in volume of a system when a differential amount of species j is added, with other species, temperature, pressure, electrical potential, and gravitational position remaining constant:

$$\overline{V}_j = \left(\frac{\partial \overline{V}}{\partial n_j} \right)_{n_i, T, P, E, h} \tag{2.6}$$

The subscript n_i on the partial derivative in Equation 2.6 means that the number of moles of each species present, other than species j, is held constant when the derivative is taken; the other four subscripts are included to remind us of the additional parameters that must be held constant.

$\overline{V}_j$ is called the *partial molal volume* of species j. (μ_j is actually the partial molal Gibbs free energy, $(\partial G/\partial n_j)_{n_i, T, P, E, h}$, as we will see in Ch. 6 and App. VI.) The partial molal volume of a substance is often nearly equal to the volume of a mole of that species, but because there is in general a slight change in total volume when substances are mixed, $\overline{V}_j$ is not exactly equal to it.[*]

---

[*] The addition of small concentrations of certain salts can actually cause an aqueous solution to contract—a phenomenon known as electrostriction—which in this special case would result in a negative value for $\overline{V}_j$.

Since work is often expressed as pressure times change in volume, we note that $\bar{V}_j P$ has the correct units for μ_j: energy per mole ($\bar{V}_j$ is the volume per mole). To help justify the form of the pressure term, let us imagine that the solution containing species j is built up by adding small volumes of species j while the system is maintained at a constant pressure P. The work done to add a mole of species j would then be the existing pressure times some volume change of the system characterizing a mole of that species, namely, $\bar{V}_j$. (In App. VI the inclusion of the $\bar{V}_j P$ term in the chemical potential of species j is justified more formally.)

The influence of electrical potential on the μ_j of an ion is expressed by the term $z_j F E$ in Equation 2.4, where z_j is an integer representing the charge number of species j, F is a constant known as the faraday (see p. 107), and E is the electrical potential. Because water is uncharged ($z_w = 0$), the electrical term does not contribute to its chemical potential. But electrical potential is of central importance when discussing ions and the origin of membrane potentials; both of these are examined in detail in Chapter 3, where we explicitly consider the $z_j F E$ term. The full expression for chemical potential given by Equation 2.4, including $z_j F E$, is often referred to as *electrochemical potential* when discussing the properties of charged particles.

Finally, Equation 2.4 includes a gravitational term $m_j g h$ expressing the amount of work required to raise an object of mass m_j per mole to a vertical height h, where g is the gravitational acceleration (close to 9.8 m s^{-2}; see App. II for details). In the case of water, which is the primary concern of this chapter, the mass per mole m_w is $0.018016 \text{ kg mol}^{-1}$ ($18.016 \text{ g mol}^{-1}$). The gravitational term in the chemical potential is not only important for the fall of rain, snow, or hail, but also it affects the percolation of water downward through porous soil and the upward movement of water in a tree. The gravitational term $m_j g h$ can have units of $(\text{kg mol}^{-1})(\text{m s}^{-2})(\text{m})$, or $\text{kg m}^2 \text{ s}^{-2} \text{ mol}^{-1}$, which is J mol^{-1}. (Conversion factors between joules, ergs, calories, and other energy units are given in App. III.)

Standard State

The additive constant term μ_j^* in Equation 2.4 is actually the chemical potential of species j for a specific reference state. From the preceding definitions of the various quantities involved, this reference state is attained when the following conditions hold: the activity of species j is unity ($RT \ln a_j = 0$); the hydrostatic pressure equals atmospheric pressure ($\bar{V}_j P = 0$);

the species is uncharged or the electrical potential is zero ($z_jFE = 0$); we are at the zero level for the gravitational term ($m_jgh = 0$); and the temperature equals the temperature of the system under consideration. Under these conditions, μ_j equals μ_j^* (Eq. 2.4).

As suggested on p. 66, an activity of unity is defined in different ways for the solute and the solvent. To describe liquid properties, such as the dielectric constant, the heat of vaporization, and the boiling point, the most convenient standard state is that of the pure solvent. For a solvent, $a_{solvent}$ equals $\gamma_{solvent}$ times $N_{solvent}$, and hence the activity is unity when the mole fraction $N_{solvent}$ is unity ($\gamma_{solvent} = 1$ for pure solvent). For example, the standard reference state for water is pure water at atmospheric pressure and at the temperature and gravitational level of the system under consideration. Water has an activity of unity when N_w as given by Equation 2.8 is one. The concentration of water on a molality basis (number of moles of a substance per kg of water for aqueous solutions) is then (1 kg)/(0.018016 kg mol^{-1}), or 55.5 molal. The accepted convention for a solute, on the other hand, is that a_j is unity when γ_jc_j equals 1 molal. For example, if γ_j equals one, a solution with a 1 molal concentration of solute j has an activity of 1 molal for that solute. Thus, the standard state for an ideal solute is when its concentration is 1 molal, in which case $RT \ln a_j$ is zero.* A special convention is used for the standard state of a gas such as CO_2 or O_2 in an aqueous solution— namely, the activity is unity when the solution is in equilibrium with a gas phase containing that gas at a pressure of one atmosphere. (At other pressures, the activity is proportional to the partial pressure of that gas in the gas phase.)

The value of the chemical potential for a solute in the standard state, μ_j^*, depends on the solvent. To be explicit, we will next argue that μ_j^* for a polar molecule is smaller for an aqueous solution (where the species readily dissolves) than it is for an organic phase such as olive oil (where it is not as soluble). Let us consider a two-phase system consisting of water with an overlying layer of olive oil, similar to that discussed in Chapter 1 in relation to measuring partition coefficients (p. 29). If we put the polar solute into our system, shake the water and olive oil together, and wait for the two phases to separate out again, the solute concentration will be higher in the bottom or water phase. Let us next analyze this event using symbols, where phase A represents water and phase B is olive oil. Waiting for the phase separation is equivalent to waiting for the attainment of equilibrium, and so μ_j^A is then

---

* Since molality involves the moles of a substance per kg of solvent, it is a legitimate SI unit.

equal to μ_j^B. In the present case this means that $\mu_j^{*,A} + RT \ln a_j^A$ equals $\mu_j^{*,B} + RT \ln a_j^B$ by Equation 2.4. But the polar solute is more soluble in water than it is in olive oil; hence $RT \ln a_j^A$ is greater than $RT \ln a_j^B$. Because we are at equilibrium, this means that $\mu_j^{*,A}$ must be less than $\mu_j^{*,B}$, as we already indicated.

Hydrostatic Pressure

Because of the rigid cell walls, fairly large hydrostatic pressures can exist in plant cells, while hydrostatic pressures in animal cells generally are relatively small. In addition to being involved in the support of a plant, hydrostatic pressures are important for the movement of water and solutes in the xylem and probably also in the phloem. The term expressing the effect of pressure on the chemical potential of water is $\bar{V}_w P$ (see Eq. 2.4), where $\bar{V}_w$ is the partial molal volume of water and P is the hydrostatic pressure in the aqueous solution in excess of the ambient atmospheric pressure. The approximate value of $\bar{V}_w$ can be estimated using the following argument. The density of water is about $1\,000$ kg m^{-3}; therefore, when 1 mol or 18.0×10^{-3} kg of water is added to water, the volume increases by 18.0×10^{-6} m^3. Actually, using the definition of $\bar{V}_w$ as a partial derivative (see Eq. 2.6), we need add only an infinitesimally small amount of water (dn_w) and observe the infinitesimal change in volume of the system (dV). We thus find that $\bar{V}_w$ for pure water is 18.0×10^{-6} m^3 mol^{-1} (18.0 cm^3 mol^{-1}). Although $\bar{V}_w$ can be influenced by the various solutes present, it is generally close to 18.0×10^{-6} m^3 mol^{-1} for a dilute solution, a value that we will use for calculations in this book.

Various units are used for expressing pressures in plant physiology and ecology (see footnote, p. 40). A pressure of one standard atmosphere, or 0.1013 MPa, can support a column of mercury 760 mm high or a column of water 10.3 m high. Research concerning the water relations of plants often uses pressures in bars. However, as indicated in Chapter 1, the SI unit for pressure is the pascal (Pa), which is 1 N m^{-2}; a quantity of convenient size for hydrostatic pressures is often the MPa (1 MPa $= 10$ bars $= 9.87$ atmospheres). (An extensive list of conversion factors for pascals, bars, atmospheres, and other pressure units is given in App. III, which also includes values for related quantities such as RT.) Since pressure is force per unit area, it is dimensionally the same as energy per unit volume, e.g., 1 Pa $= 1$ N m^{-2} $= 1$ J m^{-3}. Since $\bar{V}_w$ has the units of m^3 mol^{-1}, the quantity $\bar{V}_w P$ and hence μ_w itself can be expressed in J mol^{-1}.

Water Activity and Osmotic Pressure

The presence of solutes in an aqueous solution tends to decrease the activity of water (a_w). As a first approximation, we can view the decrease in a_w when more and more solutes are added as simply a dilution effect. In other words, the concentration (or mole fraction) of water becomes lower when the water molecules are displaced by those of the solute. As its activity thus decreases, we would expect a lowering of the chemical potential of water, as does indeed occur (see the $RT \ln a_j$ term in μ_j). We also know that the presence of solutes can lead to an osmotic pressure (Π) in the solution. An increase in the concentration of solutes raises the osmotic pressure, indicating that Π and a_w change in opposite directions. In fact, the osmotic pressure and water activity are related, a fundamental definition of Π being

$$RT \ln a_w = -\bar{V}_w \Pi \qquad (2.7)$$

where the subscript w refers to water. As solutes are added, a_w decreases from its value of unity for pure water, $\ln a_w$ is therefore negative, and Π (defined by Eq. 2.7) is positive, which is consistent with the arguments presented above. Equation 2.7 also indicates that $RT \ln a_w$, a term in the chemical potential of water (see Eq. 2.4), can be replaced by $-\bar{V}_w \Pi$.

Unfortunately, there is much variation in the use of the terms "osmotic pressure" and "osmotic potential," as well as the algebraic sign of the quantity. Osmotic pressures were originally measured using an osmometer (Fig. 2.6), a device employing a membrane that in the ideal case is permeable to water but not to any of the solutes present. When pure water is placed on one side of the membrane, and some solution on the other, there is a net diffusion of water toward the side with the solutes. To counteract this tendency and establish equilibrium, a hydrostatic pressure is necessary on the solution side. This pressure is often called the *osmotic pressure*. It depends, of course, on the presence of the solutes. Does an isolated solution have an osmotic pressure? In the sense of requiring an applied hydrostatic pressure to maintain equilibrium, the answer is no. Yet the same solution when placed in an osmometer can manifest an osmotic pressure. Thus, some physical chemistry texts say that a solution has an osmotic *potential*—i.e., it could show an osmotic pressure if placed in an osmometer—and that osmotic potentials are a property characteristic of solutions (see Morris). Many plant physiology books (e.g., Salisbury and Ross) have defined the *negative* of the same quantity as the osmotic potential, so that the water potential (a quantity we will introduce shortly) will depend directly on the osmotic

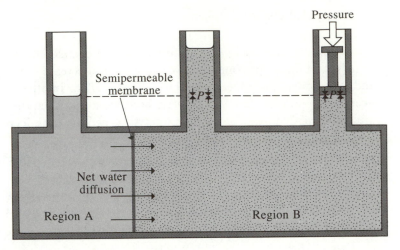

Figure 2.6
Schematic diagram indicating the principle behind an osmometer in which a semipermeable membrane (permeable to water, but not to solutes) separates pure water (region A) from water containing solutes (region B). Water tends to diffuse to where it is less concentrated, in this case into region B. This causes the solution to rise in the open central column until at equilibrium the hydrostatic pressure (P) at the base of this column is equal to the osmotic pressure (Π) of the solution. Alternatively, we can apply a hydrostatic pressure P to the right-hand column to prevent a net diffusion of water into region B, this P again being equal to Π.

potential, rather than on its negative. We will refer to Π, defined by Equation 2.7, as the osmotic pressure (see Van Holde) and to its negative as the osmotic potential. (The osmotic potential will be symbolized by Ψ_Π, a component of the water potential Ψ; see p. 78.)

A direct and convenient way of evaluating osmotic pressure and water activity as defined by Equation 2.7 is to use another *colligative* property. The four colligative properties of a solution—those that depend on the concentration of solutes regardless of their nature, at least in a dilute enough solution—are the freezing point depression, the boiling point elevation, the lowering of the vapor pressure of the solvent, and the osmotic pressure. Thus, if the freezing point of cell sap is measured, $\Pi^{cell\ sap}$ can be calculated using the freezing point depression of $1.86°C$ for a 1 molal solution together with the *Van't Hoff relation*.

The Van't Hoff Relation

For many purposes in plant physiology it is convenient to relate osmotic pressure directly to the concentration of solutes instead of expressing Π in terms of the water activity, a_w, as is done in Equation 2.7. In general, the greater the concentration of solutes, the more negative is $\ln a_w$ and the larger is the osmotic pressure. Thus, some way of expressing a_w in terms of the properties of the solutes would seem to be an appropriate approach for expressing Π as a function of the various solute concentrations. The ensuing derivation not only will show how a_w can be so expressed, but will also serve to indicate the many approximations necessary to achieve a rather simple expression for Π.

The activity of pure water is unity, while in general a_w equals $\gamma_w N_w$, where γ_w is the activity coefficient of water and N_w is its mole fraction. The *mole fraction* of a species is the ratio of the number of moles of that species to the total number of moles of all species in the system. Thus, N_w is given by

$$N_w = \frac{n_w}{n_w + \sum_j n_j} = \frac{n_w + \sum_j n_j - \sum_j n_j}{n_w + \sum_j n_j} = 1 - \frac{\sum_j n_j}{n_w + \sum_j n_j} \qquad (2.8)$$

where n_w is the number of moles of water, n_j is the number of moles of solute j, and the summation $\sum_j$ is over all solutes in the system being considered. Equation 2.8 also expresses the familiar relation that N_w equals 1 minus the mole fraction of solutes. For an ideal solution, γ_w equals 1. This condition of unit activity coefficient is approached for dilute solutions, in which case n_w is much greater than $\sum_j n_j$. (The expression $n_w \gg \sum_j n_j$ can be taken as one one way of defining a dilute solution, where the symbol $\gg$ means "is much greater than.") Using Equation 2.8 and assuming a dilute solution, we obtain the following relations for $\ln a_w$:

$$\ln a_w \simeq \ln N_w = \ln\left(1 - \frac{\sum_j n_j}{n_w + \sum_j n_j}\right) \cong -\frac{\sum_j n_j}{n_w + \sum_j n_j} \cong -\frac{\sum_j n_j}{n_w} \qquad (2.9)$$

The penultimate step in Equation 2.9 is based on the series expansion of a logarithm: $\ln(1 - x) = -x - x^2/2 - x^3/3 - \ldots$, a series that converges rapidly for $|x| \ll 1$. The last two steps employ the approximation $\left(n_w \gg \sum_j n_j\right)$ relevant to a dilute solution. Equation 2.9 is thus restricted to dilute ideal solutions. Nevertheless, it is a useful expression, clearly indicating that, in the absence of solutes $\left(\sum_j n_j = 0\right)$, $\ln a_w$ is zero and a_w is unity, while presence of solutes decreases the activity of water from the value of one for pure water.

To obtain a rather familiar form for expressing the osmotic pressure, we can incorporate the approximation for $\ln a_w$ given by Equation 2.9 into the definition of osmotic pressure given by Equation 2.7 ($RT \ln a_w = -\bar{V}_w \Pi$), which yields

$$\Pi_s \cong -\frac{RT}{\bar{V}_w}\left(-\frac{\sum_j n_j}{n_w}\right) = RT \sum_j \frac{n_j}{\bar{V}_w n_w} = RT \sum_j c_j \qquad (2.10)$$

where $\bar{V}_w n_w$ is the total volume of water in the system (essentially the total volume of the system for a dilute aqueous solution), $n_j/\bar{V}_w n_w$ is the number of moles of species j in that volume and is therefore the concentration of species $j (c_j)$, and the summations are over all solutes.

Osmotic pressure is often expressed by Equation 2.10, known as the Van't Hoff relation, but this is justified only in the limit of *dilute ideal* solutions. As we have already indicated (p. 67), an ideal solution has ideal solutes dissolved in an ideal solvent. Since in writing the string of equalities in Equation 2.9 we assumed that γ_w was essentially unity, our subsequently derived expression (Eq. 2.10) strictly applies only when water acts as an ideal solvent ($\gamma_w = 1.00$). To emphasize that we are neglecting any factors which cause γ_w to deviate from unity and thereby affect the measured osmotic pressure (such as the interaction between water and colloids, which we will discuss shortly), Π_s instead of Π has been used in Equation 2.10, and we will follow this convention throughout. The effect of solute concentration on osmotic pressure as described by Equation 2.10 is portrayed in Figure 2.7 and was first clearly recognized by the botanist Pfeffer in 1877. Both the measurement of osmotic pressures and a recognition of their effects have been crucial for an understanding of the water relations of plants.

Relatively dilute solutions have fairly large osmotic pressures (Fig. 2.7). The cellular fluid expressed from young leaves of plants like pea or spinach often contains about 0.3 mol osmotically active particles per kg of water (about 300 mol m^{-3}). This fluid is referred to as being 0.3 osmolal by analogy with molality, which refers to the total concentration. For example, 0.1 molal $CaCl_2$ is nearly 0.3 osmolal since most of the $CaCl_2$ is dissociated (Fig. 2.7). We note that molality (moles of solute/kg solvent) is a concentration unit that is independent of temperature and suitable for colligative properties; e.g., the freezing point depression is 1.86°C for a 1 osmolal aqueous solution. But mol m^{-3} or molarity (moles of solute/litre of solution, M) is often a more convenient unit. Below 0.1 M, molarity and molality are nearly the same for a small solute in an aqueous solution, while at high

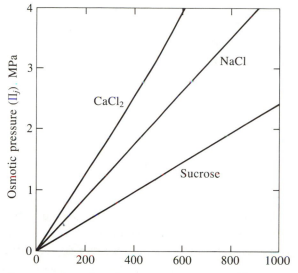

Figure 2.7
Relationship between concentration and osmotic pressure for a
nonelectrolyte (sucrose) and two readily dissociating salts (NaCl
and CaCl$_2$). The different slopes indicate the different degrees of
dissociation for the three substances. The observed linearity (except
at the higher concentrations of CaCl$_2$) is in agreement with the
Van't Hoff relation (Eq. 2.10). See Weast (Ch. 1 reference).

concentrations the molarity can be considerably less than the molality. At
20°C, RT is 0.002437 m^3 MPa mol^{-1}. (See App. III for various values of
RT.) By Equation 2.10, the osmotic pressure of the 0.3 osmolal solution is

$$\Pi_s = (0.002437 \text{ m}^3 \text{ MPa mol}^{-1})(300 \text{ mol m}^{-3}) = 0.73 \text{ MPa}$$

The osmotic pressure of cell sap pressed out of mature leaves of most angiosperms is 0.6 to 3 MPa (Altman and Dittmer); for comparison, sea water
has a Π_s of 2.5 MPa (Ψ_{Π_s} of -2.5 MPa).

Matric Potential

For certain applications in plant physiology, another term is frequently
included in the chemical potential of water, namely $\bar{V}_w\tau$, where τ is the
matric potential, or matric pressure. The use of this term is sometimes

convenient for dealing explicitly with interactions occurring at interfaces, although these interfacial forces can also be adequately represented by their contributions to Π or P (see also Passioura). In other words, the matric potential does not represent a new or different contribution to μ_w—but it can be used as a bookkeeping device for handling interfacial interactions. To help make this statement more meaningful, we will briefly consider the influence of liquid-solid interfaces on the chemical potential of water at the surfaces of colloids.[*]

When water molecules are associated with interfaces such as those provided by the surfaces of colloidal particles suspended in an aqueous solution, they have less tendency either to react chemically in the bulk solution or to escape into a surrounding vapor phase. The presence of interfaces thus lowers the thermodynamic activity of the water (a_w), especially near the surfaces of the colloids. The presence of solutes also lowers water activity (Eq. 2.10). As a useful first approximation, we can consider that these two different effects lowering water activity are additive in a solution containing both ordinary solutes and colloids. Osmotic pressure (Π) depends on the actual activity of water regardless of the reason for the departure of a_w from unity; i.e., Π still equals $-(RT/\overline{V}_w) \ln a_w$, as it must by Equation 2.7. Incorporating these various concepts (and recalling that $a_w = \gamma_w N_w$), we can write the following relation:

$$\Pi = -\frac{RT}{\overline{V}_w} \ln \gamma_w N_w = -\frac{RT}{\overline{V}_w} \ln \gamma_w - \frac{RT}{\overline{V}_w} \ln N_w = \tau + \Pi_s \quad (2.11)$$

where τ is a matric pressure resulting from the water-solid interactions at the surfaces of the colloids.

We note that Π_s in Equation 2.11 is the osmotic pressure of all solutes, including the *concentration* of the colloids, as represented by Equation 2.10 $\left(\Pi_s \cong RT \sum_j c_j\right)$. In other words, in the derivation of Π_s we were actually dealing only with $-(RT/\overline{V}_w) \ln N_w$, since γ_w was then set equal to unity. For an exact treatment when many interfaces are present—e.g., in the cytosol of a typical cell—we cannot set γ_w equal to unity, since the activity of water, and hence the osmotic pressure (Π), is strongly affected by the proteins and other colloids present. In such a case, Equation 2.11 suggests

---

[*] "Colloid" is a generic term for solid particles approximately 0.002 to 1 μm in diameter suspended in a liquid, e.g., proteins, ribosomes, and even some membrane-bounded organelles.

a simple way in which matric pressure may be related to the reduction of the activity coefficient of water caused by the interactions at interfaces. (Note that, if we let $\tau = -(RT/\overline{V}_w) \ln \gamma_w$, then γ_w less than unity means a positive τ.) Equation 2.11 should not be viewed as a relation defining matric potentials for all situations; rather, it is one example where it might be useful to represent interfacial interactions by a separate term that can be added to Π_s, the effect of the solutes on Π.

Although Π and a_w may be the same throughout some system, both Π_s and τ in Equation 2.11 may vary. For example, water activity in the bulk of the solution may be predominantly lowered by the presence of the solutes, whereas at or near the surface of colloids the main factor decreasing a_w from unity could be the interfacial attraction and binding of water. Such interfacial interactions do not change the mole fraction of water, but they do reduce its activity coefficient, γ_w.

Other areas of plant physiology where matric potentials have been invoked are descriptions of the chemical potential of water in soil or in cell walls. Soil matric potentials will be mentioned in Chapter 9, while cell walls will be further considered at the end of this chapter. The interfacial interactions between water and the many surfaces lining the interstices of the cell wall lead to a tension in the cell wall water. Such a tension is simply a negative hydrostatic pressure; i.e., P is less than zero. Although it is not necessary to do so, P is sometimes used to refer only to positive pressures, while a negative P in cell wall water has been called a positive matric potential. Some books define this same negative hydrostatic pressure in the water in cell wall interstices or between soil particles as a negative matric potential. It is far more straightforward and consistent to refer to the quantity that reduces μ_w in such pores as a negative P.

Water Potential

From the definition of chemical potential (Eq. 2.4) and the formal expression for osmotic pressure (Eq. 2.7), we can write the chemical potential of water (μ_w) as

$$\mu_w = \mu_w^* - \overline{V}_w\Pi + \overline{V}_w P + m_w gh \tag{2.12}$$

where the electrical term ($z_j FE$) is not included, since water carries no net charge ($z_w = 0$). The quantity $\mu_w - \mu_w^*$ has proved to be of considerable importance for discussing the water relations of plants. It represents the work involved in moving one mole of water from some point in a system (at constant pressure and temperature) to a pool of pure water at atmospheric pressure, at the same temperature as the system under consideration, and at the zero level for the gravitational term. A difference between two locations in the value of $\mu_w - \mu_w^*$ indicates that water is not in equilibrium, so there will be a tendency for water to flow toward the region where $\mu_w - \mu_w^*$ is lower.

A quantity proportional to $\mu_w - \mu_w^*$ commonly used in studies of plant water relations is the so-called *water potential*, Ψ. This is defined as

$$\Psi = \frac{\mu_w - \mu_w^*}{\overline{V}_w} = P - \Pi + \rho_w gh \qquad (2.13a)$$

which follows directly from Equation 2.12 plus the identification of $m_w/\overline{V}_w$ with the density of water, ρ_w. Equation 2.13a indicates that an increase in hydrostatic pressure raises the water potential, while an increase in osmotic pressure lowers it. We will consider the values of Ψ on both sides of cellular and subcellular membranes in the next two sections.

We have already mentioned that work must be performed to raise an object in the gravitational field of the earth. Thus, vertical position also affects the chemical potential, and consequently the term $\rho_w gh$ must be included in the water potential given by Equation 2.13a, at least when water moves an appreciable distance vertically in the gravitational field. The magnitude of $\rho_w g$ is 0.0098 MPa m^{-1}; if water moves 10 m vertically upward in a tree, the gravitational contribution to the water potential is increased by about 0.1 MPa (1 bar). For certain applications in this text, such as considerations of chemical reactions or the crossing of membranes, there is very little or no change in vertical position, and the gravitational term can then be omitted from μ_j and Ψ.

As we alluded to when introducing osmotic pressure, a number of conventions are used to describe the osmotic and other water potential terms. One such convention used by many plant physiologists and ecologists is to define Ψ as follows:

$$\Psi = \Psi_P + \Psi_\Pi + \Psi_h \qquad (2.13b)$$

where Ψ_P $(= P)$ is called the hydrostatic potential or pressure potential, Ψ_Π $(= -\Pi)$ is the osmotic potential, and Ψ_h $(= \rho_w gh)$ is the gravitational potential. Although uniformity of expression is a cherished ideal, one should not be too alarmed that various conventions are used, since persons from

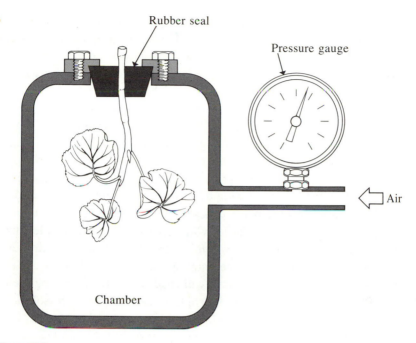

Figure 2.8

Schematic diagram of a "pressure bomb," which can be used to measure the xylem pressure, P^{xylem}, averaged over the material placed in the chamber. To make a measurement, a severed part of a plant is placed in the chamber with its freshly cut end protruding through a rubber seal. The air pressure (P^{air}) in the chamber is then gradually increased until it just causes the exudation of xylem sap at the cut end. At this stage, the resulting pressure of the sap, which equals $P^{xylem} + P^{air}$, is zero, and so P^{xylem} equals $-P^{air}$, which is read on the pressure gauge. If the xylem osmotic pressure can be ignored, P^{xylem} would be approximately equal to Ψ^{xylem}, which can be the same as the water potential of the other tissue in the chamber (if water equilibration has been achieved). See Scholander et al. or Turner.

many fields and backgrounds have contributed to our understanding of plants. Because Ψ is so important in understanding plant water relations, a method for measuring it is illustrated in Figure 2.8.

CENTRAL VACUOLE AND CHLOROPLASTS

The water relations of both the large central vacuole and the chloroplasts will be considered next, using the water potential defined above. We will focus our attention on the situation where there is no net flow of water across the limiting membranes surrounding these subcellular compartments and for the special case of nonpenetrating solutes. Hence, we will be considering the idealization of semipermeable membranes (see Fig. 2.6). The more general and biologically realistic case of penetrating solutes will be discussed in Chapter 3 after a consideration of the properties of solutes and the introduction of concepts from irreversible thermodynamics.

The central vacuole occupies up to 90% of a mature plant cell, which means that under such circumstances most of the cellular water is in the vacuole. The vacuolar volume is generally about 10^4 μm^3 for the mesophyll cells in a leaf, but can be much larger in certain algal cells. Because it is nearly as large as the cell containing it, special procedures are required to remove the vacuole from a plant cell without rupturing its surrounding membrane, the tonoplast (see Matile). Chloroplasts are much smaller than the central vacuole, often having volumes near 30 μm^3 in vivo (although the sizes vary considerably with the plant species). When chloroplasts are carefully isolated, suitable precautions will ensure that their limiting membranes remain intact. Such intact chloroplasts can be placed in solutions having various osmotic pressures, and the resulting movement of water into or out of the isolated plastids can be precisely measured.

Water Relations of the Central Vacuole

To predict whether and in what direction water will move, we need to know the value of the water potential in the various compartments under consideration. At equilibrium, the water potential is the same in all communicating phases, such as those separated by membranes. For example, when water is in equilibrium across the tonoplast, the value of the water potential is the same in the vacuole as it is in the cytosol. No force then drives water across this membrane, and no net flow of water occurs into or out of the vacuole.

The tonoplast most likely does not have an appreciable difference in hydrostatic pressure across it. A higher internal hydrostatic pressure would cause an otherwise slack (folded) tonoplast to be mechanically pushed outward. The observed lack of such motion thus indicates that ΔP is zero across a typical tonoplast. If the tonoplast were taut, ΔP would cause a stress in the membrane, analogous to the cell wall stresses discussed in the last chapter; namely, the stress would be $r\,\Delta P/2t$ for a spherical vacuole. But the tensile strength of biological membranes is low—most membranes rupture before a stress of even 0.1 MPa develops in them. For a tonoplast

7 nm thick with a maximum stress of 0.1 MPa surrounding a spherical vacuole 14 μm in radius, the hydrostatic pressure difference across the tonoplast would be

$$\Delta P = \frac{2t\sigma}{r} = \frac{(2)(7 \times 10^{-9} \text{ m})(0.1 \text{ MPa})}{(14 \times 10^{-6} \text{ m})} = 10^{-4} \text{ MPa}$$

Thus, P may well have essentially the same value in the cytosol and the vacuole. With this simplifying assumption, and for the equilibrium situation, where Ψ is the same in the two phases, Equation 2.13a ($\Psi = P - \Pi + \rho_w gh$, where $\Delta h = 0$ across a membrane) gives

$$\Pi^{\text{cytosol}} = \Pi^{\text{vacuole}} \tag{2.14}$$

where Π^{cytosol} is the osmotic pressure in the cytosol and Π^{vacuole} that in the central vacuole.

The vacuole appears to be a relatively homogeneous aqueous phase, while the cytoplasm is a more complex phase containing many colloids and membrane-bounded organelles. Since the vacuole contains few colloidal or other interfaces, any matric potential in it is probably negligible compared with the osmotic pressure resulting from the vacuolar solutes. Expressing osmotic pressure by Equation 2.11 ($\Pi = \Pi_s + \tau$) and assuming that τ^{vacuole} is negligible, we can replace Π^{vacuole} in Equation 2.14 by Π_s^{vacuole}; i.e., the decrease in vacuolar water activity is due almost solely to the presence of solutes. This also means that water is acting like an ideal solvent ($\gamma_w \cong 1$). On the other hand, water activity in the cytosol presumably is considerably lowered both by solutes and by the presence of interfaces. Unlike the case for the vacuole, individual water molecules in the cytosol are never very far away from proteins, organelles, or other colloids, so that interfacial interactions can appreciably affect a_w^{cytosol}. In terms of Equation 2.11, this means that both Π_s^{cytosol} and τ^{cytosol} contribute to Π^{cytosol}. Hence, Equation 2.14 indicates that $\Pi_s^{\text{cytosol}} + \tau^{\text{cytosol}}$ equals Π_s^{vacuole}. Since τ^{cytosol} is a positive quantity, we conclude that at equilibrium the osmotic pressure in the vacuole due to solutes, Π_s^{vacuole}, must be larger than Π_s^{cytosol}. Equation 2.14 thus leads to the prediction that the vacuole has a higher concentration of osmotically active solutes than does the cytosol.

There are various possible roles for the large central vacuole in plant cells. Because of the large central vacuole, the cytoplasm occupies a thin layer around the periphery of the cell (see Fig. 1.1); therefore, for its volume, the cytoplasm has a relatively large surface area, across which diffusion

can readily occur. Vacuoles also provide large, relatively simple compartments in which hydrostatic pressures lead to the cellular turgidity necessary
for the growth and support of a plant. The vacuole can also act as a storage
reservoir for toxic products or metabolites; e.g., the nighttime storage of
organic acids like malic acid takes place in the vacuoles of Crassulacean
acid metabolism plants (to be discussed in Ch. 8). Also, certain secondary
chemical products such as phenolics, alkaloids, tannins, glucosides, and
flavonoids like the anthocyanins can accumulate in vacuoles (see Matile).

Boyle–Van't Hoff Relation

The volume of a chloroplast or other membrane-bounded body changes in
response to variations in the osmotic pressure of the external solution, Π°.
This is a consequence of the properties of membranes, which generally allow
water to move readily across them, at the same time restricting the passage
of certain solutes, such as sucrose. For example, if Π° outside a membrane-
bounded aqueous compartment were raised, we would expect that water
would flow out across the membrane to the region of lower chemical potential
and thereby decrease the compartment volume, while in general there would
be very little movement of solutes in such a case. This differential permeability
leads to the "osmometric behavior" characteristic of many cells and organelles. The conventional expression quantifying this volume response to
changes in the external osmotic pressure is the *Boyle–Van't Hoff relation*:

$$\Pi^\circ(V - b) = RT \sum_j \varphi_j n_j = RTn \qquad (2.15)$$

where Π° is the osmotic pressure of the external solution; V is the volume
of the cell or organelle; b is the so-called nonosmotic volume (osmotically
inactive volume), and it is frequently considered to be the volume of a solid
phase within volume V that is not penetrated by water; n_j is the number
of moles of species j within $V - b$; φ_j is the osmotic coefficient, generally
construed as a correction factor indicating the relative osmotic effect of
species j; and $n = \sum_j \varphi_j n_j$ is the apparent number of osmotically active moles
in $V - b$ (see Dick, Giese, or Nobel).

In this chapter we will derive the Boyle–Van't Hoff relation using the
chemical potential of water, and in Chapter 3 we will extend the treatment
to penetrating solutes by employing irreversible thermodynamics. Although
the Boyle–Van't Hoff expression will only be used to interpret the osmotic

responses of chloroplasts, the equations that will be developed are quite general and can be applied equally well to data obtained with mitochondria, whole cells, or other membrane-surrounded bodies.

The Boyle–Van't Hoff relation can apply to the equilibrium situation, when equality of the water potential has been achieved across the membranes surrounding chloroplasts. When Ψ^i equals Ψ^o, net water movement across the membrane ceases, and the volume of a chloroplast is steady. (The superscript i refers to the inside of the cell or organelle, the superscript o to the outside.) If we were to measure the chloroplast volume under such conditions, the external solution would generally be at atmospheric pressure ($P^o = 0$). By Equation 2.13a ($\Psi = P - \Pi$, when the gravitational term is ignored), the water potential in the external solution is then

$$\Psi^o = -\Pi^o \qquad (2.16)$$

Like Ψ^o (Eq. 2.16), the water potential inside the chloroplasts, Ψ^i, also depends on osmotic pressure. But the internal hydrostatic pressure (P^i) may be different from atmospheric pressure and should be included in the expression for Ψ^i. In addition, macromolecules and solid-liquid interfaces inside the chloroplasts can lead to a matric potential, τ^i. To allow for this possibility, the internal osmotic pressure (Π^i) will be considered to be the sum of the solute and the interfacial contributions, in the manner expressed by Equation 2.11. Using Equation 2.13a, Ψ^i is therefore

$$\Psi^i = P^i - \Pi^i = P^i - \Pi_s^i - \tau^i \qquad (2.17)$$

We will now re-express Π_s^i and then equate Ψ^o to Ψ^i. For a dilute solution inside the chloroplasts, Π_s^i equals $RT \sum_j n_j^i/(\bar{V}_w n_w^i)$ by Equation 2.10. Equating the water potential outside the chloroplast (Eq. 2.16) to that inside (Eq. 2.17), the condition for water equilibrium across the limiting membranes of the chloroplast becomes

$$\Pi^o = RT \frac{\sum\limits_j n_j^i}{\bar{V}_w n_w^i} + \tau^i - P^i \qquad (2.18)$$

Although this expression may look a little frightening, it states only that at equilibrium the water potential is the same on the two sides of the membranes and that we can explicitly recognize various possible contributors to the two Ψ's involved.

To appreciate the refinements that this thermodynamic treatment introduces into the customary expression describing the osmotic responses of cells and organelles, Equation 2.18 should be compared with Equation 2.15, the conventional Boyle–Van't Hoff relation. The volume of water inside the chloroplast is $\bar{V}_w n_w^i$, since n_w^i is the number of moles of internal water and $\bar{V}_w$ is the volume per mole of water. This factor in Equation 2.18 can be identified with $V - b$ in Equation 2.15. Instead of being designated the "nonosmotic volume," b is more appropriately called the "nonwater volume," for it includes the volume of the internal solutes including that of colloids. In other words, the total volume (V) minus the nonwater volume (b) equals the volume of internal water ($\bar{V}_w n_w^i$). We also note that the possible hydrostatic and matric contributions included in Equation 2.18 are neglected in the usual Boyle–Van't Hoff relation. In short, although certain approximations and assumptions are incorporated into Equation 2.18 (e.g., that solutes do not cross the limiting membranes, and that we have dilute solutions), it is still a more satisfactory expression for describing osmotic responses than is the conventional Boyle–Van't Hoff relation.

Osmotic Responses of Chloroplasts

To illustrate the use of Equation 2.18 in interpreting osmotic data, we will consider osmotic responses of pea chloroplasts suspended in external solutions of various osmotic pressures. Since it is customary to plot the volume V versus the reciprocal of the external osmotic pressure, $1/\Pi^o$, certain algebraic manipulations are needed to put Equation 2.18 into a more convenient form for discussing such osmometric studies. By transferring $\tau^i - P^i$ to the left side of Equation 2.18 and then multiplying both sides by $\bar{V}_w n_w^i/(\Pi^o - \tau^i + P^i)$, $\bar{V}_w n_w^i$ can be shown to equal $RT \sum_j \gamma_j^i n_j^i/(\Pi^o - \tau^i + P^i)$. The measured chloroplast volume V can be represented by $\bar{V}_w n_w^i + b$, i.e., as the sum of the aqueous and nonaqueous contributions. Hence, V should be equal to $RT \sum_j \gamma_j^i n_j^i/(\Pi^o - \tau^i + P^i) + b$, if the modified form of the Boyle–Van't Hoff relation is obeyed. Figure 2.9 indicates that the volume of pea chloroplasts varies linearly with $1/\Pi^o$ over a considerable range of external osmotic pressures. Consequently, $\tau^i - P^i$ in both Equation 2.18 and the above expression must be either negligibly small for pea chloroplasts, compared with the various values of Π^o employed, or perhaps proportional to Π^o. For simplicity, we will consider only the observed proportionality between V and $1/\Pi^o$. In other words, we will assume that $V - b$ (or $\bar{V}_w n_w^i$) equals $RT \sum_j n_j^i/\Pi^o$ for pea chloroplasts. We will return to such consider-

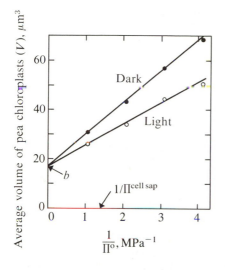

Figure 2.9
Volumes of chloroplasts at various external osmotic pressures, Π^o. Pea chloroplasts were isolated from plants in the light or the dark as indicated. [Source: P. S. Nobel, *Biochimica et Biophysica Acta* *172*:134–143 (1969). Data reprinted by permission.]

ations in Chapter 3, where we will further refine the Boyle–Van't Hoff relation to include the more realistic case in which solutes can cross the limiting membranes.

The relatively simple measurement of the volumes of pea chloroplasts for various external osmotic pressures can yield a considerable amount of information about the organelles. If we measure the volume of the isolated chloroplasts at the same osmotic pressure as in the cytosol, we can determine the chloroplast volume in the plant cell. Cell "sap" expressed from young pea leaves can have an osmotic pressure of 0.70 MPa. Such sap comes mainly from the central vacuole, but since we expect $\Pi^{cytosol}$ to be essentially equal to $\Pi^{vacuole}$ (Eq. 2.14), $\Pi^{cell\ sap}$ is about the same as $\Pi^{cytosol}$ (some uncertainty exists, inasmuch as during extraction the cell sap can come into contact with water in the cell walls). At an external osmotic pressure of 0.70 MPa (indicated by an arrow in Fig. 2.9), pea chloroplasts have a volume of 29 μm^3 when isolated from illuminated plants, and 35 μm^3 when isolated from plants in the dark. Since these volumes occur at approximately the same osmotic pressure found in the cell, they are presumably reliable estimates of pea chloroplast volumes in vivo.

The data in Figure 2.9 indicate that light affects the chloroplast volume in vivo. Although the basis for the size changes is not fully understood, chloroplasts in many plants do have a larger volume in the dark than in the light. The decrease in volume upon illumination of the plants is observed by both phase contrast and electron microscopy as a flattening or decrease in thickness of the chloroplasts. This flattening, which amounts to about 20% of the thickness for pea chloroplasts in vivo, is in the vertical direction for the chloroplast depicted in Figure 1.9. We also note that the slopes of the osmotic response curves in Figure 2.9 are equal to $RT \sum_j n_j^i$.

The slope is 50% greater for pea chloroplasts from plants in the dark than from those in the light, suggesting that chloroplasts in the dark contain more osmotically active particles. In fact, illumination causes an efflux of K^+ and Cl^- from the chloroplasts.

The intercept on the ordinate in Figure 2.9 is the chloroplast volume theoretically attained in an external solution of infinite osmotic pressure—a $1/\Pi^\circ$ of zero is the same as a Π° of infinity. For such an infinite Π°, all of the internal water would be removed ($n_w^i = 0$), and the volume, which is actually obtained by extrapolation, would be that of the nonaqueous components of the chloroplasts.* Thus, the intercept on the ordinate of a V-versus-$1/\Pi^\circ$ plot corresponds to b in the conventional Boyle–Van't Hoff relation (Eq. 2.15). This intercept (indicated by an arrow in Fig. 2.9) equals $17 \ \mu m^3$ for chloroplasts both in the light and in the dark. (The extra solutes in a chloroplast in the dark correspond to less than $0.1 \ \mu m^3$ of solids.) Using the chloroplast volumes obtained for a Π° of 0.70 MPa and $17 \ \mu m^3$ for the nonwater volume b, we find that the fractional water content of pea chloroplasts in the dark is

$$\frac{V - b}{V} = \frac{(35 \ \mu m^3 - 17 \ \mu m^3)}{(35 \ \mu m^3)} = 0.51$$

or 51%. In the light it is 41%. These relatively low water contents in the organelles are consistent with the extensive internal lamellar system and the abundance of CO_2 fixation enzymes in chloroplasts (mentioned in Ch. 1). Thus, osmometric responses of cells and organelles can be used to provide information on their fractional water content.

WATER POTENTIAL AND PLANT CELLS

In this section we will shift our emphasis from a consideration of the water relations of subcellular bodies to those of whole cells, and extend the development to include the case of water fluxes. Whether water enters or leaves a plant cell, and how much, depends on the water potential outside compared with that inside. The external water potential Ψ° can often be varied experimentally, and the direction as well as the magnitude of the resulting water

---

*Some uncertainty exists for water that is tightly bound to proteins and other substances. Some of this water presumably remains bound even at the hypothetical infinite osmotic pressures considered here. Such water should not be counted as part of the internal water ($\bar{V}_w n_w^i$).

movement will give information about Ψ^i. Moreover, the equilibrium value of Ψ^o can be used to estimate the internal osmotic pressure Π^i.

A loss of water from plants—indeed, sometimes even an uptake—occurs at cell-air interfaces. As we would expect, the chemical potential of water in cells compared with that in the adjacent air determines the direction for net water movement at such locations. Thus, we must obtain an expression for the water potential in a vapor phase, and then relate this Ψ_{wv} to Ψ for the liquid phases in a cell. We will specifically consider the factors influencing the water potential at the plant cell-air interface, namely, in the cell wall. We will find that $\Psi^{\text{cell wall}}$ is dominated by a negative hydrostatic pressure resulting from surface tension effects in the cell wall pores.

Incipient Plasmolysis

For usual physiological conditions, a positive hydrostatic pressure exists inside a plant cell; i.e., it is under turgor. By suitably adjusting the solutes in an external solution bathing such a turgid cell, it is possible to reduce P^i to zero and thereby to obtain an estimate of Π^i, as the following argument will indicate (see Fig. 2.10).

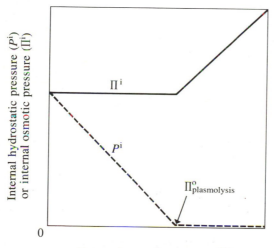

Figure 2.10

Responses of an initially turgid cell ($P^i > 0$) placed in pure water ($\Pi^o = 0$) to changes in the external osmotic pressure (Π^o). At the point of incipient plasmolysis (plasmalemma just beginning to pull away from the cell wall), the internal hydrostatic pressure (P^i) is reduced to zero. Constancy of cell volume and water equilibrium are assumed at each step as Π^o is raised.

Let us place the cell in pure water ($\Pi^o = 0$) at atmospheric pressure ($P^o = 0$). Ψ^o is then zero (ignoring the gravitational term, $\Psi = P - \Pi$, Eq. 2.13a), while Ψ^i is $P^i - \Pi^i$ (the inner phase can be the cytosol). If the cell can be assumed to be in equilibrium with this external solution ($\Psi^o = \Psi^i$), P^i must then equal Π^i (Fig. 2.10). Suppose that Π^o is now gradually raised from its initial zero value—for example, by adding solute to the external solution. If the cell remains in equilibrium, then Π^o equals $\Pi^i - P^i$ ($P^o = 0$, since the external solution is at atmospheric pressure). As Π^o is increased, P^i will tend to decrease, while Π^i usually does not change very much. More precisely, because the cell wall is quite rigid, the cell will not appreciably change its volume in response to small changes in P^i. (Since the cell wall has elastic properties—see Ch. 1—some water flows out as P^i decreases and the cell shrinks, a point which we will return to shortly.) If the cell volume can be assumed not to change in response to changes in P^i and no internal solutes leak out, Π^i will remain constant. As the external osmotic pressure is raised, Π^o will eventually become equal to Π^i. In such a plant cell at equilibrium, P^i is 0, which means that no internal hydrostatic pressure is exerted against the cell wall when Π^o equals Π^i (Fig. 2.10). The cell will thus lose its turgidity. If Π^o is increased even further, an appreciable amount of water will flow out of the cell, plasmolysis will occur as the plasmalemma pulls away from the rigid cell wall, and the internal osmotic pressure will increase (the same amount of solutes in less water). Ignoring for the moment any overall volume changes of the cell, we find that the condition under which water just begins to move out of the cell—referred to as the point of *incipient plasmolysis* and illustrated in Figure 2.10—is

$$\Pi^o_{plasmolysis} = \Pi^i \tag{2.19}$$

Equation 2.19 suggests that only a relatively simple measurement ($\Pi^o_{plasmolysis}$) is needed to estimate the osmotic pressure (Π^i) occurring inside an individual plant cell (see Stadelmann).

The existence of an internal hydrostatic pressure within a plant cell leads to stresses in its cell wall and a resulting elastic deformation or strain, as indicated in Chapter 1. The decrease of P^i to 0—which occurs in taking a turgid plant cell to the point of incipient plasmolysis—must therefore be accompanied by a contraction of the cell as the wall stresses are removed. This decrease in volume means that some water will actually flow out of the cell before the point of incipient plasmolysis is reached. If no internal solutes leave as the cell shrinks, then the osmotic pressure inside will increase (the same amount of solutes in a smaller volume). As a useful first approximation, we can assume that the change in osmotic pressure reciprocally follows the change in volume; i.e., the product of Π^i and the cellular volume is approximately constant. With this assumption, the osmotic pressure in the cell—described by Equation 2.19, and determined by using the technique of measuring the point of incipient plasmolysis—can be corrected to its original value by using the ratio of the initial volume to the final volume of the cell. The change in volume is only a few percent for some plant cells, in which case fairly accurate estimates of Π^i can be obtained from plasmolytic studies alone.

Measurements of $\Pi^o_{plasmolysis}$ often give values near 0.7 MPa for cells in storage tissues like onion bulbs or carrot roots and in young leaves of pea or spinach. These values of the external osmotic pressure provide information on various contributors to the water potential inside the cell, as Equation 2.19 indicates. For the case in which water is in equilibrium within the plant cell at the point of incipient plasmolysis, the term Π^i in Equation 2.19 is the osmotic pressure both in the cytosol and in the vacuole. Under certain circumstances Π^i can be replaced by $\Pi^i_s + \tau^i$ (Eq. 2.11), where Π^i_s is the osmotic pressure contributed by the internal solutes and τ^i is the matric potential. This relation was invoked for example when the implications of Equation 2.14 ($\Pi^{cytosol} = \Pi^{vacuole}$) were discussed, and the various arguments presented at that time extend to the present case. Specifically, since the possible matric potential in the vacuole is probably negligible, $\Pi^o_{plasmolysis}$ should be a rather good estimate of $\Pi^{vacuole}_s$. Moreover, $\tau^{cytosol}$ is most probably significant, as the water activity in the cytosol can be lowered by the many interfaces present there. Thus, $\Pi^o_{plasmolysis}$ is an upper limit for $\Pi^{cytosol}_s$. We see, therefore, that determination of the external osmotic pressure at the point of incipient plasmolysis provides information on the osmotic pressure existing in different compartments within the plant cell.

Chemical Potential and Water Potential of Water Vapor

Water molecules in an aqueous solution continually escape into a surrounding gas phase. At the same time, water molecules also condense back into the liquid. The rates of escape and condensation depend on the chemical activities of water in the gas and in the liquid phases, the two rates becoming equal at equilibrium. The gas phase adjacent to the given solution then contains as much water as it can hold at that temperature and still remain in equilibrium with the liquid. It is thus *saturated* with water vapor for the particular solution under consideration. The partial pressure exerted by the water vapor in equilibrium with pure water is known as the *saturation vapor pressure*.

The vapor pressure at equilibrium depends on the temperature and the solution, but it is independent of the relative or absolute amounts of liquid and vapor. When air adjacent to pure water is saturated with water vapor (100% relative humidity), the gas phase then has the maximum water vapor pressure possible at that temperature—unless it is supersaturated, a non-equilibrium situation. This saturation vapor pressure of pure water increases markedly with temperature; e.g., it goes from 0.61 kPa at 0°C to 7.37 kPa at 40°C (see App. II, where the saturation vapor pressure is also expressed in mbar; also see Fig. 8.4). Thus, heating air at constant pressure causes the relative humidity to drop dramatically (which helps explain the desiccating environment often found in heated buildings during cold weather).

As solutes are added to the liquid phase and the activity of its water is thereby lowered, water molecules have less of a tendency to leave the solution. Hence, the water vapor pressure in the gas phase at equilibrium becomes less—this is one of the colligative properties of solutions that we mentioned earlier. In fact, for dilute solutions the actual partial pressure of water vapor (P_{wv}) at equilibrium depends linearly on the mole fraction of water (N_w) in the liquid phase. This is *Raoult's law*, which is also mentioned in Appendix VI. For pure water, N_w equals one and P_{wv} has its maximum value, namely P_{wv}^*, the saturation vapor pressure.

The chemical potential of water vapor μ_{wv}, which depends on the partial pressure of water vapor P_{wv}, can be represented as follows:

$$\mu_{wv} = \mu_{wv}^* + RT \ln \frac{P_{wv}}{P_{wv}^*} + m_{wv}gh \qquad (2.20)$$

where P_{wv}^* is the saturation vapor pressure in equilibrium with pure liquid water at atmospheric pressure and at the same temperature as the system under consideration, and m_{wv} is the mass per mole of water vapor, which is the same as the mass per mole of water, m_w. (The form of the pressure term in Eq. 2.20 is derived in App. VI.) To handle deviations from ideality, P_{wv} is replaced by $\gamma_{wv}P_{wv}$, where γ_{wv} is the activity coefficient (more properly, the "fugacity" coefficient) of water vapor. If water vapor obeyed the ideal gas law—$P_jV = n_jRT$, where P_j is the partial pressure of species j in a volume V containing n_j moles of that gas—γ_{wv} would equal unity. This is an extremely good approximation for situations of interest in biology.

Let us next consider the various parameters at the surface of pure water ($a_w = 1$) at atmospheric pressure ($P = 0$) in equilibrium with its vapor. In such a case, P_{wv} equals P_{wv}^*, and we can take the zero level of the gravitational term as the surface of the liquid. Hence, μ_w equals μ_w^*, and μ_{wv} is μ_{wv}^*. Since there is no tendency for a net gain or loss of water molecules from the liquid phase at equilibrium ($\mu_w = \mu_{wv}$), we conclude that the constant μ_w^* must equal the constant μ_{wv}^* when the two phases are at the same temperature.

Let us now consider the water potential of water vapor in a gas phase such as air, Ψ_{wv}. Using a definition of Ψ analogous to that in Equation 2.13a $[\Psi = (\mu_w - \mu_w^*)/\overline{V}_w]$ and defining μ_{wv} by Equation 2.20, we have

$$\Psi_{wv} = \frac{\mu_{wv} - \mu_w^*}{\overline{V}_w} = \frac{\mu_{wv} - \mu_{wv}^*}{\overline{V}_w}$$

$$= \frac{RT}{\overline{V}_w} \ln \frac{P_{wv}}{P_{wv}^*} + \rho_w gh$$

$$= \frac{RT}{\overline{V}_w} \ln \frac{\% \text{ relative humidity}}{100} + \rho_w gh \tag{2.21}$$

where $m_{wv}/\overline{V}_w$ (the same as $m_w/\overline{V}_w$) has been replaced by the density of water, ρ_w. We note that $\overline{V}_w$, not $\overline{V}_{wv}$, is used in the definition of Ψ_{wv} in Equation 2.21. This is necessary since the fundamental term representing free energy per mole is the chemical potential—we want to compare $\mu_j - \mu_j^*$ for the two phases when predicting changes at an interface—and thus the proportionality factor between $\mu_w - \mu_w^*$ or $\mu_{wv} - \mu_{wv}^* (= \mu_{wv} - \mu_w^*)$ and the more convenient terms, Ψ or Ψ_{wv}, must be the same in each case, namely, $\overline{V}_w$ ($\overline{V}_w$ should not vary with location in the system). Equation 2.21 also indicates that P_{wv}/P_{wv}^* equals the % relative humidity/100; relative humidity is a readily measured quantity and has been extensively used in studying the water relations of plants.

What happens to P_{wv} as we move vertically upward from pure water at atmospheric pressure in equilibrium with water vapor in the air? If we let h be zero at the surface of the water, Ψ equals zero ($\Psi = P - \Pi + \rho_w gh$, Eq. 2.13a), and since the water vapor is by supposition in equilibrium with the liquid phase, Ψ_{wv} is also zero. Equation 2.21 indicates that, as we go vertically upward in the gas phase, $\rho_w gh$ must make an increasingly positive contribution to Ψ_{wv}. Since we are at equilibrium, the other term in Ψ_{wv}, $(RT/\overline{V}_w) \ln (P_{wv}/P_{wv}^*)$, must make a compensating negative contribution. Thus, at equilibrium P_{wv} must decrease with altitude. We can further appreciate this conclusion simply by noting that gravity attracts the molecules of water vapor and other gases toward the earth, so air pressure is higher at sea level than it is on the top of a mountain. Rather great heights are needed before the partial pressure of water vapor is reduced significantly. In particular, let us consider a 7% reduction in P_{wv} from P_{wv}^* (its value at the water surface at equilibrium at atmospheric pressure), to $0.93\, P_{wv}^*$. Using Equation 2.21 and values at 20°C in Appendix II, we find that the height necessary is

$$h = -\frac{1}{\rho_w g}\frac{RT}{\overline{V}_w}\ln\frac{P_{wv}}{P_{wv}^*}$$

$$= -\frac{1}{(0.0098\ \text{MPa m}^{-1})}\,(135\ \text{MPa})\ln\left(\frac{0.93\ P_{wv}^*}{P_{wv}^*}\right) = 1\,000\ \text{m}$$

In other words, not until we reached a height of 1 km would the partial pressure of water vapor at equilibrium decrease by 7% from its value at sea level, indicating that the gravitational term generally has relatively little influence on Ψ_{wv} over the distances involved for a particular plant.

Plant-Air Interface

Water equilibrium across the plant-air interface occurs when the water potential in the leaf cells equals that of the surrounding atmosphere. (This presupposes that the leaf and the air are at the same temperature, an aspect that we will reconsider in Ch. 8.) To measure Ψ^{leaf}, the leaf can be placed in a closed chamber and the relative humidity adjusted so that there is no net water gain or loss by the leaf. Such a determination is experimentally rather demanding, since small changes in relative humidity have large effects on Ψ_{wv}, as we will now show.

Extremely large negative values are possible for Ψ_{wv}. In particular, $RT/\overline{V}_w$ at 20°C is 135 MPa (1 350 bars). In the expression for Ψ_{wv} (Eq. 2.21) this relatively large factor multiplies $\ln$ (% relative humidity/100), and a wide range of relative humidities can occur in the air. By Equation 2.21, with $\rho_w g h$ ignored, a relative humidity of 100% corresponds to a water potential in the vapor phase of 0 ($\ln 1 = 0$). This Ψ_{wv} would be in equilibrium with pure water at atmospheric pressure, which also has a water potential of 0. For a relative humidity of 99%, Ψ_{wv} given by Equation 2.21 would be

$$\Psi_{wv} = (135\ \text{MPa})\ln\left(\frac{99}{100}\right) = -1.36\ \text{MPa}$$

Hence, going from 100% to 99% relative humidity corresponds to a decrease in water potential of 1.36 MPa (Table 2.1). Small changes in relative humidity do indeed reflect large differences in the water potential in air! Finally, we note that a relative humidity of 50% at 20°C leads to a Ψ_{wv} of -94 MPa.

One important consequence of the large negative values of Ψ_{wv}^{air} is easily

Table 2.1
Magnitudes of certain water vapor parameters useful for understanding water movement at the plant-air interface. Data are for 20°C. Ψ_{wv} is calculated from Equation 2.21 ignoring the gravitational term. We note that $P_{wv}/P^*_{wv} \times 100 =$ relative humidity in %, kPa $\times$ 10 = mbars, and MPa $\times$ 10 = bars.

Relative humidity (%)	P_{wv} (kPa)	Ψ_{wv} (MPa)
100.0	2.33	0.00
99.6	2.32	-0.54
99.0	2.31	-1.36
96.0	2.24	-5.51
90.0	2.09	-14.2
50.0	1.17	-93.6
0.0	0.00	$-\infty$

appreciated when values of the water potential that may occur in plant leaves are considered. The actual value of Ψ^{leaf} depends on the ambient conditions as well as the plant type and its physiological status. The range of values for most mesophytes is -0.3 to -3 MPa, with -0.5 MPa typical for leaves of a garden vegetable such as lettuce. From Equation 2.21, the relative humidity corresponding to a water potential of -0.5 MPa is 99.6%.[*] Even during a rainstorm the relative humidity of the air rarely exceeds 99%. Since relative humidity is lower than 99.6% under most natural conditions, water is continually being lost from a leaf having an internal water potential of -0.5 MPa. (Again, it should be emphasized that throughout this discussion the leaf is assumed to be at the same temperature as the air—a restriction that we will remove in Ch. 8.) A few desert plants have a Ψ^{leaf} as low as -5.5 MPa (-55 bars). Even in this case of adaptation to arid climates, however, water still tends to leave the plant unless the relative humidity is above 96% (Table 2.1).[†]

---

[*] Such an extremely high value for the relative humidity in equilibrium with a Ψ^{leaf} of -0.5 MPa indicates why it is difficult to determine Ψ^{leaf} by measuring the Ψ^{air}_{wv} for which no water is gained or lost by the leaf. See Figure 2.8, Boyer, and Turner for ways to measure Ψ and Ψ_{wv}.

[†] Structural modifications are generally more important than a low Ψ^{leaf} for adapting to xerophytic conditions. We will discuss this topic in Chapters 7 and 8.

Pressure in the Cell Wall Water

Plant cells come into contact with air at the cell walls bounding the intercellular air spaces (see Fig. 1.2). Thus, the water potential in the cell walls must be considered with respect to Ψ_{wv} in the adjacent gas phases. The main contributor to Ψ in cell wall water is often the negative hydrostatic pressure arising from surface tension effects at the numerous air-liquid interfaces of the cell wall interstices. In turn, $P^{\text{cell wall}}$ can be related to the geometry of the cell wall pores and the contact angles that occur.

The magnitude of the negative hydrostatic pressure that develops in cell wall water can be analyzed by considering the pressure that occurs near the surface in a liquid within a cylindrical pore—the argument is basically the same as the one presented earlier in this chapter in discussing capillary rise. One of the forces acting on the fluid in a narrow pore of radius r is the result of surface tension, σ. This force equals $2\pi r\sigma \cos \alpha$ (see Fig. 2.3), where α is the contact angle, a quantity that can be essentially zero for wettable walls. Since the adhesive forces at the wall are transmitted to the rest of the fluid by means of cohesion, a tension or negative hydrostatic pressure develops in the fluid. The total force resisting the surface adhesion can be regarded as this tension times the area over which it acts, πr^2, and hence the force is $\pi r^2 \times (-P)$. Equating the two forces gives the following expression for the pressure developed in the fluid contained within a cylindrical pore:

$$P = -\frac{2\sigma \cos \alpha}{r} \tag{2.22}$$

Adhesion of water at interfaces generally creates negative hydrostatic pressures in the rest of the fluid (Eq. 2.22 describes this P near the air-water interface, where the gravitational term can be ignored). Such negative hydrostatic pressures arising from interfacial interactions have sometimes been treated in plant physiology as positive matric pressures, a convention that we mentioned earlier.

The strong water-wall adhesive forces, which are transmitted throughout the cell wall interstices by water-water hydrogen bonding, conceivably can greatly reduce the water potential in the cell wall, as the following calculation indicates. At 20°C the surface tension of water is 7.28×10^{-8} MPa m (App. II), the voids between the fibers in the cell wall are often about 10 nm across ($r = 5$ nm), and $\cos \alpha$ can equal one for wettable walls. For water in such cylindrical pores, Equation 2.22 indicates that P could be

$$P = -\frac{(2)(7.28 \times 10^{-8} \text{ MPa m})(1)}{(5 \times 10^{-9} \text{ m})} = -29 \text{ MPa}$$

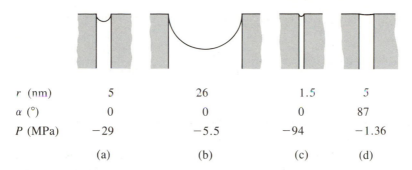

r (nm)	5	26	1.5	5
α (°)	0	0	0	87
P (MPa)	-29	-5.5	-94	-1.36
	(a)	(b)	(c)	(d)

Figure 2.11

Influence of radius r and contact angle α on the hydrostatic pressure P near the surface in cylindrical pores open to air.

(see Fig. 2.11a). This is an estimate of the negative hydrostatic pressure or tension that *could* develop in the aqueous solution within cell wall interstices of typical dimensions, supporting the contention that $\Psi^{\text{cell wall}}$ can be markedly less than zero. Moreover, the previous discussion of the tensile strength of water (p. 58) indicates that the hydrogen bonding in water can withstand such a tension. We will next consider the actual pressures that develop in the cell wall water.

Let us begin by considering a Ψ_{wv} of -5.5 MPa, which corresponds to a relative humidity of 96% (Table 2.1). According to Equation 2.22, water in cylindrical cell wall pores with radii of 26 nm would have a P of -5.5 MPa (see Fig. 2.11b), and hence it would be in equilibrium with air of 96% relative humidity in the intercellular spaces of the leaf. If the relative humidity near the cell wall surface were decreased, water in such interstices could be lost, but it would still remain in the finer pores, where a more negative P or larger tension can be created (Figs. 2.11a and c). Hence, if the plant material is exposed to air of moderate to low relative humidity—as can happen for an herbarium specimen—then larger and larger tensions resulting from interfacial interactions develop in the cell walls as the plant material dries out. Water is then retained only in finer and finer pores of the wall, as it must be according to Equation 2.22.

We will now show the effect of the availability of water adjacent to that in wettable cell walls on the actual contact angle in the interstices and on $P^{\text{cell wall}}$. Suppose that pure water in mesophyll cell wall interstices 10 nm across is in equilibrium with water vapor in the intercellular air spaces where the relative humidity is 99%. As we calculated on p. 92, Ψ_{wv} for this relative humidity is -1.36 MPa at 20°C. Hence, at equilibrium the (pure) water in the cell wall interstices must be under a hydrostatic pressure of -1.36 MPa ($\Psi = P - \Pi + \rho_w gh$, Eq. 2.13a). From Equation 2.22 we can calculate the contact angle where P can be -1.36 MPa for pores 5 nm in radius:

$$\cos \alpha = -\frac{rP}{2\sigma} = -\frac{(5 \times 10^{-9} \text{ m})(-1.36 \text{ MPa})}{(2)(7.28 \times 10^{-8} \text{ MPa m})} = 0.0467$$

and so α is $87°$. For the wettable cell walls, α *can* be $0°$, in which case $\cos \alpha$ is unity and the maximum negative pressure does develop, namely -29 MPa for the pores 5 nm in radius. On the other hand, when water is available, it can be pulled into the interstices by such possible tensions and thereby cause α for wettable walls to increase toward $90°$, which leads to a decrease in $\cos \alpha$ and consequently a decrease in the absolute value of the pressure developed. In the present example, α is $87°$, P is -1.36 MPa, and the pore 5 nm in radius is nearly filled (Fig. 2.11d). Thus, the water in the interstices can be nearly flush with the cell wall surface. In fact, the large tensions that could be present in the cell wall generally do not occur in living cells, since water is usually available within the plant and is "pulled" into the interstices, thus nearly filling them, depending of course on their dimensions (see Eq. 2.22).

The only contribution to the water potential of the cell wall that we have been considering is the negative pressure that develops because of air-liquid interfaces, as can be the case for pure water. When the gravitational term in Equation 2.13a is ignored, $\Psi^{\text{cell wall}}$ equals $P^{\text{cell wall}} - \Pi^{\text{cell wall}}$. The bulk solution in the cell wall pores generally has an average osmotic pressure of 0.3 to 1.5 MPa. The local $\Pi^{\text{cell wall}}$ next to the Donnan phase (Ch. 3, p. 134) is considerably greater because of the many ions present there; by way of compensation, $P^{\text{cell wall}}$ is also higher near the Donnan phase than it is in the bulk of the cell wall water.

Water Flux

When the water potential inside a cell differs from that outside, the water is no longer in equilibrium and we can expect a net water movement toward the region of lower water potential. Our attention here will be specifically focused on water flow into and out of plant cells. This volume flux density of water, J_{V_w}, is assumed to be proportional to the difference in water potential ($\Delta\Psi$) across the membrane or membranes restricting the flow. The proportionality factor indicating the permeability to water flow at the cellular level is expressed by a *water conductivity coefficient*, L_w:

$$J_{V_w} = L_w \Delta\Psi = L_w(\Psi^{\text{o}} - \Psi^{\text{i}}) \tag{2.23}$$

In Equation 2.23, J_{V_w} is the volume flow of water per unit area of the barrier per unit time. It can have units of $m^3 \; m^{-2} \; s^{-1}$, or $m \; s^{-1}$, which is a velocity. In fact, J_{V_w} is the average velocity of water moving across the barrier being considered.[*] L_w can have units of $m \; s^{-1} \; MPa^{-1}$, in which case the water potentials would be expressed in MPa (Experimentally, L_w is usually the same as L_P, a coefficient also describing water conductivity, which we will formally introduce in Ch. 3 in a rather different manner.)

When Equation 2.23 is applied to cells, Ψ^o is the water potential in the external solution, and Ψ^i usually refers to the water potential in the vacuole. L_w then indicates the conductivity for water flow across the cell wall, the plasmalemma, and the tonoplast, all in series. For a group of barriers in series the overall water conductivity coefficient of the pathway, L_w, is related to the conductivities of the individual barriers by

$$1/L_w = \sum_j 1/L_{w_j} \qquad (2.24)$$

where L_{w_j} is the water conductivity coefficient of barrier j. To see why this is so, let us go back to Equation 2.23, $J_{V_w} = L_w \Delta \Psi$. When the barriers are in series, J_{V_w} is the same across each one, and so J_{V_w} equals $L_{w_j} \Delta \Psi_j$, where $\Delta \Psi_j$ is the drop in water potential across barrier j, and $\sum_j \Delta \Psi_j = \Delta \Psi$. We therefore obtain the following string of equalities: $J_{V_w}/L_w = \Delta \Psi = \sum_j \Delta \Psi_j = \sum_j J_{V_w}/L_{w_j} = J_{V_w} \sum_j 1/L_{w_j}$; hence, $1/L_w$ must equal $\sum_j 1/L_{w_j}$, as indicated in Equation 2.24. (We also note that $1/L_w$ corresponds to a resistance, and the resistance of a group of resistors in series is the sum of the resistances, $\sum_j 1/L_{w_j}$.)

We shall now consider a membrane separating two solutions that differ only in osmotic pressure. This will help us to relate L_w to the permeability coefficient of water, P_w, and also to view Fick's first law $[J_j = P_j(c_j^o - c_j^i),$ Eq. 1.8] in a slightly different way. The appropriate form for Fick's first law when describing the diffusion of the solvent water is $J_{V_w} = P_w (N_w^o - N_w^i)$ (see p. 67, where we used N_j instead of c_j in the activity of a solvent). By Equation 2.23, J_{V_w} also equals $L_w (\Psi^o - \Psi^i)$, which becomes $L_w (\Pi^i - \Pi^o)$ in our case (recall that $\Psi = P - \Pi + \rho_w gh$, Eq. 2.13a). Thus, a volume flux

---

[*] To help see why this is so, consider a volume element of cross-sectional area A, extending back from the barrier for a length equal to the average water velocity $\bar{v}_w$ multiplied by dt. In time dt this volume element, $A\bar{v}_w dt$, would cross area A of the barrier, and for such a period the volume flow of water per unit area and per unit time, J_{V_w}, would be $(A\bar{v}_w dt)/(A\, dt)$, which is $\bar{v}_w$. See also Figure 3.3.

density of water toward a region of higher osmotic pressure (Eq. 2.23) is equivalent to J_{V_w} toward a region where water is less concentrated in the sense of having a lower mole fraction (Fick's first law applied to water). Let us pursue this one step further. Using Equations 2.8 through 2.10, we can obtain the following string of equalities: $N_w \cong 1 - \sum_j n_j/n_w \cong 1 - \bar{V}_w\Pi_s/RT$.

Therefore, $N_w^o - N_w^i$ equals $(1 - \bar{V}_w\Pi_s^o/RT) - (1 - \bar{V}_w\Pi_s^i/RT)$, or $\bar{V}_w(\Pi_s^i - \Pi_s^o)/RT$. When we incorporate this last relation into our two expressions for the volume flux density of water, we obtain $J_{V_w} = L_w(\Pi^i - \Pi^o) = P_w(N_w^o - N_w^i) = P_w\bar{V}_w(\Pi_s^i - \Pi_s^o)/RT$ (Π^i is here the same as Π_s^i). Hence, we obtain the following relationship:

$$L_w = \frac{P_w\bar{V}_w}{RT} \qquad (2.25)$$

which states that the water conductivity coefficient (L_w) is proportional to the water permeability coefficient (P_w).

Next, we will estimate a possible maximum value for L_w. P_w is generally at most 10^{-4} m s^{-1} for plasmalemmas (p. 30), RT is 2.437×10^3 m^3 Pa mol^{-1} at 20°C (App. II), and $\bar{V}_w$ is 1.8×10^{-5} m^3 mol^{-1}. Thus, L_w can be

$$L_w = \frac{(10^{-4} \text{ m s}^{-1})(1.8 \times 10^{-5} \text{ m}^3 \text{ mol}^{-1})}{(2.437 \times 10^3 \text{ m}^3 \text{ Pa mol}^{-1})} = 7 \times 10^{-13} \text{ m s}^{-1} \text{ Pa}^{-1}$$

When P_w is less than 10^{-4} m s^{-1}, L_w will be proportionally less than 7×10^{-13} m s^{-1} Pa^{-1}.

When the value of the water conductivity coefficient is known, the water potential difference necessary to give an observed water flux can be calculated using Equation 2.23. For the internodal cells of *Nitella* and *Chara*, L_w for water entry is high, about 10^{-12} m s^{-1} Pa^{-1} (see Dainty). For convenience of calculation, we will consider cylindrical cells 100 mm long and 1 mm in diameter as an approximate model for such algal cells (see Fig. 3.9). The surface area across which the water flux occurs is $(2\pi r)(l)$, where r is the cell radius and l is the cell length. Thus, the area equals $(2\pi)(0.5 \times 10^{-3} \text{ m})(100 \times 10^{-3} \text{ m})$, or 3.14×10^{-4} m^2. [The area of each end of the cylinder (πr^2) is much less than $2\pi rl$; in any case, a water flux from the external solution across them would not be expected, since they are in contact with other cells, not the bathing solution; see Fig. 3.9]. The volume of a cylinder is $\pi r^2 l$, which equals $(\pi)(0.5 \times 10^{-3} \text{ m})^2(100 \times 10^{-3} \text{ m})$, or 7.9×10^{-8} m^3, in the present case. Internodal cells of *Nitella* and *Chara* grow relatively slowly—a change in volume of about 1% per day being a possible growth rate for the fairly mature cells which were used in determining L_w. This growth rate means that the water content increases by about 1% of the volume per day (1 day $= 8.64 \times 10^4$ s). The volume flux density for water is the rate of volume increase divided by the surface area across which water enters. Using the indicated value of L_w and Equation 2.23, we can then calculate the drop in water potential involved:

$$\Delta\Psi = \frac{J_{V_w}}{L_w} = \frac{\frac{1}{A}\frac{dV}{dt}}{L_w} = \frac{\frac{1}{(3.14 \times 10^{-4} \text{ m}^2)}\frac{(0.01)(7.9 \times 10^{-8} \text{ m}^3)}{(8.64 \times 10^4 \text{ s})}}{(10^{-12} \text{ m s}^{-1} \text{ Pa}^{-1})} = 29 \text{ Pa}$$

In other words, the internal water potential (Ψ^i) needed to sustain the water influx accompanying a growth of about 1% per day is only 29 Pa less than the outside water potential (Ψ^o). *Nitella* and *Chara* can grow in pond water, which is a dilute aqueous solution often having a water potential near -7 kPa (-0.07 bar). Thus, Ψ^i need be only slightly more negative than -7 kPa to account for the influx of water accompanying a growth of 1% per day.

Kinetics of Volume Changes

We will now indicate how both cell wall and membrane properties influence the kinetics of swelling or shrinking for plant cells. When the water potential outside a cell or group of cells is changed, water movement will be induced. A useful expression describing the time constant for the resulting volume change is

$$t_e = \frac{V}{A\,L_w\,(\varepsilon + \Pi^i)} \tag{2.26}$$

where V is the initial volume of a cell, A is the area across which water enters or leaves, ε is the volumetric elastic modulus (Eq. 1.17), and Π^i is the initial internal osmotic pressure (see Dainty or Zimmermann, both Ch. 1 references; and Zimmermann and Steudle). The time constant t_e represents the time to complete all but $1/e$ or 37% of the volume shift. (The halftime, which is the time required to shift halfway from the initial to final volume, equals ln 2 times t_e.) Equation 2.26 indicates that t_e for the swelling or shrinking is low for a cell with a small V/A (a large available surface area per unit volume), a high L_w (which usually means a high plasmalemma permeability to water), or a high ε (a very rigid cell wall). Figure 2.12 shows a device that can be used to measure ΔV and ΔP, thus allowing a determination of ε [see Eq. 1.17, $\varepsilon = \Delta P/(\Delta V/V)$] as well as other water-relations parameters.

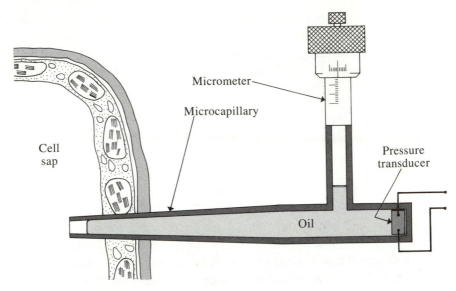

Figure 2.12
Schematic diagram of "pressure probe," an apparatus that can be used to measure P, L_w (or L_P, see Ch. 3), and ε for individual plant cells. The intracellular hydrostatic pressure is transmitted to the pressure transducer via an oil-filled microcapillary introduced into the cell. Volume can be changed by adjusting the micrometer and observing the motion of the interface between the cell sap and the oil (in the drawing the cell is greatly enlarged relative to the rest of the apparatus). See Hüsken et al. for details.

Let us next estimate the time constant for volume changes that might occur for a mesophyll cell (Fig. 1.2). We will assume that it has a volume of 10^4 μm^3 and that water enters from a neighboring cell over an area of 100 μm^2, so that V/A is 100 μm. We will let L_w be 10^{-13} m s^{-1} Pa^{-1}, ε be 20 MPa, and Π^i be 1 MPa (usually Π^i is much less than ε, and so Π^i is often ignored when calculating t_e). From Equation 2.26, the time constant is

$$t_e = \frac{(100 \times 10^{-6}\ \text{m})}{(10^{-13}\ \text{m s}^{-1}\ \text{Pa}^{-1})(21 \times 10^6\ \text{Pa})} = 50\ \text{s}$$

Hence, the volume changes caused by water transport between adjacent cells can be rather rapid, especially if their contacting area is large and if L_w and ε are fairly high. Time constants for volume changes have important implications for the water exchanges underlying stomatal opening (discussed in Ch. 8) and capacitance effects (discussed in Ch. 9).

Problems

2.1. The rise of water in a long capillary 2 mm in diameter with wettable walls is 15 mm. (a) If the capillary is tilted 45° from the horizontal, what is the vertical height of the rise? Assume that the contact angle does not change upon tilting. (b) If sucrose is added to the solution so that the density becomes 1 200 kg m^{-3} (1.2 g cm^{-3}), what is the height of the rise? (c) If the capillary extends only 7.5 mm above the main surface of the water, what is the contact angle in the capillary? (d) If the wall of the capillary is so treated that the contact angle becomes 60°, what is the height of the rise? (e) What is the rise of water in a capillary similar to the original one, but with a 1 μm radius? (f) In which of the 5 cases is the greatest weight supported by capillary (surface tension) forces?

2.2. A neutral solute at equilibrium has a concentration of 0.1 molal on one side of a barrier permeable only to water and a c_{solute} of 1 molal on the other side. Let the partial molal volume of the solute be 40×10^{-6} m^3 mol^{-1}, the temperature be 20°C, and the activity coefficients of the solute and the water be unity, except in (c). (a) What is the hydrostatic pressure difference across the barrier? (b) What would it be if the barrier were permeable to solute only? (c) If γ_{solute} is 0.5 on the more concentrated side and if other conditions are unchanged, what is ΔP across the barrier permeable to the solute only? (d) What would happen at equilibrium if the barrier were permeable to both water and solute? (e) If P on the 0.1 molal side is the same as atmospheric pressure, what is the chemical potential of the solute there?

2.3. An ideal solution contains 80 g of sorbitol (molar mass of 0.182 kg mol^{-1}) in 1 kg of water at 20°C. (a) What are N_w and a_w in the solution? (b) What is the osmotic pressure of the sorbitol solution? (c) By what percent is the activity of water reduced for an osmotic pressure of 1 MPa at 20°C compared with the value of a_w for pure water? (d) If we assume that activity coefficients are unity, what concentration of a solute corresponds to a Π of 1 MPa at 20°C? (e) A 0.25 mol m^{-3} solution of a particular polymer has a measured osmotic pressure at 20°C of 0.01 MPa. What is the osmotic pressure predicted by the Van't Hoff relation (Eq. 2.10)? Explain any discrepancies. (f) In the vacuole of a certain cell, the mole fraction of water is 0.98, the hydrostatic pressure is 0.8 MPa, and the temperature is 20°C. Assuming activity coefficients are unity, what is the water potential in the vacuole?

2.4. A tank 10 m tall and open at the top is filled with an ideal solution at 20°C. We will assume that the system is in equilibrium and that the zero level for the gravitational term is at the top of the tank, where Ψ is -0.600 MPa. (a) What is the water potential 0.1 m below the surface and at the bottom of the tank? (b) What is Π at the upper surface, 0.1 m below the surface, and at the bottom of the tank? (c) What are P and the gravitational term at the three levels in (b)? (d) What relative humidity would be in equilibrium with the water in the tank?

2.5. Chloroplasts are isolated from a plant cell whose cytosol has an osmotic pressure of 0.4 MPa at 20°C. When the chloroplasts are placed in solutions at 20°C containing an impermeant solute, the volumes are 36 μm^3 at an external pressure of 0.33 MPa, 28 μm^3 at 0.5 MPa, and 20 μm^3 at 1.0 MPa. Assume that the activity coefficients are unity. (a) What is the volume of the chloroplasts in the plant cell?

(b) What is the nonaqueous volume per chloroplast? (c) What volume fraction of the chloroplast is occupied by water in vivo? (d) What is the amount of osmotically active particles per chloroplast?

2.6. A spherical algal cell 1 mm in diameter has a water conductivity coefficient L_w equal to 10^{-12} m s^{-1} Pa^{-1}. Let the internal osmotic pressure be 1.0 MPa, the internal hydrostatic pressure be 0.6 MPa at 20°C, and the volumetric elastic modulus be 5 MPa. (a) What is the initial net volume flux density of water into or out of the cell when it is placed in pure water at atmospheric pressure? (b) What is the time constant for the change in (a)? (c) What is the water flux density at the point of incipient plasmolysis? (d) What is the water flux density when the external solution is in equilibrium with a gas phase at 97% relative humidity? (e) Assume that the water in the cell walls is in equilibrium with the internal cellular water. What are cos α and the contact angle at the air-water interface for cylindrical cell wall pores 20 nm in diameter? Assume that $\Pi^{\text{cell wall}}$ is negligible.

References

Adamson, A. W. 1976. *Physical Chemistry of Surfaces*, 3rd ed. Wiley, New York.

Altman, P. L., and D. S. Dittmer, eds. 1966. *Environmental Biology*. Federation of American Societies for Experimental Biology, Bethesda, Maryland.

Bikerman, J. J. 1970. *Physical Surfaces*. Academic Press, New York.

Boyer, J. S. 1969. Measurement of the water status of plants. *Annual Review of Plant Physiology 20*:351–364.

Bull, H. B. 1964. *An Introduction to Physical Biochemistry*. F. A. Davis, Philadelphia.

Dainty, J. 1963. Water relations of plant cells. *Advances in Botanical Research 1*: 279–326.

Davies, J. T., and E. K. Rideal. 1963. *Interfacial Phenomena*, 2nd ed. Academic Press, New York.

Dick, D. A. T. 1966. *Cell Water*. Butterworths, London.

Edsall, J. T., and J. Wyman. 1958. *Biophysical Chemistry*, Vol. I. Academic Press, New York.

Eisenberg, D., and W. Kauzmann. 1969. *The Structure and Properties of Water*. Oxford University Press, Oxford.

Giese, A. C. 1973. *Cell Physiology*, 4th ed. Saunders, Philadelphia.

Hayward, A. T. J. 1971. Negative pressure in liquids: Can it be harnessed to serve man? *American Scientist 59*:434–443.

Hüsken, D., E. Steudle, and U. Zimmermann. 1978. Pressure probe technique for measuring water relations of cells in higher plants. *Plant Physiology 61*:158–163.

Matile, P. 1978. Biochemistry and function of vacuoles. *Annual Review of Plant Physiology 29*:193–213.

Morris, J. G. 1968. *A Biologist's Physical Chemistry*. Addison-Wesley, Reading, Massachusetts.

Nobel, P. S. 1969. The Boyle–Van't Hoff relation. *Journal of Theoretical Biology 23*:375–379.

Oertli, J. J. 1971. The stability of water under tension in the xylem. *Zeitschrift für Pflanzenphysiologie 65*: 195–209.

Passioura, J. B. 1980. The meaning of matric potential. *Journal of Experimental Botany 31*: 1161–1169.

Pauling, L. 1970. *General Chemistry*, 3rd ed. W. H. Freeman, San Francisco.

Salisbury, F. B., and C. Ross. 1978. *Plant Physiology*, 2nd ed. Wadsworth, Belmont, California.

Scholander, P. F., H. T. Hammel, E. A. Hemmingsen, and E. D. Bradstreet. 1964. Hydrostatic pressure and osmotic potential in leaves of mangroves and some other plants. *Proceedings of the National Academy of Sciences, United States 52*: 119–125.

Schuster, P., G. Zundel, and C. Sandorfy, eds. 1976. *The Hydrogen Bond*, 3 volumes. North-Holland, Amsterdam.

Slatyer, R. O. 1967. *Plant-Water Relationships*. Academic Press, New York.

Stadelmann, E. J. 1966. Evaluation of turgidity, plasmolysis, and deplasmolysis of plant cells. In *Methods in Cell Physiology*, Vol. II, D. M. Prescott, ed. Academic Press, New York. Pp. 143–216.

Turner, N. C. 1981. Techniques and experimental approaches for the measurement of plant water status. *Plant and Soil 58*: 339–366.

Van Holde, K. E. 1971. *Physical Biochemistry*. Prentice-Hall, Englewood Cliffs, New Jersey.

Zimmermann, M. H., and C. L. Brown. 1971. *Trees: Structure and Function*. Springer-Verlag, New York.

Zimmerman, U., and E. Steudle. 1978. Physical aspects of water relations of plant cells. *Advances in Botanical Research 6*: 45–117.

CHAPTER 3

Solutes

In this chapter we turn our attention to the properties of solutes. We will compare chemical potentials in the aqueous phases on the two sides of a membrane or across some other region to predict the direction of passive solute fluxes as well as the driving forces leading to such motion. We will also show how the fluxes of charged species can account for the electrical potential differences observed across biological membranes.

Many solute properties are intertwined with those of the ubiquitous solvent, water. For example, the osmotic pressure term in the chemical potential of water is mainly due to the decrease of the water activity caused by the solutes ($RT \ln a_w = -\bar{V}_w \Pi$, Eq. 2.7). Also, the movement of water through the soil to the root and then to the xylem can influence the entry of dissolved nutrients, while the subsequent distribution of these nutrients throughout the plant depends on water movement in the xylem (and the phloem in some cases). In contrast with water, however, solute molecules can carry a net positive or negative electrical charge. For such charged particles, the electrical term must be included in their chemical potential, which leads to a consideration of electrical phenomena in general and an interpretation of the electrical potential differences across membranes in particular. Whether an observed ionic flux of some species into or out of a cell can be accounted for by the passive process of diffusion depends on the differences in both the concentration of that species and the electrical potential between

105

the inside and the outside. Ions can also be actively transported across membranes, in which case metabolic energy is involved in moving the solutes.

When both solutes and water traverse the same barrier, we should replace the classical thermodynamic approach with one based on irreversible thermodynamics. The various forces and fluxes are then viewed as interacting with each other, which means that the movement of water across a membrane influences the movement of solutes, and vice versa. This more general approach helps us quantitatively describe the observation that the osmotic pressure difference *effective* in causing a volume flux across a membrane permeable to both water and solutes is less than the *actual* osmotic pressure difference across that membrane.

CHEMICAL POTENTIAL OF IONS

We have already introduced the concept of chemical potential in Chapter 2. Since a substance spontaneously tends to move to regions where its chemical potential is lower, this quantity is useful for analyzing passive movements of solutes. Using a linear combination of the various contributors, we expressed the chemical potential of any species j by Equation 2.4 as follows: $\mu_j = \mu_j^* + RT \ln a_j + \bar{V}_j P + z_j FE + m_j gh$. Since water is uncharged ($z_w = 0$), the electrical term does not enter into its chemical potential. For ions, however, $z_j FE$ becomes important. In fact, for charged solutes under most conditions of biological interest, differences in the electrical term are usually far larger than changes in the terms for hydrostatic pressure or gravity. In particular, for movements across membranes, Δh is zero, and so we will omit the gravitational term in this chapter. Changes in the $\bar{V}_j P$ term can also be generally ignored because they are relatively small, as we will indicate below. Therefore, the chemical potential generally used when dealing with ions is $\mu_j^* + RT \ln a_j + z_j FE$, commonly referred to as the *electrochemical* potential. This emphasizes the role played by electrical potentials, but the expression "chemical potential" represents the sum of *all* the different contributors affecting a particular species, and so it is not really necessary to use a special term like "electrochemical" potential when dealing with ions.

Electrical Potential

The difference in electrical potential E between two locations is a measure of the amount of electrical work involved in moving a charge from one location to the other. Specifically, the work in joules (J) equals the charge in coulombs (C) times the potential difference in volts (V). The zero level for electrical potential is actually arbitrary. Since both the initial and the final electrical potential must be expressed relative to the same zero level, the arbitrariness is of no consequence when the electrical potential difference (final minus initial potential) is determined.

The charge carried by an ion of species j is a positive or negative integer z_j (the *charge number*) times the charge on a proton. For instance, z_j is $+1$ for K^+ and -2 for sulfate (SO_4^{2-}). The electrical charge carried by a proton is commonly called the *electronic charge*, since it is the same in magnitude as the charge on an electron, although opposite in sign. A proton has a charge of 1.6021×10^{-19} coulomb, abbreviated C; thus, a mole (Avogadro's number) of protons would have a charge equal to (6.022×10^{23} protons mol^{-1})(1.6021×10^{-19} C/proton), or 9.65×10^4 C mol^{-1}. Such a unit, consisting of Avogadro's number of electronic charges (i.e., one mole of single, positive charges), is called the faraday, F. This quantity, which appears in the electrical term of the chemical potential (see Eq. 2.4), equals 9.65×10^4 C mol^{-1} or 9.65×10^4 J mol^{-1} V^{-1}.

To illustrate the rather large contribution that the electrical term can make to differences in the chemical potential of a charged substance compared to the effect of changes in hydrostatic pressure, let us consider a small monovalent cation ($z_j = +1$) with a partial molal volume ($\bar{V}_j$) equal to 3.0×10^{-5} m^3 mol^{-1} (30 cm^3 mol^{-1}). We might ask what electrical potential difference (ΔE) makes the same contribution to the chemical potential of such an ion as a hydrostatic pressure drop (ΔP) of 0.1 MPa (1 bar; note that 1 MPa $= 10^6$ J m^{-3})? Thus, ΔE would be

$$\Delta E = \frac{\bar{V}_j \Delta P}{z_j F} = \frac{(3.0 \times 10^{-5} \text{ m}^3 \text{ mol}^{-1})(0.1 \times 10^6 \text{ J m}^{-3})}{(1)(9.65 \times 10^4 \text{ J mol}^{-1} \text{ V}^{-1})}$$

$$= 3.1 \times 10^{-5} \text{ V} = 0.031 \text{ mV}$$

This is an extremely small electrical potential difference. For comparison, the actual difference in electrical potential across a biological membrane is often near 100 mV, which is (100)/(0.031) or 3 200 times larger than the potential difference calculated here. Stated another way, 100 mV across a membrane would make the same contribution to the chemical potential difference as a drop of 320 MPa of hydrostatic pressure from one side of the barrier to the other for the above case of a monovalent cation having a $\bar{V}_j$ of 3.0×10^{-5} m^3 mol^{-1}. Most hydrostatic pressures encountered in plant or cell physiology are less than 1 MPa. Thus, hydrostatic pressure drops across membranes are usually negligible in comparison with the electrical contributions to the chemical potential differences of ions from one side of the membranes to the other.

Electroneutrality and Membrane Capacitance

Another consequence of the relatively large magnitude of electrical effects is the general occurrence of *electroneutrality*. Thus, in most aqueous regions that are large compared with atomic dimensions, the total electrical charge carried by the cations is essentially equal in magnitude to that carried by the anions. What is the situation for plant cells? Do we ever have an excess of negative or positive charges in cells or organelles? If there were a net charge in some region, such as near a membrane, an electrical potential difference would exist from one part of the region to another. Can we relate the size of the electrical potential difference to the net charge?

To relate charges and ΔE's, we need to introduce a new term, *capacitance*. Electrical capacitance is simply a coefficient of proportionality between a net charge and the resulting electrical potential difference. A high capacitance means that the region has the *capacity* to have many uncompensated charges separated across it without at the same time developing a large electrical potential difference across that region (see Fig. 3.1). The magnitude of such an electrical potential difference, ΔE, is related to the capacitance of the region (C) as follows:[*]

$$Q = C \, \Delta E \tag{3.1}$$

where Q is the net charge. The unit for capacitance is the *farad* (F), which equals 1 C V^{-1}. The capacitances of most biological membranes are approximately the same per unit area—about 10 mF m^{-2} (1 μF cm^{-2}; see Hope and Walker, or Lüttge and Pitman).

For convenience in estimating the net charge inside a cell that would lead to a typical membrane potential, we will consider a spherical cell of radius r. Suppose that the uncanceled or net charges are uniformly distributed in space at a concentration, c. For a conductor—a body in which electrical charges can freely move, such as an aqueous solution—the uncanceled or

---

[*] In keeping with most biological literature, E in electrical equations throughout this book refers to electrical potentials, not electric field intensities, as it does in many physics texts.

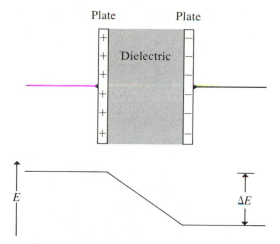

Figure 3.1
A parallel plate capacitor. Each plate represents a conductor in which charges
can freely move. Thus, each plate has a particular electrical potential, reflecting
its uncompensated positive or negative charges. Between the plates is a region
(often called the dielectric) that charges cannot cross; as a result, an electrical
potential difference ΔE occurs across it. The higher the capacitance, the greater
the charge that can be on the plates for a given ΔE across the dielectric. Capacitance
is proportional to the dielectric constant introduced on p. 59 (see Eq. 2.3 and
following discussion). Membranes act as dielectrics (D near 3 for the lipid phase)
separating the aqueous conducting phases on either side.

net charges would not remain uniformly distributed. Rather, they repel each
other and would therefore collect at the inner surface of the sphere. The
quantity c is therefore the hypothetical concentration of the net charge if it
were uniformly distributed throughout the interior of the sphere. The charge
Q within the sphere of radius r is then $(4/3)\pi r^3 cF$. (The concentration c can
be expressed in mol m^{-3}, and here refers to the net concentration of uncompensated singly charged particles; F, which is expressed in C mol^{-1}, is
necessary to convert from concentration to electrical charge.) The capacitance
C of the sphere is $4\pi r^2 C'$, where $4\pi r^2$ is the surface area of a sphere and C'
is the capacitance per unit area. After substituting these values of Q and C
into Equation 3.1 and rearranging, we obtain the following expression for
the electrical potential difference in the case of a spherical capacitor:

$$\Delta E = \frac{rcF}{3C'} \tag{3.2}$$

Equation 3.2 gives the electrical potential difference from the center of the sphere to just outside its surface. For a conductor, the internal uncompensated charges are found near the surface. In that case, ΔE actually occurs close to the bounding surface (such as a membrane) surrounding the spherical body under consideration. Equation 3.2 indicates that the electrical potential difference developed is directly proportional to the average concentration of net charge enclosed and inversely proportional to the capacitance per unit area of the sphere.

To apply Equation 3.2 to a specific situation, let us consider a spherical cell with a radius of 30 μm and an electrical potential difference across the membrane (inside negative) of -100 mV, a value close to that occurring for many cells. If the membrane capacitance per unit area has a typical value of 10 mF m^{-2} (10^{-2} C V^{-1} m^{-2}), to what net charge concentration in the cell does this potential drop correspond? Using Equation 3.2, we obtain

$$c = \frac{3C' \Delta E}{rF} = \frac{(3)(10^{-2} \text{ C V}^{-1} \text{ m}^2)(-0.1 \text{ V})}{(30 \times 10^{-6} \text{ m})(9.65 \times 10^4 \text{ C mol}^{-1})}$$

$$= -1.0 \times 10^{-3} \text{ mol m}^{-3} \, (-1 \, \mu\text{M})$$

The sign of the charge is negative, indicating more internal anions than cations. We see that the average concentration of the net (uncanceled) charge leading to a considerable electrical potential difference is rather small, an important point which we will consider next.

It is instructive to compare the net charge concentration averaged over the volume of the cell (c in Eq. 3.2) with the total concentration of anions and cations in a plant cell. Specifically, since the positive and negative ions in plant cells can each have a total concentration near 100 mol m^{-3} (0.1 M), the above-calculated excess of 10^{-3} mol m^{-3} is only about one extra negative charge per 10^5 anions. Expressed another way, the total charge of the cations inside such a cell equals or compensates that of the anions to within about one part in 100 000. When cations are taken up by a cell to any appreciable extent, either anions must accompany them or cations must be released from inside the cell, or both. Otherwise, marked departures from electrical neutrality would occur in some region, and sizable electrical potential differences would build up.

Activity Coefficients of Ions

We will now turn our attention from the electrical term in the chemical potential to the activity term—specifically, to the chemical activity itself. As indicated in Chapter 2, the activity of species j, a_j, is the thermodynamically effective concentration. For charged particles in an aqueous solution, this activity can differ appreciably from the actual concentration c_j—a fact that has not always been adequately recognized in dealing with ions. By Equation 2.5, a_j equals $\gamma_j c_j$, where γ_j is the activity coefficient of species j.

A quantitative description of the dependence of the activity coefficients of ions on the concentration of the various species in a solution was developed by Debye and Hückel in the 1920's (see Bull; Edsall and Wyman; MacInnes; or Robinson and Stokes for further details on the Debye–Hückel theory). In a local region around a particular ion, the electrostatic forces describable by relations such as Equation 2.3 constrain the movement of other ions. As the concentration increases, the average distance between the ions becomes less, which facilitates ion-ion interactions. Equation 2.3 [electrical force $= Q_1 Q_2/(4\pi\varepsilon_0 D r^2)$] indicates, for example, that the electrostatic interaction between two charged particles varies inversely as the square of their separation, so electrical forces greatly increase as the ions get closer together. When ions of opposite sign attract each other, the various other interactions of both ions are restricted, thus lowering their thermodynamically effective concentration or activity.

A simple and approximate form of the Debye–Hückel equation appropriate for estimating the values of activity coefficients of ions in relatively dilute aqueous solutions at 25°C is

$$\log \gamma_\pm = \frac{0.51 \, z_+ z_- \sqrt{\frac{1}{2} \sum_j c_j z_j^2}}{32 + \sqrt{\frac{1}{2} \sum_j c_j z_j^2}} \tag{3.3}$$

where z_+ is the charge number of the cation, z_- that of the anion, concentrations are expressed in mol m^{-3} (numerically equal to mM), and the summations are over all charged species (log is the common logarithm, to the base 10; see App. IV). Thermodynamics cannot give the activity coefficient of a single ion, because we cannot have a solution of one type of ion by itself in which to measure or to calculate γ_+ or γ_-. Thus, activity coefficients of ions always occur as the products of those of cations and anions, and hence $\gamma_\pm$ in Equation 3.3 represents the mean activity coefficient of some cation-anion pair with charge numbers z_+ and z_-. Since z_- is negative, Equation 3.3 indicates that $\log \gamma_\pm$ is also negative, and $\gamma_\pm$ is less than one. The activities of ions in dilute aqueous solutions are thus less than their concentrations, as we would indeed expect.

To estimate $\gamma_\pm$ under conditions approximating those that might occur in a plant or animal cell, let us consider an aqueous solution containing 100 mol m^{-3} (100 mM) of both monovalent cations and anions and

25 mol m^{-3} of both divalent cations and anions. For this solution, $\frac{1}{2}\sum c_j z_j^2$, known as the *ionic strength*, is as follows:

$$\frac{1}{2}\sum c_j z_j^2 = \frac{1}{2}[(100 \text{ mol m}^{-3})(1)^2 + (100 \text{ mol m}^{-3})(-1)^2$$
$$+ (25 \text{ mol m}^{-3})(2)^2 + (25 \text{ mol m}^{-3})(-2)^2]$$
$$= 200 \text{ mol m}^{-3}$$

From Equation 3.3, we can calculate log $\gamma_\pm$ for the monovalent ions as

$$\log \gamma_\pm = \frac{(0.51)(1)(-1)\sqrt{200}}{32 + \sqrt{200}} = -0.156$$

This corresponds to a mean activity coefficient of only 0.70, a value considerably less than one.

The activity coefficient of a particular ionic species depends on all the ions in the solution, as indicated by the ionic strength terms in Equation 3.3. Therefore, even when some particular ionic species is itself dilute, its activity coefficient can still be appreciably less than one because of the many electrostatic interactions with other ions. The departure from unity for activity coefficients is even greater for divalent and trivalent ions than for monovalent ions, as the $z_+ z_-$ factor in Equation 3.3 indicates. Thus, although activity coefficients of ions are often set equal to unity for convenience, this is obviously not always justified. A practical difficulty arising under most experimental situations is that c_j is much easier to determine than a_j, especially for compartments like the cytosol or the interior of a chloroplast. In those circumstances where a_j has been replaced by c_j, caution must be exercised in the interpretations or conclusions. The activity coefficients for nonelectrolytes and water are generally much closer to unity than are those for ions; hence the assumption involved in replacing a_j by c_j for such neutral species is not as severe as it is for the charged substances. For instance, γ_{sucrose} is about 0.96 for 300 mol m^{-3} (0.3 M) sucrose. An all-inclusive theory for activity coefficients is quite complicated and beyond the scope of this text. Equation 3.3 is presented only to illustrate quantitatively that activity coefficients of ions can be appreciably less than 1.00 under conditions that may occur in a plant cell (see Pytkowicz).

Nernst Potential

Having considered the electrical and activity terms in some detail, we will now turn to the role of these quantities in the chemical potential of ions.

Specifically, we will consider the deceptively simple yet extremely important relationship between the electrical potential difference across a membrane and the accompanying distribution of ions across it at equilibrium.

When ions of some species j are in equilibrium across a membrane, its chemical potential outside (o) is the same as inside (i); i.e., μ_j^o equals μ_j^i. A difference in the hydrostatic pressure term almost always makes a negligible contribution to the chemical potential differences of ions across membranes, so $\bar{V}_j P$ can be omitted from μ_j in the present case. With this approximation and the definition of chemical potential (Eq. 2.4 without the gravitational term), the condition for equilibrium of ionic species j across the membrane is given by:

$$\mu_j^* + RT \ln a_j^o + z_j F E^o = \mu_j^* + RT \ln a_j^i + z_j F E^i \qquad (3.4)$$

The term μ_j^* in Equation 3.4 is a constant referring to the same standard state of species j in the aqueous solutions on both sides of the membrane, and so it can be canceled from the equation.

Upon solving Equation 3.4 for the electrical potential difference $E^i - E^o$ across the membrane at $equilibrium$, we obtain the following important relationship:

$$E_{N_j} = E^i - E^o = \frac{RT}{z_j F} \ln \frac{a_j^o}{a_j^i} = 2.303 \frac{RT}{z_j F} \log \frac{a_j^o}{a_j^i} \qquad (3.5a)$$

which at 25°C becomes

$$E_{N_j} = \frac{25.7}{z_j} \ln \frac{a_j^o}{a_j^i} \text{ mV} = \frac{59.2}{z_j} \log \frac{a_j^o}{a_j^i} \text{ mV} \qquad (3.5b)$$

The electrical potential difference E_{N_j} in Equation 3.5 is called the $Nernst$ $potential$ of species j; i.e., $E^i - E^o = \Delta E = E_{N_j}$ in this case, where the subscript N stands for Nernst, who first derived this relation about 1900.* We derived it by assuming equality of the chemical potentials of some charged species on two sides of a membrane, but the Nernst potential can be considered in a broader context (discussed below). The natural logarithm (ln) is often replaced by 2.303 log; and 2.303 RT/F by 58.2 mV at 20°C, 59.2 mV at 25°C (as above), and 60.2 mV at 30°C (Apps. II, III, and IV).

Equation 3.5, the $Nernst$ $equation$, is an equilibrium statement showing how the internal and external activities of ionic species j are related to the

---

* Nernst made many important contributions to the understanding of the physical chemistry of solutions; he was awarded the Nobel Prize in chemistry in 1920.

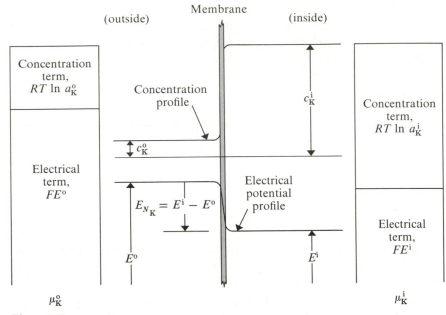

Figure 3.2
Equilibrium of K^+ across a membrane. When K^+ is in equilibrium across some membrane, the side with the higher concentration must be at a lower electrical potential for its chemical potential to be unchanged ($\mu_K^o = \mu_K^i$) in crossing the membrane (μ_K^* is the same on the two sides of the membrane). The electrical potential difference across the membrane is then the Nernst potential for K^+, E_{N_K}. Note that E^o and E^i must be expressed relative to some arbitrary baseline for electrical potentials.

electrical potential difference across a membrane (see Fig. 3.2). At equilibrium, a tenfold difference in activity of a monovalent ion across some membrane is energetically equivalent to a 59 mV difference in electrical potential (at 25°C). Hence, a relatively small electrical potential drop can balance a large difference in activity across a membrane. For some calculations, γ_j^o/γ_j^i is set equal to one (a less stringent assumption than setting both γ_j^o and γ_j^i equal to one). Under this condition, a_j^o/a_j^i in Equation 3.5 becomes the ratio of the concentrations, c_j^o/c_j^i ($a_j = \gamma_j c_j$, Eq. 2.5). Such a substitution may be justified when the ionic strengths on the two sides of a membrane are approximately the same, but it can lead to errors when the outside solution is much more dilute than the internal one, as would occur for *Chara* or *Nitella* in pond water.

   Throughout the rest of this book we will represent the *actual* electrical potential drop existing across a membrane, $E^i - E^o$, by E_M, where the subscript M refers to membrane. Hence, for both equilibrium and non-equilibrium situations, we have $E^i - E^o = \Delta E = E_M$. When a particular ionic species j is in equilibrium across some membrane, E_M equals E_{N_j}, the Nernst potential for that species. However, regardless of the actual electrical potential difference existing across a membrane (E_M), a Nernst potential for an individual ionic species j can always be *calculated* from Equation 3.5 by using the ratio of the outside to the inside activity (a_j^o/a_j^i) or concentration (c_j^o/c_j^i), e.g., $E_{N_j} = (RT/z_j F) \ln (c_j^o/c_j^i)$.

If some ionic species cannot penetrate the membrane or is actively transported across it, E_{N_j} can differ markedly from E_M. In fact, the minimum amount of energy needed to transport ions across a membrane is proportional to the difference between E_{N_j} and E_M (see Eq. 3.26). The aqueous phases designated as inside and outside can have more than one membrane intervening between them. For instance, the vacuolar sap and an external solution—the regions often considered experimentally—have two membranes separating them, the plasmalemma and the tonoplast. The thermodynamic arguments remain the same in the case of multiple membranes, with E_M and E_{N_j} referring to the electrical potential differences between the two regions actually under consideration, regardless of how many membranes occur between them.

Example of E_{N_K}

Data obtained on K^+ for the large internodal cells of the green freshwater alga *Chara corallina* are convenient for illustrating the use of the Nernst equation (see Vorobiev). The K^+ activity in the medium bathing the cells, a_K^o (where the element symbol K as a subscript is the conventional notation for K^+), was 0.096 mol m^{-3} (0.096 mM), while a_K^i in the vacuole was 48 mol m^{-3}. (See App. I for a comment on the units of activity.) By Equation 3.5 (with a factor of 58.2, since measurement was at 20°C), the Nernst potential for K^+ was

$$E_{N_K} = \frac{(58.2 \text{ mV})}{(1)} \log \left(\frac{0.096 \text{ mol m}^{-3}}{48 \text{ mol m}^{-3}} \right) = -157 \text{ mV}$$

The measured electrical potential of the vacuole relative to the external solution, E_M, was -155 mV—very close to the calculated Nernst potential for K^+. Thus, K^+ in this case may have been in equilibrium between the external solution and the vacuole. The K^+ concentration in the vacuole of these *Chara* cells was 60 mol m^{-3} (60 mM). The activity coefficient for K^+

in the vacuole (γ_K^i) equals a_K^i/c_K^i ($a_j = \gamma_j c_j$, Eq. 2.5), so γ_K^i was (48 mol m^{-3})/ (60 mol m^{-3}), or 0.80, while the K$^+$ activity coefficient in the bathing solution (γ_K^o) was about 0.96. For this example, where there are large differences in the internal and the external concentrations, the ratio γ_K^o/γ_K^i is (0.96)/(0.80), or 1.20, which differs appreciably from 1.00. If concentrations instead of activities had been used in Equation 3.5, E_{N_K} would be -162 mV, which is somewhat lower than the measured potential of -155 mV. Calculating from the concentration ratio, the suggestion that K$^+$ was in equilibrium from the bathing solution to the vacuole could not have been made with much confidence, if at all.

Equilibrium does not require that the various forces acting on a substance are zero; rather, it requires that they cancel each other. In the above example for *Chara*, the factors that tend to cause K$^+$ to move are the differences both in its activity (or concentration) and the electrical potential across the membranes. The activity of K$^+$ was much higher in the vacuole than in the external solution (see Fig. 3.2). The activity term in the chemical potential therefore represents a driving force on K$^+$ directed from the inside of the cell to the bathing solution. The electrical potential is lower inside the cell, as is indicated in Figure 3.2. Hence, the electrical driving force on the positively charged K$^+$ tends to cause its entry into the cell. At equilibrium these two tendencies for movement in opposite directions are balanced and no net K$^+$ flux occurs. As indicated above, the electrical potential difference existing across a membrane when K$^+$ is in equilibrium is the Nernst potential for K$^+$, E_{N_K}. Generally, c_K^i and a_K^i for both plant and animal cells are much higher than c_K^o and a_K^o, and K$^+$ is often close to equilibrium across the cellular membranes. From these observations, we can expect that the interiors of cells are usually at negative electrical potentials compared with the outside solutions, as is indeed the case. In general, the chemical potentials of ions are not equal in all regions of interest, so passive movements toward lower μ_j's occur—a topic we will turn to next.

FLUXES AND DIFFUSION POTENTIALS

Fluxes of many different solutes are continually occurring across biological membranes. Inward fluxes move mineral nutrients into cells while certain products of metabolism flow out of cells. The primary concern of this section will be the passive fluxes of ions toward lower chemical potentials. Although the mathematical expressions on the next few pages are rather formidable, the underlying approach is really quite straightforward. First, we indicate that the passive flux density of some species is directly proportional to the driving force causing the movement. Next, the driving force is expressed in terms of the relevant components of the chemical potential. We then examine the consequences of electroneutrality when there are simultaneous passive fluxes of more than one type of ion. This leads us to an expression describing the electrical potential difference across a membrane in terms of the properties of the ions penetrating it.

Before discussing the relation between fluxes and chemical potentials, we will briefly consider, one at a time, the fluxes already mentioned or which may be familiar from other contexts. In Chapter 1 we discussed Fick's first law of diffusion (Eq. 1.1), which says that the flux density of (neutral) species j equals $-D_j \partial c_j/\partial x$, where we can consider that the force is the negative gradient of the concentration. In Chapter 9 we will use Darcy's law (Eq. 9.6) and Poiseuille's law (Eq. 9.10), both of which indicate that the volume flux density of a solution is proportional to $-\partial P/\partial x$. Ohm's law, which describes electrical effects, can be written ΔE equals IR, where I is the current (charge moving per unit time) and R is the resistance across which the electrical potential difference is ΔE. The current per unit area A is the charge flux density, J_c, which equals $-(1/\rho) \partial E/\partial x$ where ρ is the electrical resistivity and the negative gradient of the electrical potential represents the driving force. We usually replace $-\partial E/\partial x$ by $\Delta E/\Delta x$, which leads to $I/A = (1/\rho) \Delta E/\Delta x$. For this to conform with the usual expression of Ohm's law ($\Delta E = IR$), $\rho \Delta x/A$ must equal R, which indeed it does. The gravitational force, $-m_j g$, is the negative gradient of the potential energy in a gravitational field; i.e., $-\partial m_j gh/\partial h = -m_j g$, where the minus sign indicates that the force is directed toward decreasing altitudes, namely, toward the center of the earth. The gravitational force leads to the various forms of precipitation as well as to the percolation of water down through the soil.

We have just considered examples of fluxes depending on each of the variable terms of the chemical potential ($\mu_j = \mu_j^* + RT \ln a_j + \bar{V}_j P + z_j FE + m_j gh$, Eq. 2.4). In particular, the activity term ($RT \ln a_j$) leads to Fick's first law, the pressure term ($\bar{V}_j P$) accounts for Darcy's law and Poiseuille's law, the electrical term ($z_j FE$) yields Ohm's law, and the gravitational term ($m_j gh$) is responsible for fluxes caused by gravity. In each case, flux density is found by experiment to be directly proportional to the appropriate driving force. We can generalize such relationships since nearly all transport phenomena can be represented by the following statement: flux density equals an appropriate force divided by some resistance (or an appropriate force times some conductance). Force is the negative gradient of a suitable potential, which for convenience we often take as the change in potential over some distance. But we have already shown that the chemical potential is an elegant way of summarizing all the factors that can contribute to the motion of a substance. It should not be surprising, therefore, that *in general* the flux density of species j is proportional to the negative gradient of its chemical potential, $-\partial\mu_j/\partial x$.

Flux and Mobility

Let us now consider a charged substance capable of crossing a particular membrane. When its μ_j depends on position, a net passive movement or flux of that species will tend to occur toward the region where its chemical potential is lower. The negative gradient of the chemical potential of species j, $-\partial\mu_j/\partial x$, acts as the driving force for this flux. (As in Ch. 1, the discussion will again apply to the one-dimensional case, so we will let $\partial\mu_j/\partial y = \partial\mu_j/\partial z = 0$.) The greater $-\partial\mu_j/\partial x$, the larger will be the flux of species j in the x-direction. As a very useful approximation, we will assume that the flux density is proportional to $-\partial\mu_j/\partial x$, where the minus sign means that a net positive flux density occurs in the direction of decreasing chemical potential. The magnitude of a flux density across some plane is also proportional to the local concentration of that species, c_j. That is, for a given driving force on species j, the amount of that substance which moves is proportional to how much of it is actually present—the more present, the greater the flux density. Thus, for the one-dimensional case of crossing a plane perpendicular to the x-axis, the flux density J_j can be expressed as

$$J_j = u_j c_j \left(-\frac{\partial\mu_j}{\partial x} \right)$$

$$= \bar{v}_j c_j \tag{3.6}$$

where u_j is a coefficient called the *mobility* of species j. A mobility is generally the ratio of some velocity to the force causing the motion. We will return to a consideration of u_j shortly.

As we have already indicated, the top line of Equation 3.6 is a representative example from the large class of expressions relating various flows to their causative forces. In this particular case, J_j is the rate of flow of moles of species j across unit area of a plane, and can be expressed in mol m^{-2} s^{-1}. Such a molar flux density of species j divided by its local concentration c_j gives the mean velocity $\bar{v}_j$ with which this species moves across the plane; when c_j is in mol m^{-3}, J_j/c_j can have units of (mol m^{-2} s^{-1})/(mol m^{-3}), or

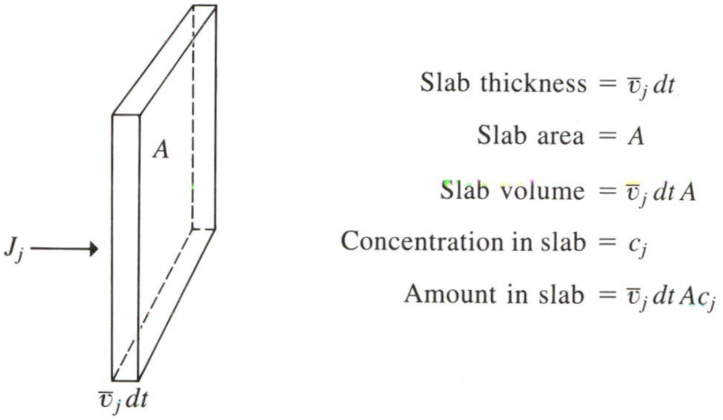

Slab thickness $= \bar{v}_j \, dt$

Slab area $= A$

Slab volume $= \bar{v}_j \, dt \, A$

Concentration in slab $= c_j$

Amount in slab $= \bar{v}_j \, dt \, A c_j$

$$\text{Flux density} = \frac{\text{(amount)}}{\text{(area)(time)}} = \frac{\bar{v}_j \, dt \, A c_j}{A \, dt} = \bar{v}_j c_j$$

Figure 3.3
Geometrical argument showing that the flux density J_j across surface area A equals $\bar{v}_j c_j$, where $\bar{v}_j$ is the mean velocity of species j. Note that $\bar{v}_j \, dt$ is an infinitesimal distance representing the slab thickness. By the definition of $\bar{v}_j$, all the molecules of species j in the slab cross A in time dt.

m s^{-1}. Perhaps this important point can be better appreciated by considering it in the following way. Suppose that the average velocity of species j moving perpendicularly toward area A of the plane of interest is $\bar{v}_j$, and consider a volume element of cross-sectional area A extending back from the membrane for a distance $\bar{v}_j \, dt$ (see Fig. 3.3). In time dt, all molecules in the volume element $\bar{v}_j \, dt \, A$ will cross area A, which means that the number of moles of species j crossing in this interval is $(\bar{v}_j \, dt \, A) \times (c_j)$, where c_j is the number of moles per unit volume. Hence, the molar flux density J_j (which is the number of moles crossing unit area per unit time) is $\bar{v}_j \, dt \, A \, c_j/ (A \, dt)$, or $\bar{v}_j c_j$ (Fig. 3.3). In other words, the mean velocity of species j moving across the plane, $\bar{v}_j$, times the number of those molecules per unit volume that can move, c_j, equals the flux density of that species, J_j, as is given by the bottom line of Equation 3.6. The upper line of Equation 3.6 indicates that this average velocity, J_j/c_j, equals the mobility of species j, u_j, times $-\partial \mu_j/\partial x$, the latter being the force on j that causes it to move. Thus, mobility is the proportionality factor between the mean velocity of motion ($\bar{v}_j = J_j/c_j$) and the causative force ($-\partial \mu_j/\partial x$). The greater the mobility of some species, the larger is its velocity in response to a given force.

The particular form of the chemical potential of species j, μ_j, that is to be substituted into Equation 3.6 depends on the application. For charged particles moving across biological membranes, the appropriate μ_j is $\mu_j^* +$

$RT \ln a_j + z_j FE$. (As we mentioned before, the $\bar{V}_j P$ term makes only a relatively small contribution to the $\Delta\mu_j$ of an ion, and so it is not included.) For treating the one-dimensional case described by Equation 3.6, μ_j must be differentiated with respect to x, $\partial\mu_j/\partial x$, which leads to $RT\, \partial \ln a_j/\partial x + z_j F\, \partial E/\partial x$ when T is constant.* The quantity $\partial \ln a_j/\partial x$ equals $(1/a_j)\, \partial a_j/\partial x$, which is $(1/\gamma_j c_j)\partial\gamma_j c_j/\partial x$ ($a_j = \gamma_j c_j$, Eq. 2.5). Using the above form of μ_j appropriate for charged solutes and this expansion of the derivative, $\partial \ln a_j/\partial x$, the net flux density of species j in Equation 3.6 can be written

$$J_j = -\frac{u_j RT}{\gamma_j}\frac{\partial\gamma_j c_j}{\partial x} - u_j c_j z_j F\frac{\partial E}{\partial x} \qquad (3.7)$$

Equation 3.7 is a general expression for the one-dimensional flux density of species j either across a membrane or in a solution in terms of two components of the driving force: the gradients in activity and electrical potential.

Before proceeding, let us examine the first term on the right side of Equation 3.7. When γ_j varies across a membrane, $\partial\gamma_j/\partial x$ can be considered to represent a driving force on species j. In keeping with common practice, we will ignore this possible force. That is, we will assume that γ_j can be treated as a constant—in any case $\partial\gamma_j/\partial x$ would be extremely difficult to measure. For constant γ_j, the first term on the right side of Equation 3.7 becomes $-u_j RT(\partial c_j/\partial x)$; i.e., it is proportional to the concentration gradient. In the absence of electrical potential gradients ($\partial E/\partial x = 0$), or for neutral solutes ($z_j = 0$), Equation 3.7 indicates that J_j equals $-u_j RT(\partial c_j/\partial x)$. But this is just the type of flux described by Fick's first law ($J_j = -D_j\partial c_j/\partial x$, Eq. 1.1), with $u_j RT$ taking the place of the diffusion coefficient, D_j. In other words, we can replace $u_j RT$ in Equation 3.7 by D_j. Since D_j equals $u_j RT$, diffusion coefficients must depend on temperature. Moreover, u_j is generally inversely proportional to the viscosity, which decreases as T increases. Thus the dependence of D_j on T can be pronounced. (For gases in air, D_j depends approximately on $T^{1.8}$; see Fuller et al.) Consequently, the temperature should be specified when the value of a diffusion coefficient is given.

Let us now resume our consideration of fluxes. In the absence of electrical effects and for constant γ_j, the net flux of species j given by Equation 3.7 equals $-D_j\partial c_j/\partial x$ when $u_j RT$ is replaced by D_j. Thus, Fick's first law ($J_j = -D_j\partial c_j/\partial x$, Eq. 1.1) is a special case of our general flux relation

---

* Actually, we must assume isothermal conditions in order for $-\partial\mu_j/\partial x$ to represent the force precisely; i.e., T is a constant here, and hence $\partial(RT \ln a_j)/\partial x = RT\, \partial \ln a_j/\partial x$.

(Eq. 3.6), where we first ignored the pressure and gravitational effects to obtain Equation 3.7 and then omitted the electrical effects. But this eventual reduction of Equation 3.6 to Fick's first law is as we should expect, since the only driving force considered when we presented Fick's first law (Ch. 1) was the concentration gradient. Such agreement between our present thermodynamic approach and the *seemingly* more empirical Fick's first law is quite important. It serves to justify the logarithmic term for activity in the chemical potential $(\mu_j = \mu_j^* + RT \ln a_j + \bar{V}_j P + z_j FE + m_j gh,$ Eq. 2.4). In other words, if the activity of species j appeared in Equation 2.4 in a form other than $\ln a_j$, Equation 3.7 would not reduce to Fick's first law under the appropriate conditions. Since Fick's first law has been amply demonstrated experimentally, such a disagreement between theory and practice would necessitate some modification in the expression used to define chemical potential.*

In contrast to the case for a neutral solute, the flux of an ion also depends on an electrical driving force, represented in Equation 3.7 by $-\partial E/\partial x$. A charged solute spontaneously tends to move in an electrical potential gradient, with a cation moving in the direction of lower or decreasing electrical potential. Of course, the concentration gradient also affects charged particles. If a certain type of ion were present in some region but absent in an adjacent one, the ions would tend to diffuse into the latter region. As such charged particles diffuse toward regions of lower concentration, an electrical potential difference is created. This electrical potential difference is referred to as a "diffusion potential," which we will discuss in detail below. The interrelationship between concentration and electrical effects is extremely important in biology. Most membrane potentials can be treated as diffusion potentials resulting from different rates of movement of the various ions across a membrane.

---

* As indicated in Chapter 2, the terms in the chemical potential can be justified or "derived" by different methods. The forms of some terms in μ_j can be readily appreciated since they follow from familiar definitions of work, e.g., the gravitational term (p. 68) and the electrical term (p. 106). The above comparison with Fick's first law indicates that the $\ln a_j$ form is the appropriate way of handling the activity term. Another derivation of the $RT \ln a_j$ term will be found in Appendix VI, together with a discussion of the pressure term for both liquids and gases. Some of these derivations incorporate conclusions from empirical observations. Moreover, the fact that the chemical potential can be expressed as a series of terms which can be added together agrees with experiment. Thus, a thermodynamic expression for the chemical potential such as Equation 2.4 can (1) summarize the results of previous observations, (2) stand the test of experiments, and (3) lead to new and useful predictions.

Diffusion Potential in a Solution

We will now use Equation 3.7 to derive the electrical potential difference created by ions diffusing down a concentration gradient in a solution containing one type of cation and its accompanying anion. This case is simpler than the biologically more realistic one to follow, and thus it may more clearly illustrate the relationship between concentration gradients and the accompanying electrical potential differences. We will assume that the cations and anions are initially placed at one side of the solution. In time, they will *diffuse* across the solution toward regions of lower concentration. In general, one ionic species will have a higher mobility, u_j, than the other. The more mobile ions will tend to diffuse faster than their oppositely charged partners, resulting in a microscopic charge separation. This slight charge separation sets up an electrical potential gradient leading to a *diffusion potential*. Using certain simplifying assumptions, we will calculate the magnitude of the electrical potential difference so created.

For convenience of analysis, let us consider a solution containing only a monovalent cation ($+$) and its monovalent anion ($-$). We will assume that their activity coefficients are constant. As our previous calculations on electrical effects have indicated, solutions are essentially neutral in regions that are large compared with atomic dimensions. Thus, c_+ equals c_-, so the concentration of either species can be designated c. Furthermore, no charge imbalance will develop in time, which means that the flux density of the cation across some plane in the solution, J_+, equals that of the anion across the same plane, J_-. (A very small charge imbalance does develop, which sets up the electrical potential gradient, but this uncompensated flux density is transitory and in any case negligible compared with J_+ or J_-.) Both flux densities, J_+ and J_-, can be expressed by Equation 3.7 and then equated to each other, which gives

$$-u_+ RT \frac{\partial c}{\partial x} - u_+ cF \frac{\partial E}{\partial x} = -u_- RT \frac{\partial c}{\partial x} + u_- cF \frac{\partial E}{\partial x} \tag{3.8}$$

where the plus sign on the right side occurs because the monovalent anion carries a negative charge ($z_- = -1$). Rearrangement of Equation 3.8 yields the following expression for the electrical potential gradient:

$$\frac{\partial E}{\partial x} = \frac{u_- - u_+}{u_+ + u_-} \frac{RT}{Fc} \frac{\partial c}{\partial x} \tag{3.9}$$

Equation 3.9 indicates that a nonzero $\partial E/\partial x$ occurs when the mobility of the cation differs from that of the anion and there is a concentration gradient. If u_- is greater than u_+, the anions move (diffuse) faster than the cations toward regions of lower concentration. As some individual anion moves ahead of its "partner" cation, an electric field is set up in such a direction as to speed up the cation and slow down the anion until they both move at the same speed, thus preserving electrical neutrality.

To obtain the difference in electrical potential produced by diffusion between planes of differing concentration, we must integrate Equation 3.9. At this point we will restrict our consideration to the *steady state* case where neither E nor c changes with time.* In going along the x-axis from region I to region II, the change in the electrical potential term in Equation 3.9 is $\int_I^{II} (\partial E/\partial x)\,dx$, which becomes $\int_I^{II} dE$ for the steady state condition[†] and so equals $E^{II} - E^I$. The integral of the concentration term is $\int_I^{II} (1/c)(\partial c/\partial x)\,dx$, which becomes $\int_I^{II} dc/c$, or $\ln(c^{II}/c^I)$. Using these two relations, integration of Equation 3.9 leads to

$$E^{II} - E^I = \frac{u_- - u_+}{u_+ + u_-} \frac{RT}{F} \ln \frac{c^{II}}{c^I} \tag{3.10a}$$

or, at 25°C,

$$E^{II} - E^I = 59.2 \frac{u_- - u_+}{u_+ + u_-} \log \frac{c^{II}}{c^I} \text{ mV} \tag{3.10b}$$

where ln has been replaced by 2.303 log, and the value 59.2 mV has been substituted for 2.303 RT/F at 25°C (App. II). In the general case, the anions have different mobilities than the cations. As the ions diffuse to regions of lower chemical potential in the solution, an electrical potential difference—

---

* At equilibrium, μ_j does not change with time or position—we are dealing with communicating regions—and there is no net flux of that solute ($J_j = 0$). When μ_j changes with position but not with time, the situation is referred to as a steady state, a condition often used to approximate problems of biological interest; J_j is then constant. Actually, it is a matter of judgment whether μ_j is constant enough in time to warrant a description of the system as being in a steady state. Similarly, constancy of μ_j for appropriate time and distance intervals is necessary before indicating that a system is in equilibrium.

† As we note in Appendix V (Eq. V.5), a total differential like dE equals $(\partial E/\partial x)\,dx + (\partial E/\partial t)\,dt$ and dc is $(\partial c/\partial x)\,dx + (\partial c/\partial t)\,dt$. For a steady state condition, both $\partial E/\partial t$ and $\partial c/\partial t$ are zero. Consequently, dE then equals $(\partial E/\partial x)\,dx$ and dc equals $(\partial c/\partial x)\,dx$.

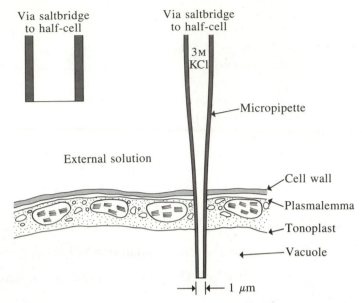

Figure 3.4
Glass micropipette filled with 3 M KCl and inserted into the vacuole of a plant cell so that the electrical potential difference across the plasmalemma and tonoplast in series can be measured. The micropipette is carefully inserted into the cell with a micro-manipulator while observing its progress with a light microscope. The micropipette tip must be strong enough to penetrate the cell wall and yet fine enough not to disturb the cell mechanically or electrically. The finer the tip, the higher is its resistance—a 1 μm tip usually has a resistance of about 10^6 ohms, which is generally acceptable. The micropipette is connected via a saltbridge containing an electrically conducting solution to a half-cell (discussed on pp. 298–299). This half-cell plus another one in electrical contact with the external solution are connected to a high input resistance device (generally at least 10^{10} ohms) so that the membrane potentials can be measured without drawing much current (see G. P. Findlay and A. B. Hope in Lüttge and Pitman, or Nobel 1974).

given by Equation 3.10 and called a diffusion potential—is set up across the section where the concentration changes from c^I to c^{II}.

An example of a diffusion potential describable by Equation 3.10 occurs at the open end of the special micropipettes used for measuring electrical potential differences across membranes (Fig. 3.4). The fine tip of the micro-pipette is open and provides an electrically conducting pathway into the cell or tissue. Ions diffusing from this fine tip give rise to a diffusion potential between the interior of the micropipette and the aqueous compartment into which the tip is inserted. To estimate the magnitude of this potential, we will assume that there is a concentration ratio of 30 across the micropipette tip, and then calculate the diffusion potentials for typical electrolytes, NaCl and KCl (see Briggs et al.). The chloride mobility u_{Cl} is about 1.52 times u_{Na}, and so the diffusion potential calculated from Equation 3.10b at 25°C is

$$E^{II} - E^{I} = (59 \text{ mV}) \left(\frac{1.52 u_{Na} - u_{Na}}{1.52 u_{N_a} + u_{Na}} \right) \log \left(\frac{1}{30} \right)$$

$$= -18 \text{ mV}$$

For KCl as the electrolyte, u_{Cl} is $1.04 u_{K}$, and the diffusion potential is only -2 mV in going from the interior of the micropipette, where the concentration of KCl is, say, 3 000 mol m^{-3} (3 M), to a region where it is 100 mol m^{-3}. Thus, from the point of view of minimizing the diffusion potential across the fine tip, KCl is a much more suitable electrolyte for micropipettes than is NaCl. In fact, KCl is employed in nearly all micropipettes used for measuring membrane potentials (generally, c_K inside cells is also higher than c_{Na}, and it does not vary as much from cell to cell as does c_{Na}). In any case, the closer u_- is to u_+, the smaller will be the diffusion potential for a given concentration ratio.

Membrane Fluxes

As is the case for diffusion potentials in open solution, membrane potentials also depend on the different mobilities of the various ions and on their concentration gradients. In this case, however, the "solution" in which the diffusion takes place toward regions of lower chemical potential is actually the membrane itself. We noted in Chapter 1 that a membrane is often the main barrier, and thus the rate-limiting step, for the diffusion of molecules into and out of cells or organelles. We would therefore expect it to be the phase across which the diffusion potential is expressed. Under biological conditions, a number of different types of ions are present, so the situation is more complex than for the single cation-anion pair analyzed above. Furthermore, the quantities of interest such as $\partial E/\partial x$ and γ_j are those within the membrane, where they are not accessible to experimental measurement. We will consider ways of dealing with these difficulties in this section.

To calculate membrane diffusion potentials, we must make certain assumptions before an integration of fluxes or gradients can be made. As a start, we will assume that the electrical potential (E) varies linearly with distance across a membrane. This means that $\partial E/\partial x$ is a constant equal to $E_M/\Delta x$, where E_M is the electrical potential difference across the membrane—i.e., $E^i - E^o = \Delta E = E_M$, as indicated on p. 114—and Δx is the membrane thickness. This assumption of a constant electric field across the membrane—the electric field equals $-\partial E/\partial x$ in our one-dimensional case—was originally suggested by Goldman in 1943 (see Goldman, or Hodgkin and Katz). It appreciably simplifies the integration conditions leading to an expression describing the electrical potential difference across membranes. As another very useful approximation, we will assume that the activity coefficient of species j, γ_j, is constant across the membrane (a less severe restriction than setting it equal to one). As noted in Chapter 1, a partition coefficient is needed to describe concentrations within a membrane, since its solvent properties differ from those of the adjoining aqueous solutions where the concentrations are actually determined. Thus, c_j in Equation 3.7 should be replaced by $K_j c_j$, where K_j is the partition coefficient for species j. Incorporating these various simplifications and conditions, we can rewrite Equation 3.7 as follows:

$$J_j = -u_j RT K_j \frac{\partial c_j}{\partial x} - u_j K_j c_j z_j F \frac{E_M}{\Delta x} \tag{3.11a}$$

We next transfer the electrical term to the left side and factor out $\dfrac{u_j z_j F E_M}{\Delta x}$:

$$\left(K_j c_j + \frac{J_j \Delta x}{u_j z_j F E_M} \right) \frac{u_j z_j F E_M}{\Delta x} = -u_j RT K_j \frac{\partial c_j}{\partial x} \tag{3.11b}$$

To transform Equation 3.11b into a form we can integrate, it is advantageous to put all terms containing the concentration on the same side of the equation. Furthermore, since the two most convenient variables are c_j and x, their differentials should appear on opposite sides of the equation. Being guided by these two objectives, and after first multiplying each term by $dx/(u_j RT)$, we can rearrange Equation 3.11 to give the following expression:

$$\frac{z_j F E_M}{RT \Delta x} dx = -\frac{K_j dc_j}{K_j c_j + \dfrac{J_j \Delta x}{u_j z_j F E_M}} \tag{3.12}$$

where $(\partial c_j/\partial x)\,dx$ has been replaced by dc_j in anticipation of the restriction to steady-state conditions.

When the term $J_j\Delta x/(u_j z_j FE_M)$ is constant, Equation 3.12 can be readily integrated from one side of the membrane to the other. The factors Δx, z_j, F, and E_M are all fixed, and only J_j and u_j need be further considered here. When the flux of species j does not change with time or position, then J_j across any plane parallel to and within the membrane is the same, and species j is neither accumulating nor being depleted in any of the regions of interest. Consequently, $\partial c_j/\partial t$ is zero, while $\partial c_j/\partial x$ is nonzero—which is the steady state condition (see footnote to p. 123). Our restriction to a steady state therefore means that J_j is constant and $(\partial c_j/\partial x)\,dx$ equals dc_j, a relation that we have already incorporated into Equation 3.12. For convenience, the mobility (u_j) of each species is also assumed to be constant within the membrane. With these restrictions, the quantity $J_j\Delta x/(u_j z_j FE_M)$ becomes a constant (b), and we can integrate Equation 3.12 from one side of the membrane (the outside, o) to the other (the inside, i). By Appendix V, $\int dx/(x + b)$ equals $\ln(x + b)$. We thus have

$$-\int_{K_j c_j^o}^{K_j c_j^i} \frac{K_j dc_j}{K_j c_j + b} = -\ln(K_j c_j + b)\Big|_{K_j c_j^o}^{K_j c_j^i} = \ln\frac{K_j c_j^o + b}{K_j c_j^i + b} \quad (3.13a)$$

Since $\int_{x^o}^{x^i} dx$ equals Δx, integration of Equation 3.12 gives

$$\frac{z_j FE_M}{RT} = \ln\frac{\left(K_j c_j^o + \dfrac{J_j\Delta x}{u_j z_j FE_M}\right)}{\left(K_j c_j^i + \dfrac{J_j\Delta x}{u_j z_j FE_M}\right)} \quad (3.13b)$$

After taking exponentials of both sides of Equation 3.13b to put it into a more convenient form, and then multiplying by $K_j c_j^i + J_j\Delta x/(u_j z_j FE_M)$, Equation 3.13b becomes

$$K_j c_j^i e^{z_j FE_M/RT} + \frac{J_j\Delta x}{u_j z_j FE_M} e^{z_j FE_M/RT} = K_j c_j^o + \frac{J_j\Delta x}{u_j z_j FE_M} \quad (3.14)$$

A quantity of considerable interest in Equation 3.14 is J_j, the net flux density of species j. This equation can be readily solved for J_j, giving

$$J_j = J_j^{in} - J_j^{out} = \left(\frac{K_j u_j z_j FE_M}{\Delta x}\right)\left(\frac{1}{e^{z_j FE_M/RT} - 1}\right)(c_j^o - c_j^i e^{z_j FE_M/RT}) \quad (3.15)$$

where J_j^{in} is the influx or inward flux density of species j, J_j^{out} is its efflux, and their difference is the net flux density. (The net flux density can represent either a net influx or a net efflux, depending on which of the unidirectional components, J_j^{in} or J_j^{out}, is larger.) We will use Equation 3.15 to derive the Goldman equation—describing the diffusion potential across membranes— and later to derive the Ussing–Teorell equation, a relation obeyed by certain passive fluxes.*

Equation 3.15 shows how the passive flux of some charged species j depends on its internal and external concentrations and on the electrical potential difference across the membrane. For most cell membranes, E_M is negative; i.e., the inside of the cell is at a lower electrical potential than the outside. For a cation (z_j a positive integer) and a negative E_M, the terms in the first two parentheses on the right side of Equation 3.15 are both negative; hence their product is positive. For an anion (z_j a negative integer) and a negative E_M, both parentheses are positive; hence their product is still positive. (For $E_M \geq 0$, the product of the first two parentheses is also positive for both anions and cations.) Thus, the sign of J_j depends on the value of c_j^o relative to that of $c_j^i e^{z_j FE_M/RT}$. When c_j^o is greater than $c_j^i e^{z_j FE_M/RT}$, the expression in the last parenthesis of Equation 3.15 is positive, and a net inward flux density of species j occurs ($J_j > 0$). Such a condition may be contrasted with Equation 1.8 $[J_j = P_j(c_j^o - c_j^i)]$, where the net flux density in the absense of electrical effects was inward when c_j^o was simply larger than c_j^i, as would adequately describe the situation for neutral solutes. But knowledge of the concentration difference alone is not sufficient to predict the magnitude or even the direction of the flux of ions; we must also consider the electrical potential difference between the two regions.

Membrane Diffusion Potential—Goldman Equation

Passive fluxes of ions, which can be described by Equation 3.15 and are caused by gradients in the chemical potentials of the various species, lead to an electrical potential difference (diffusion potential) across a membrane.

---

* Although the mathematical manipulations necessary to get from Equation 3.11 to Equation 3.15 are lengthy and cumbersome, the resulting expression is extremely important for our understanding of both membrane potentials and the passive fluxes of ions. Moreover, throughout this text we have generally presented the actual steps involved in a particular derivation to avoid statements such as "it can *easily* be shown" that such and such follows from so and so—expressions that can be most frustrating and are often untrue.

We can determine the magnitude of this electrical potential difference by considering the contributions from all ionic fluxes across the membrane and the condition of electroneutrality. Certain assumptions are needed, however, to keep the equations manageable. Under usual biological situations not all anions and cations can easily move through the membranes. Many divalent cations do not readily enter or leave plant cells passively, meaning that their mobility in the membranes is small. Such ions usually do not make a large enough contribution to the fluxes into or out of plant cells to influence markedly the diffusion potentials across the membranes. Thus, we will omit them in the present analysis, which is nevertheless rather complicated.

For many plant cells, the total ionic flux consists mainly of movements of K^+, Na^+, and Cl^- (see Dainty 1962, Gutknecht and Dainty, or Hope). These three ions generally have fairly high concentrations in and near plant cells, and we would therefore expect them to make substantial contributions to the total ionic flux density. More specifically, a flux density of species j depends on the product of its concentration and its mobility ($J_j = -u_j c_j \partial \mu_j / \partial x$, Eq. 3.6); thus ions having relatively high local concentrations or moving in the membrane fairly easily (high u_j) will tend to be the major contributors to the total ionic flux density. In some cases there may be a sizable flux density of H^+ or OH^- (which can have high u_j's) as well as of other ions, the restriction here to three ions being partially in the interest of algebraic simplicity. However, the real justification for considering only K^+, Na^+, and Cl^- is that the diffusion potentials calculated for the passive fluxes of these three ions across certain membranes are in good agreement with the measured electrical potential differences.

Our previous electrical calculations (p. 110) indicated that aqueous solutions are essentially neutral, a condition that we will again invoke here. In other words, because of the large effects resulting from small amounts of uncanceled charge, the net charge actually needed to cause the electrical potential difference across a membrane (E_M) is negligible compared with the existing concentrations of ions. Furthermore, the steady-state fluxes of the ions across the membrane will not change this condition of electrical neutrality, which means that no net charge is transported by the algebraic sum of the various charge movements across the membrane, i.e., $\sum_j z_j J_j = 0$. When the bulk of the ionic flux density consists of K^+, Na^+, and Cl^- movements, this important condition of electroneutrality can be described by equating the cationic flux densities ($J_K + J_{Na}$) to the anionic one (J_{Cl}), which leads to the following relation (as mentioned above, the various ions are indicated by using only their element symbols as subscripts):

$$J_K + J_{Na} - J_{Cl} = 0 \qquad (3.16)$$

Equation 3.16 describes the net ionic flux densities leading to the electrical potential difference across a membrane. After substituting the expressions for the various J_j's into Equation 3.16, we will solve the resulting equation for the diffusion potential across a membrane, E_M.

To obtain a useful expression for E_M in terms of measurable parameters, it is convenient to introduce a permeability coefficient for species j, P_j. In Chapter 1, such a permeability coefficient was defined as $D_j K_j / \Delta x$, where D_j is the diffusion coefficient of species j, K_j is its partition coefficient, and Δx is the membrane thickness (Eq. 1.9). Upon comparing Equation 3.7 with Equation 1.1 ($J_j = -D_j \partial c_j / \partial x$), we can see that $u_j RT$ takes the place of the diffusion coefficient of species j, D_j, as pointed out on p. 120. The quantity $K_j u_j RT / \Delta x$ can thus be equated to $K_j D_j / \Delta x$, which is the permeability coefficient P_j of species j for some particular membrane. In this way, the unknown mobility of species j in a given membrane, the unknown thickness of the membrane, and the partition coefficient for the solute can all be replaced by one parameter, which describes the permeability properties of that solute crossing the particular membrane under consideration.

With all the preliminaries out of the way, let us now derive the expression for the diffusion potential across a membrane for the case where most of the ionic flux density is due to K^+, Na^+, and Cl^- movements. Using the permeability coefficients of the three ions and substituting in the net flux density of each species as defined by Equation 3.15, Equation 3.16 becomes

$$P_K \left(\frac{1}{e^{FE_M/RT} - 1} \right) (c_K^o - c_K^i e^{FE_M/RT}) + P_{Na} \left(\frac{1}{e^{FE_M/RT} - 1} \right)$$

$$\times (c_{Na}^o - c_{Na}^i e^{FE_M/RT}) + P_{Cl} \left(\frac{1}{e^{-FE_M/RT} - 1} \right) (c_{Cl}^o - c_{Cl}^i e^{-FE_M/RT}) = 0$$

$$(3.17)$$

where z_K and z_{Na} have been replaced by 1, z_{Cl} has been replaced by -1, and FE_M/RT has been canceled from each of the terms for the three net flux densities. To simplify this unwieldly expression, the quantity $1/(e^{FE_M/RT} - 1)$ can be canceled from each of the three terms in Equation 3.17—note that $1/(e^{-FE_M/RT} - 1)$ in the last term is the same as $-e^{FE_M/RT}/(e^{FE_M/RT} - 1)$.

Equation 3.17 then assumes a much more manageable form:

$$P_K c_K^o - P_K c_K^i e^{FE_M/RT} + P_{Na} c_{Na}^o - P_{Na} c_{Na}^i e^{FE_M/RT}$$

$$- P_{Cl} c_{Cl}^o e^{FE_M/RT} + P_{Cl} c_{Cl}^i = 0 \quad (3.18)$$

After solving Equation 3.18 for $e^{FE_M/RT}$ and taking logarithms, we obtain the following expression for the electrical potential difference across a membrane:

$$E_M = \frac{RT}{F} \ln \frac{(P_K c_K^o + P_{Na} c_{Na}^o + P_{Cl} c_{Cl}^i)}{(P_K c_K^i + P_{Na} c_{Na}^i + P_{Cl} c_{Cl}^o)} \quad (3.19)$$

Equation 3.19 is generally known as the *Goldman*, or *constant field*, equation. As mentioned on p. 126, the electric field equals $-\partial E/\partial x$, which Goldman in 1943 set equal to a constant (here $-E_M/\Delta x$) to facilitate the integration across a membrane (see Goldman for a consideration of the constant field situation in a general case).* In 1949 Hodgkin and Katz applied the general equation derived by Goldman to the specific case of K^+, Na^+, and Cl^- diffusing across a membrane (see Hodgkin and Katz), and Equation 3.19 is sometimes referred to as the Goldman–Hodgkin–Katz equation.

Equation 3.19 gives the diffusion potential existing across a membrane. We derived it by assuming independent passive movements of K^+, Na^+, and Cl^- across a membrane in which $\partial E/\partial x$, γ_j, J_j, and u_j are all constant. We used the negative gradient of its chemical potential as the driving force for the net flux density of each ion. Thus, Equation 3.19 gives the electrical potential difference arising from the different tendencies of K^+, Na^+, and Cl^- to *diffuse* across a membrane to regions of lower chemical potential. When other ions cross some particular membrane in appreciable amounts, they will also make a contribution to its membrane potential. However, the inclusion of divalent and trivalent ions in the derivation of an expression for E_M complicates the algebra considerably (e.g., if Ca^{2+} is also considered, Eq. 3.18 has fourteen terms on the left side instead of six, and the equation becomes a quadratic in powers of $e^{FE_M/RT}$). But the flux densities of such ions are often rather small, in which case Equation 3.19 can be adequate for describing the membrane potential.

---

* The assumption of a constant electric field in the membrane is actually not essential for obtaining Equation 3.19; we could invoke Gauss's law and perform a more difficult integration.

Application of the Goldman Equation

In certain cases, the quantities in Equation 3.19—namely, the permeabilities and the internal and external concentrations of K^+, Na^+, and Cl^-—have all been measured. The validity of the Goldman equation can then be checked. In other words, we can compare the predicted diffusion potential with the actual electrical potential difference measured across the membrane.

As a specific example, we will use the Goldman equation to evaluate the membrane potential across the plasmalemma of *Nitella translucens*. The concentrations of K^+, Na^+, and Cl^- in the external bathing solution and in the cytosol of *Nitella translucens* are given in Table 3.1. The ratio of the permeability of Na^+ to that of K^+, P_{Na}/P_K, is about 0.18 for *Nitella translucens* (see MacRobbie 1962). The plasmalemma of *Nitella* is much less permeable to Cl^- than it is to K^+, probably only 0.1% to 1% as much. Thus, for purposes of calculation, we will let P_{Cl}/P_K be 0.003. Using these relative permeability coefficients and the concentrations given in Table 3.1, the Goldman equation (Eq. 3.19) predicts the following membrane potential ($RT/F = 25.3$ mV at 20°C, App. II):

$$E_M = (25.3 \text{ mV}) \ln \frac{(P_K)(0.1 \text{ mM}) + (0.18\ P_K)(1.0 \text{ mM}) + (0.003\ P_K)(65 \text{ mM})}{(P_K)(119 \text{ mM}) + (0.18\ P_K)(14 \text{ mM}) + (0.003\ P_K)(1.3 \text{ mM})}$$

$$= -140 \text{ mV}$$

Thus, we expect the cytosol to be electrically negative with respect to the external bathing solution, as is indeed the case. In fact, the measured value of the electrical potential difference across the plasmalemma of *Nitella translucens* is -138 mV at 20°C (Table 3.1). This excellent agreement between the observed electrical potential difference and that calculated from the Goldman equation supports the contention that the membrane potential is in fact a diffusion potential. This can be fairly easily checked by varying the external concentration of K^+, Na^+, and/or Cl^- and then seeing whether the membrane potential changes in accordance with Equation 3.19.

The different ionic concentrations on the two sides of a membrane help set up the passive ionic fluxes creating the diffusion potential, but the actual contribution of a particular ionic species to E_M also depends on the ease with which that type of ion can cross the membrane. Based on the relative permeabilities and concentrations, the major contribution to the electrical potential difference across the plasmalemma of *Nitella* comes from the K^+ flux, with Na^+ and Cl^- playing secondary roles. If the Cl^- terms are omitted

Table 3.1

Concentrations, potentials, and fluxes of various ions for *Nitella translucens* **in the light and the dark.** The superscript o refers to concentrations in the external bathing solution, superscript i refers to the cytosol. The Nernst potentials (E_{N_j}) were calculated from Equation 3.5 using concentration ratios and a numerical factor of 58.2 mV since the temperature was 20°C. The potential across the plasmalemma (E_M) was -138 mV. The fluxes indicated for the dark refer to values soon after removing the illumination from the cells.

					Light		Dark	
Ion	c_j^o (mol m^{-3}	c_j^i = mM)	E_{N_j} (mV)	$c_j^o/c_j^i e^{z_j F E_M/RT}$	J_j^{in}	J_j^{out} (nmol m^{-2} s^{-1})	J_j^{in}	J_j^{out}
Na$^+$	1.0	14	-67	17	5.5	5.5	5.5	1.0
K$^+$	0.1	119	-179	0.20	8.5	8.5	2.0	(8.5)
Cl$^-$	1.3	65	99	0.000085	8.5	8.5	0.5	—

SOURCES: E. A. C. MacRobbie, *Journal of General Physiology* 45:861–878 (1962) and R. M. Spanswick and E. J. Williams, *Journal of Experimental Botany* 15:193–200 (1964). Data reprinted by permission.

from Equation 3.19 (i.e., if P_{Cl} is set equal to zero), the calculated membrane potential is -153 mV, compared with -140 mV when Cl$^-$ is included. This rather small difference between the two potentials is a reflection of the relatively low permeability coefficient for chloride crossing the plasmalemma of *Nitella*, and so the Cl$^-$ flux has less effect on E_M than does the K$^+$ flux. In fact, its relatively high permeability and its high concentrations insure that K$^+$ will have a major influence on membrane potentials.

Restriction to the three ions indicated has proved to be adequate for treating the diffusion potential across certain membranes, and the Goldman or constant field expression in the form of Equation 3.19 has found widespread application. But changes in the amount of Ca^{2+} or H$^+$ in the external medium do cause some deviations from the predictions of Equation 3.19 for the electrical potential difference across the plasmalemma of *Nitella*. Thus, the diffusion potential is influenced by the properties of the particular membrane being considered, and ions other than K$^+$, Na$^+$, and Cl$^-$ may have to be included in specific cases. To allow for the influence of the passive flux of H$^+$ on E_M, for example, we could include $P_H c_H^o$ in the numerator of the logarithm in the Goldman equation (Eq. 3.19) and $P_H c_H^i$ in the denominator (note that a movement of H$^+$ in one direction has the same effect on E_M as a movement of OH$^-$ in the other). In fact, H$^+$ can be the most important ion influencing the electrical potential difference across certain membranes (see Lüttge and Higinbotham; Lüttge and Pitman; or Smith and Raven). But the main mechanism for the H^+ effect on E_M may be via an electrogenic pump (discussed on pp. 137–139), not the passive diffusion described by the Goldman equation. Such a pump generally would actively transport or extrude H^+ from the cytosol out across the plasma-lemma, thereby lowering the membrane potential and raising the cytosolic pH (or perhaps helping to maintain it near pH 7).

Donnan Potential

Another type of electrical potential difference encountered in biological systems is associated with immobile or fixed charges in some region adjacent to an aqueous phase containing small mobile ions. This is referred to as a *Donnan potential*. When a plant cell is placed in a KCl solution, for example, a Donnan potential arises between the interior of the cell wall and the bulk of the bathing fluid. The electrical potential difference arising from electro-static interactions at such a solid–liquid interface can be regarded as a special type of diffusion potential, as we will show shortly.

Pectin and other macromolecules in the cell wall have a large number of immobile carboxyl groups (—COOH) from which hydrogen ions dissoci-ate. This gives the cell wall a net negative charge, as indicated in Chapter 1 (see p. 36). Cations such as Ca^{2+} are electrostatically attracted to the negatively charged cell wall, and the overall effect is an exchange of H^+ for Ca^{2+} and other cations.* The region containing the immobile charges—such as carboxyl groups in the case of the cell wall—is generally referred to as the *Donnan phase* (see Fig. 3.5). At equilibrium, a distribution of op-positely charged ions electrostatically attracted to these immobile charges occurs between the Donnan phase and the adjacent aqueous one. This sets up an ion concentration gradient, and thus a Donnan potential is created between the center of the Donnan phase and the bulk of the solution next to it. The sign of the electrical potential in the Donnan phase relative to the surrounding electrolyte solution is the same as the sign of the charge of the immobile ions. For example, because of the presence of dissociated carboxyl groups, the electrical potential in the cell wall is negative with respect to an external solution. Membranes also generally act as charged Donnan phases. In addition, Donnan phases occur in the cytoplasm, where

---

* Such an attraction of positively charged species to the cell wall can increase the local concentration of solutes to about 600 mol m^{-3} (0.6 M), and consequently a greater osmotic pressure can exist in the cell wall than in surrounding aqueous solutions (see Dainty and Hope, or Sentenac and Grigon).

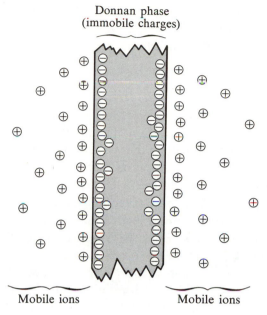

Donnan phase
(immobile charges)

Mobile ions   Mobile ions

Figure 3.5
Spatial distribution of positively
charged mobile ions ($\oplus$) occurring
on either side of a Donnan phase
in which are embedded immobile
negative charges ($\ominus$).

the immobile charges are due to large proteins and other large polymers (RNA, DNA) that have many carboxyl and phosphate groups from which protons can dissociate, leaving the macromolecules with a net negative charge. Because the net electrical charge attracts counterions, locally higher osmotic pressures can occur. Actually, all small ions in the immediate vicinity of a Donnan phase are affected, including H^+, i.e., the local pH.

Figure 3.5 illustrates a negatively charged, immobile Donnan phase with mobile, positively charged ions on either side. For example, layers containing cations are often formed in the aqueous solutions on each side of biological membranes, which generally act as Donnan phases with a net negative charge at physiological pH's.* At equilibrium no net movement of the ions occurs, so the chemical potentials of each of the mobile ions (e.g., K^+, Na^+, Cl^-, Ca^{2+}) have the same values up close to the Donnan phase as they do in the surrounding aqueous phase. The electrical potential difference (Donnan potential) can then be calculated by assuming constancy of the chemical potential. But this is exactly the same principle that we used in deriving the Nernst potential, $E_{N_j} = E^{II} - E^I = (RT/z_j F) \ln (a_j^I/a_j^{II})$, Equation 3.5, between two aqueous compartments. In fact, since the argument

---

* The electrical potential drops or Donnan potentials on either side of a membrane (cf. Fig. 3.5) are in opposite directions and are *assumed* to cancel each other when a diffusion potential across the membrane is calculated.

again hinges on the constancy of the chemical potential, the equilibrium distribution of any ion from the Donnan phase to the aqueous phase extending away from the barrier must satisfy the Nernst potential (E_{N_j}) for that particular ion.

The electrical potential difference known as the Donnan potential can also be regarded as a special case of a diffusion potential. To see why, let us imagine that the mobile ions are tending to diffuse away from the fixed immobile ions of the opposite sign. We can assume that the mobile ions are initially in the same region as the immobile ones. In time, some of the mobile ions will tend to diffuse away. This tendency, based on thermal agitation, causes a slight charge separation, thus setting up an electrical potential difference between the Donnan phase and the bulk of the adjacent solution. For the case of a single species of mobile cations and the anions fixed in the membrane (both assumed to be monovalent), the diffusion potential across that part of the aqueous phase next to the membrane can be described by Equation 3.10, $E^{II} - E^{I} = [(u_- - u_+)/(u_+ + u_-)](RT/F) \ln (c^{II}/c^{I})$, which we derived for diffusion toward regions of lower chemical potential in a solution. Fixed anions have zero mobility $(u_- = 0)$; hence $(u_- - u_+)/(u_+ + u_-)$ here is $-u_+/u_+$, or -1. Equation 3.10 then becomes $E^{II} - E^{I}$ equals $-(RT/F)$ $\ln (c^{II}/c^{I})$, which is simply the Nernst potential (Eq. 3.5) for monovalent cations $[-\ln (c^{II}/c^{I}) = \ln (c^{I}/c^{II})]$. Thus, the Donnan potential can also be regarded as a type of diffusion potential occurring as the mobile ions tend to diffuse away from the charges of opposite sign, which remain fixed in the Donnan phase.

ACTIVE TRANSPORT

The expression "active transport" implies that energy derived from metabolic processes is used to move a solute across a membrane toward a region of higher chemical potential. There are three different aspects to this description of active transport: a supply of energy, movement, and an increase in chemical potential. Although it is not really necessary, the term "active transport" has conventionally been restricted to the case of movement in the energetically uphill direction, and we will follow this restriction here.

By itself, a difference in chemical potential of a certain species across a membrane does not necessarily imply that active transport of that species is occurring. For example, if the solute does not penetrate the membrane, the solute is unable to attain equilibrium across it, and μ_j would not be expected to be the same on the two sides. For ions actually moving across some membrane, the ratio of the influx to the efflux of a particular ionic species provides information that helps to indicate whether or not active transport is taking place. A very simple but often effective approach for determining whether fluxes are active or passive is to remove possible energy sources. For photosynthesizing plant tissue, this can mean comparing the fluxes in the light with those in the dark. (In addition, we should check whether the permeability of the membrane changes or other alterations occur upon illumination.) Compounds or treatments that disrupt metabolism also can be useful for ascertaining whether metabolic energy is being used for the active transport of various solutes.

We will begin our discussion by showing how active transport can directly affect membrane potentials. We will then compare the temperature dependencies of metabolic reactions with those for diffusion processes across a barrier to show that a marked enhancement of solute influx caused by increasing the temperature does not necessarily indicate that active transport is taking place. Next, we will consider a much more reliable criterion for deciding whether fluxes are passive or not—namely, the Ussing–Teorell, or flux ratio, equation. After examining a specific case where active transport is involved, we will discuss a relationship often invoked for describing the concentration dependence of ion uptake.

Electrogenicity

One of the possible consequences of actively transporting a certain ionic species into a cell or organelle is the development of an excess of electrical charge inside the membrane-surrounded entity. If for some ionic species the active transport process involves an accompanying ion of opposite charge or an equal release of a similarly charged ion, the total charge in the cell or organelle is unaffected. However, if the charge of the actively transported ion is not directly compensated for, the process is *electrogenic*; i.e., it tends to generate an electrical potential difference across the membrane. An electrogenic uptake of an ion that produces a net transport of charge into some cell or organelle will thus affect its membrane potential (see Higinbotham and Anderson, Higinbotham et al., Hope, MacRobbie 1977, and Spanswick). We can appreciate this effect by referring to Equation 3.1 ($Q = C \Delta E$), which indicates that the difference in electrical potential across a membrane, ΔE, equals Q/C, where Q is the net charge enclosed within the cell organelle and C is the membrane capacitance. The initial movement of net charge across a membrane by active transport leads to a fairly rapid increase or decrease in the potential difference across the membrane.

To be specific, we will consider a spherical cell of radius r into which there is an electrogenic influx of chloride caused by active transport, $J_{a.t.Cl}^{in}$. The amount of charge transported in time t across the surface of the sphere (area $= 4\pi r^2$) is $J_{a.t.Cl}^{in}\, 4\pi r^2 t$. This active uptake of Cl^- increases the internal concentration of negative charge by the amount moved in divided by the cellular volume, or $J_{a.t.Cl}^{in}\, 4\pi r^2 t/(4\pi r^3/3)$, which is $3J_{a.t.Cl}^{in}\, t/r$. Let us suppose that $J_{a.t.Cl}^{in}$ is 10 nmol m^{-2} s^{-1} and that the cell has a radius of 30 μm. In 1 s the concentration of Cl^- actively transported in would be

$$c_{a.t.Cl} = \frac{3J_{a.t.Cl}^{in}\, t}{r} = \frac{(3)(10 \times 10^{-9}\,\text{mol m}^{-2}\,\text{s}^{-1})(1\,\text{s})}{(30 \times 10^{-6}\,\text{m})}$$

$$= 1.0 \times 10^{-3}\,\text{mol m}^{-3} \qquad (1\,\mu\text{M})$$

Assuming a membrane capacitance of 10 mF m^{-2}, we calculated on p. 110 that such a cell has 1.0×10^{-3} mol m^{-3} uncompensated negative charge when the interior is 100 mV negative with respect to the external solution. If no change were to take place in the other ionic fluxes, the electrogenic uptake of Cl^- into this cell would cause its interior to become more negative at the rate of 100 mV s^{-1}. The charging of the membrane capacitance is indeed a rapid process.

The initiation of an electrogenic process causes an adjustment of the passive fluxes across the membrane. In particular, the net charge actively brought in is very soon electrically compensated by appropriate passive movements of it and other ions into or out of the cell. The actual electrical potential difference across the membrane then results from the diffusion potential caused by these new passive fluxes plus a contribution from the electrogenic process involving the active transport of some charged species. We can represent the electrical potential difference generated by the active transport of species j, $E_{a.t.j}$, as follows:

$$E_{a.t.j} = z_j F J_{a.t.j} R_j^{memb} \qquad (3.20)$$

where z_j is the charge number of species j, F is the faraday, $J_{a.t.j}$ is the flux density of species j caused by active transport, and R_j^{memb} is the membrane resistance for the specific pathway along which species j is actively transported. $FJ_{a.t.j}$ in Equation 3.20—which is really a form of Ohm's law—is the charge flux density and can have units of (C mol^{-1}) (mol m^{-2} s^{-1}), or C m^{-2} s^{-1}, which is ampere m^{-2}, or current per unit area. Hence, R_j^{memb} can be expressed in ohm m^2, so that $z_j F J_{a.t.j} R_j^{memb}$ in Equation 3.20 can have units of ampere ohm, i.e., V, the proper unit for electrical potentials. For many of the plant cells and ions that have been studied, R_j^{memb} is from

0.1 to 1 ohm m^2 (see Hope or Spanswick). The sign of z_j accounts for the algebraic sign of the electrical potential change in Equation 3.20.

Let us now see how large $E_{a.t.j}$ given by Equation 3.20 might be for representative electrogenic "pumps." The electrical potential difference created by the active transport of 10 nmol m^{-2} s^{-1} of a monovalent cation across a membrane resistance of 0.3 ohm m^2 would be

$$E_{a.t.j} = (1)(9.65 \times 10^4 \text{ C mol}^{-1})(10 \times 10^{-9} \text{ mol m}^{-2} \text{ s}^{-1})(0.3 \text{ ohm m}^2)$$

$$= 3 \times 10^{-4} \text{ V} \quad (0.3 \text{ mV})$$

This is a rather small electrical potential difference. When the membrane resistance for the active transport pathway is higher, or the net flux by the electrogenic pump is greater, the potential created by the active transport process would of course be larger. For certain plant cells and some animal cells, R_j^{memb} is about 1 ohm m^2 and $J_{a.t.j}$ may be 200 nmol m^{-2} s^{-1}. Equation 3.20 indicates that the potential difference created by such an electrogenic pump for a monovalent cation would be approximately 0.019 V (19 mV). Measurements in the presence and absence of metabolic inhibitors have indicated that an electrogenic efflux of Na$^+$ or, more likely, H$^+$ can lead to an effect of more than 50 mV on E_M (see Higinbotham and Anderson). In any case, the actual electrical potential difference across a membrane can be obtained by adding the potential drop caused by active transport of uncompensated charge (Eq. 3.20) and that caused by passive fluxes (predicted by the Goldman equation, Eq. 3.19). (See Spanswick 1981.)

Boltzmann Energy Distribution and Q$_{10}$, a Temperature Coefficient

In general, most metabolic reactions are markedly influenced by temperature, while processes such as light absorption are insensitive to temperature. What temperature dependence should we expect for diffusion? Can we decide whether the movement of some solute into a cell is by active transport or by passive diffusion once we know how the fluxes depend on temperature? To answer such questions, we need an expression describing the distribution of energy among the molecules as a function of temperature in order to determine what fraction of the molecules has the requisite energy for a particular process. In aqueous solutions the relevant energy is generally the kinetic energy of motion of the species involved. Hence, we will begin by relating the distribution of kinetic energy among molecules to temperature.

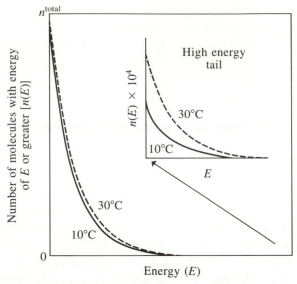

Figure 3.6
Boltzmann energy distributions at 10°C (———) and 30° (-----). The inset is a continuation of the right-hand portion of the graph with the scale of the abscissa (E) unchanged and that of the ordinate $[n(E)]$ expanded by 10^4. The difference between the two curves is extremely small over most of the energy range. Although very few molecules are in the "high energy tail," there are many more such molecules at the higher temperature.

The topics introduced here are important for a basic understanding of many aspects of biology—all the way from biochemical reactions to the consequences of light absorption.

Very few molecules possess extremely large kinetic energies. In fact, the probability that a molecule has a given kinetic energy E decreases exponentially as E increases. The precise statement of this is known as the *Boltzmann energy distribution*, which describes the frequency with which specific kinetic energies are possessed by molecules at equilibrium at a given temperature (see Bull, or Edsall and Wyman). A convenient form for the Boltzmann energy distribution (Fig. 3.6) at temperature T is as follows:

$$n(E) = n_{\text{total}}e^{-E/kT} \qquad \text{molecule basis} \qquad (3.21a)$$

where $n(E)$ is the number of molecules possessing an energy of E or more out of the total number of molecules, n_{total}, and k is Boltzmann's constant.

The quantity $e^{-E/kT}$, which by Equation 3.21a equals $n(E)/n_{total}$, is often referred to as the *Boltzmann factor*. Equation 3.21a and Figure 3.6 indicate that the number of molecules with an energy of zero or greater, $n(0)$, equals $n_{total}e^{-0/kT}$ or $n_{total}e^{-0}$, which is simply n_{total}, the total number of molecules present.

Because of collisions based on thermal motion, energy is continually being gained or lost by individual molecules in a random fashion. Hence, a wide range of kinetic energies is possible, although very high energies are less probable; see Equation 3.21a. As the temperature is raised, not only does the average energy per molecule become higher, but the relative number of molecules in the "high energy tail" of the exponential Boltzmann distribution also increases substantially (see Fig. 3.6).

So far in this section we have considered energy per *molecule*, while in many cases it is more convenient to consider energy per *mole*. To change the Boltzmann distribution from a molecule to a mole basis, we multiply Boltzmann's constant k [energy/(molecule K)] by Avogadro's number N (molecules/mole), which gives us the gas constant R [energy/(mole K)]; i.e., R equals kN.* If $n(E)$ and n_{total} are numbers of moles and E is energy per mole, we simply replace k in the Boltzmann energy distribution (Eq. 3.21a) by R:

$$n(E) = n_{total}e^{-E/RT} \qquad \text{mole basis} \qquad (3.21b)$$

For diffusion across a membrane, the appropriate Boltzmann energy distribution indicates that the number of molecules with a kinetic energy of U or greater per mole resulting from velocities in some particular direction is proportional to $\sqrt{T}e^{-U/RT}$ (see Davson and Danielli). A minimum kinetic energy (U_{min}) is often necessary to diffuse past some barrier or to cause some specific reaction. In such circumstances, any molecule with a kinetic energy of U_{min} or greater has sufficient energy for the particular process. For the Boltzmann energy distribution appropriate to this case, the number of such molecules is proportional to $\sqrt{T}e^{-U_{min}/RT}$. (These expressions having the factor $\sqrt{T}$ actually only apply to diffusion in one dimension, e.g., for mole-

---

* In this text we will use two analogous sets of expressions: (1) molecule, mass of molecule, photon, electronic charge, k, kT; and (2) mole, molecular weight, mole, faraday, R, RT (see App. II for numerical values of k, R, kT, and RT). As indicated above, a quantity in the second set, which is more appropriate for most of our applications, is Avogadro's number N (6.022×10^{23}) times the corresponding quantity in the first set. We also note that 1 electron volt (eV), an energy unit often used on atomic and molecular levels, equals 1.602×10^{-19} J and that 1 eV/molecule = 96.55 kJ mol^{-1} = 23.06 kcal mol^{-1} (see App. III).

cules diffusing across a membrane.) At a temperature 10°C higher, the number is proportional to $\sqrt{(T + 10)}e^{-U_{min}/[R(T + 10)]}$. The ratio of these two quantities is called the Q_{10}, or *temperature coefficient* of the process:

$$Q_{10} = \frac{\text{rate of process at } T + 10°C}{\text{rate of process at } T} = \sqrt{\frac{T + 10}{T}}\, e^{10U_{min}/[RT(T+10)]} \quad (3.22)$$

To obtain the form of the exponential given in Equation 3.22, we note that

$$\frac{-U_{min}}{R(T + 10)} + \frac{U_{min}}{RT} = \frac{-U_{min}T + U_{min}(T + 10)}{RT(T + 10)} = \frac{10U_{min}}{RT(T + 10)}$$

A Q_{10} near unity is characteristic of those passive processes with no energy barrier to surmount, i.e., where U_{min} equals 0. On the other hand, most enzymatic reactions take place only when the reactants have a considerable kinetic energy, so such processes tend to be quite sensitive to temperature. A value of 2 or greater for Q_{10} is often considered to indicate the involvement of metabolism, as occurs for active transport of a solute into a cell or organelle. However, Equation 3.22 indicates that *any* process having an appreciable energy barrier can have a large temperature coefficient.

The Q_{10} for a particular process can be indicative of the minimum kinetic energy required (U_{min}), and vice versa. A membrane often represents an appreciable energy barrier for the diffusion of charged solutes—U_{min} for passive ion movement across can be 50 kJ mol^{-1} (11.9 kcal mol^{-1}, or 0.52 eV molecule^{-1}). By Equation 3.22, this would lead to the following temperature coefficient for a T of 20°C:

$$Q_{10} = \sqrt{\frac{(303\text{ K})}{(293\text{ K})}}\, e^{(10\text{ K})(50 \times 10^3\text{ J mol}^{-1})/[(8.3143\text{ J mol}^{-1}\text{ K}^{-1})(293\text{ K})(303\text{ K})]}$$

$$= 1.02e^{0.68} = 2.01$$

Therefore, the passive uptake of this ion would double with only a 10°C increase in temperature. This example indicates that a passive process can have a rather high Q_{10} if there is an appreciable energy barrier, implying that a large Q_{10} for ion uptake does not necessarily indicate active transport.

A kinetic energy of 50 kJ mol^{-1} or greater is possessed by only a small fraction of the molecules in the Boltzmann energy distribution (Eq. 3.21b). For instance, at 20°C the Boltzmann factor is

$$e^{-E/RT} = e^{-(50 \times 10^3\text{ J mol}^{-1})/[(8.3143\text{ J mol}^{-1}\text{ K}^{-1})(293\text{ K})]} = 1.2 \times 10^{-9}$$

As T is raised, the fraction of molecules in the high energy part of the Boltzmann distribution increases greatly (see Fig. 3.6). Many more molecules thus have the requisite kinetic energy, U_{min}, and consequently can take part in the process being considered. In particular, at 30°C the Boltzmann factor becomes 2.4×10^{-9}. Hence, the Boltzmann factor for a U_{min} of 50 kJ mol^{-1} essentially doubles for a 10°C rise in temperature, consistent with our Q_{10} calculation for this case.

Activation Energy and Arrhenius Plots

The existence of an energy barrier requiring a minimum energy U_{min} to cross is related to the concept of *activation energy*, which refers to the minimum amount of energy necessary for some reaction to take place. In the case of a membrane, we can evaluate U_{min} by determining how the number of molecules diffusing across the membrane varies with temperature, e.g., by invoking Equation 3.22. In fact, the required kinetic energy (U_{min}) corresponds to the activation energy for crossing the barrier. For a chemical reaction, we can also experimentally determine how the process is influenced by temperature. If we represent the activation energy per mole by A, the rate constant for such a reaction should vary with temperature as follows:

$$\text{rate constant} = Be^{-A/RT} \qquad (3.23)$$

where B is essentially a constant.*

Equation 3.23 is referred to as the *Arrhenius equation*. It was originally proposed on experimental grounds by Arrhenius at the end of the 19th century and only subsequently interpreted theoretically. A plot of the logarithm of the rate constant (or rate of some reaction) versus $1/T$ is commonly known as an *Arrhenius plot* (see Fig. 3.7); by Equation 3.23, ln (rate constant) equals $\ln B - A/RT$. Hence, the slope of an Arrhenius plot $(-A/R)$ can be used to determine the activation energy. Many reactions of importance in biochemistry have large values for A and are therefore extremely sensitive to temperature. Moreover, an enzyme greatly increases the rate of the reaction which it catalyzes by reducing the value of the activation energy needed (see Eq. 3.23); i.e., many more molecules then have enough energy to get over the energy barrier separating the reactants from

---

* Actually, B can depend on temperature. For diffusion in one dimension as discussed above, the number of molecules with an energy of at least U_{min} per mole is proportional to $\sqrt{T}e^{-U_{min}/RT}$; B would then be proportional to $\sqrt{T}$.

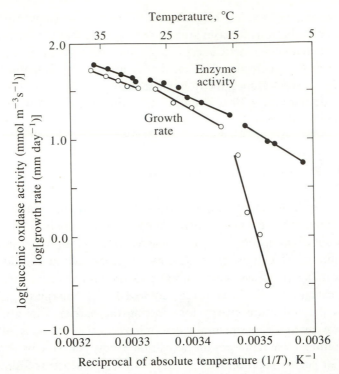

Figure 3.7
Arrhenius plots of mitochondrial succinate oxidase activity and growth
rate of the hypocotyl plus radicle of *Vigna radiata*. [Data are replotted
from J. K. Raison and E. A. Chapman, *Australian Journal of Plant
Physiology 3*:291–299 (1976). Used by permission.]

the products. Thus, the related concepts of activation energy and energy
barrier have many applications in biology (see Morris or Stein).

Let us next consider the activation energies for certain cases of diffusion.
For diffusion in water at 20°C, A is about 17 kJ mol^{-1} for K$^+$ and 21 kJ
mol^{-1} for mannitol. After replacing U_{min} by A in Equation 3.22, we can
calculate that the appropriate Q_{10} for such diffusion would be about 1.3.
The activation energy for the passive efflux of K$^+$ from many cells is about
63 kJ mol^{-1} near 20°C, which corresponds to a Q_{10} of 2.5 (see Stein). We
again conclude that a purely passive process, such as diffusion across
membranes, can have a marked temperature dependence.

Arrhenius plots have been used to identify those processes that change
markedly over the range of temperatures for which chilling-sensitive plants
such as corn, cotton, cucumber, rice, soybean, and tomato are injured (see

Lange et al.; Lyons et al.; or Simon, Ch. 1 reference). Such plants are often severely injured by exposure to temperatures that are low but far above freezing, e.g., near 10°C; prolonged exposure can even result in death. Visible symptoms include wilting, surface pitting of the leaves, and loss of chlorophyll. Damage at the cellular level caused by chilling can be manifested by loss of cytoplasmic streaming, metabolic dysfunction leading to the accumulation of toxic products, and enhanced membrane permeability. Figure 3.7 shows that the temperature dependence of the rate of growth of the chilling-sensitive *Vigna radiata* (mung bean) can change near 15°C and near 28°C. But the activation energy for a mitochondrial enzyme (succinate oxidase) also changes at these two temperatures, from 87 kJ mol^{-1} below 15°C to 44 kJ mol^{-1} above 28°C (Fig. 3.7). Moreover, changes in the organization of membrane lipids in both mitochondria and chloroplasts of mung bean occur at these two temperatures (see Raison and Chapman). In fact, transitions in the physical properties of membranes (such as a change to a more fluid state), which actually can occur over a range of temperatures, may underlie the changes in slope seen in Figure 3.7. Such changes in the membranes can affect the catalytic properties of enzymes located in them, the permeability of solutes, and the general regulation of cellular metabolism, which then obviously would affect plant growth.

Ussing–Teorell Equation

We will now consider ways of distinguishing between active and passive fluxes that are more reliable than determining Q_{10}'s. One of the most useful physicochemical criteria for deciding whether a particular ionic movement across a membrane is active or passive is the application of the *Ussing–Teorell*, or *flux ratio*, equation. For ions moving passively, this expression shows how the ratio of the influx to the efflux depends on the internal and the external concentrations of that species and on the electrical potential difference across the membrane. If the Ussing–Teorell equation is satisfied, passive movements can account for the observed flux ratio, and so active transport of the ions need not be invoked. We can readily derive this expression by considering how the influx and the efflux could each be determined experimentally, as the following arguments indicate.

For measuring the unidirectional inward component, or influx, of a certain ion (J_j^{in}), the plant cell or tissue can be placed in a solution containing a radioactive isotope of species j. Initially, none of the radioisotope is inside the cells, so the internal specific activity for this isotope equals zero at the beginning of the experiment. (As for any radioisotope study, only some of the molecules of species j are radioactive. This particular fraction is known as the *specific activity*, and must be determined for both c_j^o and c_j^i in the present experiment.) Since none is inside, the initial unidirectional outward component, or efflux, of the radioisotope (J_j^{out}) is zero, and the initial net flux density (J_j) of the isotope indicates J_j^{in}. From Equation 3.15, this influx of the radioisotope can be represented by $(K_j u_j z_j F E_M / \Delta x) \left[1/(e^{z_j F E_M / RT} - 1) \right] c_j^o$. After the isotope has entered the cell, some of it will start coming out. Therefore, only the initial flux will give an accurate measure of the influx of the radioisotope of species j.

Once the radioactivity has built up inside to a substantial level, we may remove the radioisotope from the outside solution. The flux of the isotope is then from inside the cells to the external solution. In this case, the specific activity for c_j^o equals zero, and c_j^i determines the net flux of the radioisotope. By Equation 3.15, this efflux of the isotope of species j differs in magnitude from the initial influx only by having the factor c_j^o replaced by $c_j^i e^{z_j F E_M / RT}$, while the quantities in the first two parentheses remain the same. The ratio of these two flux densities—each of which can be separately measured—takes on the following relatively simple form:

$$\frac{J_j^{in}}{J_j^{out}} = \frac{c_j^o}{c_j^i e^{z_j F E_M / RT}} \tag{3.24}$$

Equation 3.24 was independently derived by both Ussing and Teorell in 1949 (see Lüttge and Higinbotham, Teorell, and Ussing), and is known either as the Ussing–Teorell equation or the flux ratio equation. It is strictly valid only for ions moving passively without interacting with other substances that may also be moving across the membrane. The present derivation makes use of Equation 3.15, which gives the passive flux of some charged species across a membrane in response to differences in the chemical potential of that species across the barrier. Equation 3.15 was derived by considering only one species at a time; hence, possible interactions between the fluxes of different species are not included in Equation 3.24. The Ussing–Teorell equation can thus be used to determine whether the observed influxes and effluxes are passive (i.e., responses to the chemical potentials of the ions on the two sides of a membrane), or whether additional factors such as interactions between species or active transport must be invoked. For example, when active transport of species j into a cell is taking place, the actual J_j^{in} would be the passive unidirectional flux density (i.e., the one predicted by Eq. 3.24) *plus* the influx due to active transport. After one further comment about Equation 3.24, we will consider a specific application of the Ussing–Teorell equation to evaluate membrane fluxes.

The ratio of the influx of species j to its efflux, as given by Equation 3.24, can readily be related to the difference in its chemical potential across a membrane, $\mu_j^o - \mu_j^i$. This difference causes the flux ratio to differ from unity, and we will return to it to estimate the minimum amount of energy needed for active transport of that ionic species across the membrane. After taking logarithms of both sides of the Ussing–Teorell equation (Eq. 3.24) and multiplying by RT, we obtain the following equalities:

$$RT \ln \frac{J_j^{\text{in}}}{J_j^{\text{out}}} = RT \ln \frac{c_j^o}{c_j^i} - z_j F E_M$$

$$= RT \ln a_j^o + z_j F E^o - RT \ln a_j^i - z_j F E^i$$

$$= \mu_j^o - \mu_j^i \tag{3.25}$$

where the membrane potential E_M has been replaced by $E^i - E^o$, in keeping with our previous convention. The derivation is restricted to the case of constant γ_j ($\gamma_j^o = \gamma_j^i$), which means that $\ln (c_j^o/c_j^i)$ equals $\ln (\gamma_j^o c_j^o/\gamma_j^i c_j^i)$, or $\ln (a_j^o/a_j^i)$, which is $\ln a_j^o - \ln a_j^i$. Finally, the $\overline{V}_j P$ term in the chemical potential is ignored for these charged species—actually, we need only assume that $\overline{V}_j P^o$ equals $\overline{V}_j P^i$—so that μ_j then equals $\mu_j^* + RT \ln a_j + z_j F E$, where μ_j^* has the same value on the two sides of the membrane. Thus, $\mu_j^o - \mu_j^i$ is $RT \ln a_j^o + z_j F E^o - RT \ln a_j^i - z_j F E^i$, as is indicated both in Equation 3.25 and the caption to Figure 3.8.

A difference in chemical potential of species j across a membrane would cause the flux ratio of the passive flux densities to differ from unity, a conclusion that follows directly from Equation 3.25. When μ_j^o equals μ_j^i, the influx would balance the efflux, and hence there would be no net passive flux density of species j across the membrane ($J_j = J_j^{\text{in}} - J_j^{\text{out}}$ by Eq. 3.15). This condition ($\mu_j^o = \mu_j^i$) is also described by Equation 3.4, which was used to derive the Nernst equation. In fact, in such a case of equality of the chemical potentials, the electrical potential difference across the membrane is the Nernst potential, as given by Equation 3.5: $E_{N_j} = (RT/z_j F) \ln (a_j^o/a_j^i)$. Thus, when E_M equals E_{N_j} for some species, J_j^{in} equals J_j^{out}, and no net passive flux density of that ion is expected across the membrane, nor is any energy expended in moving the ion from one side of the membrane to the other. When Equations 3.24 and 3.25 are not satisfied for some species, such ions are not moving across the membrane passively, or perhaps not moving independently from other fluxes. One way this may occur is for the various

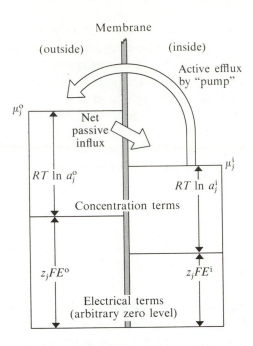

Figure 3.8
Diagram illustrating the situation for an ion not in equilibrium across a membrane. Since μ_j^o is greater than μ_j^i, there is a net passive flux density into the cell. In the steady state, this net influx is balanced by an equal efflux caused by active transport of species j out of the cell. We note that $\mu_j^o - \mu_j^i$ equals $z_j F(E_{N_j} - E_M)$, where $z_j F E_{N_j}$ is $RT \ln a_j^o - RT \ln a_j^i$ and $-z_j F E_M$ is $z_j F E^o - z_j F E^i$.

fluxes to be interdependent, a condition describable by irreversible thermo-dynamics (see p. 161). Another way is through active transport of the ions, whereby energy derived from metabolism is used to move solutes to regions of higher chemical potential.

Example of Active Transport

The above criteria for deciding whether or not active transport of certain ions is taking place can be illustrated by using data obtained with the large internodal cells of *Nitella translucens* (Table 3.1). All the parameters appearing in the Ussing–Teorell equation have been measured for Na^+, K^+, and Cl^- using *Nitella*. For experimental purposes, this fresh water alga is often placed in a dilute aqueous solution containing 1 mM NaCl, 0.1 mM KCl, plus 0.1 mM $CaCl_2$ (1 mM = 1 mol m^{-3}), which indicates the values for all three c_j^o's (this solution is not unlike the pond water in which *Nitella* grows, and it is often referred to as "artificial pond water"). The concentrations of Na^+, K^+, and Cl^- measured in the cytosol, the c_j^i's, are given in the third column of Table 3.1. Assuming that activities can be replaced by concentrations, we can calculate the Nernst potential across the plasmalemma from these concentrations by using Equation 3.5, $E_{N_j} = (58.2/z_j) \log (c_j^o/c_j^i)$ in mV (the numerical factor is 58.2 since the measurements were at 20°C). We thus get

$$E_{N_{Na}} = \frac{(58.2 \text{ mV})}{(1)} \log \frac{(1.0 \text{ mM})}{(14 \text{ mM})} = -67 \text{ mV}$$

Similarly, E_{N_K} is -179 mV and $E_{N_{Cl}}$ is 99 mV (Table 3.1). Direct measurement of the electrical potential difference across the plasmalemma (E_M) gives -138 mV, as indicated earlier in discussing the Goldman equation. Since E_{N_j} differs from E_M in all three cases, none of these ions is in equilibrium across the plasmalemma of *Nitella*.

A difference between E_M and E_{N_j} for a particular species indicates departure from equilibrium for that ionic species; it also tells in what compartment μ_j is higher. Specifically, if the membrane potential is algebraically more negative than the calculated Nernst potential, the chemical potential in the inner aqueous phase (here the cytosol) is lower for a cation (Fig. 3.8) but higher for an anion, compared to the values in the external solution— consider the effect of z_j in the electrical term of the chemical potential, $z_j FE$. Since E_M (-138 mV) is more negative than $E_{N_{Na}}$ (-67 mV), Na$^+$ is at a lower chemical potential in the cytosol than outside in the external solution. Analogously, we find that K$^+$ (with $E_{N_K} = -179$ mV) has a higher chemical potential inside, while Cl$^-$ (with $E_{N_{Cl}} = 99$ mV) is at a much higher chemical potential inside. If these ions can move across the plasmalemma, this suggests an active transport of K$^+$ and Cl$^-$ into the cell and an active extrusion of Na$^+$ from the cell, as schematically indicated in Figure 3.9.

We can also consider the movement of the various ions into and out of *Nitella* in terms of the Ussing–Teorell equation to help determine whether active transport need be invoked to explain the fluxes. The Ussing–Teorell equation predicts that the quantity on the right side of Equation 3.24 $(c_j^o/c_j^i e^{z_j FE_M/RT})$ should equal the ratio of influx to efflux of the various ions, if the ions are moving passively in response to gradients in their chemical potential. Using the values given in Table 3.1 and noting that RT/F is 25.3 mV at 20°C (App. II), we find that this ratio for Na$^+$ is:

$$\frac{J_{Na}^{in}}{J_{Na}^{out}} = \frac{(1.0 \text{ mM})}{(14 \text{ mM})e^{(1)(-138 \text{ mV})/(25.3 \text{ mV})}} = 17$$

Similarly, the expected flux ratio is 0.20 for K$^+$ and 0.000085 for Cl$^-$ (values given in Table, 3.1, column 5). However, the observed influxes in the light

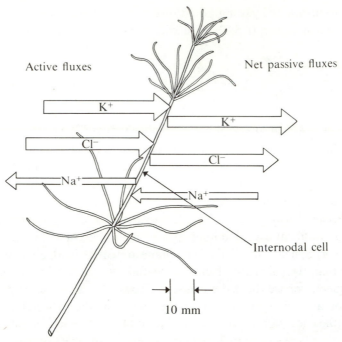

Figure 3.9
In the steady-state condition in the light the three active fluxes
across the plasmalemma of the large internodal cell of *Nitella*
are balanced by net passive K^+ and Cl^- effluxes and a net
passive Na^+ influx.

equal the effluxes for each of these three ions (Table 3.1, columns 6 and 7).
Equal influxes and effluxes are quite reasonable for mature cells of *Nitella*,
which are essentially in a steady-state condition. On the other hand, if
J_j^{in} equals J_j^{out}, the flux ratios given by Equation 3.24 are not satisfied for
Na^+, K^+, or Cl^-. In fact, as suggested by Figure 3.9, active transport of K^+
and Cl^- in and Na^+ out accounts for the marked deviations from the Ussing–
Teorell equation for *Nitella*.

As we mentioned earlier, another approach for studying active transport
is to remove the supply of energy. In the case of *Nitella*, cessation of illumination causes an appreciable decrease in the Na^+ efflux, in the K^+ influx,
and in the Cl^- influx (Table 3.1, columns 8 and 9). But these are the three
fluxes that are toward regions of higher chemical potential for the particular
ions involved, and thus we may reasonably expect all three to be active. On

the other hand, some fluxes remain essentially unchanged upon placing the cells in the dark (values in Table 3.1, columns 8 and 9, refer to the fluxes soon after extinguishing the light, not the steady-state fluxes). For instance, the Na^+ influx and K^+ efflux are initially unchanged when the *Nitella* cells are transferred from the light to the dark; i.e., these unidirectional fluxes toward lower chemical potentials do not depend on energy derived from photosynthesis. (For reasons still unclear, the energetically downhill efflux of Cl^- apparently increases in the dark. This J_{Cl}^{out} is not included in Table 3.1; see Hope and Walker.)

The passive diffusion of the ions toward regions of lower chemical potential helps create the electrical potential difference across a membrane, and active transport is instrumental in maintaining the asymmetrical ionic distributions that sustain the passive fluxes. Thus, the passive and the active fluxes are interdependent in the ionic relations of cells, and both are crucial for the generation of the observed diffusion potentials. Moreover, active and passive fluxes can occur simultaneously in the same direction. For example, we calculated that J_{Na}^{in} should equal 17 times J_{Na}^{out}, if both flux densities were passive ones obeying the Ussing–Teorell equation. Since J_{Na}^{in} is passive and equal to 5.5 nmol m^{-2} s^{-1}, we expect a passive efflux of Na^+ equaling (5.5)/(17), or 0.3 nmol m^{-2} s^{-1}. Thus, the active component of the Na^+ efflux in the light may be $5.5 - 0.3$, or 5.2, nmol m^{-2} s^{-1}. At cessation of illumination, J_{Na}^{out} decreases from 5.5 to 1.0 nmol m^{-2} s^{-1} (Table 3.1). Extinguishing the light removes photosynthesis as a possible energy source for active transport, but respiration could still supply energy in the dark. This helps explain why J_{Na}^{out} in the dark does not decrease all the way to 0.3 nmol m^{-2} s^{-1}, the value predicted for the passive efflux.

Energy for Active Transport

Suppose that the chemical potential of some species is higher outside than inside a cell, as is illustrated in Figure 3.8. The minimum amount of energy needed to transport a mole of that species from the internal aqueous phase on one side of some membrane to the external solution on the other is the difference in chemical potential of that solute across the membrane, $\mu_j^o - \mu_j^i$ (here, $\mu_j^o > \mu_j^i$). As we noted in considering Equation 3.25, the quantity $\mu_j^o - \mu_j^i$ for ions is $RT \ln (a_j^o/a_j^i) - z_j F E_M$. Since the Nernst potential E_{N_j} is $(RT/z_j F) \ln (a_j^o/a_j^i)$ (Eq. 3.5), we can express the difference in chemical potential across the membrane as follows:

$$\mu_j^o - \mu_j^i = z_j F(E_{N_j} - E_M) \tag{3.26a}$$

or

$$\mu_j^i - \mu_j^o = z_j F(E_M - E_{N_j}) \tag{3.26b}$$

We will use Equation 3.26b to discuss the active transport of ions *into* cells (i.e., when $\mu_j^i > \mu_j^o$).

Using the Nernst potentials of Na^+, K^+, and Cl^- for *Nitella translucens* and the value of E_M given in Table 3.1, we can calculate $z_j(E_M - E_{N_j})$ for transporting these ions across the plasmalemma of this alga. Such a quantity is $(+1)[(-138\ mV) - (-67\ mV)]$ or $-71\ mV$ for Na^+, $+41\ mV$ for K^+, and $+237\ mV$ for Cl^-. By Equation 3.26, these values for $z_j(E_M - E_{N_j})$ mean that Na^+ is at a higher chemical potential in the external bathing solution, while K^+ and Cl^- are at higher chemical potentials inside the cell, as we concluded above (see also Figs. 3.8 and 3.9). By Equation 3.26a, the minimum energy required to actively transport or "pump" Na^+ out across the plasmalemma of the *Nitella* cell is

$$\mu_{Na}^o - \mu_{Na}^i = (1)(9.65 \times 10^{-2}\ kJ\ mol^{-1}\ mV^{-1})[-67\ mV - (-138\ mV)]$$

$$= 6.9\ kJ\ mol^{-1}$$

Similarly, to pump K^+ inward requires $4.0\ kJ\ mol^{-1}$. The active extrusion of Na^+ from certain algal cells may be linked to the active uptake of K^+, ATP being implicated as the energy source for this coupled exchange process (see Lüttge and Higinbotham, or Raven). As we will discuss later (Ch. 6), the hydrolysis of ATP under biological conditions usually releases at least $40\ kJ\ mol^{-1}$ ($10\ kcal\ mol^{-1}$). For the case of a *Nitella* cell this is more than sufficient energy per mole of ATP hydrolyzed to pump one mole of Na^+ out and one mole of K^+ in. The transport of Cl^- inward takes a minimum of $23\ kJ\ mol^{-1}$ according to Equation 3.26, which is a fairly large amount of energy. Although the actual mechanism involved in actively transporting Cl^- into *Nitella* or other plant cells is not fully understood at the present time, exchanges with OH^- or co-transport with H^+ may be involved (see Lüttge and Pitman, or MacRobbie 1977). The involvement of proton chemical potential differences across membranes in chloroplast and mitochondrial bioenergetics will be discussed in Chapter 6.

The active uptake of K^+ and Cl^- together with an active extrusion of Na^+, as for *Nitella*, actually occurs for many plant cells (see Anderson; Gutknecht and Dainty; Lüttge and Higinbotham; and MacRobbie 1971). We might ask, why does a cell actively transport K^+ and Cl^- in and Na^+ out? Although no definitive answer can be given to such a question, we shall *speculate* on possible reasons, based on the principles we have been considering.

Let us imagine that a membrane-bounded cell containing negatively charged proteins is placed in an NaCl solution, the latter possibly reflecting primeval conditions when life on earth originated. When Na^+ and Cl^- are both in equilibrium across the membrane, E_M equals $E_{N_{Cl}}$ and $E_{N_{Na}}$. Using concentrations (instead of activities) in Equation 3.5, $\log (c_{Na}^o/c_{Na}^i)$ then equals $-\log (c_{Cl}^o/c_{Cl}^i)$, or $c_{Na}^o/c_{Na}^i = c_{Cl}^i/c_{Cl}^o$, and hence $c_{Na}^o c_{Cl}^o = c_{Na}^i c_{Cl}^i$. For electroneutrality in an external solution containing only NaCl, c_{Na}^o equals c_{Cl}^o. Since $a^2 = bc$ implies that $2a \leq b + c,$* we conclude that $c_{Na}^o + c_{Cl}^o \leq c_{Na}^i + c_{Cl}^i$. But the proteins, which cannot diffuse across the membrane, also make a contribution to the internal osmotic pressure (Π^i), so Π^o would be less than Π^i. When placed in an NaCl solution, water would therefore tend to enter such a membrane-bounded cell containing proteins, causing it to swell without limit. An outwardly directed active transport of Na^+ would lower c_{Na}^i and thus prevent excessive osmotic swelling of primitive cells.

An energy-dependent uptake into a plant cell tends to increase Π^i, leading to a rise in P^i. This higher internal hydrostatic pressure favors cell enlargement and consequently cell growth. For a plant cell surrounded by a cell wall, we might therefore expect an active transport of some species into the cell, e.g., Cl^-. (Animal cells do not have to push against a cell wall in order to enlarge, and do not generally have an active uptake of Cl^-.) Enzymes have evolved that operate efficiently when exposed to relatively high concentrations of K^+ and Cl^-. In fact, many actually require K^+ for their activity, so that an inwardly directed K^+ pump is probably necessary for biochemistry as we know it. The presence of a substantial concentration of such ions insures that electrostatic effects adjacent to a Donnan phase can in large part be screened out—otherwise, a negatively charged substrate, e.g., an organic acid or a phosphorylated sugar, might be electrostatically repelled from the catalytic site on an enzyme (proteins are generally negatively charged at cytoplasmic pH's). Once active transport has set up certain concentration differences across a membrane, the membrane potential is an inevitable consequence of the tendency of such ions to passively diffuse toward regions of lower chemical potential. The roles such diffusion potentials actually play in the physiology of plant cells are open to question. It is known, however, that they are essential for the transmission of electrical impulses in excitable cells of animals and certain plants.

---

* To show this, consider that $0 \leq (\sqrt{b} - \sqrt{c})^2 = b + c - 2\sqrt{bc}$, or $2\sqrt{bc} \leq b + c$; hence, if $a^2 = bc$, then $2a \leq b + c$. Here, we let $c_{Na}^o = c_{Cl}^o = a$, $c_{Na}^i = b$, and $c_{Cl}^i = c$.

Carriers, Solute Uptake, and the Michaelis–Menten Formalism

Although active transport is clearly of common occurrence, its actual mechanism—including the means whereby metabolic energy is used—remains uncertain. The possible involvement of a "carrier" molecule in the active transport of solutes across plant cell membranes was first suggested by Osterhout in the early 1930's. This carrier, it was proposed, would selectively bind certain molecules and then carry them across the membrane. Carriers can provide a cell with the specificity or selectivity needed to control the entry and exit of the various types of solutes encountered. Thus, certain metabolites can be specifically taken into the cell, while photosynthetic and waste products can be selectively moved out across the membranes. Also, the active transport of certain inorganic nutrients into epidermal cells in the root allows a plant to obtain and accumulate these solutes from the soil. Even though the mechanism for physically binding and moving the solutes through the membrane is still unknown, the carrier concept has found widespread application in the interpretation of experimental observations (see Epstein, Higinbotham, Lüttge and Higinbotham, Neame and Richards, Stein, or Yariv et al.).

Before proceeding to a consideration of the effects of carriers—or "transporters" (or simply "porters") as they are increasingly referred to—let us speculate on their possible structures (see Harrison and Lunt, Ch. 1 reference; or Wiskich). Most researchers agree that transporters are generally proteins. Small proteins could bind some substance on one side of a membrane, thermally diffuse across the membrane, and then somehow release the substance on the other side. Such mobile carriers could bind a single substance, or they could bind two different substances, like the proton-solute *symport** portrayed in Figure 3.10a. Candidates for transport by a proton symport in plants include inorganic ions like Na^+ and metabolites like sugars and amino acids. Much evidence indicates that many substances actually move in trans-membrane pores or channels, which could be protein-lined. Such channels could have a series of binding sites, where the molecule or molecules transported must go from site to site across the membrane (Fig. 3.10b). As another alternative, the substance to be transported might first bind to a site accessible from one side of the membrane only. Following a conformational change of the protein involved, the substance might subsequently be exposed to the solution on the other side of the membrane (Fig. 3.10c). For mechanisms as speculative as these, there are a number of

---

* A symport is a porter that causes two different substances to move in the same direction across a membrane (see Fig. 3.10a); compare "antiport," pp. 321 and 329.

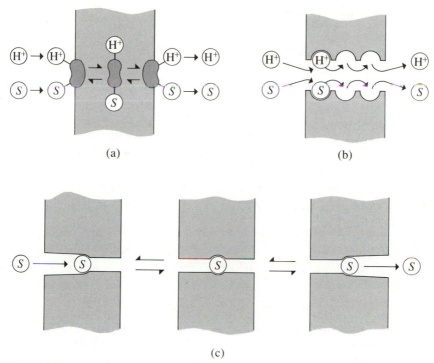

Figure 3.10
Hypothetical structures indicating possible mechanisms for transporters: (a) mobile carrier or porter acting as a symport for protons (H^+) and some transported solute (S), (b) series of binding sites in channel across membrane acting as a symport for H^+ and S, and (c) three sequential conformations of a channel that leads to unidirectional movement of solute.

ways that metabolic energy could be involved—e.g., using ATP to lead to protein conformational changes or for active transport of H^+ to maintain its chemical potential difference across the membrane so that solute transport could be coupled to the passive, energetically downhill flows of protons.

Based primarily on results from competition studies, certain solutes are apparently bound to, or associated with, a particular transporter. When an ion of some species is attached to this transporter, another similar ion (of the same or a different species) competing for the same binding site cannot also be bound. For example, the similar monovalent cations K^+ and Rb^+ appear to bind in a competitive fashion to the same site on some transporter. For some cells, one and the same carrier might transport Na^+ out of the cell and K^+ in—the so-called sodium-potassium pump alluded to above. Ca^{2+} and Sr^{2+} may compete with each other for binding sites on another common carrier. Two other divalent cations, Mg^{2+} and Mn^{2+}, are apparently transported by a single carrier which is different from the one for Ca^{2+} and Sr^{2+}. The halides (Cl^-, I^-, and Br^-) may also be transported by a single carrier.

One of the most important variables in the study of carrier-mediated uptake is the external concentration. As the external concentration of a solute increases, the rate of uptake generally reaches an upper limit. We may then presume that all binding sites on the carriers for that particular solute have become filled or saturated. In particular, the rate of active uptake of species j, J_j^{in}, is often proportional to the external concentration of that solute, c_j^o, over the lower range of concentrations, but as c_j^o is raised a maximum rate $J_{j\,max}^{in}$, is eventually reached. We can describe this kind of behavior by the following equation:

$$J_j^{in} = \frac{J_{j\,max}^{in}\, c_j^o}{K_j + c_j^o} \qquad (3.27a)$$

where K_j is a constant characteristic of species j crossing a particular membrane; it is expressed in the units of concentration. For the uptake of many ions into roots and other plant tissues, $J_{j\,max}^{in}$ is 30 to 300 nmol m^{-2} s^{-1}. Often, two different K_j's are observed for the uptake of one and the same ion into a root. The lower one is generally between 6 and 100 μM (6 and 100 mmol m^{-3}), which is in the concentration range of many ions in soil water, while the other K_j can be above 10 mM (see Epstein).

Equation 3.27a is similar in appearance to the Michaelis–Menten equation used for describing enzyme kinetics in biochemistry, $V = V_{max}s/(K_M + s)$ (see Epstein, Lehninger, or White et al.). The substrate concentration s in the latter relation is analogous to c_j^o in Equation 3.27a, and the enzyme reaction velocity V, to J_j^{in}. The term in the Michaelis–Menten equation equivalent to K_j in Equation 3.27a is the substrate concentration for half-maximal velocity of the reaction, K_M (the Michaelis constant). The lower the K_M, the greater is the reaction velocity at low substrate concentrations. Likewise, a low value for K_j indicates that the ion or other solute is more readily bound to some carrier and then transported across the membrane. The physiological consequence of a low K_j is that species j is actually favored or selected for active transport into the cell, even when its external concentration is relatively low.

The two most common ways of graphing data on solute uptake that fit Equation 3.27a are illustrated in Figure 3.11. Figure 3.11 shows that, when the external concentration of species j, c_j^o, is equal to K_j, then J_j^{in} equals $\frac{1}{2}J_{j\,max}^{in}$, as we can see directly from Equation 3.27a; i.e., J_j^{in} is then $J_{j\,max}^{in}K_j/$

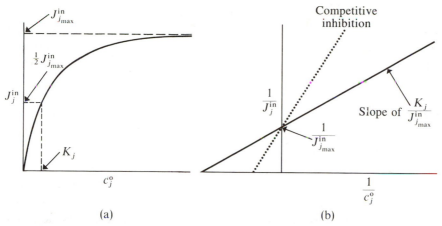

Figure 3.11

Relationship between the external solute concentration (c_j^o) and the rate of influx (J_j^{in}) for active uptake according to a Michaelis–Menten type of kinetics, as given by Equation 3.27: (a) linear plot, and (b) double reciprocal plot.

($K_j + K_j$), or $\frac{1}{2}J_{j\,max}^{in}$. Thus, K_j is the external concentration at which the rate of active uptake is half-maximal—in fact, the observed values of K_j have proved to be convenient parameters for describing the uptake of various solutes. If two different solutes compete for the same site on some carrier, then J_j^{in} for species j will be decreased by the presence of the second solute; this is known as *competitive inhibition* of species j. In the case of competitive inhibition, the asymptotic value for the active influx, $J_{j\,max}^{in}$, is not affected, since in principle we can raise c_j^o high enough to eventually obtain the same maximum rate for the active uptake of species j. But the half-maximum rate occurs at a higher concentration, which means that K_j is raised if a competing solute is present.

For many purposes, it is advantageous to plot the experimental data in such a way that a linear relationship is obtained when Equation 3.27a is satisfied for the active uptake of species j. Taking reciprocals of both sides of Equation 3.27a, we note that

$$\frac{1}{J_j^{in}} = \frac{K_j + c_j^o}{J_{j\,max}^{in}c_j^o} = \frac{K_j}{J_{j\,max}^{in}c_j^o} + \frac{1}{J_{j\,max}^{in}} \tag{3.27b}$$

When $1/J_j^{in}$ is plotted against $1/c_j^o$ (Fig. 3.11b), Equation 3.27b yields a straight line with a slope of $K_j/J_{j\,max}^{in}$ and an intercept on the ordinate of $1/J_{j\,max}^{in}$. This latter method of treating the experimental results has proved to be rather convenient, and has found widespread application in studies of solute uptake by plant tissues, especially into roots. Since K_j increases while $J^{in}_{j\,max}$ remains the same for competitive inhibition, the presence of a competing species causes the slope of the line (i.e., $K_j/J^{in}_{j\,max}$) to be greater, while the intercept on the y-axis ($1/J^{in}_{j\,max}$) is unchanged (see Fig. 3.11b).

Certain compounds structurally unrelated to species j can inhibit the functioning of the carriers. A *noncompetitive* inhibitor does not bind to the site used for transporting species j across the membrane, and K_j is unaffected. Since a noncompetitive inhibitor lowers the maximum rate of active influx ($J^{in}_{j\,max}$), both the slope and the y-axis intercept in Figure 3.11b are changed. Finally, an *uncompetitive* inhibitor does not affect the binding of the solute, but rather inactivates the complex after the solute has been bound; the slope in a double reciprocal plot is unchanged, but the intercept is increased (see Lehninger).

Facilitated Diffusion

The basis of the mathematical form of Equation 3.27 for describing uptake is the competitive binding of solutes to a limited number of carriers or transporters. In other words, active processes involving metabolic energy do not have to be invoked; if a solute were to *diffuse* across the membrane only when bound to a carrier, the expression for the influx could also be Equation 3.27. This passive entry of a solute mediated by a carrier is termed *facilitated diffusion*.

Since facilitated diffusion is so important in biology and yet is often misunderstood, we will briefly elaborate upon it. Certain molecules passively enter cells more readily than would be expected from consideration of their molecular structure or from observations with analogous substances (e.g., from a comparison of P_j's), so some mechanism is apparently *facilitating* their entry. The net flux density is still toward lower chemical potentials, and hence is still in the same direction as ordinary diffusion.* To help explain facilitated diffusion, transporters are proposed to act as shuttles for a net passive movement of the specific molecules across the membrane toward regions of lower energy. Instead of the usual diffusion across the barrier—based on random thermal motion of the solutes—transporters select out and bind certain molecules and then release them on the other side of the membrane without the intervention of metabolism. Such facilitation of entry

---

* The term "diffusion" usually refers to net thermal motion toward regions of lower concentration. It is used here with a somewhat broader meaning— namely, net motion toward regions of lower chemical potential.

may also be regarded as a special means of lowering the activation energy needed for the solute to cross the energy barrier represented by the membrane. Thus, transporters facilitate the influx of solutes in the same way that enzymes facilitate biochemical reactions.*

There are certain general characteristics of a facilitated diffusion system. As already mentioned, the net flux is toward lower chemical potentials. (According to the usual definition, p. 136, active transport is in the energetically uphill direction; conceivably, it can employ the same carriers used in facilitated diffusion.) Facilitated diffusion causes fluxes to be larger than those expected for ordinary diffusion. Furthermore, the transporters can exhibit a high degree of selectivity; i.e., they can be specific for certain types of molecules, while not binding closely related species. In addition, carriers in facilitated diffusion become saturated when the external concentration of the species transported is raised sufficiently, a behavior consistent with Equation 3.27. Finally, since carriers can exhibit competition phenomena, the flux density of a species entering the cell by facilitated diffusion can be reduced when structurally similar molecules are added to the external solution. (Such molecules would compete for sites on the carrier and thereby reduce the binding and the subsequent transfer of the original species into the cell.)

For convenience, we have been discussing facilitated diffusion *into* a cell, but exactly the same principles apply for exit, as well as for fluxes at the organelle level. Let us assume that a transporter for K^+ exists in the membrane of a certain cell and that it is being used as a shuttle for facilitated diffusion. Not only would the carrier lead to an enhanced net flux density toward the side with the lower chemical potential, but also both the unidirectional fluxes J_K^{in} and J_K^{out} could be increased over the values predicted for ordinary diffusion. This increase in the unidirectional fluxes by a carrier is often called *exchange diffusion*. In such a case, the molecules are interacting with a membrane component, —namely, the carrier—and hence the Ussing–Teorell equation—Equation 3.24, $J_j^{in}/J_j^{out} = c_j^o/(c_j^i e^{z_j FE_M/RT})$—would not be obeyed, since it presupposes that there are no interactions with other substances. In fact, observation of departures from predictions of the Ussing–Teorell equation is often how cases of exchange diffusion are actually discovered.†

---

* As is true for all carrier mechanisms, the exact molecular details have not been worked out. We are thus using the term "carrier" or "transporter" in a phenomenological or operational sense.

† The term "exchange diffusion" has another usage in the literature—namely, to describe the carrier-mediated movement of some solute in one direction across a membrane in exchange for a *different* solute being transported in the opposite direction. Again, the Ussing–Teorell equation is not obeyed.

Both active and passive fluxes across the cellular membranes can occur concomitantly; these movements depend on concentrations in rather different ways. For passive diffusion, the unidirectional component J_j^{in} is proportional to c_j^o, as indicated by Equation 1.8 for neutral solutes $[J_j = P_j(c_j^o - c_j^i)]$, and by Equation 3.15 for ions. This proportionality strictly applies only over the range of external concentrations for which the permeability coefficient is essentially independent of concentration, and the membrane potential must not change in the case of charged solutes. Nevertheless, ordinary passive influxes do tend to be proportional to the external concentration, while an active influx or the special passive influx known as facilitated diffusion—either of which can be described by a Michaelis–Menten type of formalism—shows saturation effects at the higher concentrations. Moreover, facilitated diffusion and active transport exhibit both a high degree of selectivity and competition phenomena, while ordinary diffusion does not.

PRINCIPLES OF IRREVERSIBLE THERMODYNAMICS

So far, we have been using classical thermodynamics—though, it may have been noticed, often somewhat illegitimately. For example, let us consider Equation 2.23 ($J_{V_w} = L_w \Delta \Psi$). Since there is a difference in chemical potential represented here by a change in water potential, we expect a net (and irreversible) flow of water from one region to another—obviously not an equilibrium situation. Strictly speaking, however, classical thermodynamics is concerned solely with equilibria, and not with movement. Indeed, classical thermodynamics might have been better named "thermostatics." Thus, we have frequently been involved in a kind of hybrid enterprise—appealing to classical thermodynamics for the driving forces, and using nonthermodynamic arguments and analogies to discuss fluxes. One of the objectives of irreversible thermodynamics is to help legalize the arguments. But, as we shall see, legalizing them brings in new ideas and considerations.

Irreversible thermodynamics uses the same parameters as classical thermodynamics—namely, temperature, pressure, free energy, activity, and so on. But these quantities are strictly defined for macroscopic amounts of matter only in equilibrium situations. How can we use them to discuss processes *not* in equilibrium, the domain of irreversible thermodynamics? This dilemma immediately circumscribes the range of validity of the theory of irreversible thermodynamics: it can deal only with "slow" processes or situations not very far from equilibrium, for only in such circumstances can equilibrium-related concepts such as temperature and free energy retain their validity—at least approximately. We have to assume from the outset that we can talk about and use classical thermodynamics parameters even in nonequilibrium situations.

Another refinement we should introduce into our theory is to recognize that the movement of one species may affect the movement of a second species. A particular flux of some solute may interact with another flux by way of collisions, each species flowing under the influence of its own force. For example, water, ions, and other solutes moving through a membrane toward regions of lower values for their respective chemical potentials can exert a frictional drag on each other. The magnitude of the flux of a solute may then depend on whether water is also flowing. In this way, the fluxes of various species across a membrane become interdependent. Stated more formally, the flux of a solute is not only dependent on the negative gradient of its own chemical potential—which is the sole driving force we have recognized up to now—but it may also be influenced by the gradient in the chemical potential of water. Again using Equation 2.23 ($J_{V_w} = L_w \Delta \Psi$) as an example, we have considered that the flow of water depends solely on the difference in its own chemical potential between two locations, and have thus far ignored any coupling to concomitant fluxes of solutes.

A quantitative description of interdependent fluxes and forces is given by irreversible thermodynamics, a subject that treats nonequilibrium situations such as those actually occurring under biological conditions.* In this brief introduction to irreversible thermodynamics we will emphasize certain underlying principles and then derive the *reflection coefficient*, a parameter important in plant and cell physiology. To keep the analysis manageable, we will restrict our attention to isothermal conditions, which approximate many biological situations where fluxes of water and solutes are considered. (For further details see Dainty 1963, De Groot and Mazur, Katchalsky and Curran, Kedem and Katchalsky, or Prigogine†.)

Fluxes, Forces, and Onsager Coefficients

In our previous discussion of fluxes, the driving force leading to the flux density of species j, J_j, was the negative gradient in its chemical potential,

---

* Nonequilibrium and irreversible are related, since a system in a *nonequilibrium* situation left isolated from external influences will spontaneously and *irreversibly* move toward equilibrium.

† Prigogine received the 1977 Nobel Prize in chemistry for his contributions to nonequilibrium thermodynamics.

$-\partial\mu_j/\partial x$. Irreversible thermodynamics takes a more general view—namely, that the flux of species j depends not only on $-\partial\mu_j/\partial x$, but also can be affected by *any* other force occurring in the system, such as the negative gradient in the chemical potential of some other species. A particular force, X_k, can likewise influence the flux of any species. Thus, the various fluxes become independent, or coupled, since they can respond to changes in any of the forces. Another premise of irreversible thermodynamics is that J_j is linearly dependent on the various forces. This means that in practice we will treat only those cases that are not too far from equilibrium. Even with this simplification, the algebra often becomes cumbersome, owing to the coupling of the various forces and fluxes.

Using a linear combination of all the forces, we can represent the flux density of species j by

$$J_j = \sum_k L_{jk}X_k = L_{j1}X_1 + L_{j2}X_2 + \cdots + L_{jj}X_j + \cdots + L_{jn}X_n \quad (3.28)$$

where the summation $\sum_k$ is over all forces (all n X_k's), and the L_{jk}'s are referred to as the *Onsager coefficients*, or the phenomenological coefficients, in this case for conductivity. The first subscript on these coefficients (j on L_{jk}) identifies the flux density J_j that we are considering; the second subscript (k on L_{jk}) designates the force—e.g., it can represent the gradient in chemical potential of species k. Each term, $L_{jk}X_k$, is thus the partial flux density of species j due to the particular force X_k. The individual Onsager coefficients in Equation 3.28 are therefore the proportionality factors indicating what contribution each force X_k makes to the flux density of species j, J_j. Equation 3.28 is sometimes referred to as the *phenomenological equation*. Phenomenological equations are used to describe observable phenomena without regard to explanations in terms of atoms or molecules. For instance, Ohm's law and Fick's laws are also phenomenological equations. They too assume linear relations between forces and fluxes.

A convenient relationship exists between the various phenomenological coefficients, namely, L_{jk} equals L_{kj}, which is known as the *reciprocity relation*. Such an equality of cross-coefficients was derived in 1931 by Onsager—awarded the Nobel Prize in chemistry in 1968—from statistical considerations utilizing the principle of "detailed balancing." The argument involves microscopic reversibility; i.e., for local equilibrium, any molecular process and its reverse will be taking place at the same average rate in that region. The Onsager reciprocity relation means that the proportionality coefficient giving the flux density of species k caused by the force on species

j equals the proportionality coefficient giving the flux density of j caused by the force on k. (Strictly speaking, conjugate forces and fluxes must be used, as they will in the case below. Also, see App. VII.) The fact that L_{jk} equals L_{kj} can be further appreciated by considering Newton's third law— equality of action and reaction. For example, the frictional drag exerted by a moving solvent on the solute is equal to the drag exerted by the moving solute on the solvent. The pairwise equality of cross-coefficients given by the Onsager reciprocity relation reduces the number of coefficients needed to describe the interdependence of forces and fluxes in irreversible thermo-dynamics and consequently leads to a simplification in solving the sets of simultaneous equations.

Water and Solute Flow

As a specific application of the principles just introduced, we will consider in some detail the important coupling of water and solute flow. The driving forces for the fluxes are the negative gradients in chemical potential, which we will assume to be proportional to the differences in chemical potential across some barrier, here considered to be a membrane. In particular, we will represent $-\partial\mu_j/\partial x$ by $\Delta\mu_j/\Delta x$, which in the present case is $(\mu_j^o - \mu_j^i)/\Delta x$. (For convenience, the thickness of the barrier, Δx, will be incorporated into the coefficient multiplying $\Delta\mu_j$ in the flux equations.) To help keep the algebra relatively simple, the development will be carried out for a single nonelectrolyte. The fluxes are across a membrane permeable to both water (w) and the single solute (s), thereby removing the restriction in Chapter 2, where membranes permeable only to water were considered. Using Equation 3.28, we can represent the flux densities as the following linear combination of the differences in chemical potential:

$$J_w = L_{ww}\Delta\mu_w + L_{ws}\Delta\mu_s \tag{3.29}$$

$$J_s = L_{sw}\Delta\mu_w + L_{ss}\Delta\mu_s \tag{3.30}$$

Equations 3.29 and 3.30 allow for the possibility that each of the flux densities may depend on the differences in both chemical potentials, $\Delta\mu_w$ and $\Delta\mu_s$. Four phenomenological coefficients are used in these two equations. But by the Onsager reciprocity relation, L_{ws} equals L_{sw}. Thus, three different coefficients (L_{ww}, L_{ws}, and L_{ss}) are needed to describe the relationship of these two flux densities to the two driving forces. This is in contrast to Equation 3.6, $J_j = u_j c_j(-\partial\mu_j/\partial x)$, where each flux density depends on but one force; accordingly, only two coefficients would then be involved in describing J_w and J_s.*

To obtain more convenient formulations for the fluxes of water and the solute, we generally express $\Delta\mu_w$ and $\Delta\mu_s$ in terms of the differences in the osmotic and the hydrostatic pressures, $\Delta\Pi$ and ΔP, since it is usually easier to measure ΔP and $\Delta\Pi$ than $\Delta\mu_w$ and $\Delta\mu_s$. The expression for $\Delta\mu_w$ is straightforward; the only possible ambiguity is in deciding on the algebraic sign. In keeping with the usual conventions for this specific case, $\Delta\mu_w$ is the chemical potential of water on the outside minus that on the inside, $\mu_w^o - \mu_w^i$. From Equation 2.12 ($\mu_w = \mu_w^* - \bar{V}_w\Pi + \bar{V}_wP + m_wgh$), $\Delta\mu_w$ is given by

$$\Delta\mu_w = -\bar{V}_w\Delta\Pi + \bar{V}_w\Delta P \tag{3.31}$$

where $\Delta\Pi$ here equals $\Pi^o - \Pi^i$ and ΔP is $P^o - P^i$ ($\Delta h = 0$ across a membrane).

To express $\Delta\mu_s$ in terms of $\Delta\Pi$ and ΔP, we will first consider the activity term, $RT \ln a_s$. The differential $RT\, d(\ln a_s)$ equals $RT\, da_s/a_s$, or $RT\, d(\gamma_sc_s)/(\gamma_sc_s)$. When γ_s is constant, this latter quantity becomes $RT\, dc_s/c_s$. By Equation 2.10 ($\Pi_s = RT \sum_j c_j$), $RT\, dc_s$ equals $d\Pi$ for a dilute solution of a single solute. Hence, $RT\, d(\ln a_s)$ can be replaced by $d\Pi/c_s$ as a useful approximation. In expressing the difference in chemical potential across a membrane we are interested in macroscopic changes, not in the infinitesimal changes given by differentials. To go from differentials (d) to differences (Δ), $RT\, d(\ln a_s)$ becomes $RT\,\Delta \ln a_s$ and so $d\Pi/c_s$ can be replaced by $\Delta\Pi/\bar{c}_s$, where $\bar{c}_s$ is essentially the mean concentration of solute s, in this case across the membrane. Alternatively, we can simply define $\bar{c}_s$ as that concentration for which $RT\,\Delta \ln a_s$ exactly equals $\Delta\Pi/\bar{c}_s$. In any case, we can replace the term $RT\,\Delta \ln a_s$ in $\Delta\mu_s$ by the equivalent term, $\Delta\Pi/\bar{c}_s$. By Equation 2.4, μ_s equals $\mu_s^* + RT \ln a_s + \bar{V}_sP$ for a neutral species, and the difference in chemical potential of the neutral solute across the membrane, $\Delta\mu_s$, becomes

$$\Delta\mu_s = \frac{1}{\bar{c}_s}\Delta\Pi + \bar{V}_s\Delta P \tag{3.32}$$

The two expressions representing the driving forces, $\Delta\mu_w$ (Eq. 3.31) and $\Delta\mu_s$

---

* If the solute were a salt dissociable into two ions, we would have to consider three components and three forces, $\Delta\mu_w$, $\Delta\mu_+$, and $\Delta\mu_-$. This would lead to nine coefficients in the three flux equations (for J_w, J_+, and J_-). Invoking the Onsager reciprocity relation, we would then have six different phenomenological coefficients to describe the movement of water, a cation, and its accompanying anion.

(Eq. 3.32), are expressed as functions of the same two pressure differences, $\Delta\Pi$ and ΔP, which are experimentally more convenient to measure.

Volume Flux Density, L_P, and σ

Now that we have appropriately expressed the chemical potential differences of water and the solute, we will direct our attention to the fluxes. Expressed in our usual units, the flux densities J_w and J_s are the moles of water and solute, respectively, moving across one square metre of membrane surface in a second. A quantity of considerable interest in plant and cell physiology is the volume flux density J_V, which is the rate of movement of the total *volume* of both water and solute across unit area of the membrane J_V has the units of volume per unit area per unit time, e.g., $m^3\ m^{-2}\ s^{-1}$, or $m\ s^{-1}$.

The molar flux density of species $j\ (J_j)$ in $mol\ m^{-2}\ s^{-1}$ times the volume occupied by each mole of it $(\bar{V}_j)$ in $m^3\ mol^{-1}$ gives the volume flow for that component (J_{V_j}) in $m\ s^{-1}$. Hence, the total volume flux density is

$$J_V = \sum J_{V_j} = \sum \bar{V}_j J_j \qquad (3.33a)$$

For solute and water both moving across a membrane, J_V is the volume flow of water plus that of solute per unit area, which in the case of a single solute can be represented as follows:

$$J_V = \bar{V}_w J_w + \bar{V}_s J_s \qquad (3.33b)$$

It is generally simpler and more convenient to measure the total volume flux density (such as that given in Eq. 3.33) than one of the component volume flux densities $(J_{V_j} = \bar{V}_j J_j)$. For instance, we can often determine the volumes of cells or organelles under different conditions and relate any changes in volume to J_V. The volume flux density of water J_{V_w} (used in Ch. 2, Eq. 2.23) is simply $\bar{V}_w J_w$

Although straightforward, the algebraic substitution necessary to incorporate the various forces and fluxes into the volume flow given by Equation 3.33b leads to a rather cumbersome expression. Nevertheless, it provides insight into the origin of the reflection coefficient, an important quantity that we will consider in detail shortly (see App. VII for an alternative approach, algebraically simpler, but less intuitive). When the fluxes of water and solute given by Equations 3.29 and 3.30, respectively, are substituted into the expression for the appropriate volume flux density (Eq. 3.33b), we obtain the following relation:

$$J_V = (\bar{V}_w L_{ww} + \bar{V}_s L_{sw}) \Delta \mu_w + (\bar{V}_w L_{ws} + \bar{V}_s L_{ss}) \Delta \mu_s \qquad (3.34)$$

As we mentioned above, L_{sw} equals L_{ws}, so that only the symbol L_{ws} need be retained here. Next, we will incorporate the chemical potential differences across the membrane for water and the solute, as given by Equations 3.31 and 3.32, respectively, into Equation 3.34. Upon rearrangement, J_V is then as follows:

$$J_V = (\bar{V}_w^2 L_{ww} + 2\bar{V}_s \bar{V}_w L_{ws} + \bar{V}_s^2 L_{ss}) \Delta P$$

$$- (\bar{V}_w^2 L_{ww} - \bar{V}_w \frac{1}{\bar{c}_s} L_{ws} - \bar{V}_s \frac{1}{\bar{c}_s} L_{ss} + \bar{V}_s \bar{V}_w L_{ws}) \Delta \Pi \qquad (3.35)$$

Equation 3.35 gives the dependence of the volume flux density on the hydrostatic and osmotic pressure differences across a membrane permeable to both water and the solute.

The quantities multiplying ΔP and $\Delta \Pi$ in Equation 3.35 are cumbersome, so we will find it convenient to define some new parameters. The factor multiplying the difference in hydrostatic pressure ΔP is called the *hydraulic conductivity coefficient*, L_P. By Equation 3.35, this hydraulic conductivity coefficient has the following value:

$$L_P = \bar{V}_w^2 L_{ww} + 2\bar{V}_s \bar{V}_w L_{ws} + \bar{V}_s^2 L_{ss} \qquad (3.36)$$

Such a coefficient describes membrane conductivity, since the larger is L_P, the greater is J_V for a given hydrostatic pressure difference across a membrane. The other unwieldly factor in Equation 3.35—the parenthesis multiplying $\Delta \Pi$—is incorporated into the *reflection coefficient*, σ. It is advantageous to define σ as the ratio of the coefficient of $\Delta \Pi$ divided by the coefficient of ΔP in the expression for the volume flow (Eq. 3.35). From the definition of L_P (Eq. 3.36), σ is therefore given by

$$\sigma = \frac{\bar{V}_w^2 L_{ww} - \bar{V}_w \dfrac{1}{\bar{c}_s} L_{ws} - \bar{V}_s \dfrac{1}{\bar{c}_s} L_{ss} + \bar{V}_s \bar{V}_w L_{ws}}{L_P} \qquad (3.37)$$

The reflection coefficient defined by Equation 3.37 is perhaps the most important parameter from irreversible thermodynamics relevant to the fluxes generally considered in plant physiology. It was introduced by Staverman in 1948 and specifically applied to biological situations by Kedem and Katchalsky in 1958 (for a more detailed development of the theory, see De Groot and Mazur, Katchalsky and Curran, Kedem and Katchalsky, or Prigogine). Since 1958, numerous applications of irreversible thermodynamics in general and reflection coefficients in particular have been made in biology. We will show that the maximum value of unity for the reflection coefficient occurs when the solute is unable to cross the membrane—the molecules are then all *reflected* from the barrier. After further considering the expression for volume flux density, we will discuss the properties of reflection coefficients. (We derive σ in a more conventional but less intuitive way in App. VII.)

Using L_P and σ as defined by Equations 3.36 and 3.37, respectively, we can rewrite the expression for the volume flux density, Equation 3.35, in the following convenient form:

$$J_V = L_P(\Delta P - \sigma \Delta \Pi) \tag{3.38}$$

We note that L_P is essentially the same as and hence usually replaces L_w, the water conductivity coefficient that we introduced in Chapter 2—e.g., in Equations 2.23 and 2.26. In the absence of a hydrostatic pressure difference across a membrane, the volume flux density J_V equals $-L_P \sigma \Delta \Pi$ by Equation 3.38. Thus, the magnitude of the dimensionless parameter σ determines the volume flux density expected in response to a difference in osmotic pressure across a membrane. It is this use which is most pertinent in plant physiology.

Many different solutes can cross a membrane under usual conditions. Each such species j can be characterized by its own reflection coefficient, σ_j, for that particular membrane. The volume flux density given by Equation 3.38 can then be generalized to

$$J_V = L_P\left(\Delta P - \sum_j \sigma_j \Delta \Pi_j\right) \tag{3.39}$$

where $\Delta \Pi_j$ is the osmotic pressure difference across the membrane for species j (e.g., $\Delta \Pi_j = RT \Delta c_j$, by Eq. 2.10). Although interactions with water are still taken into account, the generalization represented by Equation 3.39 introduces the assumption that the solutes do not interact with each other as they cross a membrane. Moreover, J_V in Equation 3.39 refers only to the movement of neutral species—otherwise we would also need a current equation to describe the flow of charge. Nevertheless, Equation 3.39 is a useful approximation of the actual situation describing the multicomponent solutions encountered by cells, and we will use it as the starting point for our general consideration of solute movement across membranes. Before discussing such movement, however, let us consider the range of values of reflection coefficients.

Values of Reflection Coefficients

A reflection coefficient characterizes some particular solute interacting with a specific membrane. In addition, σ_j depends on the solvent on either side of the membrane, as the subscripts w for water in Equation 3.37 clearly indicate—water is the only solvent we will consider here. Two extreme conditions can describe the passage of solutes: impermeability, which leads to the maximum value of unity for the reflection coefficient, and nonselectivity, where σ_j is zero. A reflection coefficient of zero may describe the movement of a solute across a very coarse barrier (one with large pores) which cannot distinguish or select between solute and solvent molecules, or it may refer to the passage through a membrane of a molecule very similar in size and structure to water itself.

Impermeability describes the limiting case where water can cross some membrane, but the solute cannot. When the solute does not cross the membrane, J_s must remain zero for any and all values of $\Delta\mu_w$ and $\Delta\mu_s$. Since J_s equals $L_{sw}\Delta\mu_w + L_{ss}\Delta\mu_s$ (Eq. 3.30), the only way J_s will always be zero is for both L_{ss} and L_{ws} to be zero. Therefore, when the membrane is impermeable to a particular solute, both L_{ss} and L_{ws} are zero. Putting these zero values for L_{ss} and L_{ws} into Equations 3.36 and 3.37, we obtain the following two relations: $L_P = \bar{V}_w^2 L_{ww}$ (Eq. 3.36) and $\sigma = \bar{V}_w^2 L_{ww}/L_P$ (Eq. 3.37). Consequently, σ then equals $\bar{V}_w^2 L_{ww}/\bar{V}_w^2 L_{ww}$, which is one. A reflection coefficient of unity thus signifies that the solute is not crossing the membrane; i.e., all of the solute is "reflected" from the membrane, and the reflection coefficient has its maximum value. When a solute can cross, L_{ss} takes on a positive value; i.e., the force on solute s causes a flux of it across the membrane. The $-L_{ss}\bar{V}_s/\bar{c}_s$ term then causes the numerator of the expression for σ (Eq. 3.37) to decrease as L_{ss} increases, while the denominator (L_P; see Eq. 3.36) tends to increase as the solute's ability to penetrate goes up. The reflection coefficient thus becomes less than one when the solute can cross the membrane, indicating that Equation 3.37 handles the general case of solute permeability. Moreover, the two terms containing L_{ws} in the numerator of the expression for the reflection coefficient (Eq. 3.37) take into account the further possibility that solute and solvent may interact when crossing the membrane; i.e., σ can also be affected by the coupling or interdependence of forces and fluxes.

When a membrane or barrier is nonselective, both water and the solute move across it at the same velocity; i.e., $v_s = v_w$. The solution is then the same on both sides of the barrier, so that $\Delta\Pi$ across the barrier is zero, which means that $\Delta\mu_w$ equals $\bar{V}_w \Delta P$ and $\Delta\mu_s$ equals $\bar{V}_s \Delta P$. (Admittedly, the idea of a solution containing a single solute is not realistic from a biological point of view, but it is convenient for illustrating the minimum value for σ.) As discussed above (p. 119), v_j is simply J_j/c_j. Using these various conditions as well as Equations 3.29 and 3.30, we obtain the following string of equalities:

$$v_w = \frac{J_w}{c_w} = \frac{L_{ww}\Delta\mu_w}{c_w} + \frac{L_{ws}\Delta\mu_s}{c_w} = \left(\frac{L_{ww}\bar{V}_w}{c_w} + \frac{L_{ws}\bar{V}_s}{c_w}\right)\Delta P$$

$$= v_s = \frac{J_s}{c_s} = \frac{L_{ws}\Delta\mu_w}{c_s} + \frac{L_{ss}\Delta\mu_s}{c_s} = \left(\frac{L_{ws}\bar{V}_w}{c_s} + \frac{L_{ss}\bar{V}_s}{c_s}\right)\Delta P$$

Next, we will let the number of moles of water per unit volume (c_w) times the volume per mole of water ($\bar{V}_w$) equal unity ($c_w\bar{V}_w \cong 1$ is another way of stating the dilute solution approximation). Also, we will identify c_s with $\bar{c}_s$. Putting these last two relations into the factors multiplying ΔP in the above string of equalities yields

$$L_{ww}\bar{V}_w^2 + L_{ws}\bar{V}_s\bar{V}_w = \frac{L_{ws}\bar{V}_w}{\bar{c}_s} + \frac{L_{ss}\bar{V}_s}{\bar{c}_s}$$

This relation indicates that the numerator on the right side of Equation 3.37 is zero, and so σ equals $0/L_p$, which is zero. Thus, for the case of nonselectivity ($v_s = v_w$), the reflection coefficient of the solute must be zero for that barrier.

Another relatively simple condition under which σ equals zero can be readily understood from Equation 3.38, $J_V = L_p(\Delta P - \sigma\Delta\Pi)$. Let us consider the rather realistic situation where ΔP is zero across a membrane. The volume flux density (J_V) is then simply $-L_p\sigma\Delta\Pi$ by Equation 3.38. For a solute having a reflection coefficient equal to zero for that particular membrane, the volume flux density would be zero. By Equation 3.33, $J_V = \bar{V}_w J_w + \bar{V}_s J_s$, a zero J_V implies that $\bar{V}_w J_w$ equals $-\bar{V}_s J_s$. In words, the volume flux density of water must be equal and opposite to the volume flux density of the solute to result in no net volume flux density. It follows that, when ΔP is zero but $\Delta\Pi$ is nonzero, the absence of a net volume flux density across a membrane permeable to both water and the single solute indicates that the reflection coefficient for that solute is zero. This condition, σ equaling zero, occurs when the volume of water flowing toward the side with the higher Π is balanced by an equal volume of solute diffusing across the membrane in the opposite direction toward the side where the solute is less concentrated (lower Π). Such a situation of zero volume flux density anticipates the concept of a "stationary state," introduced in the next section.

Solutes interacting with biological membranes exhibit properties ranging from impermeability ($\sigma_j = 1$) all the way to nonselectivity ($\sigma_j = 0$). Substances retained in or excluded from plant cells have reflection coefficients quite close to unity for the cellular membranes. For instance, the σ_j's for sucrose and amino acids are usually near 1.0 for plant cells. Methanol and ethanol enter cells very readily and have reflection coefficients of approximately 0.3 for some *Chara* and *Nitella* internodal cells (see Dainty and Ginzburg, or Lüttge and Higinbotham). On the other hand, σ_j can essentially equal zero for solutes crossing porous barriers, such as those presented by cell walls, or for molecules penetrating very readily across membranes, such as D_2O (^2H_2O). As for a permeability coefficient, the reflection coefficient of a species is the same for traversal in either direction across a membrane.

For many small neutral solutes not interacting with carriers in a membrane, the reflection coefficients are correlated with the partition coefficients (see Diamond and Wright, or Wright and Diamond). For example, when K_j for nonelectrolytes is less than about 10^{-4}, σ_j is generally close to one. Thus compounds that do not readily enter the lipid phase of a membrane (low K_j) also do not cross the membrane easily (σ_j near 1). When the partition coefficient is one or greater, the solutes can enter the membrane in appreciable amounts, and σ_j is generally close to zero. Considering the intermediate case, the reflection coefficient of some species j can be near 0.5 for a small nonelectrolyte having a K_j of about 0.1 for the membrane lipids, although individual molecules differ depending on their molecular weight, branching, and atomic composition. Neglecting frictional effects with other solutes, we see that the intermolecular interactions affecting partition coefficients are similar to those governing the value of reflection coefficients, and there is therefore a correlation between the permeability coefficient of a solute ($P_j = D_jK_j/\Delta x$) and its reflection coefficient for the same membrane. In fact, as Figure 3.12 illustrates, there is even a correlation between the reflection coefficients of a series of nonelectrolytes determined with animal membranes and the permeability coefficients of the same substances measured for plant membranes.* Thus, as the permeability coefficient goes from

---

* Exceptions occur for certain solutes, e.g., those that hydrogen-bond to membrane components.

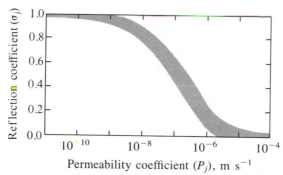

Figure 3.12
Correlation between the reflection coefficients for a series of non-electrolytes (determined using rabbit gallbladder epithelium) and the permeability coefficients for the same compounds (measured by R. Collander using *Nitella mucronata*). Most of the measurements are in the area indicated. [The curve is adapted from J. M. Diamond and E. M. Wright, *Annual Review of Physiology 31*:581–646 (1969). Used by permission.]

very small to very large values, the reflection coefficient decreases from unity (describing relative impermeability) to zero (for the opposite extreme of nonselectivity). In the next section we will specifically consider some of the consequences of reflection coefficients' differing from unity for solutes crossing cellular and organelle membranes.

Osmotic pressures play a key role in plant physiology, and so σ_j's are important parameters for quantitatively describing the solute and water relations of plants. In particular, the incorporation of reflection coefficients allows the role of osmotic pressure to be precisely stated. In Poiseuille's law [$J_V = -(r^2/8\eta)\,\partial P/\partial x$, Eq. 9.10], which can adequately describe movement in the xylem and possibly in the phloem—as well as in veins, arteries, and household plumbing—the flow is driven by the hydrostatic pressure gradient. An alternative view of the same situation is that σ equals zero for the solutes. In that case, osmotic pressures would have no direct effect on the movement described by Poiseuille's law; i.e., we would have a case of nonselectivity. At the opposite extreme of impermeability, σ is unity, and so Equation 3.38 becomes $J_V = L_P(\Delta P - \Delta\Pi) = L_P\Delta\Psi$ (recall that the water potential, Ψ, can be $P - \Pi$, Eq. 2.13a). But this is similar to Equation 2.23, $J_{V_w} = L_w\Delta\Psi$, which we obtained when only water fluxes were considered. This correspondence is not unexpected, since when the solutes are impermeant J_V is the same as J_{V_w} and L_w equals L_P (see p. 167). The real usefulness of reflection coefficients comes when σ_j is not at one of its two extremes of zero and unity. For such cases—intermediate between nonselectivity and impermeability—the volume flux density does not depend on the full osmotic pressure difference across the barrier, but it would be invalid to completely ignore the osmotic contribution of species j. (In App. VII, an expression involving reflection coefficients is developed for the flux density of a solute, J_s.)

SOLUTE MOVEMENT ACROSS MEMBRANES

We can profitably reexamine certain aspects of the movement of solutes into and out of cells and organelles by using the more general equations developed from irreversible thermodynamics. One particularly important situation amenable to relatively uncomplicated analysis is when the total volume flux density J_V is zero, an example of a *stationary state*. This stationary state, in which the volume of the cell or organelle does not change over the time period of interest, can be brought about by having the net volume flux density of water in one direction across the membrane equal the net volume flux density of solutes in the opposite direction. A stationary state is therefore not the same as a steady state or an equilibrium condition for the cell or organelle; it represents a situation occurring only at a particular time or under some special experimental arrangement. In fact, μ_j can depend on both position and time for a stationary state, but only on position for a steady state, and on neither position nor time at equilibrium. When a net amount of water is moving into a cell, $\Delta\mu_w$ is nonzero (we are not at equilibrium) and, in general, μ_w^i will be increasing with time (we are not even in a steady state). But we still might have a stationary state, where the volume is not changing. Our restriction here to cases of zero net volume flux density considerably simplifies the algebra and emphasizes the role played by reflection coefficients. Moreover, a stationary state of no volume change often characterizes the experimental situation under which the Boyle–Van't Hoff relation (Eqs. 2.15 and 2.18) or the expression describing incipient plasmolysis (Eq. 2.19) is invoked. Thus, the derivation of both of these relationships will be reconsidered in terms of irreversible thermodynamics, and we will discuss the role of reflection coefficients.

What is the relationship between the internal and the external P's and Π's in the case of a stationary state? To answer this, we must first express the consequences of a stationary state in symbols. Our point of departure is Equation 3.39, $J_V = L_P\left(\Delta P - \sum_j \sigma_j \Delta \Pi_j\right)$, where the stationary state condition of zero net volume flux density ($J_V = 0$) leads to the following equalities: $0 = \Delta P - \sum_j \sigma_j \Delta \Pi_j = P^o - P^i - \sum_j \sigma_j(\Pi^o_j - \Pi^i_j)$. Here, the osmotic pressures are for the effect of solutes on water activity, while in the general case both osmotic contributions from solutes (Π_s) and matric pressures resulting from the presence of interfaces (τ) might occur. Volume measurements for osmotic studies involving incipient plasmolysis or the Boyle–Van't Hoff relation are generally made when the external solution is at atmospheric pressure ($P^o = 0$) and when there are no external interfaces ($\tau^o = 0$). The stationary state condition of J_V equal to zero then leads to

$$\sum_j \sigma_j \Pi^o_j = \sigma^o \Pi^o = \sum_j \sigma_j \Pi^i_j + \tau^i - P^i \qquad \cdot \quad (3.40)$$

where the possibilities of interfacial interactions and hydrostatic pressures within the cell or organelle are explicitly recognized by the inclusion of τ^i and P^i. Equation 3.40 applies when the solutes are capable of crossing the barrier, as when molecules interact with real—not idealized—biological membranes. Equation 3.40 also characterizes the external solutes by an average reflection coefficient, σ^o, and the total external osmotic pressure, $\Pi^o = \sum_j \Pi^o_j$, the latter being relatively easy to measure.

The Influence of Reflection Coefficients on Incipient Plasmolysis

In Chapter 2, we used classical thermodynamics to derive the condition for incipient plasmolysis ($\Pi^o_{plasmolysis} = \Pi^i$, Eq. 2.19), which occurs when the internal pressure P^i inside a plant cell just becomes zero. This derivation was made assuming equilibrium of water, i.e., equal water potential, across a membrane impermeable to solutes. But the assumptions of water equilibrium and impermeability are often not valid. We can now rectify this situation using an approach based on irreversible thermodynamics.

Measurements of incipient plasmolysis can be made when there is zero volume flux density ($J_V = 0$) and for a simple external solution ($\tau^o = 0$) at atmospheric pressure ($P^o = 0$). In this case, Equation 3.40 would be the appropriate underlying expression from irreversible thermodynamics, instead of the less realistic condition of water equilibrium used previously.

For this stationary state condition, we obtain the following expression describing incipient plasmolysis ($P^i = 0$) when the solutes can cross the cell membrane:

$$\sigma^o \Pi^o_{\text{plasmolysis}} = \sum_j \sigma_j \Pi^i_j + \tau^i \tag{3.41}$$

Since the value of σ^o depends on the particular external solutes present, Equation 3.41 (a corrected version of Eq. 2.19) indicates that the external osmotic pressure Π^o at incipient plasmolysis can vary with the particular solute placed in the solution surrounding the plant cells. Let us suppose that solute i cannot penetrate the membrane, so σ_i equals 1, a situation often true for sucrose. We will suppose that another solute, j, can enter the cells ($\sigma_j < 1$), as is the case for many small nonelectrolytes. If we are at the point incipient plasmolysis for each of these two solutes—each taken in turn as the sole species in the external solution—$\sigma_i \Pi^o_{i \text{ plasmolysis}}$ equals $\sigma_j \Pi^o_{j \text{ plasmolysis}}$ by Equation 3.41. But solute i is unable to penetrate the membrane ($\sigma_i = 1$). Hence we obtain the following relationships:

$$\sigma_j = \frac{\Pi^o_{i \text{ plasmolysis}}}{\Pi^o_{j \text{ plasmolysis}}} = \frac{\text{effective osmotic pressure of species } j}{\text{actual osmotic pressure of species } j} \tag{3.42}$$

where the effective and actual osmotic pressures will be discussed presently.

Equation 3.42 suggests a straightforward way of describing σ_j. Since (by supposition) species j can cross the cell membrane, σ_j must be less than one. Therefore, by Equation 3.42, $\Pi^o_{j \text{ plasmolysis}}$ is greater than $\Pi^o_{i \text{ plasmolysis}}$, where the latter refers to the osmotic pressure of the impermeable solute solute at the point of incipient plasmolysis. In words, a higher external osmotic pressure is needed to cause plasmolysis if that solute is able to enter the plant cell. The "actual osmotic pressure" indicated in Equation 3.42 is defined either by Equation 2.7, $\Pi = -(RT/\bar{V}_w) \ln a_w$, where $\Pi = \sum_j \Pi_j$, or by Equation 2.10, $\Pi_j = RTc_j$. When the membrane is impermeable to the solute, σ_j equals one, and the apparent (effective) osmotic pressure of species j equals the actual Π^o_j. Effective osmotic pressure takes into consideration the fact that many solutes can move across biological membranes ($\sigma_j < 1$), and hence that the $\Delta\Pi_j$ effective in leading to a net volume flux density is reduced from its actual value—see Equation 3.39, $J_V = L_P\left(\Delta P - \sum_j \sigma_j \Delta\Pi_j\right)$. In summary, the reflection coefficient of species j indicates how effectively the osmotic pressure of that solute can be exerted across a particular membrane or other barrier.

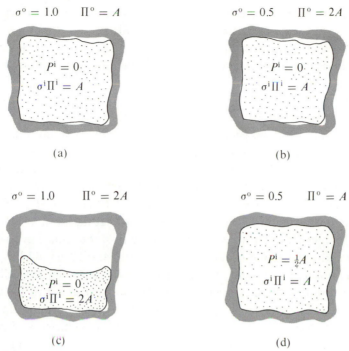

Figure 3.13

Diagrammatic cell showing a cell wall (shaded region) and a plasma-lemma (thin line) for various external osmotic pressures: (a) point of incipient plasmolysis in the presence of an impermeable solute, (b) point of incipient plasmolysis with a permeating solute, (c) extensive plasmolysis, and (d) cell under turgor.

We can use the condition of incipient plasmolysis to evaluate specific reflection coefficients. It is experimentally difficult to replace one external solution by another with no changes in the tissue taking place, or with none of the previous solution adhering to the cell (see Dainty 1963). Also, although it is easy in principle and involves only the use of a light microscope, determination of when the plasmalemma just begins to pull away from the cell wall is a rather subjective judgment. Nevertheless, the use of Equation 3.42 provides a simple way of considering individual reflection coefficients for various solutes entering plant cells.

To indicate relationships between reflection coefficients, osmotic pressures, and plasmolysis, let us consider Figure 3.13. The cell in the upper left of the diagram is at the point of incipient plasmolysis ($P^i = 0$) for a nonpenetrating solute ($\sigma^o = 1.0$) in the external solution ($\Pi^o = A$, where A is a constant);

i.e., the plasmalemma is just beginning to pull away from the cell wall. Equation 3.41 indicates that $\sigma^i \Pi^i$ would then also equal A. If we place the cell in a second solution containing a penetrating solute ($\sigma^o < 1$), Equation 3.42 indicates that the external solution must have a higher osmotic pressure for the cell to remain at the point of incipient plasmolysis. For instance, for a second solute with a σ_j of 0.5, the external osmotic pressure at the point of incipient plasmolysis would be $2A$ (see Fig. 3.13b). Thus, when the external solute can enter, Π^o is less effective in balancing the internal osmotic pressure or in leading to a flow of water (see Eq. 3.42).

On the other hand, if Π^o were $2A$ for an impermeable external solute, extensive plasmolysis of the same cell would occur, as illustrated in Figure 3.13c. Since $\sigma^o \Pi^o$ is $2A$ in this case, Equation 3.40 ($\sigma^o \Pi^o = \sigma^i \Pi^i - P^i$, when $\tau^i = 0$) indicates that $\sigma^i \Pi^i$ must also be $2A$, which means that essentially half of the internal water has left the cell. Finally, if the reflection coefficient were 0.5 and the external osmotic pressure were A, $\sigma^o \Pi^o$ would be $\frac{1}{2}A$, and we would not be at the point of incipient plasmolysis. In fact, the cell would be under turgor with an internal hydrostatic pressure equal to $\frac{1}{2}A$, at least until the concentration of the penetrating solute began to build up inside. We must therefore take into account the reflection coefficients of external (and internal) solutes to describe conditions at the point of incipient plasmolysis and, by extension, to predict the direction and the magnitude of volume fluxes across membranes.

Extension of the Boyle–Van't Hoff Relation

In Chapter 2 we derived the Boyle–Van't Hoff relation assuming that the water potential was the same on both sides of the cellular or organelle membrane under consideration. Not only were equilibrium conditions imposed on water, but we implicitly assumed that the membrane was impermeable to the solutes. However, zero net volume flux density ($J_V = 0$) is a better description of the experimental situation where the Boyle–Van't Hoff relation is applied. This condition of no volume change during the measurement is another example of a stationary state, so the Boyle–Van't Hoff relation will be reexamined from the point of view of irreversible thermodynamics. In this way, we can remove two of the previous restrictions, equilibrium for water and impermeability of solutes (see Nobel 1969).

When molecules can cross the membranes bounding cells or organelles, the reflection coefficients for both internal and external solutes should be included in the Boyle–Van't Hoff relation. Since σ° is less than one when the external solutes can penetrate, the effect of the external osmotic pressure on J_V is reduced. Likewise, the reflection coefficients for solutes within the cell or organelle will lessen the contribution of the internal osmotic pressure of each solute. Replacing Π_j^i by $RTn_j^i/(\bar{V}_w n_w^i)$ (Eq. 2.10) in Equation 3.40 leads to the following Boyle–Van't Hoff relation for the stationary state condition ($J_V = 0$ in Eq. 3.39):

$$\Pi^\circ = RT \frac{\sum_j \sigma_j n_j^i}{\sigma^\circ \bar{V}_w n_w^i} + \frac{\tau^i - P^i}{\sigma^\circ} \tag{3.43}$$

Note that the reflection coefficients of a membrane for both internal and external solutes enter into this extension of the expression relating volume and external osmotic pressure. As indicated in Chapter 2, the quantity $V - b$ in the conventional Boyle–Van't Hoff relation, $\Pi^\circ(V - b) = RT\sum_j \varphi_j n_j$ (Eq. 2.15), can be identified with $\bar{V}_w n_w^i$. In comparing Equation 2.15 with Equation 3.43, it is evident that the osmotic coefficient of species j, φ_j, can be equated to σ_j/σ° as an explicit recognition of the permeation properties of solutes, both internal and external. In fact, failure to recognize the effect of reflection coefficients on φ_j has led to misunderstandings of osmotic responses.

As we discussed in Chapter 2, the volume of pea chloroplasts (as well as other organelles and many cells) responds linearly to $1/\Pi^\circ$ (Fig. 2.9), indicating that $\tau^i - P^i$ in such plastids may be negligible compared with the external osmotic pressures used. To analyze experimental observations, it is convenient to replace $\sigma^\circ\Pi^\circ$ by $\sigma_x\Pi_x^\circ + \alpha$, where Π_x° is the contribution to the external osmotic pressure of solute x whose reflection coefficient (σ_x) is being considered, and α is the sum of $\sigma_j\Pi_j^\circ$ for all other external solutes. We can represent $RT \sum_j \sigma_j n_j^i$ by β, and replace $\bar{V}_w n_w^i$ by $V - b$. Making these substitutions into Equation 3.43, we obtain the following relatively simple form for testing osmotic responses in the case of penetrating solutes:

$$\sigma_x\Pi_x^\circ + \alpha = \frac{\beta}{V - b} \tag{3.44}$$

If we vary Π_x° and measure V, we can then use Equation 3.44 to obtain the reflection coefficients for various nonelectrolytes in the external solution.

Reflection Coefficients of Chloroplasts

When the refinements introduced by reflection coefficients are taken into account, we can use osmotic responses of cells and organelles to describe quantitatively the permeability properties of their membranes. As a specific application of Equation 3.44, we note that the progressive addition of hydroxymethyl groups in a series of polyhydroxy alcohols causes the reflection coefficients to increase steadily from 0.00 to 1.00 for pea chloroplasts (Table 3.2). In this regard, the lipid : water solubility ratio decreases in going

Table 3.2
Reflection coefficients of chloroplasts from *Pisum sativum* for alcohols. The reflection coefficients here apply to the pair of membranes surrounding the organelles, this being the barrier to solute entry or exit encountered in the plant cell.

Substance	σ_x	Substance	σ_x
Methanol	0.00	Adonitol	1.00
Ethylene glycol	0.40	Sorbitol	1.00
Glycerol	0.63	Mannitol	1.01
Meso-erythritol	0.90	Sucrose	1.00

SOURCE: C.-t. Wang and P. S. Nobel, *Biochimica et Biophysica Acta 241*: 200–212 (1971). Data reprinted by permission.

from methanol to ethylene glycol to glycerol to erythritol to adonitol; i.e., the partition coefficient K_x decreases. Since the permeability coefficient P_x equals $D_x K_x / \Delta x$ (Eq. 1.9), we expect a similar decrease in P_x as hydroxymethyl groups are added. Figure 3.12 clearly shows that, as the permeability coefficient decreases, the reflection coefficient generally increases. Consequently, the increase in σ_x of alcohols as —CHOH groups are added can be interpreted as simply a lowering of K_x. (As we go from methanol to adonitol, D_x also decreases, perhaps by a factor of 3, while K_x decreases about a thousandfold, so changes in K_x are the predominant influence on P_x and σ_x in this case.) The reflection coefficients of six-carbon polyhydroxy alcohols, such as sorbitol and mannitol, are essentially unity for pea chloroplasts. This indication of relative impermeability suggests that these compounds would serve as suitable osmotica in which to suspend chloroplasts, as is indeed the case. From these examples, we see that the rather esoteric concepts of irreversible thermodynamics can be applied in a relatively simple manner to gain insights into the physiological attributes of membranes.

Problems

3.1. At the beginning of this chapter we calculated that an average concentration of 1 mmol m^{-3} (1 μM) excess monovalent anions can lead to a -100 mV potential change across the surface of a spherical cell 30 μm in radius. (a) If the same total amount of charge were all concentrated in a layer 3 nm thick at the surface of the sphere, what would be its average concentration there? (b) If 10^7 sulfate ions are added inside the sphere, what is the new potential difference across the surface? (c) Approximately how much electrical work in joules is required to transport the 10^7 sulfate ions across the sulface of the cell?

3.2. For purposes of calculation, let us assume that an external solution is 1 mol m^{-3} (1 mM) KCl, while the solution inside a cell is 160 mol m^{-3} KCl at 20°C. (a) If K$^+$ is in equilibrium across the membrane and activity coefficients are unity, what is the electrical potential difference across the membrane? (b) If K$^+$ is in equilibrium and the mean activity coefficients are calculated from Equation 3.3, what would the membrane potential be? (c) If 1 mol m^{-3} K$_3$ATP—which fully dissociates to 3K$^+$ and ATP^{-3}—is added inside the cell, and if K$^+$-ATP^{-3} can be considered to act as an ion pair, what are $\gamma_{K\text{-}ATP}$ and a_{ATP}?

3.3. Consider a cell with a membrane potential E_M of -118 mV at 25°C. Suppose that the external solution contains 1 mM KCl, 0.1 mM NaCl, and 0.1 mM MgCl$_2$, while the internal concentration of K$^+$ is 100 mM, that of Ca^{2+} is 1 mM, and that of Mg^{2+} is 10 mM (1 mM = 1 mol m^{-3}). Assume that activity coefficients are unity. (a) Are K$^+$ and Mg^{2+} in equilibrium across the membrane? (b) If Na$^+$ and Ca^{2+} are in equilibrium, what are their concentrations in the two phases? (c) If Cl$^-$ is 177 mV away from equilibrium, such that the passive driving force on it is outward, what is its Nernst potential, and what is the internal concentration of Cl$^-$? (d) What is J_{Cl}^{in}/J_{Cl}^{out} for passive fluxes? (e) What are $\Delta\mu_{Cl}$ and $\Delta\mu_{Mg}$ across the membrane?

3.4. A 10 mM KCl solution at 25°C is placed outside a cell formerly bathed in 1 mM KCl (1 mM = 1 mol m^{-3}). (a) Assuming that some of the original solution adheres to the cell and that the ratio of mobilities (u_{Cl}/u_K) is 1.04, what diffusion potential would be present? (b) Assume that equilibrium is reached in the bathing solution after a sufficient lapse of time. The membrane may contain many carboxyl groups (—COOH) whose H$^+$'s will dissociate. The ensuing negative charge will attract K$^+$, and its concentration near the membrane may reach 200 mM. What type of and how large a potential would be associated with this situation? (c) Suppose that 10 mM NaCl is also in the external solution (with the 10 mM KCl), and that internally there is 100 mM K^+, 10 mM Na^+, and 100 mM Cl^-. Assume that P_{Na}/P_K is 0.20 and P_{Cl}/P_K is 0.01. What diffusion potential would be expected across the membrane? (d) What would E_M be if P_{Cl}/P_K were 0.00? If P_{Na}/P_K and P_{Cl}/P_K were both 0.00? Assume that other conditions are as in (c).

3.5. Consider an illuminated spherical spongy mesophyll cell 40 μm in diameter containing 50 spherical chloroplasts 4 μm in diameter. (a) Some monovalent anion produced by photosynthesis has a steady state net flux density out of the chloroplasts of 10 nmol m^{-2} s^{-1}. If this photosynthetic product is not changed in any of the cellular compartments, what is the net passive flux density out of the cell in the steady state? (b) If the passive flux density of the above substance into the cell at 25°C is 1 nmol m^{-2} s^{-1}, what is the difference in its chemical potential across the cellular membrane? (c) Suppose that, when the cell is placed in the dark, the influx and the efflux both become 0.1 nmol m^{-2} s^{-1}. If the plasmalemma potential is -118 mV (inside negative) and the same concentration occurs on the two sides of the membrane, what can be said about the energetics of the two fluxes? (d) If 1 ATP is required per ion transported, what is the rate of ATP consumption in (c)? Express your answer in μmol s^{-1} per m^3 of cellular contents.

3.6. The energy of activation for crossing biological membranes can represent the energy required to break hydrogen bonds (p. 51) between certain nonelectrolytes and the solvent water; e.g., to enter a cell the solute must first dissolve in the lipid phase of the membrane, and thus the hydrogen bonds with water must be broken. (a) What will be the Q_{10} for the influx of a solute that forms one H-bond per molecule with water if we increase the temperature from 10°C to 20°C? (b) How many hydrogen bonds would have to be ruptured per molecule to account for a Q_{10} of 3.2 under the conditions of (a)?

3.7. Suppose that transporters in the plasmalemma can shuttle K^+ and Na^+ into a cell. We will let the Michaelis constant K_j be 0.010 mM for the K^+ transporter and 1.0 mM for the Na^+ transporter (1 mM = 1 mol m^{-3}), while the maximum influx of either ion is 10 nmol m^{-2} s^{-1}. (a) What is the ratio of influxes, J_K^{in}/J_{Na}^{in}, when the external concentration of each ion is 0.010 mM? (b) What is J_K^{in}/J_{Na}^{in} when the external concentration of each ion is 100 mM? (c) If the entry of K^+ is by facilitated diffusion only, what is the rate of K^+ entry when c_K^o is 0.1 mM and ATP is being hydrolyzed at the rate of 10 mmol m^{-2} s^{-1}?

3.8. Consider a cell whose membrane has a hydraulic conductivity coefficient L_P of 10^{-12} m s^{-1} Pa^{-1}. Initially, no net volume flux density occurs when the cell is placed in a solution having the following composition: sucrose ($\Pi_j^o = 0.2$ MPa, $\sigma_j = 1.00$), ethanol (0.1 MPa, 0.30), and glycerol (0.1 MPa, 0.80). The external solution is at atmospheric pressure, while P^i is 0.5 MPa. Inside the cell the osmotic pressure caused by glycerol is 0.2 MPa, sucrose and ethanol are initially absent, and other substances having an osmotic pressure of 1.0 MPa are present. (a) What is the mean reflection coefficient for the external solution? (b) What is the mean reflection coefficient for the internal solutes other than glycerol? (c) Suppose that some treatment makes the membrane nonselective for all solutes present. Is there then a net volume flux density? (d) If another treatment makes the membrane impermeable to all solutes present, what would be the net volume flux density?

3.9. Consider a cell at the point of incipient plasmolysis in an external solution containing 0.3 molal sucrose, an impermeable solute. The concentration of glycine that just causes plasmolysis is 0.4 molal. Assume that no water enters or leaves the cell during the plasmolytic experiments. (a) What is the reflection coefficient of glycine for the cellular membrane? (b) Suppose that chloroplasts isolated from such a cell have the same osmotic responses as in Problem 2.5. What is the volume of such chloroplasts in vivo? Assume that activity coefficients are unity and that the temperature is 20°C. (c) Suppose that chloroplasts are isolated in 0.3 molal sucrose, which has a reflection coefficient of 1.00 for the chloroplasts. If 0.1 mole of glycine is then added per kg of water in the isolation medium, and if the chloroplast volume is 23 μm^3, what is the reflection coefficient of glycine for the chloroplast membranes? (d) What is the external concentration of glycerol ($\sigma_j = 0.60$) in which the chloroplasts have the same initial volume as in 0.3 molal sucrose? What is the chloroplast volume after a long time in the glycerol solution?

References

Anderson, W. P. 1972. Ion transport in the cells of higher plant tissues. *Annual Review of Plant Physiology 23*: 51–72.

Briggs, G. E., A. B. Hope, and R. N. Robertson. 1961. *Electrolytes and Plant Cells*. F. A. Davis, Philadelphia.

Bull, H. B. 1964. *An Introduction to Physical Biochemistry*. F. A. Davis, Philadelphia.

Dainty, J. 1962. Ion transport and electrical potentials in plant cells. *Annual Review of Plant Physiology 13*: 379–402.

Dainty, J. 1963. Water relations of plant cells. *Advances in Botanical Research 1*: 279–326.

Dainty, J., and B. Z. Ginzburg. 1964. The reflection coefficient of plant cell membranes for certain solutes. *Biochimica et Biophysica Acta 79*: 129–137.

Dainty, J., and A. B. Hope. 1959. Ionic relations of cells of *Chara australis*. *Australian Journal of Biological Sciences 12*: 395–411.

Davson, H., and J. F. Danielli. 1952. *The Permeability of Natural Membranes*, 2nd ed. Cambridge University Press, Cambridge.

De Groot, S. R., and P. Mazur. 1969. *Non-Equilibrium Thermodynamics*. North-Holland, Amsterdam.

Diamond, J. M., and E. M. Wright. 1969. Biological membranes: the physical basis of ion and nonelectrolyte selectivity. *Annual Review of Physiology 31*: 581–646.

Edsall, J. T., and J. Wyman. 1958. *Biophysical Chemistry*, Vol. I. Academic Press, New York.

Epstein, E. 1972. *Mineral Nutrition of Plants: Principles and Perspectives*. Wiley, New York.

Fuller, E. N., P. D. Schettler, and J. C. Giddings. 1966. A new method for prediction of binary gas-phase diffusion coefficients. *Industrial & Engineering Chemistry 58*: 19–27.

Goldman, D. E. 1943. Potential, impedance, and rectification in membranes. *Journal of General Physiology 27*: 37–60.

Gutknecht, J., and J. Dainty. 1968. Ionic relations of marine algae. *Oceanography and Marine Biology, An Annual Review 6*:163–200.

Higinbotham, N. 1974. Conceptual developments in membrane transport, 1924–1974. *Plant Physiology 54*:454–462.

Higinbotham, N., J. S. Graves, and R. F. Davis. 1970. Evidence for an electrogenic ion transport pump in cells of higher plants. *Journal of Membrane Biology 3*:210–222.

Higinbotham, N., and W. P. Anderson. 1974. Electrogenic pumps in higher plant cells. *Canadian Journal of Botany 52*:1011–1021.

Hodgkin, A. L., and B. Katz. 1949. The effect of sodium ions on the electrical activity of the giant axon of the squid. *Journal of Physiology 108*:37–77.

Hope, A. B. 1971. *Ion Transport and Membranes*. Butterworths, London; University Park Press, Baltimore.

Hope, A. B., and N. A. Walker. 1975. *The Physiology of Giant Algal Cells*. Cambridge University Press, Cambridge.

Katchalsky, A., and P. F. Curran. 1967. *Nonequilibrium Thermodynamics in Biophysics*. Harvard University Press, Cambridge.

Kedem, O., and A. Katchalsky. 1958. Thermodynamic analysis of the permeability of biological membranes to nonelectrolytes. *Biochimica et Biophysica Acta 27*: 229–246.

Lange, O. L., P. S. Nobel, C. B. Osmond, and H. Ziegler, eds. 1981. *Physiological Plant Ecology. Encyclopedia of Plant Physiology, New Series*, Vol. 12A. Springer–Verlag, Berlin.

Lehninger, A. L. 1975. *Biochemistry; The Molecular Basis of Cell Structure and Function*, 2nd ed. Worth, New York.

Lüttge, U., and M. G. Pitman. 1976. *Transport in Plants* II, *Part A, Cells*; *Part B, Tissues and Organs. Encyclopedia of Plant Physiology, New Series*, Vol. 2. Springer-Verlag, Berlin.

Lüttge, U., and N. Higinbotham. 1979. *Transport in Plants*. Springer-Verlag, New York.

Lyons, J. M., D. Graham, and J. K. Raison, eds. 1979. *Low Temperature Stress in Crop Plants: The Role of the Membrane*. Academic Press, New York.

MacInnes, D. A. 1961. *The Principles of Electrochemistry*, rev. 2nd ed. Dover, New York.

MacRobbie, E. A. C. 1962. Ionic relations of *Nitella translucens. Journal of General Physiology 45*:861–878.

MacRobbie, E. A. C. 1971. Fluxes and compartmentation in plant cells. *Annual Review of Plant Physiology 22*:75–96.

MacRobbie, E. A. C. 1977. Functions of ion transport in plant cells and tissues. *International Review of Biochemistry, Plant Biochemistry II*, D. H. Northcote, ed. *13*:211–247.

Morris, J. G. 1968. *A Biologist's Physical Chemistry*. Addison-Wesley, Reading, Massachusetts.

Neame, K. D., and T. C. Richards. 1972. *Elementary Kinetics of Membrane Carrier Transport*. Wiley, New York.

Nobel, P. S. 1969. The Boyle–Van't Hoff relation. *Journal of Theoretical Biology 23*:375–379.

Nobel, P. S. 1974. *Introduction to Biophysical Plant Physiology*. W. H. Freeman, San Francisco.

Prigogine, I. 1967. *Thermodynamics of Irreversible Processes*, 3rd ed. Interscience, New York.

Pytkowicz, R. M. 1979. *Activity Coefficients in Electrolyte Solutions*. CRC Press, Cleveland.

Raison, J. K., and E. A. Chapman. 1976. Membrane phase changes in chilling-sensitive *Vigna radiata* and their significance to growth. *Australian Journal of Plant Physiology 3*:291–299.

Raven, J. A. 1969. Action spectra for photosynthesis and light-stimulated ion transport processes in *Hydrodictyon africanum*. *New Phytologist 68*:45–62.

Robinson, R. A., and R. H. Stokes. 1970. *Electrolyte Solutions*, rev. 2nd ed. Butterworths, London.

Sentenac, H., and C. Grignon. 1981. A model for predicting ionic equilibrium concentrations in cell walls. *Plant Physiology 68*:415–419.

Smith, F. A., and J. A. Raven. 1979. Intracellular pH and its regulation. *Annual Review of Plant Physiology 30*:289–311.

Spanswick, R. M. 1973. Electrogenesis in photosynthetic tissues. In *Ion Transport in Plants*, W. P. Anderson, ed. Academic Press, London.

Spanswick, R. M. 1981. Electrogenic ion pumps. *Annual Review of Plant Physiology 32*:267–289.

White, A., P. Handler, E. L. Smith, R. L. Hill, and I. R. Lehman. 1978. *Principles of Biochemistry*, 6th ed. McGraw-Hill, New York.

Wiskich, J. T. 1977. Mitochondrial metabolite transport. *Annual Review of Plant Physiology 28*:45–69.

Wright, E. M., and J. M. Diamond. 1969. Patterns of non-electrolyte permeability. *Proceedings of the Royal Society (Series B) 172*:227–271.

Yariv, J., I. Z. Steinberg, A. J. Kalb, R. Goldman, and E. Katchalski. 1972. Permease as a rotatory carrier. *Journal of Theoretical Biology 35*:459–465.

CHAPTER 4

Light

Through a series of nuclear reactions taking place within the sun, nuclear mass is converted into energy in accordance with Einstein's famous relation, $E = mc^2$. By such conversion of mass to energy the sun maintains an extremely high surface temperature and thus radiates a great amount of energy into space. Some of this radiant energy is incident on the earth, only a small fraction of which is absorbed by plants. This absorption initiates a flow of energy through the biosphere (plants and animals and that portion of the earth they inhabit).

The first step in the utilization of sunlight for this energy flow is the conversion of its radiant energy into various forms of chemical energy by the primary processes of photosynthesis. This chemical energy may then be stored in the plants, mainly in the form of carbohydrates. The energy thus stored may then be acquired by animals—directly by herbivores, indirectly by carnivores, or both directly and indirectly by omnivores like us. However the energy is acquired, its ultimate source is the solar radiation trapped by photosynthesis. Without this continuous energy input from the sun, plants and animals would drift toward equilibrium, with the consequent cessation of life.

Sunlight also regulates certain activities of plants and animals by acting as a trigger. The energy to carry out these activities is supplied by metabolic reactions, not directly by the light itself. Examples of light acting as a trigger are vision, phototaxis, phototropism, and phytochrome regulation of plant processes.

In this chapter we will be primarily concerned with the physical nature of light and the mechanism of light absorption by molecules. We will discuss how molecular states excited by light absorption can promote endergonic (energy-requiring) reactions or be dissipated by other de-excitation processes. In Chapter 5 we will consider the photochemistry of photosynthesis and in Chapter 6 the bioenergetics of energy conversion, especially that taking place in organelles. In Chapter 7 we will demonstrate how the net energy input by radiation is dissipated by a leaf.

To help introduce the topic of light, we will briefly review certain historical developments in the understanding of its nature (see Ditchburn, Resnick and Halliday, or Seliger and McElroy for details). In 1666 Newton showed that a prism could disperse white light into many different colors, suggesting that such radiation was a mixture of many components. Soon thereafter, Huygens proposed that the propagation of light through space could be by wave motion. In the early 1800's Young attributed interference properties to the wave character of light. But a wave theory of light was not generally accepted until about 1850, when Foucault demonstrated that light travels more slowly in a dense medium such as water than in a rarefied medium such as air, one of the predictions of the wave theory. In 1887 Hertz discovered that light striking the surface of a metal could cause the release of electrons from the solid—the so-called "photoelectric" effect. But he also found that wavelengths above a certain value could not eject any electrons at all, no matter what the total energy in the light beam. This important result was contradictory to the then-accepted wave theory of light. In an important departure from wave theory, Planck proposed in 1901 that radiation was particle-like; i.e., light was describable as consisting of discrete packets, or quanta, each of a specific energy. In 1905 Einstein explained the photoelectric effect of Hertz as a special example of the particle nature of light, indicating that the absorption of a light quantum of sufficiently short wavelength by an electron in the metal could supply enough energy to cause the ejection or release of that electron, while, if the wavelengths were longer, then the individual quanta were not energetic enough to eject any electrons. The intriguing wave-particle duality of light has subsequently become described in a consistent manner through the development of quantum mechanics. (Although this text does not require a background in quantum mechanics, some knowledge of this field is essential for a comprehensive understanding of light.) Both wave and particle attributes of light are necessary for a complete description of radiation, and we will consider both aspects in this chapter.

WAVELENGTH AND ENERGY

Light is often defined as that electromagnetic radiation perceivable by the human eye. Although such a definition may be strictly correct, the word *light* is frequently used to refer to a wider range of electromagnetic radiation. In this section we will discuss the range of electromagnetic radiation important in plant and animal physiology, including the subdivisions into various wavelength intervals. The wavelength of light will be related to its energy. After noting various conventions used to describe radiation, we will briefly consider some of the characteristics of solar irradiation reaching the earth.

Light Waves

Our concern here is with the regular and repetitive changes in the intensity of the minute electric and magnetic fields that indicate the passage of a light wave (see Fig. 4.1). Light can travel in a solid (e.g., certain plastics), in a liquid (water), in a gas (air), and even in a vacuum (the space between the sun

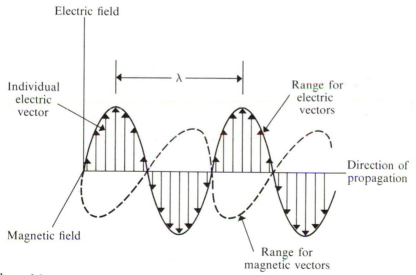

Figure 4.1
Light can be represented by an electromagnetic wave corresponding to oscillations of the local electric and magnetic fields. The oscillating electric vectors at a particular instant in time are indicated by arrows in the diagram. A moment later, the entire pattern of electric and magnetic fields will shift in the direction of propagation of the wave. A wavelength (λ) is the distance between two successive points of the same phase along the wave.

and the earth's atmosphere). One way to characterize light is by its wave-length—the distance between successive points of the same phase, such as between two successive peaks of a wave train (Fig. 4.1). A wavelength, then, is the distance per *cycle* of the wave. The preferred unit for wavelengths of light is the nanometer (nm)—10^{-9} m—and it will be used in this text (1 nm = 0.001 μm = 10 Å).

The wavelength regions of major interest in biology are the ultraviolet, the visible, and the infrared (see Table 4.1). Wavelengths below about 400 nm are referred to as ultraviolet (UV)—meaning on the other side of or beyond violet in the sense of having a shorter wavelength. The visible region extends from approximately 400 nm to 740 nm and is subdivided into various bands such as blue, green, and red (Table 4.1). These divisions are based on the subjective color experienced by humans. The infrared (IR) region has wavelengths longer than those of the red end of the visible spectrum, up to approximately 40 μm (40 000 nm).

Besides wavelength, we can also characterize a light wave by its frequency of oscillation, v, and by the magnitude of its velocity of propagation, v (i.e., v is the speed of light). These three quantities are related as follows:

$$\lambda v = v \tag{4.1}$$

where λ is the wavelength. In a vacuum the speed of light for all wavelengths is a constant, generally designated as c, which experimentally equals 299 800 km s^{-1}, or about 3.00×10^{8} m s^{-1} (see Resnick and Halliday, or Monk, for measurements of the speed of light). Light passing through a medium other than a vacuum has a speed less than c. For example, light with a wavelength of 589 nm in a vacuum is decreased in speed 0.03% by air, 25% by water, and 40% by dense flint glass. The wavelength undergoes a decrease in magnitude equal to the decrease in the speed of propagation in these various media; the unchanging property of a wave propagating through different media is the frequency. Note that v is the frequency of the oscillations of the local electric and magnetic fields of light that are illustrated in Figure 4.1.

For light, $\lambda_{\text{vacuum}} v$ equals v_{vacuum} by Equation 4.1, where v_{vacuum} is the constant c. Therefore, if we know the wavelength in a vacuum, we can calculate the frequency. In fact, the wavelengths given for light generally refer to values in a vacuum, as is the case for columns 2 and 3 in Table 4.1 (λ's in air differ only slightly from the magnitudes listed). As a specific example, let us select a wavelength in the blue region of the spectrum, e.g.,

Table 4.1

Definitions and characteristics of the various wavelength regions of light. The ranges of wavelengths leading to the sensation of a particular color are somewhat arbitrary and vary with individuals. Both frequencies and energies in the table refer to the particular wavelength indicated in column 3 for each wavelength interval. Wavelength magnitudes are those in a vacuum.

Color	Approximate wavelength range (nm)	Representative wavelength (nm)	Frequency (cycles s^{-1}, or hertz)	Energy (kJ mol^{-1})
Ultraviolet	below 400	254	11.80×10^{14}	471
Violet	400 to 425	410	7.31×10^{14}	292
Blue	425 to 490	460	6.52×10^{14}	260
Green	490 to 560	520	5.77×10^{14}	230
Yellow	560 to 585	570	5.26×10^{14}	210
Orange	585 to 640	620	4.84×10^{14}	193
Red	640 to 740	680	4.41×10^{14}	176
Infrared	above 740	1400	2.14×10^{14}	85

460 nm. By Equation 4.1, the frequency of this blue light is

$$v = \frac{(3.00 \times 10^8 \text{ m s}^{-1})}{(460 \times 10^{-9} \text{ m/cycle})} = 6.52 \times 10^{14} \text{ cycles s}^{-1}$$

Since v does not change from medium to medium, it is often desirable to describe light by its frequency, as has been done in column 4 of Table 4.1.*

Energy of Light

In addition to its wave-like properties, light can also exhibit particle-like properties, as in the case of the photoelectric effect mentioned above. Thus, light can act as if it were divided (or quantized) into discrete units, which we call *photons*. The light energy (E_λ) carried by a photon is

$$E_\lambda = hv = hc/\lambda_{\text{vacuum}} \qquad \text{photon basis} \qquad (4.2a)$$

---

* For purposes of calculation, the wavelength of light is generally expressed in nm, not nm/cycle, and v is considered to have units of s^{-1} (not cycles s^{-1}).

where h is a fundamental physical quantity called *Planck's constant*. By Equation 4.2a, a photon of light has an energy directly proportional to its frequency, and inversely proportional to its wavelength in a vacuum. For most applications in this book, we will describe light by its wavelength λ. To emphasize that the light energy of a photon depends on its wavelength, we have used the symbol E_λ in Equation 4.2a. A *quantum* (plural: quanta) refers to the light *energy* carried by a photon; i.e., $h\nu$ represents a quantum of electromagnetic energy. The terms "photons" and "quanta" are sometimes used interchangeably, a usage that is generally clear but, strictly speaking, not correct.

The introduction of the constant h by Planck in the early 1900's represented a great departure from the accepted wave theory of light. It substantially modified the classical equations describing radiation and provided a rational basis for determining the energy of a given number of photons. Since frequency has the units of reciprocal time, Equation 4.2a indicates that the dimensions of Planck's constant are energy × time, e.g., 6.626×10^{-34} J s (see App. II). We note that hc is 1 240 eV nm (App. II), and so we can readily calculate the energy per photon of blue light at 460 nm using Equation 4.2a:

$$E_\lambda = \frac{(1\ 240 \text{ eV nm})}{(460 \text{ nm})} = 2.70 \text{ eV}$$

Instead of energy per photon, we are generally interested in the energy per Avogadro's number N (6.022×10^{23}) of photons (N photons is sometimes called an einstein, but the preferred SI unit is mole). On a mole basis, Equation 4.2a becomes

$$E_\lambda = Nh\nu = Nhc/\lambda_{\text{vacuum}} \qquad \text{mole basis} \qquad (4.2b)$$

Let us consider blue light of 460 nm, which has a frequency of 6.52×10^{14} cycles s^{-1} (Table 4.1). Using Equation 4.2b, we calculate that the energy per mole of 460-nm photons is

$$E_\lambda = (6.022 \times 10^{23} \text{ mol}^{-1})(6.626 \times 10^{-34} \text{ J s})(6.52 \times 10^{14} \text{ s}^{-1})$$

$$= 260 \text{ kJ mol}^{-1}$$

Alternatively, we can calculate the energy by dividing Nhc (119 600 kJ mol^{-1} nm) by the wavelength—strictly speaking, the wavelength in a vacuum (see Table 4.1).

Quanta of visible light represent relatively large amounts of energy. The hydrolysis of ATP, the main currency for chemical energy in biology, generally yields about 40 to 50 kJ mol^{-1} (10 to 12 kcal mol^{-1}) under physiological conditions (see Ch. 6), while, as we just calculated, blue light has about five to six times as much energy/mole of photons. Quanta of ultraviolet light represent even higher energies than those of visible light; e.g., 254 nm photons have 471 kJ mol^{-1} of radiant energy (Table 4.1). This is greater than the carbon-carbon bond energy of 348 kJ mol^{-1} or the oxygen-hydrogen bond energy of 463 kJ mol^{-1}. The high quantum energy of UV radiation underlies its mutagenic and bactericidal action, since it is energetic enough to cause disruption of certain covalent bonds.

The *photoelectric* effect, where light leads to the ejection of electrons from the surface of a metal, clearly illustrates the distinction between light energy and the energy of its quanta. Millikan found that 175 kJ mol^{-1} quanta, representing wavelengths of 683 nm or less, can lead to a photoelectric effect for sodium (see Resnick and Halliday).* For wavelengths beyond 683 nm, however, no matter how much light energy is absorbed, electrons are not ejected from the surface. Hence, a quantum of a specific energy may be necessary for a certain reaction. Measurement of the total light energy does not indicate how many photons are involved or what their individual energies are unless the wavelength distribution is known.

Absorption of radiation by an atom or molecule leads to a more energetic state of the absorbing species. Such energetic states can also be produced by collisions resulting from the random thermal motion of the molecules. The higher the temperature, the greater is the average kinetic energy of the atoms and molecules, and the greater is the probability of achieving a relatively energetic state by collision. The number of molecules having a particular kinetic energy can often be approximated by the Boltzmann energy distribution (see p. 141). By Equation 3.21b $[n(E) = n_{total}e^{-E/RT}]$, the fraction of atoms or molecules having a kinetic energy of molecular Brownian motion of E or greater at thermal equilibrium equals the Boltzmann factor $e^{-E/RT}$, where we have expressed energy on a mole basis (see footnote, p. 141). Based on the kinetic theory of gases, it can be shown that the *average* kinetic energy of translational motion for molecules in a gas phase is $(3/2)RT$, which equals 3.72 kJ mol^{-1} at 25°C. One may ask what fraction of the molecules exceed

---

* For his experimental work on the photoelectric effect, Millikan was awarded the Nobel Prize in physics in 1923. Others whose contributions to the understanding of radiation led to Nobel prizes in physics include Wien in 1911 for heat radiation laws, Planck in 1918 for the quantum concept, Stark in 1919 for spectral properties, Einstein in 1921 in part for discovering the photoelectric effect, Bohr in 1922 for atomic radiation, Franck and Hertz in 1925 for atom-electron interactions, and Pauli in 1945 for atomic properties.

this energy? The Boltzmann factor becomes $e^{-(3\,RT/2)/(RT)}$, or $e^{-1.5}$, which is 0.22, and so 22% of the molecules have higher kinetic energies. At what temperature would 44% have such kinetic energies? The Boltzmann factor then equals 0.44, and so T would be

$$T = \frac{1}{-\ln(0.44)} \frac{E}{R} = \frac{1}{(0.82)} \frac{(3\,720 \text{ J mol}^{-1})}{(8.3143 \text{ J mol}^{-1} \text{ K}^{-1})}$$

$$= 546 \text{ K} \qquad (273°\text{C})$$

Raising the temperature clearly increases the fraction of molecules with higher energies. However, temperature can be raised to only a limited extent under physiological conditions.

Let us next consider a kinetic energy E of 260 kJ mol^{-1}, as is possessed by blue light of 460 nm. The fraction of molecules possessing at least this energy is $e^{-E/RT}$, which equals $e^{-(260 \text{ kJ mol}^{-1})/(2.48 \text{ kJ mol}^{-1})}$, or only 3×10^{-46} at 25°C! For comparison, the total number of atoms in the entire biomass of all living plants and animals is about 3×10^{41}. The chance that a particular molecule can gain the equivalent of 260 kJ mol^{-1} by means of thermal collisions is therefore vanishingly small. Hence, the absorption of blue light can lead to energetic states that otherwise would simply not occur at temperatures encountered by plants and animals. Absorption of the relatively high quantum energy of light thus promotes the attainment of very improbable energetic states—this is a key point in the understanding of photobiology.

Illumination, Photon Flux Density, and Irradiance

For many purposes in studying plants, we need to know the total amount of incident light. There are three common classes of intruments for measuring such fluxes: (1) photometers, which measure the available illuminating power, a quantity related to the wavelength sensitivity of the human eye; (2) quantum meters, which measure the number of photons; and (3) radiometers, which measure the total energy of the radiation.

By definition, photometers do not appreciably respond to radiation in the infrared or the ultraviolet. They are "light" meters in the sense of human vision; i.e., they respond to photons in the visible region, in the same way as the light meter on a camera. A candle is a unit of luminous intensity, originally based on a standard candle or lamp. The present international unit is called a *candela* (often still referred to a "candle"), which is defined as the total light intensity emitted from 1.67 mm^2 of a blackbody radiator (one that radiates maximally; see p. 197) at the melting temperature of pure platinum (2 042 K). The total light emitted in all directions by a source of 1 candela is 4π lumens; 1 lumen m^{-2} is the photometric illuminance unit, lux. Measurement in lux is adequate for certain purposes where human vision is involved, but it is not recommended for studies with plants (see McCree).

For many purposes in plant studies, it is important to know the photon flux density. For instance, the rate of photosynthesis depends on the rate of absorption of photons (not the rate of absorption of energy). Some instruments are sensitive only to photons that might be useful photosynthetically (e.g., wavelengths from 400 to 700 nm), the so-called *photosynthetically active radiation* (or PAR), which is also referred to as the *photosynthetic photon flux density* (or PPFD). Photon flux density is commonly expressed in μmol m^{-2} s^{-1}.

Radiometers—e.g., thermocouples, thermopiles, or thermistors that have been treated so as to absorb all wavelengths—respond to radiant energy and so are sensitive to irradiation in the ultraviolet and infrared as well as in the visible. Readings are expressed in energy per unit area and time, e.g., J m^{-2} s^{-1}, which is W m^{-2} (where W is the abbreviation for watts; conversion factors for radiometric units are given in App. III). If the irradiance or radiant energy flux density at a specified wavelength is measured in radiometric units, the value can be converted to a photon flux density by using the energy carried by individual photons (see Eq. 4.2).* In general, radiometric units

---

* Various terms are used to describe radiation. Definitions of the more common ones follow. *Emittance* is the flux density at the emitting surface, e.g., radiant emittance (expressed in J m^{-2} s^{-1} = W m^{-2}). *Illuminance* is the luminous flux density or "illumination" at some surface (lux). *Intensity* refers to a property of a source; e.g., light intensity designates the rate of light emission for a photometric source (lumens per unit solid angle, or candelas), and radiant intensity is the power emitted per unit solid angle (W steradian^{-1}). *Irradiance* (commonly termed *irradiation*) is the radiant energy flux density received on some surface (W m^{-2}). *Radiance* is the rate of radiant energy emission (power emitted) per unit solid angle per unit area (W steradian^{-1} m^{-2}). *Radiant flux* is the radiant energy emitted or received per unit time (W). *Radiant flux density* is the radiant flux per unit area (W m^{-2}). The term *fluence* is sometimes used for amount per area, and so radiant energy per unit area would be the energy fluence (J m^{-2}). *Fluence rate* is therefore the fluence per unit time; e.g., the photon fluence rate (mol m^{-2} s^{-1}) is the same as photon flux density. (See Bell and Rose; Gates; Incoll et al., Ch. 1 reference; McCree; or Monteith.)

A solid angle is the three-dimensional counterpart of the more familiar planar angle. It can be interpreted as a region of space defined by a series of lines radiating from a point, the lines forming a smooth surface (as a conical surface viewed from its apex). The region delimits a certain surface area A on a sphere of radius r; the solid angle in steradians then equals A/r^2. For instance, a hemisphere viewed from its center subtends 2π steradians.

cannot be unambiguously converted to photometric units, or vice versa.

In choosing a lamp to be used in a controlled-environment chamber for growing plants, we usually must consider both the total energy emitted and the photon flux density in the photosynthetically active part of the spectrum. Tungsten lamps are comparatively poor sources of PAR, since about 90% of their radiant energy is emitted in the infrared—the actual amount depends on the operating temperature of the filaments—creating appreciable cooling problems when large numbers of such lamps are used. Since typical fluorescent lamps emit only about 10% of their energy in the IR, cooling problems are less severe than with tungsten lamps. On the other hand, the wavelength distribution of the sun's photons in the visible region (Fig. 4.2) is matched far better by tungsten lamps than by fluorescent ones. A close match can be important, since the wavelength distribution influences the relative amounts of the two forms of phytochrome discussed at the end of this chapter (see Bickford and Dunn, Kaufman and Christensen, Langhans, Seliger and McElroy, and Tibbitts and Kozlowski).

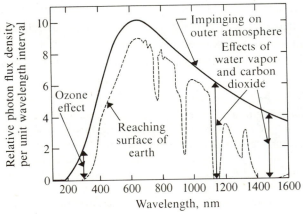

Figure 4.2
Wavelength distributions of the sun's photons incident on the earth's atmosphere and its surface. The curve for the solar irradiation on the atmosphere is an idealized one based on Planck's radiation distribution formula. The spectral distribution and the amount of solar irradiation reaching the earth's surface depend on clouds, other atmospheric conditions, altitude, and the sun's angle in the sky. The pattern indicated by the lower curve is appropriate at sea level on a clear day with the sun nearly overhead. (For further details see Bickford and Dunn, Gates, Monteith, and Seliger and McElroy.)

Sunlight

Essentially all energy for life originates in the form of electromagnetic radiation from the sun. In radiometric units the radiant flux density of solar irradiation (irradiance) perpendicularly incident on the earth's atmosphere—the so-called "solar constant"—is about $1\,360$ W m^{-2} (see Monteith or Gates).[*] In Chapter 6 we will consider the solar constant in terms of the annual photosynthetic yield, and in Chapter 7 in terms of the energy balance of a leaf. Based on the solar constant and averaged for all latitudes, the mean total daily amount of radiant energy from the sun incident on a horizontal surface just outside the earth's atmosphere is 29.4 MJ m^{-2} day^{-1}, of which atmospheric conditions permit an average of only 58% or 17.0 MJ m^{-2} day^{-1} to reach the earth's surface (the amount at the earth's surface on a cloudless day in the summer can be about 30 MJ m^{-2} day^{-1}). The instantaneous irradiance in the visible region at noon on a cloudless day with the sun overhead can be 420 W m^{-2} (about 100 000 lux). If we represent sunlight by yellow light of 570 nm, which by Table 4.1 carries 210 kJ mol^{-1}, then 420 W m^{-2} in the visible region from the sun is about (420 J m^{-2} s^{-1})/ (210 000 J mol^{-1}), or 2.0×10^{-3} mol m^{-2} s^{-1} (2 000 μmol m^{-2} s^{-1}) at the earth's surface.

Figure 4.2 shows the relative number of the sun's photons impinging on the earth's atmosphere and reaching its surface as a function of wavelength. About 5% of the photons incident on the earth's atmosphere are in the ultraviolet (below 400 nm), 28% in the visible, and 67% in the infrared (beyond 740 nm). Most of the ultraviolet fraction of sunlight incident on the atmosphere is prevented from reaching the surface of the earth by ozone (O_3) present in the stratosphere, 20 to 30 km above the earth's surface (see Gates).[†] Ozone absorbs some visible radiation (especially near 600 nm), and effectively screens out the shorter ultraviolet rays by absorbing strongly below 300 nm. Much of the infrared from the sun is absorbed by atmospheric water vapor and CO_2 (see Fig. 4.2). Water absorbs strongly near 900 nm and 1 100 nm, and above 1 200 nm, having a major infrared absorption band at

---

[*] The solar constant varies from the average by up to $\pm 3.5\%$ because the earth's orbit is elliptical. The value given is for the mean distance between the earth and sun (the earth is closest to the sun on 3 January, at 1.471×10^8 km, and furthest away on 4 July, at 1.521×10^8 km). There are additional variations in solar irradiation based on changes in solar activity, as occur for sun spots.

[†] The absorption of sunlight by this O_3 leads to a pronounced heating of the upper atmosphere.

1 400 nm (1.4 μm). Although the amount of water vapor in the air varies with latitude, longitude, and season, the mean vapor concentration is equivalent to a path of liquid water approximately 20 mm thick (see Gates, and Seliger and McElroy). Because of the substantial absorption of UV and IR by atmospheric gases, the solar irradiation at the earth's surface has a larger fraction in the visible region than that incident on the outer atmosphere. In the example in Figure 4.2, about 2% of the photons at the earth's surface are in the ultraviolet, 45% in the visible, and 53% in the infrared.

The radiation environment below the surface of a body of water is quite different from that on land. For instance, absorption by water causes most of the IR in sunlight to be removed after penetrating less than 1 m in lakes or oceans. This, coupled with greater scattering at shorter wavelengths, tends to cause a greater fraction of the photons to be in the visible region at greater depths. But water attenuates the visible region too, so even at the wavelengths for greatest penetration (slightly less than 500 nm) and in the clearest oceans, only about 1% of the solar photon flux density incident on the water surface penetrates to 200 m (see Jerlov).

We can readily appreciate that the wavelength distribution of photons reaching the earth's surface profoundly influences life. For example, the appreciable absorption of UV by ozone reduces the potential hazard of mutagenic effects caused by this short wavelength irradiation. On the other hand, prior to the advent of much ozone in the upper atmosphere, UV irradiation from the sun would have been a potent factor affecting genetic processes. Even now, exposure to the UV in sunlight can inhibit photo-synthesis and decrease leaf expansion (see Caldwell for a discussion of plant responses to UV radiation). As another example, the peak near 680 nm for photons reaching the earth's surface (Fig. 4.2) coincides with the red absorption band of chlorophyll. Vision also utilizes the wavelength region where most of the sunlight reaches the earth. It seems reasonable to suppose that there was selective pressure for the evolution of photochemical systems able to use the most abundant wavelengths.

Planck's and Wien's Formulae

The shape of the curve depicting the wavelength distribution of photons incident upon the earth's atmosphere can be closely predicted using *Planck's radiation distribution formula*. This expression indicates that the relative *photon* flux density per unit wavelength interval is proportional to $\lambda^{-4}/$

($e^{hc/\lambda kT}$ − 1), where T is the temperature of the radiation source. Such a formula applies exactly to a perfectly efficient emitter, a so-called *blackbody*. A blackbody is a convenient idealization describing an object that absorbs *all* wavelengths—it is uniformly "black" at all wavelengths—and emits in accordance with Planck's radiation distribution formula. This formula is a good approximation for describing radiation from the sun—T is the surface temperature (about 5 800 K)—and so it was used to obtain the upper curve in Figure 4.2. For the basic photometric source, the candela, Planck's formula is appropriate with T equal to 2 042 K (see p. 193). Also, the radiation from a tungsten lamp of a few hundred watts can be fairly well described by Planck's radiation distribution formula using a T of 2 900 K; i.e., the curve for the relative photon flux density from a tungsten lamp has the same shape as the solid line in Figure 4.2, although it is shifted toward longer wavelengths, since the temperature of a tungsten filament is less than that of the sun's surface (see Bickford and Dunn, Kaufman and Christensen, or Seliger and McElroy). Planck's radiation distribution formula indicates that any object with a temperature greater than 0 K will emit electromagnetic radiation.

If we know the temperature of a blackbody, we can predict at what wavelength the radiation from it will be maximal. To derive such an expression, we differentiate Planck's radiation distribution formula with respect to wavelength and set the derivative equal to zero (in App. V we discuss the minima and maxima of curves). The relation thus obtained is known as *Wien's displacement law*, which can be stated as follows: the wavelength position for maximum *photon* flux density, λ_{max}, times the temperature of the source, T, equals 3.6×10^6 nm K. Since the surface of the sun is about 5 800 K, Wien's displacement law predicts that

$$\lambda_{max} = \frac{(3.6 \times 10^6 \text{ nm K})}{(5\,800 \text{ K})} = 620 \text{ nm}$$

as the upper line in Figure 4.2 indicates. For a tungsten lamp operating at a temperature of 2 900 K (half the temperature of the sun's surface), the position for maximum photon flux density shifts to 1 240 nm (a doubling of λ_{max} compared with the sun). This is consistent with the fact that tungsten lamps emit primarily infrared radiation. From Wien's displacement law, the λ_{max} for a maximum photon flux density from a body at 298 K would occur at 1.2×10^4 nm, which is 12 μm, i.e., far into the IR. Plants emit such infrared radiation, and this is a crucial aspect of their overall energy balance, as we will consider in Chapter 7.

When considering the number of photons available for absorption by pigment molecules, as is relevant for discussing photosynthesis, we generally find it most convenient to use the spectral distribution of *photons* per unit wavelength interval (see Fig. 4.2). On the other hand, for applications such as describing the energy gain by leaves exposed to sunlight, we are usually more interested in the spectral distribution of *energy* per unit wavelength interval. To recast Figure 4.2 on an energy basis, we need to divide each point on the curves by its wavelength—the ordinate then becomes "Relative energy flux density per unit wavelength interval." Planck's radiation distribution formula indicates that the relative *energy* flux density per unit wavelength interval is proportional to $\lambda^{-5}(e^{hc/\lambda kT} - 1)$. Wien's displacement law becomes: λ_{max} for maximum *energy* output $\times\ T = 2.9 \times 10^6$ nm K (see Ditchburn, Gates, or Resnick and Halliday for further details).

ABSORPTION OF LIGHT BY MOLECULES

In our discussion of light utilization we will first consider the absorption process itself. Only light that is actually absorbed can be effective in producing a chemical change, a principle embodied in the Grotthus–Draper law of photochemistry. This is true whether radiant energy is to be converted to some other form and then stored or used as a trigger. Another important principle of photochemistry is the Stark–Einstein law, which specifies that each absorbed photon activates but one molecule. Einstein further postulated that all the energy of the light quantum is transferred to a single electron during the absorption event, resulting in the movement of this electron to a higher energy state. To help understand light absorption, we should first consider some of the properties of electrons. We will discuss the fate of the excited electron in the next section. (See Calvert and Pitts, Cowan and Drisko, Depuy and Chapman, or Turro for a more thorough treatment of the topics in this section.)

Role of Electrons in Absorption Event

From a classical viewpoint, an electron is a charged particle that moves in some orbit around an atomic nucleus. Its energy depends both on the location of the orbit in space and on how fast the electron moves in its orbit. The increase in energy of an electron upon absorbing a photon could be used either to transfer that electron into an orbit which lies at a higher energy than the original orbit or to cause the electron to move more rapidly about the nucleus than it did before the light arrived. The locations of various possible electronic orbits and the speeds for electrons moving in them are both limited to certain discrete, or "allowed," values, a phenomenon that has been interpreted by quantum mechanics. This means that the energy of an electron in an atom or molecule can change only by certain specific amounts. Light of appropriate wavelength will have the proper energy to cause the electron to move from one possible energetic state to another. For light absorption to occur, therefore, the energy of a photon as given by Equation 4.2 must equal the difference in energy between some allowed excited state of the atom or molecule and the initial state, the latter usually being the ground (lowest energy) state.

During light absorption an interaction occurs between the electromagnetic field of the light and some electron. Because electrons are charged particles, they experience a force when they are in an electric field. The oscillating electric field of light (Fig. 4.1) thus represents periodic driving force acting on electrons. This electric field—a vector having a certain direction in space, such as along the vertical axis in Figure 4.1—causes or induces the electrons to move. If the frequency of the electromagnetic radiation is such as to cause a large sympathetic oscillation or beating of some electron, that electron is said to be in *resonance* with the light wave. Such a resonating electron corresponds to an *electric dipole* (local separation of positive and negative charge in the molecule), as the electron is forced to move first in a certain direction and then in the opposite one in response to the oscillating electric field of light. The displacement of the electron back and forth requires energy—in fact, it may take the entire energy of the photon, in which case the quantum is captured or absorbed. The direction and magnitude of the induced electric dipole will depend on the resisting, or restoring, forces on the electron provided by the rest of the molecule. These restoring forces depend on the other electrons and the atomic nuclei in the molecule; consequently, they are not the same in different types of molecules. Therefore, the actual electric dipoles that can be induced in a particular molecule are characteristic of that molecule, which helps explain why each molecular species has its own unique absorption spectrum.

The probability that light will be absorbed depends on both its wavelength and the relative orientation of its electromagnetic field with respect to the possible induced oscillations of the electrons in the molecules. Absorption of a photon without ejection of an electron from the absorbing species can take place only if the following two conditions exist: (1) the photon has the proper energy to get to a discrete excited state of the molecule, i.e., has a specific wavelength (see Eq. 4.2); and (2) the electric field vector associated with the light (Fig. 4.1) has a component parallel to the direction of some potential electric dipole in the molecule so that an electron can be induced to oscillate. In other words, the electric field of light must be in the correct direction so that it is able to exert a force on some electron. It must exert this force in the direction of a potential electric dipole so that the electron can be induced to move and thus be able to accept the energy of the photon. The probability for absorption is actually proportional to the square of the cosine of the angle between the electric field vector of light and the direction of the induced electric dipole in the molecule. Within these limits set by the wavelength and the orientation, light energy can be captured by the molecule, placing it in an excited state.

Electron Spin and State Multiplicity

Light absorption is affected by the arrangement of electrons in an atom or molecule, which depends among other things on a property of the individual electrons known as their *spin*. We can consider that each electron is a charged particle spinning about an axis in much the same way as the earth spins about its axis. Such rotation has an angular momentum, or spin, associated with it. The magnitude of the spin of all electrons is the same; but since spin is a vector quantity, it can have different directions in space. For an electron, only two orientations are actually found—the spin of the electron is aligned either parallel or antiparallel to the local magnetic field, i.e., either in the same direction or in the opposite direction as the magnetic field. Even in the absence of an externally applied magnetic field, such as that of the earth or some electromagnet, there is a local internal magnetic field provided by both the moving charges in the nucleus and the motion of the electrons. So a magnetic field always exists with which the electron spin can be aligned.

Angular momentum and hence spin have units of energy $\times$ time. It proves convenient to express the spin of electrons in units of $h/(2\pi)$, where h is Planck's constant, which itself has units of energy $\times$ time, e.g., J s. In units of $h/(2\pi)$, the projection along the magnetic field of the spin for a single electron is either $+\frac{1}{2}$ (e.g., when it is parallel to the local magnetic field), or $-\frac{1}{2}$ (when the spin is in the opposite direction or antiparallel). The net spin of an atom or molecule is the vector sum of the spins of all the electrons, each individual electron having a spin of either $+\frac{1}{2}$ or $-\frac{1}{2}$. The magnitude of this net spin is given the symbol S, an extremely important quantity in spectroscopy.

When discussing the spectroscopic properties of various molecules, we will find it convenient to introduce a new term known as the *spin multiplicity*.

The spin multiplicity of an electronic state is defined as $2S + 1$, where S is the magnitude of the net spin for the whole atom or molecule. For example, if S equals zero—which means that the spin projections of all the electrons taken along the magnetic field cancel each other—then $2S + 1$ is one, and the state is called a *singlet*. On the other hand, if S equals one, the state is a *triplet* ($2S + 1$ then equals 3). Singlets and triplets are the two most important spin multiplicities encountered in biological systems. When referring to an absorbing species, the spin multiplicity is generally indicated by S for singlet and T for triplet, as we will do here. When S equals $\frac{1}{2}$, as could occur if there were an odd number of electrons in a molecule, $2S + 1$ is 2; such *doublets* occur for free radicals (molecules with a single unpaired electron—such molecules consequently are generally quite reactive).

Electrons are found only in certain "allowed" regions of space; the particular locus in which some electron can move is referred to as its *orbital*. In the 1920's Pauli noted that, when an electron is in a given atomic orbital, a second electron having its spin in the same direction is excluded from that orbital. This led to the enunciation of the *Pauli exclusion principle* of quantum mechanics: when two electrons are in the same orbital, their spins must be in opposite directions. When a molecule has all of its electrons paired in orbitals with their spins in opposite directions, the total spin of the molecule is zero ($S = 0$), and the molecule is in a singlet state ($2S + 1 = 1$; see Fig. 4.3a).

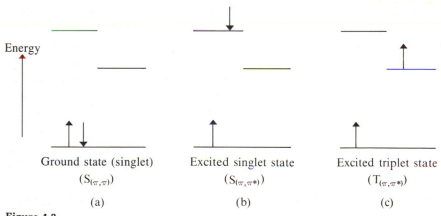

Ground state (singlet)	Excited singlet state	Excited triplet state
$(S_{(\pi,\pi)})$	$(S_{(\pi,\pi*)})$	$(T_{(\pi,\pi*)})$
(a)	(b)	(c)

Figure 4.3
Effect of light on a pair of electrons in a molecular orbital. The arrows indicate the directions of the electron spins with respect to the local magnetic field. (a) In the ground state, the two electrons in a filled orbital have their spins in opposite directions. (b) The absorption of a photon can cause the molecule to go to an excited *singlet* state where the spins of the electrons are still in opposite directions. (c) In an excited *triplet* state, the spins of the two electrons are in the same direction.

The ground, or unexcited, state of most molecules is such a singlet; i.e., all the electrons are in pairs in the lower energy orbitals. When some electron is excited to an unoccupied orbital, two spin possibilities exist. The spins of the two electrons (which are now in different orbitals) may be in opposite directions, as they were when paired in the ground state (Fig. 4.3b); this electronic configuration is still a singlet state. Or the two electrons may have their spins in the same direction—a triplet state (Fig. 4.3c). (Since the electrons are in different orbitals, their spins can be in the same direction without violating the Pauli principle.) An important rule—first enunciated by Hund, based on empirical observations, and later explained using quantum mechanics—is that the level with the greatest spin multiplicity has the lowest energy. Thus, an excited triplet state is lower in energy than the corresponding excited singlet state, as is illustrated in Figure 4.3c.

Molecular Orbitals

For a discussion of the light absorption event involving the interaction of an electromagnetic wave with some electron, it is easiest to visualize the electron as a small point located at some specifiic position in the atom or molecule. For a classical description of the energy of an electron, we can imagine the electron moving in some fixed trajectory or orbit about the nucleus. However, to describe the role of electrons in binding together atoms to form a molecule, it is most convenient to imagine the electron as spatially distributed like a cloud of negative charge surrounding the nuclei of adjacent atoms in the molecule. In this last description, involving probability considerations introduced by quantum mechanics, we say that the electrons are located in *molecular orbitals*. Such considerations lead us into a different way of looking at things. Our common experience yields relations such as Newton's laws of motion, which describe events on a scale much larger than atomic or molecular. In contrast, the main application of quantum mechanics is to molecular, atomic, and subatomic dimensions. At this scale our intuition often fails us. Moreover, many things that otherwise seem to be absolute, such as the position of an object, turn out to be describable only on a relative basis or as a *probability*. We cannot say that an electron is located at such and such a place. Instead, we must be satisfied with knowing only the probability of finding an electron in some region about a nucleus.

A logical way to begin our discussion of molecular orbitals is to consider the probabilities of finding electrons in given regions about a single atomic nucleus. The simplest atom is that of hydrogen, which has a single electron.

If we were to determine the probability of finding that electron in the various regions of space about its nucleus, we would find that it spends most of its time fairly close to the nucleus (within about 0.1 nm), and that the probability distribution is *spherically symmetrical* in space. In other words, the chance of finding the electron is the same in all directions about the nucleus. In atomic theory, this spherically symmetrical distribution of the electron about an H nucleus is called an *s* orbital, and the electron is referred to as an *s* electron.

The next simplest atom is that of helium; He has two *s* electrons moving in the *s* orbital about its nucleus. By the Pauli exclusion principle these two electrons must have their spins in opposite directions. What happens for lithium, which contains three electrons? Two of its electrons are in the same type of orbital as that of He. This orbital is known as the *K* shell, and represents the spherically symmetrical orbital closest to the nucleus. The third electron is excluded from the *K* shell but occurs in another orbital with spherical symmetry whose probability description indicates that on the average its electrons are farther away from the nucleus than is the case for the *K* shell. Electrons in this new orbital are still *s* electrons, but they are in the *L* shell. Beryllium, which has an atomic number of four, can have two *s* electrons in its *K* shell and two more in its *L* shell. What happens when we have 5 electrons moving about a nucleus, as for an uncharged boron atom? The fifth electron is in an orbital that is not spherically symmetrical about the nucleus. Instead, the probability distribution has the shape of a dumbbell, although it is still centered about the nucleus (similar in *appearance* to the distributions depicted in Fig. 4.4b). This new orbital is referred to as a *p* orbital and electrons in it are called *p* electrons. The probability of finding a *p* electron in various regions of space is greatest along some axis through the nucleus and vanishingly small at the nucleus itself.

How do all these various atomic orbitals relate to the spatial distribution of electrons in molecules? A molecule contains more than one atom (except for "molecules" like helium or neon), and certain of the electrons can move between the atoms—this interatomic motion is of course crucial for holding the molecule together. Fortunately, the spatial localization of electrons in molecules can be described using suitable linear combinations of the spatial distributions represented by various atomic orbitals centered about the nuclei involved. In fact, molecular orbital theory is concerned with giving the correct quantum-mechanical or wave-mechanical description of the probability of finding electrons in various regions of space in *molecules* by using the already established descriptions for the probability of finding electrons in *atomic* orbitals.

Some of the electrons in molecules are localized about a single nucleus, while others are *delocalized*, or shared between nuclei. For the delocalized electrons, the combination of the various atomic orbitals used to describe the spatial positions of the electrons in three dimensions is consistent with a sharing of the electrons between adjacent nuclei. In other words, the molecular orbitals of the delocalized electrons spatially overlap more than one nucleus. This sharing of electrons is responsible for the chemical bonds that prevent the molecule from separating into its constituent atoms; i.e., the negative electrons moving between the positive nuclei hold the molecule together by attracting the nuclei of different atoms. Moreover, these de-localized electrons are the ones usually involved in light absorption by molecules.

The lowest energy molecular orbital is a σ orbital, which is analogous in symmetry properties to an s atomic orbital.* Nonbonding, or lone-pair, electrons contributed by atoms like oxygen or nitrogen occur in n orbitals and retain their atomic character in the molecule. These n electrons are essentially physically separate from the other electrons in the molecule and do not take part in the bonding between nuclei. We shall devote most of our attention to electrons in π molecular orbitals, which are the molecular equivalent of p electrons in atoms. These π electrons are delocalized in a bond joining two or more atoms. They are of prime importance in light absorption. In fact, most photochemical reactions and spectroscopic properties of biological importance result from the absorption of photons by the π electrons.

The excitation of a π electron by light absorption can lead to an excited state of the molecule in which the electron moves into a π^* orbital, the asterisk referring to an excited, or high-energy, molecular orbital. The probability distributions for electrons in both π and π^* orbitals are illustrated in Figure 4.4 (the circumscribed regions indicate where the electrons are most likely to be found). Figure 4.4a shows a π orbital delocalized between two nuclei; the same clouds of negative charge electrostatically attract both nuclei. Such sharing of electrons in the π orbital helps join the atoms together, and so we refer to this type of molecular orbital as *bonding*. As shown in Figure 4.4b, electrons in a π^* orbital do not help join atoms together; rather, they tend to decrease bonding between atoms in the molecule, since the clouds of negative charge in adjacent atoms are located in a position to repel each other. A π^* orbital is therefore referred to as *antibonding*. The decrease in bonding in going from a π to a π^* orbital results in a less stable

---

* The spatial distribution of the electron cloud for a σ orbital is actually cylindrically symmetrical about the internuclear axis for the pair of atoms involved in the σ bond. It can be constructed by linear combinations of s atomic orbitals.

(a)   (b)

Figure 4.4

Typical π and π* orbitals, indicating the spatial distribution about the nuclei (⊕) where the greatest probability of finding the electrons occurs: (a) π orbital (bonding), and (b) π* orbital (antibonding).

(higher energy) electronic state for the molecule, so that a π* orbital is at a higher energy than a π orbital.

In photochemistry, the energy required to move an electron from the attractive (bonding) π orbital to the antibonding π* orbital is obtained by the absorption of a photon of the appropriate wavelength. For molecules such as chlorophylls and carotenoids, which we will consider later, the π* orbitals are often only a few hundred kJ mol^{-1} higher in energy than the corresponding π orbitals. In such molecules, the absorption of visible light corresponding to photons with energies from 160 to 300 kJ mol^{-1} (Table 4.1) can lead to the excitation of π electrons into the π* orbitals.

Photoisomerization

Light energy can cause molecular changes known as *photoisomerizations*. The three main types of photoisomerization are (1) *cis-trans* isomerization about a double bond, (2) a double bond shift (possibly occurring in the interconversion of the two forms of phytochrome that we will consider at the end of the chapter), and (3) a molecular rearrangement involving changes in carbon–carbon bonds, e.g., ring cleavage or formation. *Cis-trans* photoisomerization—which illustrates the consequences of the different spatial distribution of electrons in π and π* orbitals—makes possible the generally restricted rotation about a double bond.* A double bond has two π electrons; light absorption leads to the excitation of one of them to a π* orbital. The

---

* In the *trans* configuration, the two large groups are on opposite sides of the double

bond between them $\left(\begin{array}{c} R \\ \diagdown \\ \diagup \end{array} C{=}C \begin{array}{c} \diagup \\ \diagdown \\ R' \end{array} \right)$, and in the *cis* form they are on the same side.

attraction between the two carbon atoms caused by the remaining π electron is mostly canceled by the antibonding, or repulsive, contribution from this π^* electron (see Fig. 4.4). Hence, the molecular orbitals of the original two π electrons which prevented rotation about the double bond have been replaced by an electronic configuration permitting relatively easy rotation about the carbon-carbon axis. Let us consider the absorption of light by a molecule in the *cis* form. When the excited π^* electron drops back to a π orbital, the two large groups can be on the same side (the original *cis* isomer) or on opposite sides (the *trans* isomer) of the double bond. Since the two isomers can have markedly different properties, such use of light to trigger their interconversion can have profound importance in biology. For example, light can cause the photoisomerization of a *cis* isomer of retinal,[*] yielding a *trans* isomer. This is the basic photochemical event triggering vision (see Cowan and Drisko).

Light Absorption by Chlorophyll

We will use chlorophyll to help illustrate some of the terms introduced above describing the absorption of light by molecules. (Chlorophyll will be considered in more detail in Ch. 5.) The principal energy levels and electronic transitions of chlorophyll are presented in Figure 4.5. Chlorophyll is a singlet in the ground state, as are nearly all pigments of importance in biology. On absorption of a photon, some π electron is excited to a π^* orbital. When this excited state is a singlet, it is represented by $S_{(\pi,\pi^*)}$ (Figs. 4.3 and 4.5). The first symbol in the subscript is the type of electron (here a π electron) that has been excited to the antibonding orbital indicated by the second symbol (here π^*). We will represent the ground state by $S_{(\pi,\pi)}$, indicating that no electrons are then in excited, or antibonding, orbitals. If the orientation of the spin of the excited π electron becomes reversed during excitation, it would be in the same direction as the spin of the electron which remained in the π orbital (see. Fig. 4.3c). (Each filled orbital contains two electrons whose spins are in opposite directions.) In this case, the net spin of the molecule in the excited state would be 1, i.e., a triplet. The excited triplet state of chlorophyll is represented by $T_{(\pi,\pi^*)}$ in Figure 4.5.

Chlorophyll has two principal excited singlet states which differ considerably in energy. One of these states, designated $S^a_{(\pi,\pi^*)}$ in Figure 4.5, can be reached by absorption of red light, e.g., a wavelength of 680 nm. The other

---

[*] Retinal is a carotenoid attached to the lipoprotein opsin; the complex is referred to as rhodopsin.

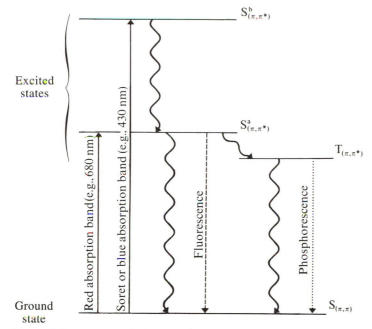

Figure 4.5

Energy level diagram indicating the principal electronic states and
some of the transitions of chlorophyll. Straight vertical lines represent
the absorption of light; wavy lines indicate radiationless transitions;
broken lines indicate those de-excitations accompanied by radiation.

state, $S_{(\pi,\pi*)}^{b}$, involves a π^* orbital that lies higher in energy and is reached
by absorption of blue light, e.g., 430 nm. The electronic transitions caused
by the absorption of photons are indicated by solid vertical arrows in Figure
4.5, while the vertical distances correspond approximately to the differences
in energy involved. Excitation of a singlet ground state to an excited triplet
state is usually only about 10^{-5} times as probable as going to an excited
singlet state, so that the transition from $S_{(\pi,\pi)}$ to $T_{(\pi,\pi*)}$ has not been indicated
for chlorophyll in Figure 4.5. To go from $S_{(\pi,\pi)}$ to $T_{(\pi,\pi*)}$, the energy of an
electron must be substantially increased *and* the orientation of its spin must
be simultaneously reversed. The coincidence of these two events is very
improbable, so very few chlorophyll molecules are directly excited to $T_{(\pi,\pi*)}$
from the ground state by the absorption of light.

We will now briefly compare the two most important excitations in
photobiology—the transitions of n and π electrons to π^* orbitals. The n
electrons have very little spatial overlap with other electrons, and they tend
to be higher in energy than the π electrons, which are reduced in energy

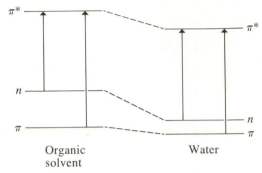

Figure 4.6
Influence of solvent on the energies of n, π, and π^* orbitals.
The indicated n-to-π^* transition takes more energy in
water than in an organic solvent, and the π-to-π^*
transition takes less.

(stabilized) by being delocalized over a number of nuclei. Therefore, the excitation of an n electron to a π^* orbital generally takes less energy than the transition of a π electron to the same antibonding orbital (see Fig. 4.6). Hence, an $S_{(n,\pi^*)}$ state usually occurs at a lower energy—a longer wavelength is required for its excitation—than does the analogous $S_{(\pi,\pi^*)}$. Another difference is that π and π^* orbitals may overlap spatially (see Fig. 4.4), while n and π^* orbitals generally do not. Consequently, the n to π^* transition is not as favored, or probable, as the π to π^* one. Thus, transitions to $S_{(\pi,\pi^*)}$ states tend to dominate the absorption properties of a molecular species (as evidenced, for example, by an absorption spectrum), compared with excitations yielding $S_{(n,\pi^*)}$ states.

As our final topic in this section, we will consider the effect of molecular environment on n, π, and π^* orbitals (Fig. 4.6). The n electrons generally interact strongly with water, e.g., by participating in hydrogen bonding, and therefore the energy level of n electrons is considerably lower in an aqueous environment than in an organic solvent. The energy of π^* electrons is also lowered by water, but to a lesser extent than for n electrons. The π orbitals are the least affected by the solvent. Thus, a transition from an n to a π^* orbital takes more energy in water than in an organic solvent (Fig. 4.6). On the other hand, the π to π^* transition takes less energy in an aqueous environment than in an organic one. As an example, let us consider the transition of chlorophyll to an excited singlet state in different solvents. The excitation to either excited singlet state takes less energy in an aqueous environment than it does in an organic solvent; i.e., when chlorophyll is in water, the required photons have a longer wavelength than when the chlorophyll is dissolved in an organic solvent like acetone. This lower energy requirement in water is consistent with our statement that the transition for chlorophyll corresponds to the promotion of a π electron to a π^* orbital.

DE-EXCITATION

The primary processes of photochemistry involve the light absorption event, which we have already discussed, together with the subsequent de-excitation reactions. We can portray such transitions on an energy level diagram, as in Figure 4.5 for chlorophyll. In this section we will discuss the various de-excitation processes, including a consideration of their rate constants and lifetimes.

One characteristic of the various excitation and de-excitation processes is the time needed for the transitions. A useful estimate of the time it takes for the absorption of a photon is the time required for one cycle of the light wave to pass by an electron. This time is the distance per cycle of the wave divided by the speed with which light travels, or λ/v, which equals $1/v$ by Equation 4.1. Hence, the time required for the absorption of a photon is approximately equal to the reciprocal of the frequency of the light, $1/v$, which is the time necessary for one complete oscillation of the electromagnetic field. During one oscillation the electric vector of light can induce an electron to move first in one direction and then in the opposite one, thereby setting up a beating or resonating of the electron. To be specific, let us consider blue light with a wavelength of 460 nm in a vacuum. From its frequency (given in Table 4.1), we can calculate that the time for one cycle is $1/(6.52 \times 10^{14}\ \text{s}^{-1})$, or 1.5×10^{-15} s. Light absorption is indeed an extremely rapid event!

Times for de-excitation reactions are usually expressed in *lifetimes*. A lifetime is the time required for the number of molecules in a given state to decrease to $1/e$, or 37%, of the initial number. Lifetimes are extremely convenient for describing first-order processes because the initial species in such processes decay (disappear) exponentially with time (see Eq. 4.8 and also App. V, pp. 563–565, for examples of first-order processes). A *half-time*, the time necessary for the number of species in a given state to decrease by 50%, is occasionally used to describe de-excitation processes; for an exponential decay, one half-time is ln 2, or 0.693, times the size of a lifetime.

Fluorescence, Phosphorescence, and Radiationless Transitions

As we would expect, the excess energy of the excited state can be dissipated by a number of competing reactions. One of the ways in which de-excitation can occur is by the emission of light known as *fluorescence* (indicated by a dashed straight line in Fig. 4.5 for the principal transitions of chlorophyll). Fluorescence describes the electromagnetic radiation emitted when a molecule goes from an excited singlet state to a singlet ground state. Fluorescence lifetimes for most organic molecules range from 10^{-9} to 10^{-6} s (see Turro).* The properties of chlorophyll fluorescence are extremely important for an understanding of the primary events of photosynthesis, as we shall see in Chapter 5.

De-excitation of an excited state often occurs without the emission of any radiation. Such transitions are termed nonradiative, or *radiationless* (indicated by wavy lines in Fig. 4.5). For a radiationless transition to a lower energy excited state or back to the ground state, the energy of the absorbed photon is eventually converted to heat, which is passed on by collisions with the surrounding molecules. Radiationless transitions can be extremely rapid from some excited singlet state to a lower energy excited singlet state in the same molecule. For example, the radiationless transition from $S^b_{(\pi,\pi*)}$ to $S^a_{(\pi,\pi*)}$ indicated for chlorophyll in Figure 4.5 takes about 10^{-12} s (see Clayton 1965). This transition is so rapid that hardly any fluorescence can be emitted from $S^b_{(\pi,\pi*)}$, so no fluorescence emission from the upper excited singlet state of chlorophyll is indicated in Figure 4.5. The excited singlet state lying at the lower energy, $S^a_{(\pi,\pi*)}$, can decay to the ground state by a radiationless transition. As another alternative, $S^a_{(\pi,\pi*)}$ can go to $T_{(\pi,\pi*)}$, also by a radiationless transition (Fig. 4.5). In fact, excited triplet states in molecules are mainly formed by radiationless transitions from excited singlet states at higher energies. The de-excitation of $T_{(\pi,\pi*)}$ to $S_{(\pi,\pi)}$ can be radiationless, or it can be by radiation known as *phosphorescence* (dotted straight line in Fig. 4.5), which we will consider next.

Phosphorescence is the electromagnetic radiation that can accompany the transition of a molecule from an excited triplet state to a ground state singlet. Since we go from a triplet to a singlet state, the net spin of the molecule must change during the emission of this radiation. The lifetimes for phosphorescence are usually from 10^{-3} to 10 s (see Calvert and Pitts), which are long compared with those for fluorescence (10^{-9} to 10^{-6} s). These relatively long times for de-excitation occur because there is both the transition from one electronic state to another and the simultaneous reversal of the electron spin—the coincidence of these two events is rather improbable. In fact, the above-mentioned low probability of forming $T_{(\pi,\pi*)}$ from $S_{(\pi,\pi)}$ by

---

* The lifetime of an excited state indicates the time course for de-excitation, not the time required for the de-excitation event per se; e.g., the time required for the emission of a photon is essentially the same as for its capture.

light absorption has the same physical basis as the long lifetime for de-excitation of the excited triplet state back to the ground state by emitting phosphorescence.

Another way for an excited molecule to emit radiation is by so-called "delayed fluorescence" (see Clayton 1980). Specifically, a molecule in the relatively long-lived excited triplet state $T_{(\pi,\pi^*)}$ can sometimes be supplied enough thermal energy by collisions to put it into the higher energy excited singlet state $S_{(\pi,\pi^*)}$. Subsequent radiation, as the molecule goes from this excited singlet state to the ground state $S_{(\pi,\pi)}$, has the characteristics of fluorescence. However, it is considerably *delayed* after light absorption compared with normal fluorescence, since the excitation spent some time in $T_{(\pi,\pi^*)}$.

Competing Pathways for De-excitation

Each excited state not only has a definite energy, but a specific lifetime as well, the length of which depends on the particular processes competing for the de-excitation of that state. In addition to fluorescence, phosphorescence, and the radiationless transitions we have just introduced, the excitation energy can also be transferred to another molecule, putting this second molecule into an excited state while the originally excited molecule returns to its ground state. As an example of yet another type of de-excitation process, an excited (energetic) electron may actually leave the molecule that absorbed the photon; this occurs for certain excited chlorophyll molecules we will discuss later. The excited state that appears to be of pivotal importance in photosynthesis is the lower excited singlet state of chlorophyll, indicated by $S_{(\pi,\pi^*)}^a$ in Figure 4.5. We will use this state to illustrate some of the possible competing reactions leading to the de-excitation of an excited singlet state, $S_{(\pi,\pi^*)}$ (see Cowan and Drisko, or Seliger and McElroy).

The absorbed quantum can be reradiated as electromagnetic energy, $h\nu$, causing the excited molecule $S_{(\pi,\pi^*)}$ to drop back to its ground state $S_{(\pi,\pi)}$:

$$S_{(\pi,\pi^*)} \xrightarrow{k_1} S_{(\pi,\pi)} + h\nu \tag{4.3}$$

Such fluorescence exponentially decays away with time after the exciting light is removed, indicating a first-order process having a rate constant k_1 in Equation 4.3 (first-order rate constants have units of s^{-1}).* The fluores-

---

* For a first-order process, the rate of disappearance of the excited state, $dS_{(\pi,\pi^*)}/dt$, is linearly proportional to the amount of $S_{(\pi,\pi^*)}$ present at any time. First-order processes are further discussed in Appendix V.

cence lifetime of $S_{(\pi,\pi^*)}$ is $1/k_1$, typically about 10^{-8} s. (We will discuss the relationship between lifetimes and rate constants shortly.) This fluorescence lifetime would be the actual lifetime of $S_{(\pi,\pi^*)}$, if no other competing de-excitation processes occurred. When the energy of the absorbed quantum is dissipated as fluorescence, no photochemical work can be done. Therefore, the fluorescence lifetime is an upper time limit within which any biologically useful reactions can be driven by the lowest excited singlet state of a molecule; i.e., if the reactions take longer than the fluorescence lifetime, most of the absorbed energy will already have been dissipated.

The next de-excitation processes we will consider are the radiationless transitions by which $S_{(\pi,\pi^*)}$ eventually dissipates its excess electronic energy as heat. Actually, two different states can be reached by radiationless transitions from $S_{(\pi,\pi^*)}$:

$$S_{(\pi,\pi^*)} \xrightarrow{k_2} S_{(\pi,\pi)} + \text{heat} \tag{4.4}$$

$$S_{(\pi,\pi^*)} \xrightarrow{k_3} T_{(\pi,\pi^*)} + \text{heat} \tag{4.5}$$

Radiationless transitions such as those indicated in Equations 4.4 and 4.5 involve de-excitations in which the excess energy is often first passed on to other parts of the same molecule. This causes the excitation of certain vibrational modes for other pairs of atoms within the molecule—vibrations like those we will discuss in conjunction with the Franck–Condon principle (see Fig. 4.7). This energy, which has become distributed over the molecule, is subsequently dissipated by collisions with other molecules in the randomizing interchanges that are the basis of temperature. When an excited molecule directly returns to its ground state by a radiationless transition (Eq. 4.4), all of the radiant energy of the absorbed light is eventually converted into the thermal energy of motion of the surrounding molecules. In Equation 4.5 only some of the excess electronic energy appears as the quantity designated "heat," which in that case represents the difference in energy between $S_{(\pi,\pi^*)}$ and $T_{(\pi,\pi^*)}$.

As for fluorescence, radiationless transitions also obey first-order kinetics. Although the dissipation of excitation energy as heat in a transition to the ground state (Eq. 4.4) is photochemically wasteful in that no biological work is performed, the transition to $T_{(\pi,\pi^*)}$ indicated by Equation 4.5 can be quite useful. The lowest excited triplet state usually lasts 10^4 to 10^8 times longer than does $S_{(\pi,\pi^*)}$, which allows time for many more intermolecular collisions. Since each collision increases the opportunity for a given reaction to occur, $T_{(\pi,\pi^*)}$ can be an important excited state in photobiology.

The absorption of light can lead to a photochemical reaction initiated by a molecule other than the one that actually absorbed the photon. This phenomenon suggests that electronic excitation can be transferred between molecules, resulting in the excitation of one and the de-excitation of the other. For instance, the excitation energy of $S_{(\pi,\pi*)}$ might be transferred to a second molecule, represented in the ground state by $S_{2(\pi,\pi)}$:

$$S_{(\pi,\pi*)} + S_{2(\pi,\pi)} \xrightarrow{k_4} S_{(\pi,\pi)} + S_{2(\pi,\pi*)} \tag{4.6}$$

This second molecule thereby becomes excited, indicated by $S_{2(\pi,\pi*)}$, while the molecule that absorbed the quantum becomes de-excited and is returned to its ground state. Such transfer of electronic excitation from molecule to molecule underlies the energy migration among the pigments involved in photosynthesis (see Ch. 5). We will assume that Equation 4.6 represents a first-order reaction, as it does for the excitation exchanges between chlorophyll molecules in vivo (in certain cases, Eq. 4.6 can represent a second-order reaction; i.e., $dS_{(\pi,\pi*)}/dt$ then equals $-k' S^2_{(\pi,\pi*)}$).

As another type of de-excitation process, $S_{(\pi,\pi*)}$ can take part in a *photochemical* reaction. For example, the excited $\pi*$ electron can be donated to a suitable acceptor. In the following equation this ejected electron is represented by e*:

$$S_{(\pi,\pi*)} \xrightarrow{k_5} D_{(\pi)} + e* \tag{4.7}$$

The electron removed from $S_{(\pi,\pi*)}$ is replaced by another one donated from some other compound; $D_{(\pi)}$ in Equation 4.7, which represents a doublet since one of the π orbitals contains an unpaired electron, then goes back to its original ground state $S_{(\pi,\pi)}$. (We will discuss electron energy and donation in Chs. 5 and 6.) Photochemical reactions of the form of Equation 4.7 serve as the crucial link in the conversion of radiant energy into chemical or electrical energy. Moreover, Equation 4.7 can be used to represent the photochemical reaction taking place at the special chlorophyll molecules P_{680} and P_{700}, which we will discuss later.

Lifetimes

Equations 4.3 through 4.7 represent five competing pathways for the de-excitation of the excited singlet state $S_{(\pi,\pi*)}$. They must all be considered when predicting the lifetime of $S_{(\pi,\pi*)}$. Assuming that each of these de-excitation processes is first-order and that no reaction leads to the formation of $S_{(\pi,\pi^*)}$, then the disappearance of the excited singlet state satisfies the following relation:

$$-\frac{dS_{(\pi,\pi^*)}}{dt} = (k_1 + k_2 + k_3 + k_4 + k_5)S_{(\pi,\pi^*)} \tag{4.8}$$

where the various k_j's in Equation 4.8 are the rate constants for the five individual decay reactions (Eqs. 4.3 through 4.7). Upon integration of Equation 4.8 (see pp. 564–565, App. V), we obtain the following expression for the time dependence of the number of molecules in the excited singlet state:

$$S_{(\pi,\pi^*)_t} = S_{(\pi,\pi^*)_0}e^{-(k_1+k_2+k_3+k_4+k_5)t} \tag{4.9}$$

where $S_{(\pi,\pi^*)_0}$ represents the number of molecules in the excited singlet state when the illumination ceases ($t = 0$), and $S_{(\pi,\pi^*)_t}$ is the number of excited singlet states remaining at a subsequent time t. Relations such as Equation 4.9—showing the amount of some state remaining at various times after illumination or other treatment—are extremely important for describing processes with first-order rate constants.

Since the lifetime of an excited state is the time required for the number of excited molecules to decrease to $1/e$ of the initial value, $S_{(\pi,\pi^*)_t}$ in Equation 4.9 equals $1/e$ $S_{(\pi,\pi^*)_0}$ when t equals the lifetime τ, i.e.,

$$S_{(\pi,\pi^*)_\tau} = e^{-1}S_{(\pi,\pi^*)_0} = S_{(\pi,\pi^*)_0}e^{-(k_1+k_2+k_3+k_4+k_5)\tau} \tag{4.10a}$$

which leads to the following relationship:

$$(k_1 + k_2 + k_3 + k_4 + k_5)\tau = 1 \tag{4.10b}$$

Equation 4.10 indicates that the greater the rate constant for *any* particular de-excitation process, the shorter will be the lifetime of the excited state.

Equation 4.10 can be generalized to include all competing reactions. This gives the following expression for the lifetime:

$$\frac{1}{\tau} = k = \sum_j k_j = \sum_j \frac{1}{\tau_j} \tag{4.11a}$$

where k_j is the first-order rate constant for the jth de-excitation process and τ_j is its lifetime ($\tau_j = 1/k_j$). Also, τ is the lifetime of the excited state, while k in Equation 4.11a is the overall rate constant for its decay. Using Equation 4.11a, we can re-express Equation 4.9 as follows (see App. V):

$$S_{(\pi,\pi*)_t} = S_{(\pi,\pi*)_0} e^{-kt} = S_{(\pi,\pi*)_0} e^{-t/\tau} \qquad (4.11b)$$

Equations 4.10 and 4.11 indicate that, when more than one de-excitation process is possible, τ is less than the lifetime of any individual competing reaction acting alone. In other words, the observed rate of decay of an excited state is faster than deactivation by any single competing reaction acting by itself.

If the rate constant for a particular reaction is much larger than for its competitors, the excited state will become de-excited predominantly by that process. As an example, we will consider an excited triplet state of a molecule that shows delayed fluorescence (see p. 211). Suppose that the lifetime for phosphorescence, τ_P, is 10^{-2} s, in which case k_P is 100 s^{-1}. When sufficient thermal energy is supplied, $T_{(\pi,\pi*)}$ can be raised in energy to an excited singlet state, which could then emit "delayed" fluorescence if $S_{(\pi,\pi*)}$ decayed to the ground state by emitting electromagnetic radiation. Suppose that the rate constant $(k_{T*\to S*})$ for the transition from $T_{(\pi,\pi*)}$ to $S_{(\pi,\pi*)}$ is 20 s^{-1}. By Equation 4.11a, k for these two competing pathways is 100 s^{-1} + 20 s^{-1} or 120 s^{-1}, which corresponds to a lifetime of $1/(120$ s$^{-1})$, or 0.008 s. Suppose now that another molecule, which can readily take on the excitation of $T_{(\pi,\pi*)}$ of the original species, is introduced into the solution—k_{transfer} might be 10^4 s^{-1}. Because of the relatively large rate constant, such a molecule "quenches" the phosphorescence and delayed fluorescence originating from $T_{(\pi,\pi*)}$ (i.e., its decay pathway predominates over the competing processes), and so it is generally referred to as a *quencher*. For the three pathways indicated, the overall rate constant is 100 s^{-1} + 20 s^{-1} + 10^4 s^{-1}, or essentially 10^4 s^{-1}. The de-excitation here is dominated by the quencher, since $k_{\text{transfer}} \gg k_P + k_{T*\to S*}$.

Quantum Yields

We often use a *quantum yield* (or *quantum efficiency*), Φ_i, to describe the de-excitation processes following absorption of light. Here Φ_i represents the fraction of molecules in some excited state that will decay by the ith de-excitation reaction out of all the possible competing pathways:

$$\Phi_i = \frac{\text{number of molecules using } i^{\text{th}} \text{ de-excitation reaction}}{\text{number of excited molecules}}$$

$$= \frac{k_i}{\sum_j k_j} = \frac{\tau}{\tau_i} \qquad (4.12)$$

Equation 4.12 states that the rate constant for a particular pathway determines what fraction of the molecules in a given excited state will use the ith de-excitation process. Hence, k_i determines the quantum yield for that de-excitation pathway. From Equation 4.11a and the definition of τ_j given above, we can also indicate such competition among pathways using lifetimes (see Eq. 4.12). The shorter the lifetime τ_i for a particular de-excitation pathway, the larger will be the fraction of the molecules using that pathway, and hence the higher will be its quantum yield. Finally, by Equations 4.11a and 4.12, the sum of the quantum yields for all the competing de-excitation pathways, $\sum_i \Phi_i$, equals unity.

To illustrate the use of Equation 4.12, let us consider the quantum yield for chlorophyll fluorescence, Φ_{Fl}. The fluorescence lifetime τ_{Fl} of the lower excited singlet state of chlorophyll in ether is 1.5×10^{-8} s, while the observed lifetime τ for de-excitation of this excited state in ether is 0.5×10^{-8} s (see Clayton 1965). By Equation 4.12, the expected quantum yield for fluorescence would be $(0.5 \times 10^{-8} \text{ s})/(1.5 \times 10^{-8} \text{ s})$, or $\frac{1}{3}$, which is consistent with the observed Φ_{Fl} of 0.33 for the fluorescence de-excitation of chlorophyll in ether.

ABSORPTION SPECTRA AND ACTION SPECTRA

The absorption of radiation causes a molecule to go from its ground state to some excited state in which one of the electrons enters an orbital of higher energy. We have so far considered that both the ground state and the excited states occur at *specific* energy levels, as is indicated by the horizontal lines in Figure 4.5 for the case of chlorophyll. This leads to a consideration of whether only a very limited number of wavelengths are absorbed. For instance, are 430 nm and 680 nm the only wavelengths absorbed by chlorophyll (Fig. 4.5)? We will find that each electronic energy level is actually divided (or split) into various sublevels that differ in energy. The largest splitting is due to *vibrational sublevels*. To help appreciate which wavelengths of light might be involved in photosynthesis or other photobiological processes, we must know something about the vibrational sublevels that occur for a given electronic state.

Vibrational sublevels are the result of the vibration of atoms in a molecule with respect to each other. This motion determines the sum of the kinetic and potential energies of the atoms, and consequently it affects the total energy of the molecule. We can describe this atomic oscillation by the accompanying change in the internuclear distance. Therefore, we will refer to an *energy level diagram* indicating the range of positions, or trajectories,

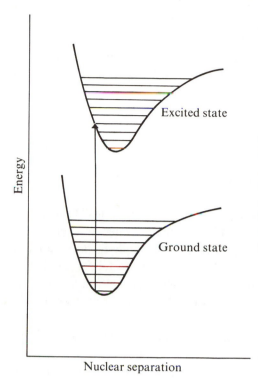

Figure 4.7
Energy curves for the ground state and an excited state showing the various vibrational sublevels. The vertical arrow represents a transition caused by the absorption of a photon and consistent with the Franck–Condon principle.

taken on by the vibrating nuclei (Fig. 4.7). The trajectories of such nuclear vibrations are quantized; i.e., only certain specific vibrations can occur. As a consequence of this quantization of the trajectories of the vibrating nuclei, only discrete energies are possible for the vibrational sublevels of a given electronic state.

At temperatures normally encountered by plants, essentially all molecules are in the ground or unexcited state. Moreover, these molecules are nearly all in the lowest vibrational sublevel of the ground state—another consequence of the Boltzmann energy distribution (see p. 140). The absorption of a photon can cause a transition of the molecule from the lowest vibrational sublevel of the ground state to one of the vibrational sublevels of the excited electronic state. The actual sublevel reached depends on the energy of the absorbed quantum. For reasons that should become clear as we proceed in the discussion of vibrational sublevels, the *probability* that a photon will be absorbed also depends on its energy. Consideration of this absorption probability as a function of wavelength leads to an *absorption spectrum* for that particular molecule. The *effect*, or consequences of light absorption on some process, when presented as a function of wavelength, leads to an *action spectrum*. At the end of this section we will discuss action spectra for a phytochrome-mediated reaction.

Vibrational Sublevels in an Energy Level Diagram

Various vibrational sublevels of both ground and excited states of a molecule are schematically indicated in Figure 4.7. In principle, such an energy level diagram can be prepared for any pigment, e.g., chlorophyll, carotenoid, or phytochrome. Energy level diagrams are extremely useful for predicting which electronic transitions are most likely to accompany the absorption of light. They can also be used to help explain why certain wavelengths predominate in the absorption process.

The abscissa in Figure 4.7 represents the distance between a pair of nuclei vibrating back and forth with respect to each other and can be obtained by imagining one nucleus to be situated at the origin of the coordinate system while the position of the other nucleus (in the same molecule) is plotted relative to this origin. The ordinate represents the total energy of the electrons plus the pair of nuclei. Figure 4.7 shows that the excited state is at a higher energy than the ground state and indicates that the two electronic states are split into many vibrational sublevels differing in energy (see Seliger and McElroy, or Turro).

Since internuclear and interelectronic repulsive forces act against the electrostatic attraction between nuclei and shared electrons, only a certain range of internuclear separations can occur for a particular bond in a molecule. As internuclear separation decreases, the nuclei repel each other more and more, and the energy of the molecule increases. Moreover, the clouds of negative charge representing electrons localized on each nucleus have a greater overlap as internuclear separation decreases, resulting in interelectronic repulsion and therefore an increase in molecular energy. These effects account for the steep rise of the energy curves in Figure 4.7 as nuclear separation becomes less (left side of the figure). On the other hand, the delocalized (bonding) electrons shared by the two nuclei resist an unlimited increase in nuclear separation, which would diminish the attractive electrostatic interaction between nuclei and electrons; such an increase in internuclear distance corresponds to a stretching of the chemical bond. Thus, the energy curve in Figure 4.7 also rises as the internuclear distance becomes greater (right side of the figure). Because of these two opposing tendencies, the range of possible nuclear separations is confined to an *energy trough*. It is within these potential energy curves — one for the ground state and another for the excited state—that the trajectories of the nuclear vibrations occur.

A horizontal line in Figure 4.7 represents the actual range of nuclear separations corresponding to a given vibrational sublevel; i.e., nuclei vibrate back and forth along the distance indicated by a horizontal line. As is evident in the figure, both ground and excited states have many vibrational sublevels differing in energy. For the upper vibrational sublevels of a given electronic state, the nuclei vibrate over longer distances (i.e., more extensive range of nuclear separations in Fig. 4.7), which also corresponds to higher vibrational energies. Since the excited state has an electron in an antibonding orbital, it usually has a greater mean internuclear separation than does the ground state. This increase in bond length is shown in Figure 4.7 by a slight displacement to greater nuclear separations, i.e., to the right, for the upper curve.

The direction of nuclear motion is reversed at the extremities of the vibrational pathways (horizontal lines in Fig. 4.7), so the velocity of the nuclei must be zero at these turning points. As the turning point at either end of the oscillation range is approached, the nuclei begin to slow down and eventually stop before reversing their direction of motion. Consequently, the nuclei spend most of their time at or near the extreme ends of their trajectory. We can thus consider that a photon is most likely to arrive at the molecule when the nuclei are at or near the extremes of their vibrational range.* Because of our probability consideration, the electronic transition resulting from the absorption of a photon (the vertical arrow in Fig. 4.7) has been initiated from one of the ends of the nuclear oscillation range for the lowest vibrational sublevel of the ground state. The arrow in Figure 4.7 begins from the lowest sublevel because nearly all ground state molecules are in the lowest vibrational sublevel at the temperatures encountered in plants, which, as we noted above, is a consequence of the Boltzmann energy distribution.

During the light absorption event, the energy of the photon is transferred to some electron in the molecule. Since the molecular orbital describing the trajectory of an electron in the excited state has a small, but finite, probability

---

* Actual quantum-mechanical calculations beyond the scope of this text lead to somewhat different conclusions, especially for the lowest vibrational sublevel. See Calvert and Pitts, or Turro.

of spatially overlapping with the nuclei, an interaction is possible between the excited electron and the nuclei over which it is delocalized. Such interactions do occur and generally take place over a rather short time period (less than 10^{-13} s). An interaction between an energetic electron and the nuclei can cause the excitation of nuclei to higher energy vibrational states. In fact, the transition represented by the arrow in Figure 4.7 corresponds to both the excitation of an electron leading to an excited state of the molecule and the subsequent excitation of the nuclei to some excited vibrational sublevel. Thus, part of the energy of the light quantum is rapidly passed on to nuclear vibrations. The length of the arrow in Figure 4.7 is proportional to the light energy (or quantum) actually added to the molecule, and therefore represents the energy distributed to the nuclei plus that remaining with the excited electron. Next, we will demonstrate which vibrational sublevel of the excited state has the highest probability of being reached by the excitation process.

The Franck–Condon Principle

Franck and Condon in 1926 enunciated a principle, based mainly on classical mechanics, to help rationalize the various bands in absorption spectra and fluorescence emission spectra. We will direct our attention to the nuclei to discuss the effect of the quantized modes of nuclear vibration, and will therefore use the energy level diagram presented in Figure 4.7. The Franck–Condon principle states that the nuclei change neither their *separation* nor their *velocity* during those transitions for which the absorption of a photon is most probable (see Seliger and McElroy, or Turro). We can use this principle to predict which vibrational sublevels of the excited state are most likely to be involved in the electronic transitions accompanying light absorption.

First, we will examine that part of the Franck–Condon principle which states that nuclei do not change their position for the most probable electronic transitions caused by light absorption. Since a vertical line in Figure 4.7 corresponds to no change in nuclear separation, the absorption of a photon has been indicated by a vertical arrow. This condition of constant internuclear distance during absorption can be satisfied more often when the nuclei are moving slowly or actually stopped at the extremes of their oscillation range. Thus, the origin of the arrow indicating an electronic transition in Figure 4.7 is at one of the turning points of the lowest vibrational sublevel of the ground state, and the tip is drawn to an extremity of one of the vibrational sublevels in the excited state (the fourth vibrational sublevel for the particular case illustrated).

The other condition embodied in the Franck–Condon principle is that a quantum has the greatest chance of being absorbed when the velocity (a vector) of the vibrating nuclei does not change. In other words, absorption is maximal when the nuclei are moving in the same direction and at the same speed in both ground and excited states. Again, this condition has the greatest statistical probability of being met when the nuclei are moving slowly or not at all, as occurs at the turning points for a nuclear oscillation. Therefore, the most probable electronic transition represented in a diagram like Figure 4.7 is a vertical line, which originates from one of the ends of the horizontal line representing the range of nuclear separations for the lowest vibrational sublevel of the ground state and goes to the end of the vibrational trajectory for some sublevel of the excited state.

Another way to view the Franck–Condon principle is to consider that the light absorption event is so rapid that the nuclei do not have a chance to move during it. The absorption of a photon requires about 10^{-15} s. In contrast, the period for one nuclear vibration back and forth along an oscillation range (like those represented by horizontal lines in Fig. 4.7) is generally somewhat more than 10^{-13} s. Thus, the nuclei cannot move any appreciable distance during the time necessary for the absorption of a photon, especially when the nuclei are moving relatively slowly near the ends of their vibrational trajectory. Also, nuclear velocity would not change appreciably in a time interval as short as 10^{-15} s (see Turro).

Since we have discussed the frequency of nuclear oscillations, we should note another aspect of this time scale that has far-reaching consequences. As the nuclei oscillate back and forth along their trajectories, they can interact with other nuclei. These encounters make possible the transfer of energy from one nucleus to another (within the same molecule or to adjacent molecules). Thus, the time for one cycle of a nuclear vibration, approximately 10^{-13} s, is an estimate of the time in which excess vibrational energy can be dissipated as heat by interactions with other nuclei. As excess energy is exchanged by such processes, the part of the molecule indicated in Figure 4.7 soon reaches the lowest vibrational sublevel of the excited state; these transitions within the same electronic state, e.g., $S_{(\pi,\pi^*)}$, are usually complete in about 10^{-12} s. Fluorescence lifetimes generally are on the order of 10^{-8} s, so an excited singlet state gets to its lowest vibrational sublevel before appreciable de-excitation can occur by fluorescence. The rapid dissipation of excess vibrational energy causes some of the energy of the absorbed photon to be released as heat. Therefore, fluorescence is generally of lower energy (longer wavelength) than the absorbed light, as we will show for chlorophyll in the next chapter.

The Franck–Condon principle predicts the most likely electronic transition caused by the absorption of light. But others do occur. These other transitions become statistically less probable, the further the conditions are from no change in nuclear position or velocity during the absorption of the photon. Since electronic transitions from the ground state to both higher and lower vibrational sublevels in the excited state occur with a lower probability than transitions to the optimal sublevel, the absorption of light is not as great at the wavelengths that excite the molecule to such vibrational sublevels. Moreover, some transitions begin from an excited vibrational sublevel of the ground state. We can calculate the fraction of the ground state molecules in the various vibrational sublevels from the Boltzmann energy distribution, $n(E) = n_{total}e^{-E/RT}$ (Eq. 3.21b). For example, $e^{-E/RT}$ is 0.22 for an E of 3.72 kJ mol^{-1}, the mean translational energy at 25°C (see p. 191).

What is the energy spacing between vibrational sublevels? For many molecules the vibrational sublevels of both the ground and the excited states are approximately 10 kJ mol^{-1} apart. Consequently, as the wavelength of incident light is increased or decreased from that for the most intense absorption, transitions involving other vibrational sublevels become important, e.g., at approximately 10 kJ mol^{-1} intervals.* This can be seen by the various peaks near the major absorption bands of chlorophyll (Fig. 5.2) or the three peaks in the absorption spectra of typical carotenoids (Fig. 5.4). In summary, the amount of light absorbed will be maximal at a certain wavelength corresponding to the most probable transition predicted by the Franck–Condon principle. Transitions from other vibrational sublevels of the ground state and to other sublevels of the excited state occur less frequently and help create an absorption spectrum that is characteristic of a particular molecule.

Absorption Bands and Absorption Coefficients

Our discussion of light absorption has so far been primarily concerned with transitions from the ground state energy level to those of excited states, which we just expanded to include the occurrence of vibrational sublevels of the electronic states. In addition, vibrational sublevels are subdivided into *rotational* states. In particular, the motion of the atomic nuclei within a

---

* An energy difference of 10 kJ mol^{-1} (about 0.1 eV molecule^{-1}, since 1 eV molecule^{-1} = 96.5 kJ mol^{-1}; App. III) between two photons corresponds to a difference in wavelength of about 25 nm near the middle of the visible region (green or yellow, Table 4.1).

molecule can be described by quantized rotational states of specific energy, leading to the subdivision of a given vibrational sublevel of an electronic state into a number of rotational *sub*sublevels. The energy increments between rotational states are generally about 1 kJ mol^{-1} (approximately 3 nm in wavelength in the visible region). The further broadening of absorption lines because of a continuum of *translational* energies of the whole molecule is generally much less, often about 0.1 kJ mol^{-1}. Moreover, interactions with the solvent or other neighboring molecules can affect the distribution of electrons in a particular molecule and, consequently, can shift the position of the various energy levels. The magnitude of the shifts caused by intermolecular interactions vary considerably and can be 5 kJ mol^{-1} or more (see Depuy and Chapman, or Seliger and McElroy).

The quanta absorbed in an electronic transition involving specific vibrational sublevels—including the range of energies due to the various rotational subsublevels and other shifts—give rise to an *absorption band*. These wavelengths represent the transition of an electron from a vibrational sublevel of the ground state to some vibrational sublevel of an excited state. A plot of the relative efficiency for light absorption as a function of wavelength is an absorption *spectrum*, which may include more than one absorption *band*. Such bands can represent transitions to or from different vibrational sublevels and to different excited states. The smoothness of the absorption bands of most pigment molecules indicates that a great range of photon energies can correspond to the transition of an electron from the ground state to some excited state (see absorption spectra in Figs. 4.11, 5.2, 5.4, and 5.6).* Because of the rather large effects that intermolecular interactions can have on electronic energy levels, we should always specify the solvent when presenting an absorption band or spectrum.

Absorption bands and spectra indicate how light absorption varies with wavelength. The absorption at a particular wavelength by a certain species is quantitatively described using an *absorption coefficient*, ε_λ (ε_λ is also referred to as an *extinction* coefficient). Because of its descriptive usefulness in plant and cell physiology, we will now derive an expression incorporating ε_λ.

Let us consider a monochromatic beam of parallel light of flux density J. Since "monochromatic" refers to light of a single wavelength, J can be

---

* To "sharpen" an absorption spectrum, it can be experimentally advantageous to reduce the translational broadening of absorption bands by substantially decreasing the temperature, e.g., by using liquid nitrogen (boiling point = $-196°C$) to cool the sample. Also, the reduction in temperature decreases the number of the absorbing molecules in excited vibrational sublevels and higher energy rotational sublevels of the ground state—we can predict the relative populations of these states from the Boltzmann factor, Equation 3.21.

expressed as either photon or energy flux densities. Some of the light may be absorbed in passing through a solution, and so the emerging beam will generally have a lower flux density—we will assume that scattering and reflection are negligible. In a small path length dx along the direction of the beam, J decreases by an amount dJ due to absorption by a species having a concentration c. Lambert is often credited with recognizing in 1768 that $-dJ/J$ is proportional to dx—actually, Bouguer had expressed this in 1729—and in 1852 Beer noted that $-dJ/J$ is proportional to c. Upon putting these two observations together, we obtain the following expression:

$$-\frac{dJ}{J} = k_\lambda c \, dx \tag{4.13}$$

where k_λ is a proportionality coefficient referring to a particular wavelength, and the minus sign indicates that the flux density is decreased by absorption. We can integrate Equation 4.13 across a solution of a given concentration, which leads to

$$\int_{J_0}^{J_b} -dJ/J = -\ln\frac{J_b}{J_0} = \int_0^b k_\lambda c \, dx = k_\lambda c b \tag{4.14}$$

where J_0 is the flux density of the incident beam, and J_b is its flux density after traversing a distance b through the solution (see Fig. 4.8).

Equation 4.14 is usually recast into a slightly more convenient form. We can replace the natural logarithm by the common logarithm ($\ln = 2.303 \log$, App. III or IV) and $k_\lambda/2.303$ by the absorption coefficient at a specific wavelength, ε_λ. Equation 4.14 then becomes

$$A_\lambda = \log\frac{J_0}{J_b} = \varepsilon_\lambda c b \tag{4.15}$$

where A_λ is the absorbance (colloquially, the "optical density") of the solution at the particular wavelength considered. Equation 4.15 is generally referred to as Beer's law (although it is also called the Beer–Lambert law, the Lambert–Beer law, and even the Bouguer–Lambert–Beer law). When more than one absorbing species is present in a solution, we can generalize Equation 4.15 to give

$$A_\lambda = \log\frac{J_0}{J_b} = \sum_j \varepsilon_{\lambda_j} c_j b \tag{4.16}$$

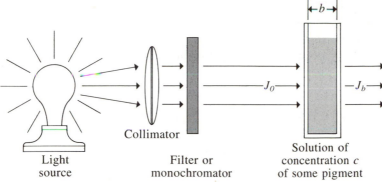

Figure 4.8
Quantities involved in light absorption by a solution as described by Beer's law, $\log (J_0/J_b) = \varepsilon_\lambda cb$ (Eq. 4.15).

where c_j is the concentration of species j, and ε_{λ_j} is its absorption coefficient at wavelength λ (see Calvert and Pitts, Clayton 1980, Eisenberg and Crothers, or Williams and Fleming for additional information).

According to Beer's law, the absorbance at some particular wavelength is proportional to the concentration of the absorbing species, to its absorption coefficient at that wavelength, and to the optical path length b (Fig. 4.8). Values of ε_λ for organic compounds range from less than 0.1 to over 10^4 m^2 mol^{-1} in the visible region.* If ε_λ at some wavelength is known for a particular species, we can determine its concentration from the measured absorbance at that wavelength by using Beer's law (Eq. 4.15).

As an application of Beer's law, we will estimate the average chlorophyll concentration in leaf cells. The palisade and spongy mesophyll cells in the leaf section portrayed in Figure 1.2 might correspond to an average thickness of chlorophyll-containing cells of about 200 μm. The maximum molar absorption coefficient ε_λ in the red or the blue bands of chlorophyll (see Figs. 4.5 and 5.2) is about 10^4 m^2 mol^{-1}. At the peaks of the absorption bands, about 99% of the incident red or blue light can be absorbed by chlorophyll in a leaf section like the one illustrated in Figure 1.2. This corresponds

---

* For laboratory absorption studies, the optical path length b is often 1 cm and c_j is expressed in mol litre^{-1} (i.e., molarity), in which case ε_{λ_j} has units of litre mol^{-1} cm^{-1} and is referred to as the *molar absorption* (or *extinction*) *coefficient* (1 litre mol^{-1} cm^{-1} = 1 M^{-1} cm^{-1} = 10^{-3} mM^{-1} cm^{-1} = 10^3 cm^2 mol^{-1} = 10^{-1} m^2 mol^{-1}). The absorbing species is usually dissolved in a solvent that does not absorb at the wavelengths under consideration.

to having an emergent flux density J_b equal to 1% of the incident flux density J_0, so that the absorbance A_λ in Equation 4.15 equals log (100/1) or 2—for simplicity, we are ignoring absorption by pigments other than chlorophyll. Using Beer's law (Eq. 4.15), we find that the average chlorophyll concentration is

$$c = \frac{A_\lambda}{\varepsilon_\lambda b} = \frac{(2)}{(10^4 \ \text{m}^2 \ \text{mol}^{-1})(200 \times 10^{-6} \ \text{m})}$$

$$= 1 \ \text{mol m}^{-3} \qquad (1 \ \text{m}\textsc{m})$$

a value characteristic of the average chlorophyll concentration in the photosynthesizing cells of many leaves.

Chlorophyll is located only in the chloroplasts, which in turn occupy about 3% to 4% (e.g., 1/30) of the volume of a mesophyll cell in the leaf of a higher plant. The average concentration of chlorophyll in chloroplasts is thus about 30 times higher than the above estimate of chlorophyll in a leaf, or approximately 30 mol m^{-3}. A typical light path across the thickness of a chloroplast (see Fig. 1.9) is about 2 μm. Using Beer's law (Eq. 4.15), we find that the absorbance of a single chloroplast in the red or blue bands is about

$$A_\lambda = (10^4 \ \text{m}^2 \ \text{mol}^{-1})(30 \ \text{mol m}^{-3})(2 \times 10^{-6} \ \text{m})$$

$$= 0.6 = \log \frac{J_0}{J_b}$$

Hence, J_b equals $J_0/[\text{antilog} \ (0.6)]$, or $0.25J_0$. Therefore, approximately 75% of the incident red or blue light at the peak of absorption bands is absorbed by a single chloroplast, which helps explain why individual chloroplasts appear rather green under a light microscope.

Conjugation

Light absorption by molecules generally involves transitions of π electrons to excited states where the electrons are in π^* orbitals. These π electrons occur in double bonds; the more double bonds in some molecule, the greater is the probability for light absorption by that species. Both the effectiveness in absorbing electromagnetic radiation and the wavelengths involved are determined by the number of double bonds in *conjugation*, where conjugation refers to the alternation of single and double bonds (e.g., C—C=C—C=C—C=C—C) along some part of the molecule. The absorption coefficient ε_λ increases with the number of double bonds in the molecule, because there are then more *delocalized* (shared) π electrons capable of interacting with light. As the number of double bonds in the conjugated system increases, the absorption bands also shift to longer wavelengths.

To help understand why the number of double bonds in conjugation affects the wavelength position for an absorption band, let us consider the shifts in energy for the various orbitals as the number of π electrons in the conjugated system changes. The various π orbitals in a conjugated system occur at different energy levels, the average energy remaining about the same as for the π orbital in an isolated double bond not part of a conjugated system (Fig. 4.9). The more double bonds in the conjugated system, the more π orbitals there are in the conjugated system, and the greater is the energy range from the lowest to the highest energy π orbital. Since the *average* energy of the π orbitals in a conjugated system does not markedly depend on the number of double bonds, the energy of the highest energy π orbital increases as the number of double bonds in conjugation increases. The π^* orbitals are similarly split into various energy levels (diagrammed in Fig. 4.9). Again, the range of energy levels about the mean for these π^* orbitals increases as the number of double bonds in the conjugated system increases. Consequently, the more π^* orbitals available in the conjugated system, the lower in energy will be the lowest of these (see Calvert and Pitts).

In Figure 4.9 we present transitions from the highest energy π orbital to the lowest energy π^* orbital for a series of molecules differing in the number of double bonds in conjugation. We immediately notice that the more delocalized π electrons there are in the conjugated system, the less the energy required for a transition. This is illustrated in Figure 4.9 by a decrease in the length of the vertical arrow (which represents an electronic transition) as the number of double bonds in conjugation increases. Moreover, the most likely or probable electronic transition in this case is the one involving the least amount of energy; i.e., the excitation from the highest energy π orbital to the lowest π^* orbital is the transition that actually predominates.*

The decrease in energy separation between the π orbitals and the π^* orbitals as the number of double bonds in conjugation increases (Fig. 4.9)

---

* The transition probability depends on the spatial overlap between the quantum-mechanical wave functions describing the trajectories of the electrons in the two states, and such overlap is relatively large between the highest energy π orbital and the lowest energy π^* one. See Orchin and Jaffé.

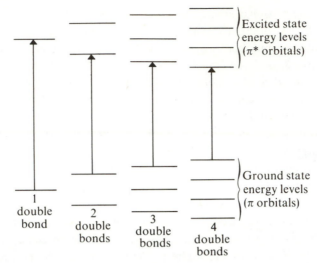

Figure 4.9
Effect on the energy levels of π and π* orbitals as the number of double
bonds in conjugation increases. The vertical arrows represent transitions caused by the absorption of photons—the required energy becomes less and less as the extent of conjugation increases.

accounts for the accompanying shift of the peaks of the absorption bands
toward longer wavelengths. For example, an isolated double bond (C—C=
C—C) generally absorbs maximally near 185 nm in the ultraviolet and has
a maximum absorption coefficient of nearly 10^3 m^2 mol^{-1}. For two double
bonds in conjugation (C—C=C—C=C—C), the absorption coefficient
doubles, and the wavelength position for maximum absorption shifts to
about 225 nm, i.e., toward longer wavelengths. As the number of double
bonds in conjugation in straight-chain hydrocarbons increases from 3 to 5
to 7 to 9, the center of the absorption band for these hydrocarbons when
they are dissolved in hexane is shifted from approximately 265 to 325 to 375
to 415 nm, respectively (see Calvert and Pitts, as well as Levine). The absorption coefficient is approximately proportional to the number of double bonds
in the conjugated system, and hence it increases to almost 10^4 m^2 mol^{-1}
for the hydrocarbon containing 9 double bonds in conjugation. For molecules
to absorb strongly in the visible region, a fairly extensive conjugated system
of double bonds is generally necessary, as is indeed the case for pigments
such as the chlorophylls and carotenoids discussed in the next chapter.

Action Spectra

The relative *effectiveness* of various wavelengths in producing a specified response is of basic importance in photobiology and leads to an *action spectrum*. An action spectrum is complementary to an absorption spectrum, the latter being the relative probability for the absorption of different wavelengths, e.g., ε_λ versus λ. When many different types of pigments are present, the action spectrum for a particular response can differ greatly from the absorption spectrum of the entire system. However, the Grotthus-Draper law cited on p. 198—only absorbed light leads to a photochemical reaction— implies that an action spectrum should resemble the absorption spectrum of the particular substance that absorbs the light responsible for the specific effect or action being considered.

To obtain an action spectrum for some particular response, we could expose the system to the same photon flux density at each of a series of wavelength intervals and measure the resulting effect or action. The action could be the amount of O_2 evolved, the fraction of seeds germinating, or some other measured change. We could then plot the responses obtained as a function of their respective wavelength intervals to see which wavelengths are most effective in leading to that "action" (see Fig. 5.10). Another way to obtain an action spectrum would be to plot the reciprocal of the number of photons required in the various wavelength intervals to give a particular response. If twice as many photons were needed at one particular wavelength compared with a second, the action spectrum would have half the height at the first wavelength, and thus the effectiveness of various wavelengths could easily be presented (see Fig. 4.12). Using the latter approach, the photon flux density would be varied until the response is the same at each wavelength interval. This is an important point; if it is to be a true action spectrum, the action or effect measured must be linear with photon flux density at each of the wavelength intervals used; i.e., we must not be approaching light saturation. When we approach light saturation at certain wavelengths, the measured action per photon is less than it should be, compared with the values at other wavelengths. Consequently, an action spectrum, such as Figure 5.10, is flattened for those wavelengths where we are approaching light saturation. In the extreme case of light saturation at *all* wavelengths, the action spectrum plotted in Figure 5.10 would be perfectly flat, since the response is then the same at each wavelength (see Clayton 1980, or Seliger and McElroy).

We can compare the action spectrum of some response with the absorption spectra of the various pigments suspected of being involved to see whether one type of pigment is responsible. If the measured action spectrum closely resembles the known absorption spectrum of some molecule, light absorbed by that molecule may be leading to the particular action we are considering. Examples where the use of action spectra have been important in understanding the photochemical aspects of plant physiology include the study of photosynthesis (Ch. 5) and investigations of the responses mediated by the pigment phytochrome, to which we now turn.

Absorption and Action Spectra of Phytochrome

Phytochrome is an important plant pigment that may be present in all eucaryotic photosynthetic organisms. It regulates photomorphogenic aspects of plant growth and development, such as seed germination, stem elongation, leaf expansion, formation of certain pigments, chloroplast development, and flowering. Many of these classical phytochrome effects can be caused by very low levels of light; e.g., 500 μmol m^{-2} of red light (the photons in the visible for $\frac{1}{4}$ second of full sunlight) can saturate most of the processes, and 0.3 μmol m^{-2} can lead to half-saturation for a very sensitive one (see Salisbury and Ross). High irradiance levels can also lead to effects mediated by phytochrome. We will direct our attention first to the structure of phytochrome and then to the absorption spectra for two of its forms. The absorption spectra will subsequently be compared with the action spectra obtained for the promotion as well as for the inhibition of seed germination of *Lactuca sativa* (lettuce).

Phytochrome consists of a protein to which is covalently bound the *chromophore*, or light-absorbing part of the pigment (Fig. 4.10a). The chromophore is a *tetrapyrrole*, as are the chlorophylls and the phycobilins that we will discuss in the next chapter. *Pyrrole* refers to a five-membered ring having 4 carbons, 1 nitrogen, and 2 double bonds, ⟦　⟧. The chromophore-protein bond (see Fig. 4.10a) may be an ester linkage to the propionic acid side chain on ring III and a thioether bond (to cysteine) to ring I. Light absorption changes double bonds in or near ring I and may also affect the protein binding, which presumably underlies the conversion of phytochrome to its physiologically active form.

The chromophore of phytochrome is highly conjugated, as the structure for P$_r$ in Figure 4.10a indicates (P stands for pigment and the subscript r indicates that it absorbs in the red region). In fact, P$_r$ has 8 double bonds in conjugation. Not all the double bonds are in the conjugated system,

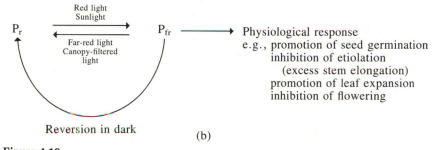

(a)

(b)

Figure 4.10
Phytochrome structure, interconversions, and associated physiological responses.
(a) Possible structure for P_r, indicating the tetrapyrrole forming the chromophore and
the protein attachment (see Brandlmeier et al., Lagarias and Rapoport, or Pratt for
details). (b) Light and dark interconversions of phytochrome, indicating some of the
reactions promoted by the physiologically active form, P_{fr}.

since only those alternating with single bonds along the molecule are part
of the main conjugation. For instance, the double bond at the top of pyrrole
ring III is a branch, or cross-conjugation, to the main conjugation and only
slightly affects the wavelength position for maximum absorption.* On the
basis of the extensive conjugation, we would expect phytochrome to absorb
in the visible region, as indeed it does.

The absorption spectra of two forms of phytochrome are presented in
Figure 4.11. P_r has a major absorption band in the red, with a peak near
667 nm (the exact location can vary with species). Upon absorption of red
light, P_r can be converted to a form having an absorption band in the far-red,
P_{fr} (Figs. 4.10 and 4.11). Although maximum absorption for P_{fr} occurs near

---

* There are semiempirical rules for predicting the wavelength position of maximum
absorption based on all double bonds that occur, including those in branches to the
main conjugated system. See Calvert and Pitts, or Williams and Fleming.

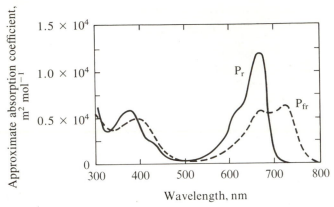

Figure 4.11
Absorption spectra for the red (P_r) and the far-red (P_{fr}) absorbing forms of phytochrome. (Data are replotted from W. L. Butler, S. B. Hendricks, and H. W. Siegelman, in *Chemistry and Biochemistry of Plant Pigments*, T. W. Goodwin, ed., Academic Press, London, 1965, pp. 197–210. Used by permission.)

725 nm, appreciable absorption occurs in the near IR up to about 800 nm. In fact, the absorption bands of both pigments are rather broad, and many different wavelengths of light can be absorbed by each of them (Fig. 4.11). As indicated in Figure 4.11, the maximum absorption coefficients are about 10^4 m^2 mol^{-1} for both P_r and P_{fr}. The plant pigments we will consider in the next chapter have similar high values of ε_λ. We should also note that the absorption spectra for P_r and P_{fr} actually overlap considerably; e.g., 660 nm light can be readily absorbed by either pigment.

We will next examine the action spectra for seed germination responses (Fig. 4.12) with regard to the known absorption properties of phytochrome. In green tissues, chlorophyll absorption in the red region markedly affects the amount of light incident on phytochrome, which makes the evaluation of action spectra more difficult. Consequently, we have chosen germination responses of seeds, which are relatively free of chlorophyll, to illustrate a possible phytochrome-mediated reaction. Figure 4.12 indicates that a pigment absorbing in the red region promotes the germination of lettuce seeds. If we compare this action spectrum with the absorption spectra given in Figure 4.11, we see that the pigment absorbing the light that promotes seed germination has an absorption spectrum essentially identical to that of the P_r form of phytochrome. This enhancement of seed germination by red light can be reversed by subsequent irradiation of the seeds with far-red

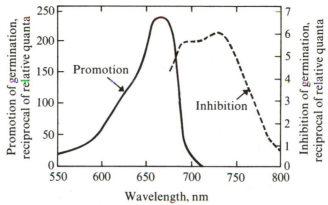

Figure 4.12
Action spectra for the promotion of lettuce seed germination and its reversal, or "inhibition." (Data are replotted from S. B. Hendricks and H. A. Borthwick, in *Chemistry and Biochemistry of Plant Pigments*, T. W. Goodwin, ed., Academic Press, London, 1965, pp. 405–436. Used by permission.)

light. In other words, the potential promotion of seed germination by red light can be stopped or at least reduced by irradiation with various longer wavelengths immediately following irradiation with the red light. The action spectrum for such "inhibition" of lettuce seed germination is also presented in Figure 4.12. This inhibition induced by far-red light requires fewest quanta at wavelengths near 725 to 730 nm, similar to the position of the peak in the absorption spectrum of P_{fr} (Fig. 4.11). In summary, both the promotion of lettuce seed germination and the reversal of this promotion are apparently controlled by two forms of phytochrome, P_r and P_{fr}, which can be reversibly interconverted by light.*

Physiological consequences of phytochrome interconversions reflect the fact that P_{fr} is the active form of the pigment (Fig. 4.10b). The absorption of light by P_r leads to its conversion to P_{fr}, and this latter form promotes the germination of the lettuce seeds. If far-red light is incident on the tissue, its absorption by P_{fr} tends to convert phytochrome to the inactive P_r form, and germination is not promoted. Other environmental factors also influence the phytochrome-mediated reactions. For instance, only in hydrated seeds is

---

* It should be noted that germination is not enhanced by red light for seeds of all species. Enhancement tends to be more common in small seeds rich in fat that come from wild plants (see Salisbury and Ross).

P_r converted to P_{fr}, presumably because water is necessary for the conformational changes of the protein. Although the photochemistry of phytochrome interconversions is fairly independent of temperature, temperature can have a marked influence on the synthesis, degradation, and physiological expression of the pigments.

To illustrate the ecological consequences of the phytochrome control of seed germination, let us consider a seed present on the surface of the soil under a dense canopy of leaves. Because of their chlorophyll, leaves in the canopy absorb red light preferentially to far red, so very little red light compared with far-red reaches the seed. Hence, the phytochrome in the seed occurs predominantly in the P_r form. However, if this seed were to become located in a patch of sunlight—as could happen because of a fire or the natural death of a shading tree—a larger fraction of the P_r would be converted to the active form, P_{fr}. Seed germination would then proceed under light conditions favorable for photosynthesis and thus for growth.

Studies using action spectra have indicated that, on a quantum basis, the maximum sensitivity is usually quite different for the opposing responses attributable to the two forms of phytochrome. (Since the expression of phytochrome action involves a multistep process including biochemical reactions, there is no reason why the two responses should have the same sensitivity.) For lettuce seed germination, about 30 times more far-red quanta (e.g., at 730 nm) are required to cause a 50% inhibition than the number of red quanta (e.g., at 660 nm) needed to promote seed germination by 50%—compare the different scales used for the oridinates in Figure 4.12. Thus, ordinary sunlight is physiologically equivalent to red light, since much of the phytochrome is converted to the active form, P_{fr}. Also, P_r has a much larger absorption coefficient in the red region than does P_{fr} (Fig. 4.11).

The P_{fr} form of phytochrome reverts spontaneously to P_r in the dark (Fig. 4.10b), except in most monocots. Reversion can take place in less than one hour and probably is not used by plants as a timing mechanism per se, although phytochrome may be involved in other aspects of photoperiodism (the influence of daylength or nightlength on plant processes). Also, the interconversions of phytochrome indicated in a simplified fashion in Figure 4.10b are actually multistep processes involving a number of intermediates whose structures and absorption characteristics are under investigation. The mechanisms of action of phytochrome are not yet fully understood, although changes in membrane permeability, in the activity of membrane-bound enzymes, and in the transcription or translation of genetic information have all been implicated. Finally, as we might expect, individual plants vary considerably in how they amplify and use the information provided by the

P_r-P_{fr} system. (See De Greef, Kendrick and Frankland, or Salisbury and Ross for further information on phytochrome.)

Our discussion of phytochrome leads us to the concept of a *photostationary state*. A photostationary state refers to the relative amounts of interconvertible forms of some pigment that occur in response to a particular steady illumination—such a state might have been called, more appropriately, a "photosteady" state. When illumination is constant, the conversion of P_r to P_{fr} will eventually achieve the same rate as the reverse reaction (see Fig. 4.10). The ratio of P_r to P_{fr} for this photostationary state depends on the absorption properties of each form of the pigment for the incident wavelengths (note that the absorption spectra overlap, Fig. 4.11), the number of photons in each wavelength interval, the kinetics of the competing de-excitation reactions, and the kinetics of pigment synthesis or degradation. If the light quality or quantity were to change to another constant condition, we would shift to a new photostationary state, where P_r/P_{fr} would in general be different. Since the ratio of P_r to P_{fr} determines the overall effect of the phytochrome system, a description in terms of photostationary states can be useful for discussing the influence of this pigment.

Problems

4.1. Consider electromagnetic radiation with the indicated wavelengths in a vacuum. (a) If λ is 400 nm, how much energy is carried by 10^{20} photons? (b) If a mol of 1 800 nm photons is absorbed by 10^{-3} m^3 (1 litre) of water at 0°C, what would be the final temperature? (Assume that there are no other energy exchanges with the external environment.) (c) A certain filter, which passes all wavelengths below 600 nm and absorbs all those above 600 nm, is placed over a radiometric device. If the meter indicates 1 W m^{-2}, what is the maximum photon flux density in μmol m^{-2} s^{-1}? (d) What is the illuminance in (c) expressed in lux (lumens m^{-2})?

4.2. Consider electromagnetic radiation having a frequency of 0.9×10^{15} cycles s^{-1}. (a) The speed of the radiation is 2.0×10^8 m s^{-1} in dense flint glass. What are the wavelengths in a vacuum, in air, and in such glass? (b) Can such radiation cause an $S_{(\pi,\pi)}$ ground state to go directly to $T_{(\pi,\pi*)}$? (c) Can such radiation cause the transition of a π electron to a $\pi*$ orbital in a molecule having six double bonds in conjugation? (d) Electromagnetic radiation is often expressed in "wave numbers," which is the frequency divided by the speed of light in a vacuum, i.e., v/c, which is $1/\lambda_{\text{vacuum}}$. What is the wave number in m^{-1} in the present case?

4.3. Suppose that the quantum yield for ATP formation—molecules of ATP formed/number of excited chlorophyll molecules—is 0.40 at 680 nm, and the rate of ATP formation is 0.20 mol m^{-2} hour^{-1}. (a) What is the minimum photon flux density in μmol m^{-2} s^{-1} at 680 nm? (b) What is the energy flux density under the conditions of (a)? (c) If light of 430 nm is used, the ground state of chlorophyll, S$_{(\pi,\pi)}$, is excited to S$_{(\pi,\pi*)}^b$. Suppose that 95% of S$_{(\pi,\pi*)}^b$ goes to S$_{(\pi,\pi*)}^a$ in 10^{-12} s and that the rest of the upper excited singlet state returns to the ground state. What is the energy conversion efficiency of 430 nm light as an energy source for ATP formation compared with 680 nm light? (d) The hydrolysis of ATP to ADP and phosphate under physiological conditions can yield about 40 kJ mol^{-1} of free energy. What wavelength of light has the same amount of energy mol^{-1}?

4.4. Assume that some excited singlet state can become de-excited by three competing processes: (1) fluorescence (lifetime = 10^{-8} s), (2) a radiationless transition to an excited triplet state (5 × 10^{-9} s), and (3) a radiationless transition to the ground state (10^{-8} s). (a) What is the lifetime of the excited singlet state? (b) What is the maximum quantum yield for all de-excitations leading directly or indirectly to electromagnetic radiation? (c) Suppose that the above molecule is inserted into a membrane, which adds a de-excitation pathway involving intermolecular transfer of energy from the excited singlet state (rate constant = 10^{12} s^{-1}). What is the new lifetime of the excited singlet state?

4.5. The *cis*-isomer of some species has an absorption coefficient of 2.0 × 10^3 m^2 mol^{-1} at 450 nm, where the spectrophotometer has a photon flux density of 10^{17} photons m^{-2} s^{-1}. (a) What concentration of the *cis*-isomer will absorb 65% of the incident 450 nm light for a cuvette (a transparent vessel used in a spectrophotometer) with an optical path length of 10 mm? (b) What is the absorbance of the solution in (a)? What would be the absorbance if the flux density at 450 nm were halved? (c) Suppose that the 450 nm light caused a photoisomerization of the *cis*-isomer to the *trans*-isomer with a quantum yield of 0.50, the other de-excitation pathway being the return to the *cis*-form. If the *trans*-isomer did not absorb at 450 nm, what would be the initial rate of decrease of the *cis*-isomer and rate of change of absorbance at 450 nm under the conditions of (a)? (d) If ε_{450} for the *trans*-isomer were 10^3 m^2 mol^{-1}, and 450 nm light led to a photoisomerization of the *trans*-isomer with a quantum yield of 0.50 for forming the *cis*-isomer, what would be the ratio of *cis* to *trans* after a long time?

4.6. (a) Suppose that the spacing in wave numbers (see Problem 4.2) between vibrational sublevels for the transition depicted in Figure 4.6 is 1.2 × 10^5 m^{-1} and that the most probable absorption predicted by the Franck–Condon principle occurs at 500 nm (the main band). What are the wavelength positions of the satellite bands that occur for transitions to the vibrational sublevels just above and just below the one for the most probable transition? (b) Suppose that the main band has a maximum absorption coefficient of 5 × 10^3 m^2 mol^{-1}, while each satellite band has an ε_λ one-fifth as large. If 20% of the incident light is absorbed at the wavelengths of either of the satellite bands, what percentage is absorbed at the main band wavelength? (c) When the pigment is placed in a cuvette with an optical path length of 5 mm, the maximum absorbance is 0.3. What is the concentration?

4.7. A straight chain hydrocarbon has eleven double bonds in conjugation. Suppose that it has three bands in the visible region, one at 450 nm ($\varepsilon_{450} = 1.0 \times 10^4$ m^2 mol^{-1}), one at 431 nm ($\varepsilon_{430} = 2 \times 10^3$ m^2 mol^{-1}), and a minor band near 470 nm ($\varepsilon_{470} \cong 70$ m^2 mol^{-1}). Upon cooling from 20°C to liquid helium temperatures, the minor band essentially disappears. (a) What is the splitting between vibrational sublevels? (b) What transition could account for the minor band? Support your answer by calculation. (c) If the λ_{max} for fluorescence is at 494 nm, what transition is responsible for the 450 nm absorption band? (d) If the double bond in the middle of the conjugated system is reduced (by adding 2 H's so that it becomes a single bond) and the rest of the molecule remains unchanged, calculate the new λ_{max} for the main absorption band and its absorption coefficient. Assume that for every double bond added to the conjugated system, λ_{max} shifts by 25 kJ mol^{-1}, and that $\varepsilon_{\lambda max}$ is directly proportional to the number of double bonds in conjugation.

References

Bell, C. J., and D. A. Rose. 1981. Light measurement and the terminology of flow. *Plant, Cell, and Environment 4*:89–96.

Bickford, E. D., and S. Dunn. 1972. *Lighting for Plant Growth*. Kent State University Press, Kent, Ohio.

Brandlmeier, T., H. Scheer, and W. Rüdiger. 1981. Chromophore content and molar absorptivity of phytochrome in the P$_r$ form. *Zeitschrift für Naturforschung 36c*: 431–439.

Caldwell, M. M. 1981. Plant response to solar ultraviolet radiation. In *Physiological Plant Ecology*, O. L. Lange, P. S. Nobel, C. B. Osmond, and H. Ziegler, eds. *Encyclopedia of Plant Physiology, New Series*, Vol. 12A. Springer-Verlag, Berlin. Pp. 169–197.

Calvert, J. G., and J. N. Pitts, Jr. 1966. *Photochemistry*. Wiley, New York.

Clayton, R. K. 1965. *Molecular Physics in Photosynthesis*. Blaisdell, New York.

Clayton, R. K. 1980. *Photosynthesis—Physical Mechanisms and Chemical Patterns*. Cambridge University Press, Cambridge.

Cowan, D. O., and R. L. Drisko. 1976. *Elements of Organic Photochemistry*. Plenum Press, New York.

De Greef, J. 1980. *Photoreceptors and Plant Development*. European Photomorphogenesis Symposium, Antwerpen University Press, Antwerp.

Depuy, C. H., and O. L. Chapman. 1972. *Molecular Reactions and Photochemistry*. Prentice-Hall, Englewood Cliffs, New Jersey.

Ditchburn, R. W. 1976. *Light*, 3rd ed. Academic Press, London.

Eisenberg, D., and D. Crothers. 1979. *Physical Chemistry with Applications to the Life Sciences*. Benjamin-Cummings, Menlo Park, California.

Gates, D. M. 1980. *Biophysical Ecology*. Springer-Verlag, New York.

Jerlov, N. G. 1976. *Marine Optics*, 2nd ed. Elsevier, Amsterdam.

Kaufman, J. E., and J. F. Christensen, eds. 1972. *IES Lighting Handbook*, 5th ed. Illuminating Engineering Society, New York.

Kendrick, R. E., and B. Frankland. 1976. *Phytochrome and Plant Growth*. The Institute of Biology's Studies in Biology, No. 68. Edward Arnold, London.

Lagarias, J. C., and H. Rapoport. 1980. Chromopeptides from phytochrome. The structure and linkage of the P_R form of the phytochrome chromophore. *Journal of the American Chemical Society 102*:4821–4828.

Langhans, R. W., ed. 1978. *A Growth Chamber Manual: Environmental Control for Plants*. Cornell University, Ithaca, New York.

Levine, J. N. 1970. *Quantum Chemistry*, Vol. I, *Quantum Mechanics and Molecular Electronic Structure*. Allyn and Bacon, Boston.

McCree, K. J. 1981. Photosynthetically active radiation. In *Physiological Plant Ecology*, O. L. Lange, P. S. Nobel, C. B. Osmond, and H. Ziegler, eds. *Encyclopedia of Plant Physiology, New Series*, Vol. 12A. Springer-Verlag, Berlin. Pp. 41–55.

Monk, G. S. 1963. *Light: Principles and Experiments*, 2nd ed. Dover, New York.

Monteith, J. L. 1973. *Principles of Environmental Physics*. American Elsevier, New York.

Orchin, M., and H. H. Jaffé. 1971. *Symmetry, Orbitals, and Spectra*. Interscience, Wiley, New York.

Pratt, L. H. 1978. Molecular properties of phytochrome. *Photochemistry and Photobiology 27*:81–105.

Resnick, R., and D. Halliday. 1977 (Part One) and 1978 (Part Two). *Physics*, 3rd ed. Wiley, New York. See also D. Halliday and R. Resnick. 1974. *Fundamentals of Physics*, Revised Printing. Wiley, New York.

Salisbury, F. B., and C. Ross. 1978. *Plant Physiology*, 2nd ed. Wadsworth, Belmont, California.

Seliger, H. H., and W. D. McElroy. 1965. *Light: Physical and Biological Action*. Academic Press, New York.

Tibbitts, T. W., and T. T. Kozlowski, eds. 1979. *Controlled Environment Guidelines for Plant Research*. Academic Press, New York.

Turro, N. J. 1978. *Modern Molecular Photochemistry*. Benjamin Cummings, Menlo Park, California.

Williams, D. H., and I. Fleming. 1973. *Spectroscopic Methods in Organic Chemistry*, 2nd ed. McGraw-Hill, London.

CHAPTER 5

Photochemistry of Photosynthesis

Photosynthesis is the largest-scale synthetic process on earth. In one year about 7.2×10^{13} kg (72 billion tons) of carbon are fixed into organic compounds by photosynthetic organisms (this is often called the *net primary productivity*; see Lieth and Whittaker). Such a large number is difficult to grasp. It corresponds to the amount of organic compounds in a solid carbohydrate "tree" 10 m high and over 100 km in diameter! Also, it equals about 1% of the world's known reserves of fossil fuels (coal, gas, and oil), or ten times the world's present annual energy consumption (see Hall and Rao). The carbon source used in photosynthesis is the 0.03% CO_2 contained in the air (about 70×10^{13} kg carbon) and the CO_2 or HCO_3^- dissolved in lakes and oceans (about $4\,000 \times 10^{13}$ kg carbon). In addition to the organic compounds, another product of photosynthesis essential for all respiring organisms is O_2. At the present rate, the entire atmospheric content of O_2 is replenished by photosynthesis every few thousand years.

Photosynthesis is not a single reaction; rather, it is composed of many individual steps that work together with remarkably high overall efficiency. We can divide the process into three stages: (1) the photochemical steps, our primary concern in this chapter; (2) electron transfer, to which is coupled the formation of ATP, which we will consider in both this and the next chapter; and (3) the biochemical reactions involving the incorporation of CO_2 into carbohydrates. Figure 5.1 summarizes the processes involved and introduces the relative amounts of the various reactants and products taking

239

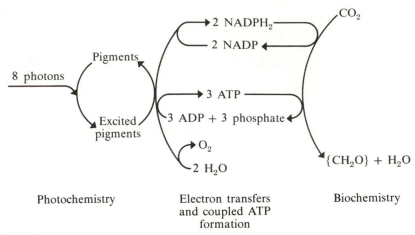

Photochemistry   Electron transfers   Biochemistry
                        and coupled ATP
                        formation

Figure 5.1
Schematic representation of the three stages of photosynthesis. The absorption of light can cause the excitation of photosynthetic pigments, which leads to the photochemical events where electrons are donated by special chlorophylls. The electrons are then transferred along a series of molecules leading to the reduction of NADP to NADPH [NADPH$_2$, which represents NADPH + H$^+$ (see p. 310), is used in the figure to help keep track of the hydrogens involved]; ATP formation is coupled to the electron transfer steps. The biochemistry of photosynthesis can proceed in the dark and requires 3 moles of ATP and 2 moles of NADPH per mole of CO$_2$ fixed into a carbohydrate, represented in the figure by {CH$_2$O}.

part in the three stages of photosynthesis. The photochemical reactions, which are often referred to as the *primary events* of photosynthesis, lead to electron transfer along a sequence of molecules, which results in the formation of NADPH and ATP.

Let us now consider the net chemical reaction for photosynthesis. In Figure 5.1, two H$_2$O's are indicated as reactants in the O$_2$ evolution step, and one H$_2$O is a product in the biochemical stage. Hence, the overall *net* chemical reaction describing photosynthesis is: CO$_2$ plus H$_2$O yields carbohydrate plus O$_2$. It is instructive to indicate the energy required to break each of the chemical bonds in these compounds, which leads us to the following representation for the net photosynthetic reaction:

$$H \overset{463}{\rule{1cm}{0.4pt}} O \overset{463}{\rule{1cm}{0.4pt}} H + O \overset{800}{=\!=} C \overset{800}{=\!=} O \longrightarrow$$

$$\underset{\frac{1}{2}(348)}{\overset{\frac{1}{2}(348)}{H \overset{413}{\rule{1cm}{0.4pt}} C \overset{350}{\rule{1cm}{0.4pt}} O \overset{463}{\rule{1cm}{0.4pt}} H}} + O \overset{498}{=\!=} O \qquad (5.1)$$

where the numbers represent the various bond energies in kJ mol^{-1} (literature values for bond energies vary somewhat; see Atkins; Pauling, Ch. 2 reference; as well as Rabinowitch and Govindjee). A C—C bond, which occurs on two sides of the carbon in {CH_2O}, has a value of 348 kJ mol^{-1}, so $\frac{1}{2}(348)$ has been indicated in the appropriate places in Equation 5.1.

The formulation of photosynthesis in Equation 5.1 fails to do justice to the complexity of the reactions, but it is useful for estimating the amount of Gibbs free energy that is stored. We will begin by considering the energy of the chemical bonds. The total bond energy of the reactants in Equation 5.1 is 2 526 kJ mol^{-1} (463 + 463 + 800 + 800), and it is 2 072 kJ mol^{-1} for the products (413 + 348 + 350 + 463 + 498). Thus the reactants H_2O and CO_2 represent the lower energy (i.e., they are more "tightly" bonded), since 454 kJ mol^{-1} is necessary for the bond changes to convert them to the products {CH_2O} plus O_2. This energy change actually represents the increase in enthalpy required, ΔH, while we are really more concerned here with the change in Gibbs free energy, ΔG (see Ch. 6 and App. VI). (For a reaction at constant temperature, ΔG equals $\Delta H - T\Delta S$, where S is the entropy. ΔG is about the same as ΔH for Eq. 5.1.) Although the actual ΔG per mole of C depends somewhat on the particular carbohydrate involved, 454 kJ is approximately the increase in Gibbs free energy per mole of CO_2 that reacts in Equation 5.1. For instance, the Gibbs free energy released when glucose is oxidized to CO_2 and H_2O is 2 872 kJ mol^{-1} (686 kcal mol^{-1}) of glucose, or 479 kJ mol^{-1} of C. In discussing photosynthesis we will frequently use this ΔG, which actually refers to standard state conditions (25°C, pH 7, 1 molal concentrations, 1 atmosphere pressure).

About 8 photons are required in photosynthesis per CO_2 fixed and O_2 evolved (see Fig. 5.1). Red light at 680 nm corresponds to 176 kJ mol^{-1} (Table 4.1), so that 8 moles of such photons have 1 408 kJ of radiant energy. Using this as the energy input and 479 kJ as the amount of energy stored per mole of CO_2 fixed, the efficiency of energy conversion by photosynthesis is (479 kJ/1 408 kJ)(100), or 34%. Actually, slightly more than 8 photons may be required per CO_2 fixed. Furthermore, the energy for wavelengths less than 680 nm, which are also used in photosynthesis, is higher than 176 kJ mol^{-1}. Both of these considerations lead to a calculated maximum efficiency of the utilization of absorbed energy of somewhat less than 34%. Nevertheless, photosynthesis is an extremely efficient energy conversion process, considering all the steps involved, each with its inherent energy losses.

Nearly all enzymes involved in the synthetic reactions of photosynthesis are also found in nonphotosynthetic tissue. Thus, the unique feature of photosynthesis is the conversion of radiant energy into chemical energy.

The present chapter will emphasize the light absorption and excitation transfer aspects of photosynthesis. We will consider the structures and absorption characteristics of photosynthetic pigments and the means by which radiant energy is trapped, transferred, and eventually used. Thus, the emphasis will be on the *photo* part of photosynthesis. In the next chapter, we will discuss energy conversion in a broader context, paying particular attention to ATP and NADPH.

CHLOROPHYLL—CHEMISTRY AND SPECTRA

Chlorophylls represent the principal class of pigments responsible for light absorption in photosynthesis and are found in all photosynthetic organisms. There are a number of different types of chlorophyll, as Tswett demonstrated in 1906 using adsorption chromatography. For instance, approximately 1 g of the chlorophylls designated *a* and *b* is present per kg fresh weight of green leaves. The empirical formulas were first given by Willstätter, beginning in 1913; Fisher established the actual structures of various chlorophylls by 1940. These two investigators, as well as Woodward, who first synthesized chlorophyll in vitro, have all received the Nobel prize for their studies on this important plant pigment. We will first consider the structure of chlorophyll *a* (abbreviated, Chl *a*), then its absorption and fluorescence characteristics.

Types and Structures

The various types of chlorophyll are identified by letters or by the taxonomic group of the plants in which they occur. One of the most important is Chl *a*. It has a relative molecular mass (molecular weight) of 893.5, and its structure is given in Figure 5.2. Chl *a* is found in all photosynthetic organisms except bacteria, i.e., in all species where O_2 evolution accompanies photosynthesis. It is a tetrapyrrole (see p. 230) having a relatively flat porphyrin "head" about 1.5 nm by 1.5 nm (15 Å by 15 Å), in the center of which a magnesium atom is co-ordinately bound (see p. 281). Attached to the head is a long-chain terpene alcohol, phytol, which is like a "tail" about 2 nm in length containing 20 carbon atoms (Fig. 5.2). This tail provides a nonpolar region which helps bind the chlorophyll molecules to chlorophyll-protein complexes in the lamellar membranes, but it makes no appreciable contribution to the optical properties of chlorophyll in the visible region. The system of rings in the porphyrin head of Chl *a* is highly conjugated, having

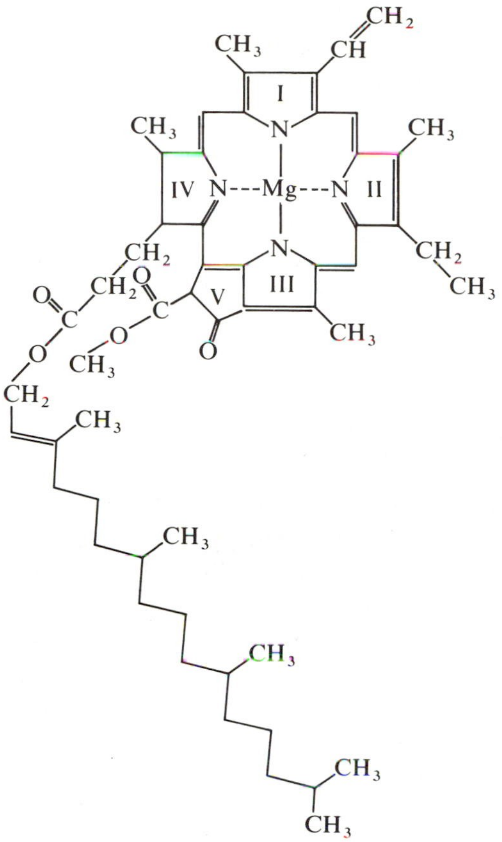

Figure 5.2
Structure of Chl *a*, illustrating the highly conjugated porphyrin "head" to which is attached a phytol "tail." The convention for numbering the various rings is also indicated.

nine double bonds in conjugation (plus three other double bonds occurring in branches to the main conjugated system). These alternating single and double bonds of the porphyrin ring provide many delocalized π electrons which can take part in the absorption of light.

A number of other chlorophyll forms structurally similar to Chl *a* occur in nature. For instance, Chl *b* differs from Chl *a* by having a formyl group (—CHO) in place of a methyl group (—CH$_3$) on ring II. Chl *b* is found in most land plants (including ferns and mosses), the green algae, and the Euglenophyta; the ratio of Chl *a* to Chl *b* in these organisms is usually about 3. Chl *b* is not essential for photosynthesis, since a barley mutant contains only Chl *a*, yet it carries out photosynthesis quite satisfactorily (see Levine).

Another type is Chl *c*, which occurs in the dinoflagellates, cryptomonads (a small class containing flagellated unicellular algae), diatoms, golden algae, and the brown algae. The common chlorophyll of bacteria is called Bchl *a*.

The purple photosynthetic bacteria contain Bchl a (or b in some species), and Bchl a plus one of two forms of Chlorobium chlorophyll occur in green photosynthetic bacteria. These bacterial pigments differ from green plant chlorophylls in that they contain two more hydrogens in the porphyrin ring. They also have different substituents around the periphery of the porphyrin ring. In addition, Chlorobium chlorophyll has the alcohol farnesol (15 C and three double bonds) in the place of phytol (20 C and 1 double bond). The pigment of principal interest in this text is Chl a (see Goodwin, Jeffrey, or Vernon and Seely for further comments on the distribution of chlorophylls among various plants).

Absorption and Fluorescence Emission Spectra

The absorption spectrum of Chl a has a blue and a red band; hence, the characteristic color of chlorophyll is green. The band in the blue part of the spectrum has a peak at 430 nm for Chl a in ether (Fig. 5.3). This band is known as the *Soret band*, which occurs in the UV, violet, or blue region for all tetrapyrroles. We will designate the wavelength position for a local maximum of the absorption coefficient in an absorption band by λ_{max}. Figure 5.3 indicates that the absorption coefficient at the λ_{max} for the Soret band of Chl a is just over 1.2×10^4 m^2 mol^{-1} (1.2×10^5 M^{-1} cm^{-1}; see footnote on p. 225). Such a high value is a consequence of the many double bonds in the conjugated system of the porphyrin ring of chlorophyll. Chl a has a major band in the red region, which has a λ_{max} at 662 nm when the pigment is dissolved in ether (Fig. 5.3).

Chl a also has a number of minor absorption bands. For instance, Chl a dissolved in ether has a small absorption band at 615 nm, which is 47 nm lower than the λ_{max} of the main red band (Fig. 5.3). Absorption of light with a wavelength of 615 nm leads to an electronic transition requiring 14 kJ mol^{-1} more energy than the main band at 662 nm (1 kJ mol^{-1} = 0.010 eV molecule^{-1}; see App. III). This extra energy is similar to the energy spacing between vibrational sublevels. In fact, this small band on the shorter wavelength (higher energy) side of the red band corresponds to electrons going to the vibrational sublevel in the excited state immediately above the sublevel for the λ_{max} at 662 nm—an aspect to which we will return shortly.

The fluorescence emission spectrum of Chl a in ether is also presented in Figure 5.3. Although chlorophyll absorbs strongly in both the red and the blue, the fluorescence is essentially all in the red region. This is because the upper singlet state of chlorophyll excited by blue light ($S^b_{(\pi, \pi^*)}$ in Fig. 4.5)

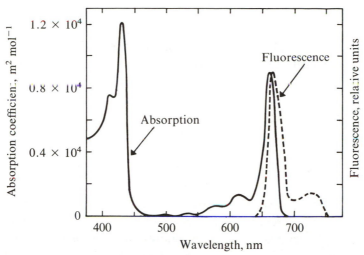

Figure 5.3

Absorption and fluorescence emission spectra of Chl *a* dissolved in ether. [Source: A. S. Holt and E. E. Jacobs, *American Journal of Botany 41*: 710–717 (1954). Data reprinted by permission.]

is extremely unstable and goes to the lower excited singlet state $S^a_{(\pi,\pi*)}$ in about 10^{-12} s, i.e., before any appreciable blue fluorescence can take place.[*] Because of such rapid energy degradation by a radiationless transition, photons absorbed by the Soret band of chlorophyll are no more effective for photosynthesis than are the lower energy photons absorbed in the red region. We can observe the red fluorescence of chlorophyll accompanying light absorption by the Soret band if we illuminate a leaf with blue or shorter-wavelength light in a darkened room. With a light microscope, we can see the red fluorescence emanating from individual chloroplasts in the leaf's cells when using such exciting light (the red fluorescence is often masked by scattering when using red exciting light, so shorter wavelengths are used in most fluorescence studies).

On the basis of the discussion in the previous chapter, we can reason that the electronic transition having a λ_{max} at 662 nm in the absorption spectrum for Chl *a* dissolved in ether corresponds to the excitation of an electron from the lowest vibrational sublevel of the ground state to some vibrational sublevel of the lower excited state. We can use the Boltzmann

---

[*] $S^b_{(\pi,\pi*)}$ and $S^a_{(\pi,\pi*)}$ are distinct excited electronic states; each has its own energy curve in a diagram like Figure 4.7.

factor $[n(E)/n_{\text{total}} = e^{-E/RT}$, Eq. 3.21b] to estimate the fraction of chlorophyll molecules in the first excited vibrational sublevel of the ground state. Since RT is 2.48 kJ mol^{-1} at 25°C (App. III), and the distance between vibrational sublevels is about 14 kJ mol^{-1} for chlorophyll, the Boltzmann factor would equal $e^{-(14 \text{ kJ mol}^{-1})/(2.48 \text{ kJ mol}^{-1})}$, or $e^{-5.65}$, which is 0.0035. Therefore, only about one in three hundred chlorophyll molecules would normally be in the first excited vibrational sublevel of the ground state when light arrives. Consequently, the absorption of a photon will most likely occur when chlorophyll is in the lowest vibrational sublevel of the ground state.

In Chapter 4 we argued that fluorescence generally occurs from the lowest vibrational sublevel of the excited singlet state. In other words, any excess vibrational energy is usually dissipated before the rest of the energy of the absorbed photon is reradiated as fluorescence. But Figure 5.3 shows that the wavelength region for most of the fluorescence is nearly coincident with the red band in the chlorophyll absorption spectrum.* In particular, the λ_{max} for fluorescence occurs at 666 nm, which is only 1 kJ mol^{-1} lower in energy than the λ_{max} of 662 nm for the red band in the absorption spectrum. The slight shift, which is much less than the distance between vibrational sublevels of 14 kJ mol^{-1} for chlorophyll, is most likely due to the loss of some rotational energy (rotational subsublevels of a vibrational sublevel are generally about 1 kJ mol^{-1} apart). Thus, the transition from the lowest vibrational sublevel of the ground state up to the lower excited state (the red absorption band) has essentially the same energy as a transition from the lowest vibrational sublevel of that excited state down to the ground state (the red fluorescence band). The only way for this to occur is to have the lowest vibrational sublevels of both ground and excited states involved in each of the transitions. Hence, the red absorption band corresponds to a transition of the chlorophyll molecule from the lowest vibrational sublevel of the ground state to the lowest vibrational sublevel of the lower excited state, as is indicated in Figure 5.4.

The participation of the lowest vibrational sublevels of both the ground state and the lower excited state of Chl a in the major red band can also be appreciated by considering the minor band adjacent to the major red band in both the absorption spectrum and the fluorescence emission spectrum (see Figs. 5.3 and 5.4). The shorter wavelength absorption band at 615 nm in ether—14 kJ mol^{-1} higher in energy than the 662 nm band— corresponds to a transition to the first excited vibrational sublevel in the

---

* The difference in energy or wavelength between absorption and fluorescence bands is often called the Stokes shift.

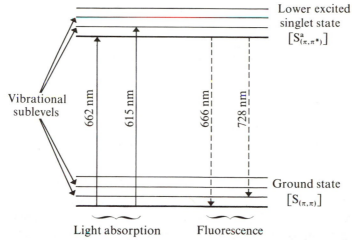

Figure 5.4
Energy level diagram, indicating the vibrational sublevels of the ground state $[S_{(\pi,\pi)}]$ and the lower excited singlet state $[S^a_{(\pi,\pi*)}]$ of Chl a. Solid vertical lines indicate the absorption of light by Chl a dissolved in ether; dashed lines represent fluorescence at the specified wavelengths. The lengths of the arrows are proportional to the amounts of energy involved in the various transitions.

lower excited state. De-excitations from the lowest vibrational sublevel of the lower excited state to excited vibrational sublevels of the ground state would correspond to fluorescence at wavelengths greater than 700 nm. In fact, a small band near 728 nm in the fluorescence emission spectrum of Chl a (Fig. 5.3) occurs about 62 nm on the long wavelength side of the main fluorescence band, indicating an electronic transition with 15 kJ mol^{-1} less energy than the 666 nm band. This far-red band most likely corresponds to fluorescence emitted as the chlorophyll molecule goes from the lowest vibrational sublevel of the lower excited state to the first excited vibrational sublevel of the ground state (Fig. 5.4). In summary, we note that excitations from excited vibrational sublevels of the ground state are uncommon, which is a reflection of the Boltzmann energy distribution, and fluorescence from excited vibrational sublevels of an excited state is difficult to observe, since radiationless transitions to the lowest vibrational sublevel are so rapid. However, transitions to excited vibrational sublevels can be quite significant (see Fig. 5.4 for Chl a).

Absorption in vivo—Polarized Light

Chl a exhibits a number of spectroscopically different forms in vivo. The various values of λ_{max} for Chl a result from interactions between the chlorophylls and surrounding molecules, such as the proteins and lipids in the lamellae as well as adjacent water molecules. In fact, probably all Chl a is conjugated to proteins in chlorophyll-protein complexes. Hydrophobic interactions among phytol tails of adjacent chlorophylls and perhaps with hydrophobic regions in the protein may help stabilize these chlorophyll–protein complexes. Because of the interactions of the porphyrin ring with the other molecules in the complex, and especially with the polar amino acids of the protein, the red bands for Chl a in vivo are shifted toward longer wavelengths than for Chl a dissolved in ether (λ_{max} at 662 nm), e.g., to 670 to 680 nm. This is an example of the pronounced effect that the solvent or other neighboring molecules can have in determining the positions of the electronic energy levels of a pigment. For completeness we should mention that the red absorption band of Chl b in vivo occurs as a "shoulder" on the short wavelength side of the Chl a red band, usually near 650 nm.

A small amount of Chl a occurs in a special site which plays a particularly important role in photosynthesis. These Chl a's have λ_{max}'s at approximately 680 nm and 700 nm and are referred to as P_{680} and P_{700}, respectively (P indicating pigment). P_{700} is apparently a dimer of Chl a molecules (i.e., 2 Chl a's acting as a unit). P_{680} may also be a dimer, although some evidence indicates that it is a single Chl a (see Barber 1977, 1978; Ciba Foundation Symposium; or Clayton).

We can describe the *bandwidth* of an absorption band by the difference in energy between photons on the two sides of the band at wavelengths where the absorption has dropped to half of that for λ_{max}. Such bandwidths of the red absorption bands of the various Chl a's in vivo are actually fairly narrow—often about 10 nm at 20°C (see Brown). At 680 nm a bandwidth of 10 nm is equivalent to 3 kJ mol^{-1}; i.e., a photon having a wavelength of 675 nm has an energy 3 kJ mol^{-1} greater than a photon with a wavelength of 685 nm. An energy of 3 kJ mol^{-1} is smaller than the spacing between vibrational sublevels of 14 kJ mol^{-1} for Chl a. Thus, a bandwidth of 3 kJ mol^{-1} results mainly from the rotational and the translational broadening of an electronic transition to a single vibrational sublevel of the excited state of Chl a.

The absorption of polarized incident light by chlorophyll in vivo can provide information on the orientations between individual chlorophyll molecules (see Clayton, Goedheer, or Vernon and Seely). (*Polarized* means that the oscillating electric vector of light, Fig. 4.1, is in some specified direction.) The electronic transition of chlorophyll to the excited singlet state responsible for the red absorption band has its electric dipole in the plane of the porphyrin ring—actually, there are two dipoles in the plane in mutually perpendicular directions. Polarized light of the appropriate wavelength with its oscillating electric vector parallel to one of the dipoles is therefore preferentially absorbed by chlorophyll—recall that the probability for absorption is proportional to the square of the cosine of the angle between the induced dipole and the electric field vector of light (see p. 200). Using polarized light, the porphyrin rings of a few percent of the Chl *a* molecules, perhaps including P_{680}, are found to be nearly parallel to the plane of the chloroplast lamellae. However, most of the chlorophyll molecules have their porphyrin heads oriented in a random fashion in the internal membranes of the chloroplasts.

Observations on the degree of polarization of fluorescence following the absorption of polarized light can tell us whether the excitation has been transferred from one molecule to another. If the same chlorophyll molecules that absorbed polarized light later emit photons when they go back to the ground state, the fluorescence would be polarized to within a few degrees of the direction of the electric vector of the incident light. However, the chlorophyll fluorescence following absorption of polarized light by chloroplasts is not appreciably polarized. We could explain this *fluorescence depolarization* if the excitation energy had been transferred from one chlorophyll molecule to another so many times that the directional aspect had become randomized, i.e., if the fluorescence emitter were randomly aligned relative to the chlorophyll molecule that absorbed the polarized light. Such fluorescence depolarization has important implications for the excitation transfer reactions of chlorophyll that we will discuss in a later section.

When unpolarized light is incident on chloroplast lamellae that have been oriented in some particular direction, the observed fluorescence is polarized. The plane of polarization is similar to the plane of the membranes, indicating that the emitting chlorophyll molecules have their porphyrin rings in about the same orientation as the membrane. But the porphyrin rings of the absorbing chlorophyll molecules are randomly oriented, as indicated above. Again, we must conclude that the excitation has been transferred from the absorbing to the emitting molecule.

OTHER PHOTOSYNTHETIC PIGMENTS

Besides chlorophyll, other molecules in photosynthetic organisms also absorb light in the visible region. These molecules pass their electronic excitations on to Chl a; they are often referred to as auxiliary or *accessory* pigments. In addition to Chl b and Chl c, two groups of accessory pigments important to photosynthesis are the *carotenoids* and the *phycobilins*. These two classes of accessory pigments can absorb yellow or green light, wavelengths where absorption by chlorophyll is not appreciable.

Fluorescence studies have provided valuable information on the sequence of excitation transfer to and from the accessory pigments. For example, light absorbed by carotenoids, phycobilins, and Chl b leads to the fluorescence of Chl a. However, light absorbed by Chl a does not lead to the fluorescence of any of the accessory pigments, suggesting that excitation energy is not transferred from Chl a to the accessory pigments. Thus, accessory pigments can increase the photosynthetic use of white light by absorbing at wavelengths where Chl a absorption is low; the excitation is then transferred to Chl a before the photochemical reactions take place.

Carotenoids

Carotenoids are found in essentially all green plants, algae, and photosynthetic bacteria (see Goodwin or Isler). In fact, the dominant pigments for plant leaves are the chlorophylls, which absorb strongly in the red and the blue regions, and the carotenoids, which absorb mostly in the blue and somewhat in the green region of the spectrum. The predominant colors reflected or transmitted by leaves are therefore green and yellow. In the autumn, chlorophyll in the leaves of deciduous plants can bleach and not be replaced, greatly reducing absorption in the red and the blue regions. The remaining carotenoids absorb only in the blue and green regions of the visible part of the spectrum, leading to the well-known fall colors of such leaves, namely, yellow, orange, and red. Animals do not synthesize carotenoids. Hence, brightly colored birds such as canaries and flamingoes, as well as many invertebrates, obtain their yellow or reddish colors from the carotenoids in the parts of the plants they eat.

Carotenoids are bound to the chlorophyll-protein complexes, of which there are a number of types in the lamellar membranes of chloroplasts (see Fig. 1.9). Carotenoids also occur in organelles known as *chromoplasts*, which are about the size of chloroplasts and are often derived from them.

Lycopene (red) is in tomato fruit chromoplasts, α- and β-carotenes (orange) occur in carrot root chromoplasts, while the various chromoplasts of flowers contain a great diversity of carotenoids (see Kirk and Tilney-Bassett for further details on chromoplasts).

Carotenoids are 40-carbon terpenoids or isoprenoids. They are composed of 8 *isoprene* units, where isoprene is a 5-carbon compound having two double bonds (CH_2=CCH=CH_2). In many carotenoids, the isoprene units

$$|$$
$$CH_3$$

on one or both ends of the molecule are part of six-membered rings (see Fig. 5.6). Carotenoids are about 3 nm long, and those involved in photosynthesis generally contain 9 or more double bonds in conjugation.

The wavelength position of the λ_{max} depends on the solvent, on the substitutions on the hydrocarbon backbone, and on the number of double bonds in the conjugated system. We can illustrate this latter point for carotenoids in *n*-hexane, where the central maxima of the three observed peaks in the absorption spectra are at 286 nm for 3 double bonds in conjugation, at 347 nm for 5, at 400 nm for 7, at 440 nm for 9, at 472 nm for 11, and at 500 nm for 13 double bonds in conjugation—the only change from molecule to molecule in this series is in the number of double bonds in conjugation (see Goodwin 1976). Thus, the greater the degree of conjugation, the longer is the wavelength representing λ_{max}, as we discussed in Chapter 4. For the 9 to 12 double bonds occurring in the conjugated systems of photosynthetically important carotenoids, the maximum absorption coefficient is greater than $10^4 \, m^2 \, mol^{-1}$.

The carotenoids that serve as accessory pigments for photosynthesis absorb strongly in the blue (425 to 490 nm, Table 4.1) and somewhat in the green (490 to 560 nm), usually having triple-banded spectra in the region from 400 to 540 nm. For β-carotene in hexane, the three bands are centered at 425, 451, and 483 nm, and another major carotenoid, lutein, has peaks at 420, 447, and 477 nm when dissolved in ethanol (absorption spectra in Fig. 5.5). The three absorption bands in each spectrum are about 17 kJ mol^{-1} apart, reasonable values for the spacing between adjacent vibrational sublevels. Hence, the triple-banded spectra characteristic of carotenoids most likely represent transitions to three adjacent vibrational sublevels in the same excited electronic state. It is not certain which vibrational sublevels are actually involved, in part because it has proved difficult to detect any carotenoid fluorescence. The spectra of the carotenoids in vivo are shifted about 20 to 30 nm toward longer wavelengths compared with absorption when the pigments are dissolved in hexane or ethanol.

Carotenoids are usually subdivided into the hydrocarbon ones, the *caro-*

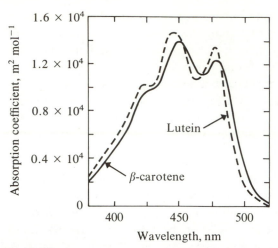

Figure 5.5
Absorption spectra for the two major carotenoids of green plants. [Data for β-carotene (in *n*-hexane) and lutein (in ethanol) are from F. P. Zscheile, J. W. White, Jr., B. W. Beadle, and J. R. Roach, *Plant Physiology 17*:331–346 (1942). Used by permission.]

tenes, and the oxygenated ones, the *xanthophylls*. The major carotene in green plants is β-carotene (absorption spectrum in Fig. 5.5, structure in Fig. 5.6); α-carotene is also abundant. The xanthophylls exhibit a much greater structural diversity than do the carotenes, since the added oxygen can be in hydroxy, keto, epoxy, or methoxy groups. The three most abundant xanthophylls in green plants are lutein (absorption spectrum in Fig. 5.5, structure in Fig. 5.6), violaxanthin, and neoxanthin, in that order, with cryptoxanthin and zeaxanthin less frequently encountered. The major carotene of algae is again β-carotene, while lutein is the most common xanthophyll, although great variation in the type and the amount of xanthophylls is characteristic of algae. For instance, golden algae, diatoms, and brown algae contain considerable amounts of the special xanthophyll, fucoxanthin, which functions as the main accessory pigment in these organisms. The distribution and the types of carotenoids in plants have important evolutionary implications and taxonomic usefulness.

In addition to functioning as accessory pigments for photosynthesis, carotenoids are also important for protecting photosynthetic organisms from destructive photooxidations which can occur in the presence of light, O_2, and certain pigments. In particular, light absorbed by chlorophyll can lead to excited states of O_2. These highly reactive states can damage chlorophyll,

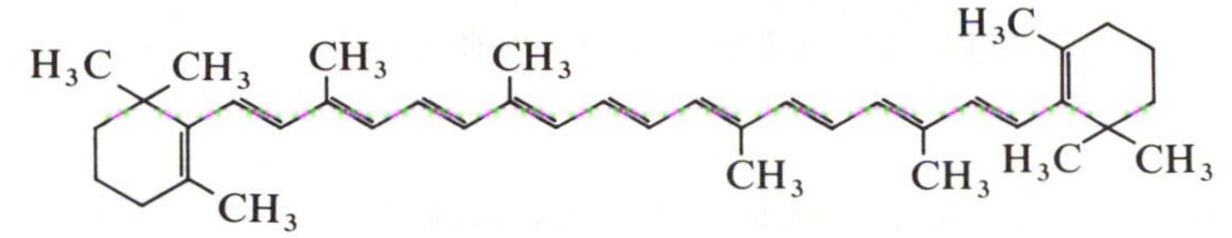

β-carotene

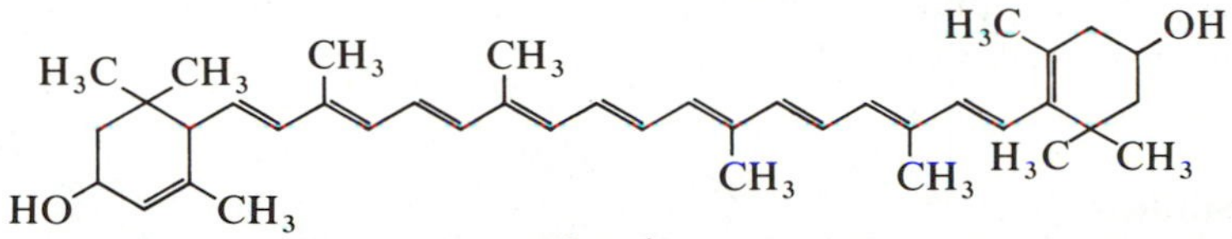

Lutein

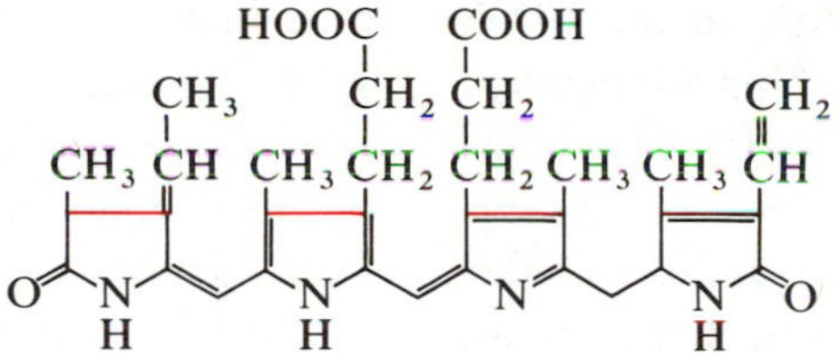

Phycoerythrobilin

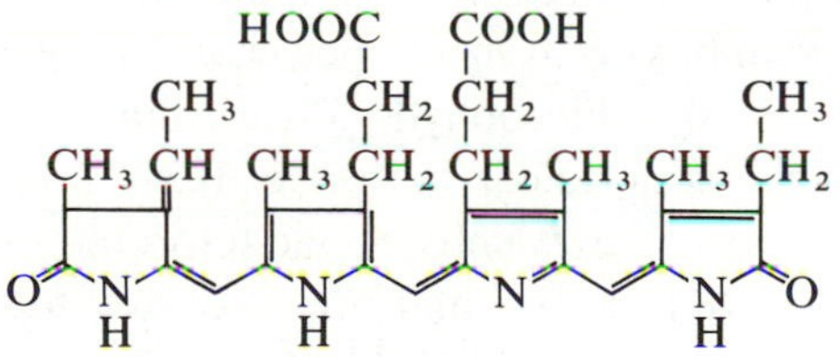

Phycocyanobilin

Figure 5.6
Structure of four important accessory pigments. The phycobilins
(lower two structures) occur covalently bound to proteins; i.e.,
they are the chromophores for phycobiliproteins.

but their interactions with carotenoids prevent harmful effects to the organism. For instance, a mutant of *Rhodopseudomonas spheroides* lacking carotenoids performs photosynthesis in a normal manner in the absence of O_2; when O_2 is introduced in the light, the bacteriochlorophyll becomes photooxidized and the bacteria are killed, a sensitivity not present in related strains containing carotenoids. In other cases, when carotenoid synthesis is inhibited, light in the presence of O_2 can be lethal to the photosynthetic organism. Since bacterial photosynthesis does not lead to O_2 evolution, it can proceed in the absence of carotenoids. However, algae and higher plants produce O_2 as a photosynthetic product, and so they must contain carotenoids to survive in the light.

Phycobilins

The other main accessory pigments important in photosynthesis are the *phycobilins*. Lemberg in the 1920's termed these molecules "phycobilins" because they resembled bile pigments. They are covalently bound to proteins, which aggregate and lead to macromolecular ensembles referred to as *phycobilisomes* (see Gantt 1980, 1981). Phycobilisomes occur associated with the outer (stromal) surfaces of the lamellar membranes in blue-green and red algae (blue-green algae are also referred to as cyanobacteria); in the cryptomonads, the biliproteins often occur packed between the photosynthetic membranes.

Like the chlorophylls, phycobilins are tetrapyrroles. However, the four pyrroles in the phycobilins occur in an open chain, not in a closed porphyrin ring, as is the case for the chlorophylls. Phycobilins have molar masses of about 0.586 kg mol^{-1} (molecular masses of 0.586 kdalton, or molecular weights of 586). They occur covalently bound to proteins that generally have molar masses of 30 to 40 kg mol^{-1} and consist of two dissimilar polypeptide subunits. Each polypeptide can bind one to four phycobilins. In vivo the proteins occur in aggregates having molar masses up to about 280 kg mol^{-1}. These stable assemblies are further organized into the phycobilisomes, which are generally 30 to 40 nm in diameter and often 5 000 to 10 000 kg mol^{-1} in molar mass. (See Dolphin; Gantt 1980, 1981; Glazer; or Goodwin 1976 for further details on phycobilins.)

Phycobilins generally have their major absorption bands in the region from 500 to 650 nm, with the Soret band occurring in the UV (Fig. 5.7).

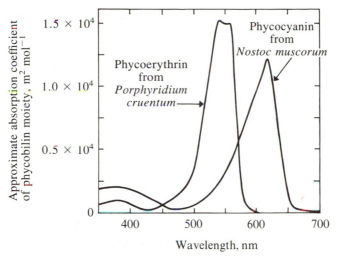

Figure 5.7

Absorption spectra of phycoerythrin from a red alga and phycocyanin from a blue-green alga. (Data are replotted from C. Ó hEocha, in *Chemistry and Biochemistry of Plant Pigments*, T. W. Goodwin, ed., Academic Press, London, 1965, pp. 175–196. Used by permission.)

These pigments are higher in concentration in many algae than are the chlorophylls and are responsible for the color of certain species. The main phycobilins are phycocyanobilin and phycoerythrobilin (structures in Fig. 5.6). Phycoerythrobilin plus the protein to which it is covalently attached is called *phycoerythrin*.* Phycoerythrin is soluble in aqueous solutions, so we can obtain absorption spectra for it under conditions similar to those in vivo. Phycoerythrin is reddish, since it absorbs green and has at least one main band between 530 and 570 nm (see absorption spectrum in Fig. 5.7). It occurs throughout the red algae, in some blue-green algae, and in the cryptomonads. *Phycocyanin* (phycocyanobilin plus protein) appears bluish, because it absorbs strongly from 610 to 660 nm (Fig. 5.7). It is the main phycobilin in the blue-green algae and is also found in the red algae and the cryptomonads.

---

* The phycobilins are covalently bound to their proteins (referred to as *apoproteins*) to form phycobiliproteins, whereas chlorophylls and carotenoids are joined to their apoproteins by weak bonds such as H-bonds and hydrophobic interactions.

As is the case for other pigments, the greater the number of double bonds in conjugation in the phycobilins, the longer are the wavelengths for λ_{max}. For example, phycoerythrobilin has 6 double bonds in the main conjugated system and absorbs maximally in the green region of the spectrum, while phycocyanobilin has 9 such double bonds and its λ_{max} occurs in the red (see the structures for these compounds in Fig. 5.6). As indicated in Figure 5.7, the maximum absorption coefficients of both phycobilins exceed 10^4 m^2 mol^{-1}.

As we indicated in Chapter 4 (p. 196), both the quantity and the quality of radiation change with depth in water, with wavelengths slightly less than 500 nm penetrating the furthest. For instance, only about 10% of the blue and the red parts of the spectrum penetrate to 50 m in clear water, so chlorophyll would not be a very useful light-harvesting pigment at that depth (see Fig. 5.3 for a chlorophyll absorption spectrum). Although there are many exceptions, changes in the spectral quality (relative amounts of various wavelengths) with depth indeed affect the distribution of photosynthetic organisms according to their pigment types. The predominant accessory pigment in green algae is Chl b, which absorbs mainly in the violet (400 to 425 nm; see Table 4.1) and the red (640 to 740 nm). Green algae as well as sea grasses and fresh water plants grow in shallow water, where the visible spectrum is little changed from that of the incident sunlight. Fucoxanthin is the major accessory pigment in brown algae, such as the kelps, and it absorbs strongly in the blue and green regions (425 to 560 nm), helping to extend the range of such plants downward to over 20 m. Marine red algae can occur at even greater depths (e.g., 100 m), and their phycoerythrin absorbs the green light (490 to 560 nm) which penetrates to such distances. Changes in spectral quality can also induce changes in the synthesis of biliproteins within a given organism. For instance, green light induces the synthesis of the green-absorbing phycoerythrin, and red light leads to the synthesis of the red-absorbing phycocyanin (see Fig. 5.7) in certain blue-green and red algae (see Glazer or Jeffrey for further details).

Interestingly enough, only two types of pigments appear to be involved in all the known photochemical reactions in plants. These are the carotenoids and the tetrapyrroles, the latter class including the chlorophylls, the phycobilins, and phytochrome. The maximum absorption coefficients for the most intense absorption bands are slightly over 10^4 m^2 mol^{-1} in each case. Cytochromes, which are involved in the electron transport reactions in chloroplasts and mitochondria, are also tetrapyrroles, as we will indicate later in this chapter. Table 5.1 summarizes the relative frequency of the main types of photosynthetic pigments that we have been discussing.

Table 5.1

Approximate relative amounts and locations of photosynthetic pigments. Data are expressed per 600 chlorophylls and are for representative leaves of green plants growing at moderate light levels (except for the phycobilins). Photosystems and the light-harvesting antenna are discussed later in this chapter; see p. 270. See Clayton, Goodwin 1976, and Thornber et al. for further details.

Pigment	Number	Location
Chl a	450	Approximately 30% in Photosystem I, 20% in Photosystem II, and 50% in light-harvesting antenna
Chl b	150	In light-harvesting antenna
P_{680}	1	Trap for Photosystem II
P_{700}	1	Trap for Photosystem I
Carotenoids	120	Some in Photosystems I and II, but most serving in light-harvesting antenna
Phycobilins	500	Covalently bound to proteins situated on the outer surface of photosynthetic membranes in red and blue-green algae; serving in light-harvesting antenna

EXCITATION TRANSFERS AMONG PHOTOSYNTHETIC PIGMENTS

Chlorophyll is at the very heart of the primary events of photosynthesis. It helps to convert the plentiful radiant energy from the sun into chemical free energy that can be stored in various ways. We will here represent light absorption, excitation transfer, and the photochemical step as chemical reactions; this will serve as a prelude to a further consideration of certain molecular details of photosynthesis.

The first step in photosynthesis is light absorption by one of the pigments. The absorption event (discussed in Ch. 4) for the various types of photosynthetic pigments described in this chapter can be represented as follows:

$$
\left.\begin{array}{c}
\text{accessory pigment} \\
\text{or} \\
\text{Chl } a \\
\text{or} \\
\text{trap chl}
\end{array}\right\} + h\nu \longrightarrow \left\{\begin{array}{c}
\text{accessory pigment*} \\
\text{or} \\
\text{Chl } a* \\
\text{or} \\
\text{trap chl*}
\end{array}\right. \qquad (5.2)
$$

where the asterisk refers to an excited state of the pigment molecule caused by the absorption of a light quantum, hv. *Trap chl* indicates a special type of Chl a (e.g., P_{680} or P_{700}) that occurs rather infrequently (see Table 5.1); we will consider its important excitation-trapping properties at the end of this section.

Since the photochemical reactions take place only at the trap chl molecules, the excitations resulting from light absorption by either the accessory pigments or the other Chl a's must be transferred to the trap chl before they can be used for photosynthesis. The relative rarity of trap chl compared with the other photosynthetic pigments means that it absorbs only a small fraction of the incident light. In fact, under natural conditions in green plants over 99% of the photons are absorbed by either the accessory pigments or Chl a. The migration of excitations from the initially excited species to the trap chl—the mechanism for which we will discuss below—can be represented as follows:

$$\text{accessory pigment*} + \text{Chl } a \longrightarrow \text{accessory pigment} + \text{Chl } a^* \quad (5.3)$$

$$\text{Chl } a^* + \text{trap chl} \longrightarrow \text{Chl } a + \text{trap chl*} \quad (5.4)$$

In other words, the direction of excitation transfer or migration is from the accessory pigments to Chl a (Eq. 5.3), and from Chl a to the special "trap" chlorophylls (Eq. 5.4) where the actual photochemical reactions take place. Hence, the overall effect of the steps described by Equations 5.2 through 5.4 is to funnel the excitations that are caused by the absorption of light to the trap chl.

A prerequisite for the conversion of radiant energy into a form that can be stored chemically is the formation of reducing and oxidizing species. The *reducing* (electron-donating) and *oxidizing* (electron-accepting) species that result from light absorption must be fairly stable and separated in such a way that they do not interact. (We will discuss oxidation and reduction in detail in Ch. 6.) If we denote the molecule that accepts an electron from the excited trap chl by A, this electron transfer step can be represented by

$$\text{trap chl*} + A \longrightarrow \text{trap chl}^+ + A^- \quad (5.5)$$

where A^- indicates the reduced state of the acceptor and trap chl$^+$ means that the special chlorophyll has lost an electron. Equation 5.5 represents a *photochemical* reaction since the absorption of a light quantum (Eq. 5.2) has led to the transfer of an electron away from a special type of chlorophyll, representing a chemical change in that molecule. The electron removed from trap chl* (Eq. 5.5) can be replaced by one coming from a donor, D, which leads to the oxidation of this latter species, D^+, and the return of the trap chl to its normal unexcited state:

$$\text{trap chl}^+ + D \longrightarrow \text{trap chl} + D^+ \qquad (5.6)$$

The generation of stable reduced (A^-) and oxidized (D^+) intermediates can be considered to complete the conversion of light energy into chemical potential energy. It proves convenient to combine Equations 5.2 through 5.6, which gives us the following relation for the net reaction describing the primary events of photosynthesis:

$$A + D + h\nu \longrightarrow A^- + D^+ \qquad (5.7)$$

In green plants and algae the light-driven change in chemical free energy represented by the conversion of $A + D$ to $A^- + D^+$ (Eq. 5.7) eventually causes chemical reactions leading to the evolution of O_2 from water, the production of a reduced compound (NADPH), and the formation of high energy phosphates (ADP + phosphate $\longrightarrow$ ATP). Such a conversion of light energy into chemical energy represented by Equation 5.7 is the cornerstone of photosynthesis.

In this section we will consider a mechanism for the transfer of an excitation from an excited molecule to another in its ground state. The concepts developed will then be applied to the specific case of the migration of an excitation among photosynthetic pigments, i.e., reactions described by Equations 5.3 and 5.4. Finally, we will discuss the trapping of the excitation in special Chl a molecules. This trapping is followed by the separation of an electron from chlorophyll, a necessary step for the generation of chemical potential energy.

Resonance Transfer of Excitation

We have already mentioned a number of different examples of excitation transfer between photosynthetic pigments. For instance, light absorbed by the accessory pigments can lead to the fluorescence of Chl a. Also, studies on the absorption of polarized light by chlorophyll in vivo, where the resulting fluorescence is not polarized, provide further evidence that excitations can migrate from molecule to molecule before the energy is emitted as radiation.

The simplest case to consider is the transfer of excitation between identical molecules; e.g., the excitation of the lower excited singlet state of chlorophyll can be passed on to a second chlorophyll molecule. This causes the deactivation of the originally excited molecule and the attainment of the lower excited singlet state in the second chlorophyll, a process described by Equation 4.6, $S_{(\pi, \pi*)} + S_{2(\pi, \pi)} \longrightarrow S_{(\pi, \pi)} + S_{2(\pi, \pi*)}$. The most widely accepted mechanism for such exchange of electronic excitation between chlorophyll molecules is *resonance transfer* (also called inductive resonance, the Förster mechanism, or weak coupling), which we will next consider qualitatively.

On the basis of our discussion in the previous chapter, we might expect that an excited molecule could induce an excited state in a second molecule in close proximity. The oscillating electric dipole representing the energetic electron in the excited state of the first molecule leads to a varying electric field. This field can cause a similar oscillation or resonance of some electron in a second molecule. A transfer of electronic excitation energy takes place when the electron oscillation in the second molecule is induced. When excitation transfer is completed, the previously excited electron in the first molecule has ceased oscillating while some electron in the second molecule is now oscillating, leading to an excited state of that molecule. Resonance transfer of excitation between molecules is thus analogous to the process by which light is originally absorbed, since an oscillation of some electron in the molecule is induced by a locally varying electric field (see p. 199). Resonance transfer of excitation is most probable when there is the proper orientation between the electric dipole in the excited molecule and the potential dipole in the second molecule *and* the energy of the original dipole is appropriate, an aspect we will consider next.

For resonance transfer of electronic excitation to occur, the energy available in the excited molecule must match the energy that can be accepted by a second molecule. The wavelengths for fluorescence indicate the energy of the excited singlet state of a molecule (at least after the very rapid radiationless transitions to the lowest vibrational sublevel of that excited state have occurred). Thus, although fluorescence itself is not involved in this type of excitation transfer, the fluorescence emission spectrum usually gives the range of energies available for transfer to a second molecule. The range of wavelengths of light that can sympathetically induce an oscillation of some electron in a second molecule is given by the absorption spectrum of that molecule (Ch. 4), and therefore the absorption spectrum shows the energies that can be accepted by a molecule. As might be expected from these two considerations, the probability for resonance transfer is high when the overlap

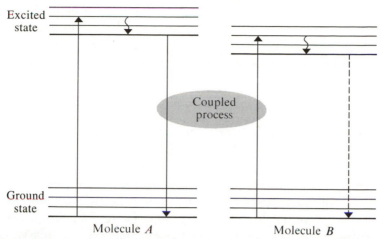

Figure 5.8

Resonance transfer of excitation from molecule A to molecule B. After light absorption by molecule A, there is a radiationless transition to the lowest vibrational sublevel of its excited state. Next, resonance transfer of the excitation takes place from A to B, causing the second molecule to go to an excited state, while molecule A returns to its ground state. Following a radiationless transition to the lowest vibrational sublevel in the excited state, fluorescence is then emitted by molecule B as it returns to its ground state. Based on the energy level diagrams (which include the vibrational sublevels for each of these two different pigments), we can conclude that generally the excitation rapidly decreases in energy after each intermolecular transfer between dissimilar molecules.

in wavelength between the fluorescence band for the excited oscillator (available energy) and the absorption band of an unexcited oscillator (acceptable energy) in a neighboring molecule is large. Since the overlap in the red region between the absorption and the fluorescence spectra of Chl a is very great (Fig. 5.3), excitations can be efficiently exchanged between Chl a molecules by resonance transfer. Figure 5.8 illustrates the various energy considerations involved in the case of resonance transfer of excitation between two different types of molecules.

The probability for resonance transfer of electronic excitation decreases as the distance between the two molecules increases. If chlorophyll molecules were uniformly distributed in three dimensions in the lamellar membranes of chloroplasts, they would have a center-to-center spacing of approximately 2 nm, an intermolecular distance over which resonance transfer of excitation can readily occur (resonance transfer is effective up to about 10 nm for chlorophyll). Thus, both the spectral properties of chlorophyll and its spacing in the lamellar membranes are conducive to an efficient migration of excitation from molecule to molecule by resonance transfer. (See Barber 1977, or Clayton, for a further discussion of resonance transfer.)

Transfers of Excitation Between Photosynthetic Pigments

In addition to the transfer from one Chl a molecule to another, excitations can also migrate by resonance transfer from the accessory pigments to Chl a. The transfers of excitations among Chl a's can be nearly 100% efficient (i.e., $k_{transfer} \gg$ rate constants for competing pathways; see p. 215), whereas the fraction of excitation transfers between dissimilar molecules is quite variable. For instance, the transfers of excitations from β-carotene to Chl a are very efficient in certain algae, but the fraction of excitation transfers from most xanthophylls to Chl a is generally low. The transfer of excitations from α-carotene and lutein to chlorophyll usually is intermediate; e.g., about 40% of the excitations of these carotenoids may be transferred to Chl a. Photons absorbed by the carotenoid fucoxanthin—found in brown algae and diatoms—can approach 100% efficiency of excitation transfer to Chl a. Also, most excitations of phycobilins and Chl b can be transferred to Chl a. In red algae, 90% of the electronic excitations produced by the absorption of photons by phycoerythrin can be passed on to phycocyanin and then to Chl a; these transfers require about 4×10^{-10} s each. (See Clayton, Goedheer, or Rabinowitch and Govindjee for efficiencies of excitation transfers between pigments.) We will next consider the direction for excitation transfer between various photosynthetic pigments and then the times involved for intermolecular excitation transfers of chlorophyll.

Some energy is generally lost by each molecule to which the excitation is transferred. Any excess vibrational or rotational energy is usually dissipated rather rapidly as heat (see Fig. 5.8). Therefore, the wavelength positions of the λ_{max} for each type of pigment involved in the sequential steps of excitation transfer tend to become longer in the direction in which the excitation migrates. (The fluorescence emission spectrum of some molecule—which, for transfer to take place efficiently, must appreciably overlap the absorption spectrum of the molecule to which the excitation is resonantly transferred—occurs at longer wavelengths than the absorption spectrum of that molecule—see Fig. 5.3 for Chl a.) Therefore, for a second molecule to become excited by resonance transfer, it should have an absorption band at longer wavelengths (lower energy) than the absorption band for the molecule from which it receives the excitation. Thus, the direction for excitation migration by resonance transfer among photosynthetic pigments is usually toward those pigments with longer λ_{max}. We can appreciate this important aspect by considering Figure 5.8. If excitation to the second excited vibrational sublevel of the excited state is the most probable transition predicted by the Franck–Condon principle for each molecule, then the excitation of molecule A requires more energy than that of B (the pigment to which the excitation is transferred). Hence, λ_{max_A} must be less than λ_{max_B} in the two absorption spectra, which is consistent with our statement that the excitation migrates toward the pigment with the longer λ_{max}.

As specific examples of the tendency for excitations to migrate toward pigments with longer λ_{max} in their absorption spectra, we will consider the transfer of excitations from the accessory pigments to Chl a. In red algae and in some blue-green algae, phycoerythrin has a λ_{max} at about 560 nm and passes excitation energy on to phycocyanin, which has an absorption maximum near 620 nm. This excitation can then be transferred to a Chl a with a λ_{max} near 670 nm. The biliprotein allophycocyanin absorbs maximally at 650 nm and apparently intervenes in the transfer of excitation between phycocyanin (λ_{max} at 620 nm) and Chl a in some blue-green and red algae. Cryptomonads contain Chl c, which also has a λ_{max} near 650 nm; for such organisms, excitation transfer may be from phycoerythrin (λ_{max} at 560 nm) to phycocyanin (620 nm) to Chl c and then to Chl a. Since some of the excitation energy is generally dissipated as heat by each molecule (see Fig. 5.8), the excitation represents less energy (longer λ) after each pigment in sequence. Consequently, the overall direction for excitation migration is essentially irreversible, a point to which we will return below.

As the nuclei vibrate back and forth after the absorption of a photon by some electron, their collisions with other nuclei every 10^{-13} s or so can lead in such short times to the dissipation of any excess energy in the excited vibrational sublevels. In addition, the radiationless transition from the upper excited singlet to the lower excited singlet of Chl a—$S^b_{(\pi, \pi*)}$ to $S^a_{(\pi, \pi*)}$ in Figure 4.5—is completed within 10^{-12} s. The time for the transfer of the excitation between two Chl a molecules in vivo is somewhat longer—about 1 to 2 $\times$ 10^{-12} s. Thus, the originally excited chlorophyll molecule usually attains the lowest vibrational sublevel of the lower excited singlet state before the excitation is transferred to another molecule. The amount of energy resonantly transferred from one Chl a to another therefore generally corresponds to the energy indicated by the fluorescence emission spectrum (Fig. 5.3). An excitation representing this amount of energy can in principle be transferred many times by resonance transfer with essentially no further degradation of the energy.

In Chapter 4 we noted that an upper time limit within which processes involving excited singlet states must occur is provided by the kinetics of fluorescence de-excitation. The lifetime for chlorophyll fluorescence from the lower excited singlet state is about 1.5×10^{-8} s. Thus, time is available for approximately 10 000 transfers of excitation among the Chl a molecules— each transfer requiring about 10^{-12} s—before the loss of the excitation by the emission of fluorescence. The number of excitation transfers between Chl a molecules actually taking place is much less than this, for reasons that will shortly become clear.

Excitation Trapping

We have already introduced the special Chl a's, P_{680} and P_{700}. These pigments tend to absorb at longer wavelengths than do the other types of Chl a; hence, the excited singlet states in P_{680} and P_{700} tend to be at lower energies. It is energetically feasible for the other Chl a's to excite such trap chl's by resonance transfer, but P_{680} and P_{700} usually do not pass the excitation back; i.e., they rapidly lose some of their excitation energy (within 10^{-12} s), and so they do not retain enough energy to re-excite the other Chl a's by resonance transfer. Therefore, the excited singlet states of other Chl a molecules can readily have their excitations passed on to the trap chl's, but not vice versa—analogous to the irreversibility of the migration of excitations from the accessory pigments to Chl a. The excitation resulting from the absorption of radiation by the various photosynthetic pigments is thereby funneled into P_{680} or P_{700}. Such collecting of excitations by one species is the net effect of Equations 5.2 through 5.4 (with the term "trap chl" replaced by P_{680} or P_{700}).

Since one of the trap chl's is present per approximately 300 chlorophylls (Table 5.1), on the average only a few hundred transfers are necessary to get an excitation from Chl a to P_{680} or P_{700}. Thus, the 10 000 possible transfers of excitation from one Chl a to another possible within the fluorescence lifetime do not actually occur. Since each excitation transfer takes approximately 10^{-12} s, 100 transfers would require about 10^{-10} s. In agreement with this, both calculations from mathematical models and ingenious experimentation have shown that over 90% of the excitations of Chl a can migrate to P_{700} (or another trap chl) in less than 10^{-9} s (see Clayton, or Rabinowitch and Govindjee).

The characteristics of fluorescence provide information on the lifetime of the excited singlet state of chlorophyll in vivo and thus on the time available for migration of excitations. Specifically, about 1% to 3% of the light absorbed by Chl a in vivo is lost by fluorescence. The amount reradiated depends on the competing de-excitation reactions, and hence is higher at higher incident light levels where the photochemical reactions become saturated. Thus, near full sunlight the quantum yield for fluorescence in vivo, Φ_{Fl}, is approximately 0.03. Equation 4.12 ($\Phi_i = \tau/\tau_i$) indicates that this quantum yield is equal to τ/τ_{Fl}, where τ is the lifetime of the excited singlet state and τ_{Fl} is its fluorescence lifetime. It is reasonable to assume that τ_{Fl} in vivo is similar to the fluorescence lifetime of Chl a in vitro, 1.5×10^{-8} s. Therefore, Equation 4.12 predicts a lifetime for the excited state of Chl a in vivo of $(0.03)(1.5 \times 10^{-8}$ s), or 0.5×10^{-9} s. This is another estimate of the average time necessary for the excitation to migrate to the trap chl (actually, not all of the Chl a readily fluoresces in vivo, and hence this is only an approximate guide to the lifetime; see Barber 1978).

P_{680} and P_{700} act as traps for excitations in chloroplast lamellae, while a special type of bacteriochlorophyll with a λ_{max} between 870 and 890 nm (depending on the species) acts in an analogous manner in bacteria. One of the useful features of such excitation traps is to have an excited singlet state lower in energy than the excited singlet states in the other pigment molecules. This lower energy state (longer λ_{max} for absorption) is a consequence of the molecular environment in which the chlorophyll molecules acting as excitation traps are located. Moreover, the longer λ_{max} ensures the directionality for the migration of excitations. Another characteristic of the trap chl's is their relative rarity. Thus, most of the photosynthetic pigments act as "antennae," or light harvesters, which collect the radiation and channel the excitations toward the trap chl, as illustrated in Figure 5.9. Processing of the excitation originally caused by light takes place only at the trap chl's. This participation in the essentially irreversible, electron-transfer process is the crucial feature of an excitation trap.

When the excitation migrates to a trap such as P_{680} or P_{700}, this special Chl a goes to an excited singlet state, as would any other Chl a. Since the trap chl cannot readily excite other chlorophylls by resonance transfer, it might become de-excited by the emission of fluorescence. However, very little fluorescence from the trap chl's is observed in vivo. This is explained by the occurrence of a photochemical event (see Eq. 5.5, trap chl* + $A \longrightarrow$ trap chl$^+$ + A^-); i.e., the rapid donation (within 10^{-10} s) of an electron to an acceptor could prevent the de-excitation of the trap chl's by fluorescence, which has a longer lifetime.

As we have just indicated, an excited trap chl can rapidly donate an electron to some acceptor molecule, which is part of the *photochemistry*

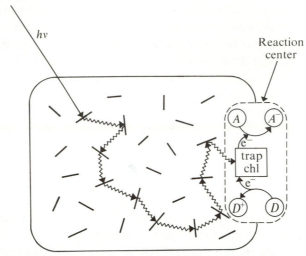

Figure 5.9
Schematic indication of a group of pigments that harvests a light
quantum (hv) and passes the excitation on to a special trap chlorophyll.
In the reaction center an electron (e^-) is transferred from the trap chl
to some acceptor (A^- in the reduced form) and then replaced by another electron coming from a suitable donor (D^+ in the oxidized form).

of photosynthesis. The donation of the electron initiates the *chemical* reactions of photosynthesis and the subsequent storage of energy in stable
chemical bonds. Moreover, once the trap chl has lost an electron, it can
take on another from some donor, indicated by D in Equation 5.6 (trap
chl$^+$ + D ⟶ trap chl + D^+) and Figure 5.9. Thus, the photochemical
reactions of photosynthesis lead to electron flow. An excitation trap such
as P_{680} or P_{700} plays a key role in the conversion of radiant energy into
forms of energy that are biologically useful. We generally refer to the trap
chl plus A and D as a *reaction center* (illustrated in Fig. 5.9). The reaction
center is the locus for the photochemistry of photosynthesis (Eq. 5.5).

GROUPINGS OF PHOTOSYNTHETIC PIGMENTS

We have discussed the absorption of light by photosynthetic pigments and
the ensuing transfers of excitation among these molecules, which leads us
to a consideration of whether there are discrete units of such pigments
acting together in some concerted fashion. Such an ensemble was presented

in Figure 5.9, where we indicated that the light-harvesting photosynthetic pigments greatly outnumber the special trap chl molecules, the latter possibly occurring in a one-to-one relationship with suitable electron acceptors and donors. In this section we will consider whether photosynthetic pigments are organized into functional groups. If they are, how many molecules make up one group? Are various groupings of pigments identical in terms of their function in photosynthesis?

Photosynthetic Units

At low light levels one CO_2 can be fixed and one O_2 evolved for approximately every eight photons absorbed by any of the photosynthetic pigments. Is one O_2 evolved for every eight photons absorbed at *high* light levels? Data to answer this question were provided in 1932 by Emerson and Arnold, who exposed the green alga *Chlorella* to a series of intense flashes of light (see Clayton, or Rabinowitch and Govindjee). These intense flashes excited practically all chlorophyll molecules and other photosynthetic pigments simultaneously. But the maximum yield in such experiments was only one evolved O_2 for every 2 000 to 2 500 chlorophyll molecules. Assuming each chlorophyll molecule absorbed one photon, this means that 300 times more photons were needed than the eight needed to produce one O_2 at low light levels. At low light levels there is sufficient time between the arrival of individual photons for the excitations of the accessory pigments and Chl *a* to be efficiently collected in a trap chl and used for the chemical reactions of photosynthesis. At high light levels, however, many photosynthetic pigments become excited at the same time, and only one excited chlorophyll out of about 300 leads to any photochemical reaction. One possible interpretation is that 300 chlorophylls could be acting together as a *photosynthetic unit*; when any one chlorophyll in this photosynthetic unit becomes excited by light and then has its excitation transferred to the reaction center, the concomitant excitation of other chlorophylls in that unit cannot be used for photosynthesis. However, the calculation really only indicates the number of chlorophylls per reaction center, and should not be taken to imply structurally definable units.

The above conclusions can also be considered in terms of Figure 5.9, where the trap chl is shown interacting with the electron acceptor *A* and donor *D*. The rate-limiting step for photosynthesis is not light absorption, excitation transfer, or photochemistry (electron donation by trap chl*), but the subsequent steps leading to O_2 evolution and CO_2 fixation. A brief intense illumination thus leads to more excitations than can be processed by the electron transfer reactions and subsequent biochemical events. In the limit of a very intense flash simultaneously exciting all photosynthetic pigments, one excitation is processed by each reaction center, but all others are dissipated by various nonphotochemical de-excitation processes, such as those discussed in the previous chapter.

Excitation Processing

The electron excitation originally caused by the absorption of a photon can be processed by the chemical reactions leading to CO_2 fixation about once every 5×10^{-3} s (5 ms). This processing time has important consequences for both the efficiency of light use at different photon flux densities and the optimal number of chlorophylls per reaction center.

The highest photon flux density normally encountered by plants occurs when the sun is directly overhead on a cloudless day, in which case the photosynthetic photon flux density (PPFD), also designated PAR (p. 193), for wavelengths from 400 to 700 nm is about $2\,000$ μmol m^{-2} s^{-1}. The average chlorophyll concentration in chloroplasts is approximately 30 mol m^{-3} (p. 226), and in passing through a chloroplast 2 μm thick about 30% of the incident PAR is absorbed. We can therefore estimate how often an individual chlorophyll molecule might absorb a photon. Specifically, $(0.3)(2\,000 \times 10^{-6}$ mol photons m^{-2} s^{-1}) or 600×10^{-6} mol photons m^{-2} s^{-1} is absorbed by (30 mol chlorophyll m^{-3})(2 $\times$ 10^{-6} m) or 60×10^{-6} mol chlorophyll m^{-2}, which is 10 mol photons (mol chlorophyll)$^{-1}$ s^{-1}. Thus, 10 photons per second are absorbed on the average by each chlorophyll in a chloroplast exposed to full sunlight.

As we have just calculated, each chlorophyll in an unshaded chloroplast could absorb a photon on the average about once every 0.1 s. When there are 300 chlorophylls per reaction center, 15 of these molecules would be excited every 5 ms ($3\,000$ excitations s^{-1} $\times$ 0.005 s). However, since the average processing time per reaction center is about 5 ms, only one of these 15 excitations could be used photochemically—the others are dissipated by nonphotochemical de-excitation reactions. Consequently, although the chemical reactions leading to CO_2 fixation will operate at their maximum rate under such conditions of high PAR, a large fraction of the electronic excitations caused by light absorption will not be used for photosynthesis.

Full midday sunlight is seldom incident on a chloroplast under natural conditions, since chloroplasts are generally shaded by other chloroplasts in the same cell, by plastids in other cells, and perhaps by overlying leaves.

Furthermore, the amount of sunlight incident on a plant is much less at sunrise or sunset and on overcast days than near noon on a clear day. For the sake of argument, let us consider that the PAR incident on a chloroplast is one tenth of that from the direct midday sun, namely 200 μmol m^{-2} s^{-1}. In this case, each chlorophyll in the chloroplast absorbs a photon once every second, which is a much longer time between the arrival of photons than the time required for processing an electronic excitation. However, when individual chlorophylls are excited every 1 s at this moderate PAR, one chlorophyll out of the 300 per reaction center would be excited on the average every 3 ms (1 s per 300 excitations means 1 000 ms/300 or 3 ms per single excitation). This excitation frequency is such that the photons can be efficiently used for photosynthesis. In other words, the photons are arriving at a rate such that the excitations produced by most of them can be used; moreover, the chemical reactions are working at their maximum capacity. Consequently, a reaction center, with its photochemistry and associated enzymatic reactions, will function very effectively at a moderate PAR.

What would happen if there were but one chlorophyll molecule per reaction center? If the chemical reactions required 5 ms as above, this single pigment molecule would be excited about once every second at a PAR of 200 μmol m^{-2} s^{-1}, and the excitation could easily be processed by the chemical reactions. But the photochemical step plus the subsequent enzymatic reactions leading to CO_2 fixation would be working at only 0.5% of capacity—(5 $\times$ 10^{-3} s)/(1 s), or 0.005, is the fraction of time they could be used. In other words, although all the absorbed photons would be used for photosynthesis, even the slowest of the chemical steps would be idle over 99% of the time.

Thus, given the 5 ms processing time for the chemical reactions, 300 chlorophylls per reaction center connected with the appropriate enzyme machinery provides a plant with a mechanism for efficiently handling the usual PAR's found in nature—both for harvesting the photons and for using the chemical reactions at a substantial fraction of their capacity. For instance, averaged over the earth's surface and the year, the mean PAR reaching the ground during the daytime on clear days is about 800 μmol m^{-2} s^{-1}. The total area of all leaves divided by the total land area is 4.3 (see Lieth and Whittaker; this ratio, called the leaf area index, will be discussed more fully in Ch. 9), and so the average PAR on a leaf is somewhat less than 200 μmol m^{-2} s^{-1}. At this PAR level, which can also occur for chloroplasts in cells on the side away from the fully sunlit side of a leaf, the rate of photon absorption is well matched to the rate of excitation processing.

In addition to interspecific variations, the ratio of chlorophyll per reaction center can depend on the PAR level present during plant development. Some algae and leaves of land plants developing under low PAR can have over 700 chlorophylls per reaction center, whereas certain leaves developing under full sunlight can have as few as 100. The ratio is fairly low in bacteria, where there are generally 40 to 100 bacteriochlorophyll molecules per reaction center (see Clayton).

Photosynthetic Action Spectra and Enhancement Effects

The electronic excitations resulting from light absorption by any photosynthetic pigment can be transferred to a trap chl and thus lead to photochemical reactions. It would therefore be reasonable to expect that the absorption spectrum of chloroplasts would fairly well match the action spectrum for photosynthesis. However, the action spectrum for CO_2 fixation or O_2 evolution and the overall absorption spectrum for the photosynthetic pigments in the same organism all do not precisely coincide. Among other things, there is a so-called "red drop," in which the photosynthetic action spectrum drops off much more rapidly in the red region beyond 690 nm than does the absorption spectrum for chlorophylls and other pigments—as illustrated in Figure 5.10 for the case of a green alga.

Emerson demonstrated in 1957 that the relatively low photosynthetic efficiency of *Chlorella* in far-red light—a red drop like that in Figure 5.10—could be increased by simultaneously using light of a shorter wavelength along with the far-red light (see Emerson). The photosynthetic rate with the

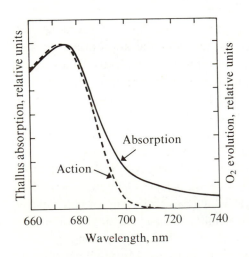

Figure 5.10
Absorption spectrum for an algal thallus and the action spectrum for its O_2 evolution, illustrating the "red drop" in photosynthesis. [Data are for *Ulva taeniata* and are taken from F. T. Haxo and L. R. Blinks, *Journal of General Physiology* 33:389–422 (1950). Used by permission.]

two beams could be 30% to 40% greater than the sum of the rates of far-red light and the shorter wavelength light used separately. We can also describe the enhancement in terms of quantum yield (defined on p. 215). For example, we could determine the quantum yield for O_2 evolution by *Chlorella* using 700 nm radiation in both the presence and the absence of 650 nm light. In such an experiment the quantum yield using the 700 nm radiation is higher when the 650 nm light is also present (see Myers). Such synergism, or enhancement, suggests that photosynthesis involves the cooperation of two distinct photochemical reactions. Light of wavelengths greater than 690 nm might mainly power only one of the two necessary reactions, and thus photosynthesis could not proceed at an appreciable rate. When shorter wavelengths are also used, however, the other necessary reaction could take place, resulting in a marked enhancement of the photosynthetic rate or quantum yield. Also, since photosynthesis is enhanced by adding a shorter wavelength to far-red illumination, a smaller percentage of light should then be reradiated as fluorescence. Indeed, the two wavelengths together do evoke less fluorescence than the two acting separately.

Two Photosystems Plus Light-Harvesting Antenna

We can use the photosynthetic enhancement effect to study the pigments in each of the two photochemical systems involved in photosynthesis. The system containing the pigments absorbing beyond 690 nm is referred to as Photosystem I, a terminology introduced by Duysens, Amesz, and Kamp in 1961 (see Duysens et al.). Much of the far-red absorption by Photosystem I is due to a type of Chl *a* with a λ_{max} near 680 nm. The special Chl *a* dimer, P_{700}, is found exclusively in Photosystem I. Therefore, light above 690 nm is absorbed mainly by this long wavelength form of Chl *a* and P_{700} in Photosystem I.

An action spectrum for the enhancement of photosynthesis in the presence of a constant irradiation with wavelengths longer than 690 nm, which are absorbed by Photosystem I, should indicate the pigments in the other system, Photosystem II. For example, the action spectrum for such photosynthetic enhancement represented in Figure 5.11 for the red alga *Porphyridium cruentum* resembles the absorption spectrum of phycoerythrin (λ_{max} near 540 nm; see Fig. 5.7). A marked increase in photosynthesis in the blue-green alga *Anacystis nidulans* occurs when far-red light is supplemented by light absorbed by phycocyanin (see Fig. 5.11 and the absorption spectrum for phycocyanin in Fig. 5.7). It is reasonable to conclude, therefore, that in such

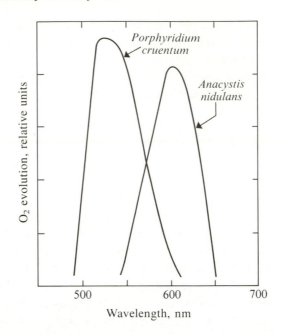

Figure 5.11
Action spectra for the enhancement of O_2 evolution for two algae. Cells were exposed to a constant red illumination (wavelengths beyond 690 nm) plus a specific photon flux density at the various wavelengths indicated on the abscissa. The ordinate represents the rate of O_2 evolution for the two beams acting together minus that produced by the shorter wavelength acting alone. The pigment absorbing the light leading to enhancement of O_2 evolution in *Porphyridium cruentum* is phycoerythrin, while phycocyanin is responsible in *Anacystis nidulans*—see absorption spectra in Figure 5.7. [Source: R. Emerson and E. Rabinowitch, *Plant Physiology 35*:477–485 (1960). Data used by permission.]

algae the phycobilins mainly funnel their excitations into Photosystem II. Studies using photosynthetic enhancement also indicate that the excitations of Chl *b* preferentially go to Photosystem II, as do those of fucoxanthin in the brown algae. In a related experimental approach, action spectra for the evocation of Chl *a* fluorescence indicate that light absorbed by the phycobilins and Chl *b* leads to the fluorescence of Chl *a* in Photosystem II, perhaps including fluorescence by P_{680}. But the results have not always been clear-cut, since photons absorbed by Chl *b* and the phycobilins sometimes lead to photochemical reactions powered by P_{700}.

Steady progress has occurred in the isolation and chemical identification of the photosystems as well as in the recognition that certain photosynthetic pigments can transfer excitations to either Photosystem I or II (see Ciba Foundation Symposium, Hatch and Boardman, and Thornber et al.). These latter pigments are referred to as being in a *light-harvesting antenna*. For instance, probably all of the Chl *b* occurs in a chlorophyll-protein complex that is part of a light-harvesting antenna (Table 5.1). The ratio of Chl *a* to Chl *b* in this complex is 1.3, and the chlorophylls are bound to proteins with molar masses of about 30 kg mol^{-1}. Some carotenoids also occur in the complex. The phycobiliproteins (organized into phycobilisomes in blue-green and red algae, which have no Chl *b*) are also part of a light-harvesting antenna. A light-harvesting antenna, which often has about the same amount of chlorophyll as the photosystems (Table 5.1), appears to form a matrix in which the photosystems are embedded. Excitations of these antenna pigments can be transferred to either photosystem, but they preferentially go to Photosystem II. Also, there is some evidence that excitations can be transferred between photosystems to a limited extent. The two photosystems are supramolecular organizations of Chl *a*, a small amount of carotenoids, and proteins. For instance, approximately 40 Chl *a*'s, one P_{700}, and a protein of about 70 kg mol^{-1} occur together as one of the components of Photosystem I. Only a single photosystem apparently occurs in photosynthetic bacteria. Bacteria are unique among photosynthetic organisms in oxidizing primarily organic acids or inorganic sulfur-containing compounds instead of water; hence, O_2 evolution does not accompany photosynthesis by bacteria.

ELECTRON FLOW

The photochemical reaction of photosynthesis involves the removal of an electron from an excited state of the special chlorophyll that acts as an excitation trap. The movement of the electron from this trap chl to some acceptor begins a series of electron transfer steps that can ultimately lead to the reduction of NADP. The oxidized trap chl, which has lost an electron, can accept another from some donor, as in the steps leading to O_2 evolution. Coupled to the electron transfer reactions in chloroplasts is the formation of ATP, which is known as "photophosphorylation." In this section we will consider some of the components of chloroplasts involved in accepting and donating electrons; a discussion of the energetics of such processes will follow in the next chapter.

The various steps of photosynthesis occur over a wide range of times (see Clayton or Kamen). For instance, the absorption of light and the transfer of excitation from both accessory pigments and Chl *a* molecules to a trap chl take from 10^{-15} to 10^{-9} s after the arrival of a photon. The photochemical event at the reaction center leads to the separation of an electron from the trap chl (Eq. 5.5), which causes a bleaching—a decrease in the absorption coefficient for wavelengths in the visible region—of this pigment. By observing the kinetics of this bleaching in the far-red region, we can determine the occurrence of photochemistry, which signals that the excitation caused by absorption of radiant energy has led to an increase in chemical energy. The donation of an electron to the oxidized trap chl (Eq. 5.6) usually occurs 10^{-7} to 10^{-4} s after the arrival of a photon, and restores the original spectral properties of the trap chl. The ensuing electron flow to some components can last into the millisecond range. Finally, the overall processing time for a reaction center plus the associated enzymes is about 5 ms per excitation.

Electron Flow Model

In 1938 Hill demonstrated that isolated chloroplasts could evolve O_2 and reduce certain exogenous compounds in the light. Oxygen evolution proceeded in the absence of CO_2, suggesting that CO_2 fixation and O_2 evolution were separate processes, contrary to the prevailing belief. Using ^{18}O-labeled H_2O and ^{18}O-labeled CO_2 in different experiments, Ruben and Kamen conclusively showed in 1941 that the evolved O_2 comes from water and not from CO_2. Subsequent studies have elucidated many of the steps intervening between O_2 evolution and CO_2 fixation in photosynthesis.

Our point of departure for the present discussion of electron flow in photosynthesis will be to indicate the overall process. Although the actual mechanism of this water oxidation is still unclear, it can be represented as follows:

$$2H_2O \longrightarrow O_2 + 4H \tag{5.8}$$

where H in Equation 5.8 represents $H^+ + e^-$. CO_2 fixation into a carbohydrate involves the reduction of carbon, four hydrogen atoms being required per carbon atom:

$$CO_2 + 4H \longrightarrow \{CH_2O\} + H_2O \tag{5.9}$$

where $\{CH_2O\}$ represents a general carbohydrate, as indicated previously (Fig. 5.1). The movement of the reductant H in Equations 5.8 and 5.9 can conveniently be followed by tracing the flow of the electrons ($H = H^+ + e^-$), which explains why our topic here is electron flow and not hydrogen move-

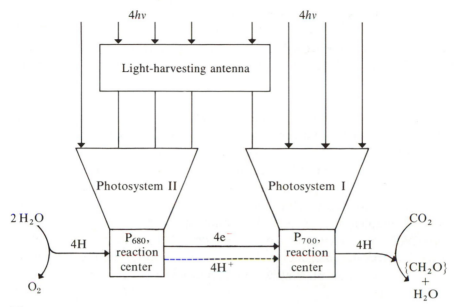

Figure 5.12
Schematic model for a series representation of the two photosystems of photosynthesis, indicating the stoichiometry of various factors involved in the reduction of a CO_2 molecule to a carbohydrate ($\{CH_2O\}$). Some of the photons (hv) are captured by the accessory pigments and Chl a in the light-harvesting antenna; these excitations are then fed into the two photosystems, but primarily to Photosystem II.

ment. Moreover, reducing a compound is chemically equivalent to adding electrons, while oxidation is the removal of electrons.

Some pathway for photosynthetic electron flow is needed so that the H's indicated as products in Equation 5.8 can become the H's required as reactants in Equation 5.9. This electron transfer pathway involves both Photosystems I and II as well as a number of other components. The oxidant involved with O_2 evolution (Eq. 5.8) is provided by Photosystem II. The primary oxidant is trap chl$^+$, which leads to the oxidation of water (see Eq. 5.6, trap chl$^+$ + $D \longrightarrow$ trap chl + D^+). The reductant required for carbon reduction (Eq. 5.9) is produced by the excited trap chl in Photosystem I (see Eq. 5.5, trap chl* + $A \longrightarrow$ trap chl$^+$ + A^-). These two photosystems are linked by a chain of components along which there is a transfer of electrons, as is illustrated in Figure 5.12.

For each photon absorbed by any of the accessory pigments or Chl a's whose excitations are funneled into a reaction center, one electron can be removed from its trap chl. Since 4 electrons are involved per O_2 derived from water, the evolution of this 1 molecule of O_2 requires the absorption of 4 photons by Photosystem II or the light-harvesting antenna feeding into it (see Eq. 5.8 and Fig. 5.12). An additional 4 photons whose excitations arrive at the trap chl of Photosystem I are required for the reduction of the 2 molecules of NADP necessary for the subsequent reduction of 1 CO_2 molecule (Eq. 5.9 and Fig. 5.12). Hence, 8 photons are needed for the evolution of 1 molecule of O_2 and the fixation of 1 molecule of CO_2. (In the next chapter we will consider how many photons are used to provide the ATP's required per CO_2 fixed.) Although there are alternatives to the series representation proposed by Hill and Bendall in 1960 (Fig. 5.12), the scheme takes into consideration the results of many investigators, and has generally become accepted as an overall description of electron flow in chloroplast grana. After introducing the concept of redox potential in the next chapter, we will portray the energetics of the series representation (see Fig. 6.4, where many of the components that we will discuss next are indicated).

Components of the Electron Transfer Pathway

We shall now turn our attention to the specific molecules that act as electron acceptors or donors in chloroplasts. A summary of the characteristics of the better-known components of the pathway is presented in Table 5.2. We will begin our discussion by considering the photochemistry at the reaction center of Photosystem II, and then proceed with a consideration of the various substances in the approximate sequence in which they are involved in electron transfer along the pathway from Photosystem II to Photosystem I. We will conclude by considering the fate of the excited electron in Photosystem I.

The electron removed from the excited trap chl in the reaction center of Photosystem II is replaced by one coming from water in the process that leads to O_2 evolution (Eq. 5.8). Using very brief flashes of light, Joliot showed in the 1960's that essentially no O_2 evolution accompanied the first or even the second flash (see Barber 1977, 1978; Clayton; or Vethuys). If the four electrons involved come from four different Photosystem II's, then a single intense flash should cause O_2 evolution, since four Photosystem II's would have been excited and the four electrons coming from each O_2 could be accepted. If two photosystems were involved, with each one accepting two electrons sequentially, then two flashes should lead to O_2 evolution. In fact, every fourth flash leads to substantial O_2 evolution, indicating that a single

Photosystem II is responsible for the four electrons involved in the evolution of each O_2 molecule; i.e., four consecutive photochemical acts are required before a molecule of O_2 can be evolved. The times for the first three steps are about 0.2 to 0.5 ms each. The last step, which leads to the release of O_2 *inside* a thylakoid (see p. 25), takes about 1 ms. (In an illuminated leaf, there are many O_2-evolving loci, and the four steps are in all different stages at any one time, thus making O_2 evolution appear to be continuous.) Manganese is apparently involved in the O_2-evolving process, presumably while bound to a protein. Six Mn's occur per reaction center in Photosystem II, and the progressive experimental removal of the four of these which are loosely bound leads to a stoichiometric reduction in the O_2-evolving ability. Chloride (or another monovalent anion of a strong acid) also is apparently necessary for O_2 evolution.

The electron from trap chl* of Photosystem II is transferred by a series of molecules making up the photosynthetic *electron transfer chain*, a term describing the pathway from Photosystem II to Photosystem I (see Fig. 5.12). All of the components involved with electron transfer in this chain of molecules may not be known, and some of the molecule-to-molecule sequences for electron transfer are uncertain. Also, what we are referring to by A and D (see Eqs. 5.5 through 5.7 and Fig. 5.9) can represent more than one molecular species. For instance, the electron removed from P_{680} may be transferred to another pigment (or a protein in bacteria) and then to a quinone in rapid succession, the latter step helping to stabilize the charge separation and thus preventing the electron from going back to P_{680}^+.

One of the first molecules to accept the electrons donated by the excited trap chl of Photosystem II is a plastoquinone. Chloroplast lamellae actually contain eight or more different types of quinones. is the structure of quinone. This becomes a semiquinone, , when one hydrogen atom is added (a semiquinone is a free radical because of the presence of an unpaired electron), and a hydroquinone, , when two are added. There is great

Table 5.2

Representative properties of some of the components involved with electron transfer in chloroplasts. The frequency of components is calculated per 600 chlorophylls (see Table 5.1). See Boardman, Clayton, and Trebst and Avron. We will discuss redox potentials in Chapter 6.

Name	Molar mass ($kg\ mol^{-1}$)	Approximate number per 600 chlorophylls	Numbers of electrons accepted or donated per molecule	Approximate midpoint redox potential in mV	Comment
Ferredoxin	11	2 to 10	1	−420	Nonheme protein with two Fe and two S atoms/molecule; accepts electrons excited from Photosystem I by way of intermediates; soluble in aqueous solutions
Ferredoxin-NADP reductase	40	≤1	...	...	An enzyme containing one flavin adenine dinucleotide/molecule (see p. 305 for structure of FAD); apparently bound to outside of lamellae
NADP-NADPH$_2$	0.743 to 0.745	30	2	−320	Soluble in aqueous solutions
Cyt b_{564}	40 to 60 per heme	2	1	−80	Apparently involved in the cyclic pathway between ferredoxin and plastoquinone pool
Cyt b_{559}	50 to 110 per heme	2	1	(55)	May act between Photosystem II and Cyt f, but data inconclusive; also called Cyt b_6

Plastoquinone A	0.748	5 to 20	113	2	Located in membrane; acts as a pool accepting electrons from Photosystem II and donating them to Photosystem I; there are about 50 quinones/600 chlorophylls in chloroplasts and only some are in the lamellar membranes and involved in electron transport
Cyt f	30 to 70 per heme	1	365	1	Associated with Photosystem I, which can oxidize it; λ_{max} for α band is at 554 nm; the relative amount of Cyt f can be considerably lower in shade plants (see Björkman)
Plastocyanin	40	$\frac{1}{2}$ to 1	360	2 (or 4)	Blue protein (reduced form is colorless) that can donate electrons to Photosystem I; contains four polypeptides and 4 Cu (2 Cu in green algae); soluble in aqueous solutions, but probably occurs on inner side of thylakoid
P_{700}	2 × 0.893	1	430	1	A Chl a dimer acting as the trap of Photosystem I
P_{680}	1 or 2 × 0.893	1	~900	1	A Chl a or a Chl a dimer acting as the trap of Photosystem II

variety among quinones because of substituents attached to the ring; e.g., plastoquinone A is

$$\text{H}_3\text{C} \overset{\displaystyle \text{O}}{\underset{\displaystyle \text{O}}{\bigbrace}} \text{H}_3\text{C} \quad -(\text{CH}_2\text{CH}=\overset{\displaystyle \text{CH}_3}{\text{C}}\text{CH}_2)_9\text{H}$$

Like chlorophyll, plastoquinone A has a nonpolar terpenoid or isoprenoid tail, which may stabilize the molecule at the proper location in the lamellar membranes of chloroplasts. When donating or accepting electrons, plasto-quinones have characteristic absorption changes in the UV near 250 to 260 nm, 290 nm, and 320 nm, which can be monitored to study their electron transfer reactions. The plastoquinones involved in electron transport can apparently be divided into two categories: (1) one or two plastoquinones, presumably bound to proteins, that rapidly receive single electrons from P_{680}; and (2) a group or pool of about 5 to 20 plastoquinones that subsequently receive two electrons from the single-electron-carrying plasto-quinones. From the plastoquinone pool, electrons may move to a blue protein known as plastocyanin or to cytochrome f (Table 5.2).

Cytochromes are extremely important components of electron transfer pathways in chloroplasts and mitochondria. They have three absorption bands in the visible region; the α, β, and γ bands. (Absorption of light by cytochromes apparently plays no part in photosynthesis.) In 1925 Keilin described three different types of cytochrome based on the spectral position of their α (long) wavelength band. Cytochromes of the a type have a λ_{max} for the α band from 600 to 605 nm, b types near 560 nm, and c types near 550 nm. All cytochromes have β bands near 515 to 530 nm. The main short wavelength band, the γ or Soret band, usually has a λ_{max} between 415 and 430 nm (see Dolphin, B. T. Storey in Davies, or Trebst and Avron for further information on cytochromes).

Cytochromes consist of an iron-containing tetrapyrrole or porphyrin known as *heme* (Fig. 5.13), which is bound to a protein. The various cyto-chromes differ in both the substituents around the periphery of the porphyrin ring and the protein to which the chromophore is attached. Cyt f^* is a c-type cytochrome occurring in chloroplasts, and contains the chromophore indicated in Figure 5.13. Many different hemoproteins of the Cyt b type are found in plants, two occurring in chloroplasts (Cyt b_{559} and Cyt b_{564}, Table

---

* f from *frons*, the Latin for leaf.

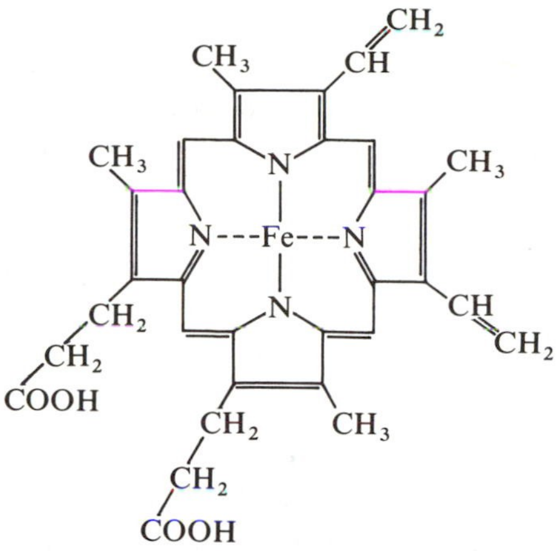

Figure 5.13

The iron-containing porphyrin known as *heme*, the chromophore for cytochrome *c*.

5.2). These various *b* cytochromes appear to have the same chromophore attached to different proteins, and thus their individual absorption properties are probably due to changes in the protein. Like the cytochromes, chlorophylls are tetrapyrroles (compare the structure of Chl *a* in Fig. 5.2), but they have a Mg atom in the center of the porphyrin ring, while cytochromes have an Fe. Furthermore, the acceptance or donation of an electron by a cytochrome involves a transition between the two states of its iron, Fe^{2+} and Fe^{3+}, whereas the electron removed from chlorophyll in the photochemical reactions of photosynthesis is apparently one of the π electrons in the conjugated system of the porphyrin ring.

Instead of the usual valence bonds, the metal atoms in chlorophylls and cytochromes should really be presented in terms of the six coordinate bonds described by ligand-field (molecular-orbital) theory. Fe has one coordinate bond to each of the four N's in the porphyrin ring (the two solid and two broken lines emanating from Fe in Fig. 5.13), one to an N in the imidazole side chain of a histidine, and another to the S in a methionine, both amino acids occurring in the protein to which the chromophore is bound (one bond is above and one is below the plane of the porphyrin in Fig. 5.13). The donation of an electron by the ferrous form of Cyt *c*, ferrocytochrome *c* (Fig. 5.13), causes the iron to go to the ferric state, Fe^{3+}. The extra positive charge on the iron in ferricytochrome *c* can either attract anions such as OH^- or be delocalized to adjacent parts of the molecule. The six coordinate bonds remain in ferricytochrome *c*, and the conjugation in the porphyrin ring is only moderately changed from that in ferrocytochrome *c*. Hence, the extensive bleaching of chlorophyll following the loss of an electron from its porphyrin ring does not occur with electron donation by the iron atom in cytochrome. Instead, the transition from Fe^{2+} to Fe^{3+} leads to less of a change in the absorption bands, since electrons in the ring conjugation are influenced only secondarily. Nevertheless, such changes in absorption upon the acceptance or donation of electrons by cytochromes can be detected and are important for monitoring electron flow in vivo.

The removal of an electron from a cytochrome (oxidation) causes its three absorption bands to become less intense, broader, and to shift toward shorter wavelengths. The absorption coefficients at λ_{max} in the reduced form are about 3×10^3 m^2 mol^{-1} for the α band, somewhat less for the β band, and over 10^4 m^2 mol^{-1} for the γ (Soret) band. Upon oxidation, the ε_λ for the α band decreases just over 50%, while smaller fractional changes generally occur in the absorption coefficients of the β and γ bands (the λ_{max} is also at shorter wavelengths in the oxidized form). Such absorption changes permit a study of the kinetics of electron transfer while the cytochrome molecules remain embedded in the internal membranes of chloroplasts or mitochondria. By monitoring the spectral changes of the chloroplast cytochromes we can observe that the electron movement along the electron transfer chain between Photosystem II and Photosystem I (Figs. 5.12 and 6.4) occurs 10^{-5} to 10^{-3} s after light absorption by some pigment molecule in Photosystem II.

Electrons that have traversed the electron transfer chain from Photosystem II to plastocyanin or Cyt f (see Table 5.2) can be accepted by the trap chl of Photosystem I, P_{700}, if the latter is in the oxidized form (see Eq. 5.6: trap $chl^+ + D \longrightarrow$ trap chl $+ D^+$). The photochemical change in P_{700} can be followed spectrophotometrically, since the loss of an electron causes a bleaching of both its Soret and red absorption bands; the subsequent acceptance of an electron restores the original spectral properties. P_{700} is bleached (oxidized) by light absorbed by Photosystem I and then restored (reduced) following the absorption of photons by Photosystem II. Thus, the electron removed from P_{700} is replaced by one coming from Photosystem II by means of the electron transfer chain (Figs. 5.12 and 6.4). Also, far-red light (above 690 nm) absorbed by Photosystem I leads to an oxidation of Cyt f, indicating that an electron from Cyt f can be donated to P_{700}.

The electron from P_{700}^* reduces the iron-containing protein, ferredoxin, via a series of intermediates. Ferredoxin contains two irons in the ferric state bound by sulfide bonds; the electron from the lower excited singlet state of P_{700} causes the reduction of one of these ferric atoms to the ferrous form. By means of the enzyme ferredoxin-NADP reductase, two molecules of ferredoxin reduce one molecule of NADP to yield $NADPH_2$ ($NADPH_2$ represents $NADPH + H^+$; we will discuss NADP and NADPH in Ch. 6).

Types of Electron Flow

We can distinguish between three different types of photosynthetic electron transfer, or flow—noncyclic, pseudocyclic, and cyclic—each one depending on the compound to which electrons are transferred from ferredoxin (see Trebst and Avron). In *noncyclic* electron flow, electrons coming originally from water are ultimately used to reduce NADP. The overall noncyclic electron flow is as follows: an electron from water goes to the trap chl^+ of Photosystem II; there it replaces a donated electron, which moves along the electron transfer chain to the oxidized P_{700} in Photosystem I; the electron from P_{700} moves to ferredoxin, and then to NADP. Such noncyclic electron flow follows essentially the same pathway as the reductant H moving from left to right in Figure 5.12 (see also Fig. 6.4).

Electrons from ferredoxin may also reduce O_2, which yields H_2O_2 and eventually H_2O ($O_2 + 2e^- + 2H^+ \rightleftharpoons H_2O_2 \rightleftharpoons H_2O + \frac{1}{2}O_2$). Since equal amounts of O_2 are evolved at Photosystem II and then consumed using reduced ferredoxin in a quite separate reaction, such electron flow is termed *pseudocyclic* (illustrated in Fig. 6.4). There is no net O_2 change accompanying pseudocyclic electron flow, although it is not a cycle in the sense of having electrons cyclically traverse a certain pathway. It is not yet clear to what extent pseudocyclic electron flow occurs in vivo, although such electron flow can readily be demonstrated with isolated chloroplasts (see H. Gimmler in Trebst and Avron).

In *cyclic* electron flow, the electron from the reduced form of ferredoxin moves back to the electron transfer chain between Photosystems I and II and eventually reduces an oxidized P_{700}. Cyt b_{564} apparently acts as an intermediate for this cyclic electron flow by intervening between ferredoxin and the plastoquinone pool (see Fig. 6.4). Cyclic electron flow does not involve Photosystem II at all, so it can be caused by far-red light absorbed only by Photosystem I—a fact often exploited in experimental studies. When far-red light absorbed by Photosystem I is used, cyclic electron flow can occur but noncyclic does not, so no NADPH is formed and no O_2 is evolved (cyclic electron flow can lead to the formation of ATP, as indicated in the next chapter). When light absorbed by Photosystem II is added to cells exposed to far-red illumination, both CO_2 fixation and O_2 evolution can proceed, and photosynthetic enhancement is achieved. Treatment of chloroplasts or plant cells with the O_2-evolution inhibitor DCMU [3-(3,4-dichlorophenyl)-1,1-dimethyl urea] also leads only to cyclic electron flow; DCMU therefore has many applications in the laboratory and is also an effective herbicide, since it markedly inhibits photosynthesis. Cyclic electron flow may occur in stromal lamellae, since they have predominantly Photosystem I activity; the two photosystems also may not be in 1-to-1 stoichiometry in the granal lamellae (each thylakoid—see p. 25—in a granum contains about 600 photosystems).

As indicated in Table 5.2, ferredoxin, the cytochromes, P_{680}, and P_{700} can accept or donate only one electron. These molecules interact with NADP, plastocyanin, and the plastoquinones, all of which can transfer two electrons at a time. The two electrons that reduce plastoquinone apparently come sequentially from the same Photosystem II. It is less clear how the two electrons move from a plastoquinone to the one-electron carrier Cyt f. The enzyme ferredoxin-NADP reductase is apparently involved in matching the one-electron chemistry of ferredoxin to the two-electron chemistry of NADP. Both the pyridine nucleotides and the plastoquinones are considerably more numerous than other molecules involved with photosynthetic electron flow (Table 5.2), and this probably has important implications for the electron transfer reactions. Moreover, NADP is soluble in aqueous solutions and so could diffuse to the NADP-ferredoxin reductase, where two electrons are transferred to it to yield $NADPH_2$ (ferredoxin and plastocyanin are also soluble in aqueous solutions).

Photophosphorylation

Three ATP molecules are generally required for the reductive fixation of one CO_2 molecule into a carbohydrate (see Fig. 5.1). Such ATP is produced by *photophosphorylation*; i.e., light absorbed by the photosynthetic pigments in the lamellar membranes leads to a flow of electrons, to which is coupled the phosphorylation of ADP. We will consider the energetics of this dehydration of ADP plus phosphate to yield ATP in the next chapter.

Photophosphorylation was first demonstrated in cell-free systems in 1954. Frenkel, working with bacterial chromatophores, and Arnon, Allen, and Whatley, using broken spinach chloroplasts, observed ATP formation in the light (see Arnon et al. and Frenkel). So far, relatively little is known at the molecular level about the enzymes involved in photophosphorylation. One difficulty is that the enzymes are localized in or on the lamellar membranes, and the energy transfer steps are apparently very sensitive to disruption of the solid system. Another difficulty is that none of the molecular species (ADP, ATP, and phosphate) can be readily determined quantitatively in vivo. It has not so far been possible to monitor the interconversions of these compounds in the chloroplasts by measuring changes in spectral properties, a technique that is quite successful for studying the acceptance or donation of electrons by cytochromes and trap chl's. Furthermore, all three molecules (ATP, ADP, and phosphate) take part in many different biochemical reactions, and this greatly complicates the interpretation of the kinetic data on photophosphorylation presently available. However, considerable progress has been made in understanding the relationship between ATP formation and proton (H^+) transport across membranes. In the next chapter we will reconsider ATP formation coupled to electron flow in both chloroplasts and mitochondria, after some of the underlying energy concepts have been introduced.

Vectorial Aspects of Electron Flow

Since the electron flow components are associated with membranes, the possibility exists for chemical asymmetries to develop (see Clayton, Hall and Rao, Trebst and Avron, or Velthuys). In fact, the electrons and their associated protons are moved in specific directions in space by the processes we have been considering, which gives the flows a vectorial nature (see Fig. 5.14).

The chlorophyll-protein complexes are oriented in the lamellar membranes in such a way that the electron transfer steps at the reaction centers lead to an outward movement of electrons. For instance, the electron donated by Photosystem II is moved to the stromal side of a thylakoid (see Figs. 1.9 and 5.14). The electron that is donated back to the trap chl (P_{680}) comes from H_2O, leading to the evolution of O_2 by Photosystem II (Eq. 5.8). The O_2 and the H^+ from this reaction are released inside the thylakoid (Fig. 5.14). Since O_2 is a small neutral molecule, it readily diffuses out across the lamellar membranes into the chloroplast stroma. But the proton (H^+) carries a charge and hence has a low partition coefficient for the membrane, and therefore it does not readily move out of the space within the thylakoid.

The electron from Photosystem II eventually reaches a plastoquinone in the plastoquinone pool, a group of plastoquinones that apparently spans the lamellar membrane. After accepting two electrons and—very important— picking up two H^+'s from outside the thylakoid, a plastoquinone delivers these to the inner side of the lamellar membrane (Fig. 5.14). The protons are then released inside the thylakoid, while the electrons are delivered to the reaction center of Photosystem I, probably by means of plastocyanin (which apparently occurs on the inner side of the lamellar membrane). Via a photochemical event, the trap chl of Photosystem I (P_{700}) donates an electron that eventually reaches ferredoxin, which occurs on the outer side of the thylakoid

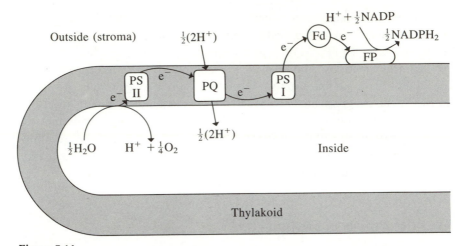

Figure 5.14
Schematic representation of reactions occurring at photosystems and certain electron transfer components, emphasizing the vectorial or unidirectional flows developed in the thylakoids of a chloroplast. The outwardly directed electron movements in the two photosystems (PS I and PS II) are photochemical reactions, where the electron donors are on the inner side and the electron acceptors are on the outer side of the membrane. The plastoquinone pool (PQ) spans the membrane, whereas the flavoprotein (FP)—ferredoxin-NADP reductase—is attached to the outside and catalyzes electron flow from ferredoxin (Fd) to NADP.

(Fig. 5.14). Ferredoxin, which is soluble in aqueous solutions, diffuses to ferredoxin-NADP reductase (a flavoprotein, Table 5.2), where two electrons are accepted by NADP, yielding $NADPH_2$. The flavoprotein that catalyzes this transfer is apparently bound on the outer side of the lamellar membranes, and so $NADPH_2$ is formed in a region where subsequent biochemical reactions can utilize this critically important molecule.

Let us now recapitulate the accomplishments of the various processes described above. O_2 is evolved inside the thylakoid and readily diffuses out. The protons from the O_2-evolving steps plus those transported by plastoquinone are released inside the thylakoid, where the membranes prevent their ready escape. [We note that in cyclic electron flow, electrons from P_{700} move to ferredoxin and thence to the plastoquinone pool (probably via Cyt b_{564} = Cyt b_6; Table 5.2), which also causes protons to be delivered inside the thylakoids.] The accumulation of protons inside the thylakoid, together with the transfer of electrons out, raises the electrical potential inside with respect to outside and also increases the internal concentration of protons, thus setting up a chemical potential gradient capable of doing work. In fact, this proton chemical potential gradient is energetically coupled to the formation of ATP (photophosphorylation), as we will see in the next chapter.

Problems

5.1. A spherical spongy mesophyll cell is 40 μm in diameter and contains 50 spherical chloroplasts 4 μm in diameter. Assume that such cells contain 1 g chlorophyll kg^{-1}, that the cell is 90% water by weight, and that the cellular density is 1 000 kg m^{-3} (1.00 g cm^{-3}). (a) What volume fraction of the cell is occupied by chloroplasts? (b) If the CO_2 fixation rate is 100 mmol (g chlorophyll)$^{-1}$ hour^{-1}, how long does it take to double the dry weight of the cell? Assume that CO_2 and H_2O are the only substances entering the cell. (c) If the ratio Chl a/Chl b is 3, what is the mean molecular weight of chlorophyll? (d) Assuming that the chlorophyll is uniformly distributed throughout the cell, what is the maximum absorbance by one cell in the red and the blue regions? Use absorption coefficients given in Figure 5.3.

5.2. Suppose that some pigment has 8 double bonds in conjugation and has a single absorption band with a λ_{max} at 580 nm, which corresponds to a transition to the fourth vibrational sublevel of the excited state (i.e., the 3rd excited vibrational sublevel; see Fig. 4.6). A similar pigment has 10 double bonds in conjugation, which causes the lowest vibrational sublevel of the excited state to move down in energy by 20 kJ mol^{-1} while the lowest vibrational sublevel of the ground state moves up in energy by 20 kJ mol^{-1} compared with the analogous levels in the other molecule. Assume that the splitting between vibrational sublevels remains at 10 kJ mol^{-1} and that the most likely transition predicted by the Franck–Condon principle for this second molecule is also to the fourth vibrational sublevel. (a) What is the λ_{max} for fluorescence by each of the two molecules? (b) Can either or both molecules readily pass their excitation on to Chl a in vivo? (c) Can the absorption of blue light by Chl a lead to excitation of either of the pigments? Give your reasoning.

5.3. Let us approximate chloroplasts by short cylinders 4 μm in diameter and 2 μm thick (i.e., 2 μm along the cylinder axis), which contain 20 mol m^{-3} chlorophyll. The chloroplasts are exposed to 40 W m^{-2} of 675 nm light parallel to the axis of the cylinder. Assume that ε_{675} is 0.60×10^4 m^2 mol^{-1} for the chlorophylls. (a) What is the absorbance at 675 nm for the chlorophyll in a single chloroplast? What is the fraction of the incident light absorbed? (b) How many μmol photons m^{-2} s^{-1} of 675 nm light will be absorbed in passing through a single chloroplast? How many chlorophyll molecules participate in this absorption? (c) Assume that a photosynthetic unit has 250 chlorophyll molecules and that 0.01 s is needed to process each excitation. How often are chlorophyll molecules excited on the average and what fraction of the absorbed photons can be processed? (d) How many moles of O_2 m^{-2} s^{-1} are evolved for each chloroplast? Assume that eight photons are needed to evolve one molecule of O_2. (e) What would be the answers to (c) for a chloroplast shaded by three overlying chloroplasts?

5.4. Chloroplasts corresponding to 10 mmol of chlorophyll m^{-3} of solution are suspended in a cuvette with a 10 mm light path. The rate of O_2 evolution is proportional to PAR up to 10 μmol photons absorbed m^{-2} s^{-1}, which gives 10^{-4} mol m^{-3} (10^{-7} M O_2) evolved s^{-1}. For very brief and intense flashes of light, the O_2 evolution is 5 $\times$ 10^{-6} mol m^{-3} per flash. (a) Using the above data, how many photons are required per O_2 evolved? (b) How many chlorophyll molecules are there in a photosynthetic unit? (c) How much time is required for the processing of an excitation by a photosynthetic unit? (d) An "uncoupler" is a compound that decreases the ATP formation coupled to photosynthetic electron flow. When such a compound is added to chloroplasts incubated at a high photon flux density, the O_2 evolution rate is higher for the first few seconds and then becomes less than a control without the uncoupler. Explain.

5.5. Suppose that the absorbance of pea chloroplasts in a cuvette with a 10 mm light path is 0.1 at 710 nm and 1.0 at 550 nm. Assume that chlorophyll is the only species absorbing at 710 nm and that no chlorophyll absorbs at 550 nm. Suppose that no CO_2 is fixed when either 550 nm or 710 nm light is used alone, but that both together lead to CO_2 fixation. (a) Is any ATP formation caused by the 550 nm or by the 710 nm light? (b) What type of pigments are absorbing at 550 nm? Are they isoprenoids or tetrapyrroles? (c) If equal but low incident photon flux densities are simultaneously used at both 550 nm and 710 nm, what is the maximum quantum yield for CO_2 fixation for each beam? (d) The initial bleaching of P_{700} at 700 nm leads to a decrease in absorbance of 10^{-5} in 10^{-6} s. What is the minimum number of moles of photons per unit area absorbed by Photosystem I that could account for this? Assume that ε_{700} is 0.8 $\times$ 10^4 m^2 mol^{-1} for the trap chl.

References

Arnon, D. I., M. B. Allen, and F. R. Whatley. 1954. Photosynthesis by isolated chloroplasts. *Nature 174*: 394–396.

Atkins, P. W. 1978. *Physical Chemistry*. W. H. Freeman, San Francisco.

Barber, J., ed. 1977. *Primary Processes of Photosynthesis. Topics in Photosynthesis*, Vol. 2. Elsevier, Amsterdam.

Barber, J. 1978. Biophysics of photosynthesis. *Reports on Progress in Physics 41*: 1157–1199.

Björkman, O. 1981. Responses to different quantum flux densities. In *Physiological Plant Ecology*, O. L. Lange, P. S. Nobel, C. B. Osmond, and H. Ziegler, eds. *Encyclopedia of Plant Physiology, New Series*, Vol. 12A. Springer-Verlag, Berlin. Pp. 57–107.

Boardman, N. K. 1970. Physical separation of the photosynthetic photochemical systems. *Annual Review of Plant Physiology 21*: 115–140.

Brown, J. S. 1972. Forms of chlorophyll in vivo. *Annual Review of Plant Physiology 23*: 73–86.

Ciba Foundation Symposium. 1979. *Chlorophyll Organization and Energy Transfer in Photosynthesis.* Excerpta Medica, Amsterdam.

Clayton, R. K. 1980. *Photosynthesis: Physical Mechanisms and Chemical Patterns.* Cambridge University Press, Cambridge.

Davies, D. D., ed. 1980. *Metabolism and Respiration.* In *The Biochemistry of Plants: A Comprehensive Treatise,* P. K. Stumpf and E. E. Conn, eds., Vol. 2. Academic Press, New York.

Dolphin, D., ed. 1979. *The Porphyrins,* 7 volumes. Academic Press, New York.

Duysens, L. M. N., J. Amesz, and B. M. Kamp. 1961. Two photochemical systems in photosynthesis. *Nature 190*:510–511.

Emerson, R. 1957. Dependence of yield of photosynthesis in long-wave red on wavelength and intensity of supplementary light. *Science 125*:746.

Frenkel, A. 1954. Light-induced phosphorylation by cell-free preparations of photosynthetic bacteria. *Journal of the American Chemical Society 76*:5568–5569.

Gantt, E. 1980. Structure and function of phycobilisomes: light harvesting pigment complexes in red and blue-green algae. *International Review of Cytology 66*:45–80.

Gantt, E. 1981. Phycobilisomes. *Annual Review of Plant Physiology 32*:327–347.

Glazer, A. N. 1977. Structure and molecular organization of the photosynthetic accessory pigments of cyanobacteria and red algae. *Molecular & Cellular Biochemistry 18*:125–140.

Goedheer, J. C. 1972. Fluorescence in relation to photosynthesis. *Annual Review of Plant Physiology 23*:87–112.

Goodwin, T. W. ed. 1976. *Chemistry and Biochemistry of Plant Pigments,* 2nd ed. Academic Press, London.

Goodwin, T. W. 1980. *The Biochemistry of the Carotenoids,* Vol. 1, *Plants,* 2nd ed. Chapman and Hall, London.

Hall, D. O., and K. K. Rao. 1981. *Photosynthesis,* 3rd ed. Edward Arnold, London.

Hatch, M. D., and N. K. Boardman, eds. 1981. *Photosynthesis.* In *The Biochemistry of Plants: A Comprehensive Treatise,* P. K. Stumpf and E. E. Conn, eds. Vol. 8. Academic Press, New York.

Hill, R., and F. Bendall. 1960. Function of the two cytochrome components in chloroplasts: a working hypothesis. *Nature 186*:136–137.

Isler, O., ed. 1971. *Carotenoids.* Birkhäuser Verlag, Basel.

Jeffrey, S. W. 1981. Responses to light in aquatic plants. In *Physiological Plant Ecology,* O. L. Lange, P. S. Nobel, C. B. Osmond, and H. Ziegler, eds. *Encyclopedia of Plant Physiology, New Series,* Vol. 12A. Springer-Verlag, Berlin. Pp. 249–276.

Kamen, M. D. 1963. *Primary Processes in Photosynthesis.* Academic Press, New York.

Kirk, J. T. O., and R. A. E. Tilney–Bassett. 1978. *The Plastids,* 2nd ed. Elsevier, Amsterdam.

Levine, R. P. 1969. The analysis of photosynthesis using mutant strains of algae and higher plants. *Annual Review of Plant Physiology 20*:523–540.

Lieth, H., and R. H. Whittaker, eds. 1975. *Primary Productivity of the Biosphere.* Ecological Studies, Vol. 14. Springer-Verlag, New York.

Myers, J. 1971. Enhancement studies in photosynthesis. *Annual Review of Plant Physiology 22*:289–312.

Rabinowitch, E., and Govindjee. 1969. *Photosynthesis.* Wiley, New York.

Thornber, J. P., J. P. Markwell, and S. Reinman. 1979. Plant chlorophyll-protein complexes: recent advances. *Photochemistry and Photobiology 29*:1205–1216.

Trebst, A. and M. Avron, eds. 1977. *Photosynthesis I: Photosynthetic Electron Transport and Photophosphorylation. Encyclopedia of Plant Physiology, New Series,* Vol. 5. Springer-Verlag, Berlin.

Velthuys, B. R. 1980. Mechanisms of electron flow in Photosystem II and toward Photosystem I. *Annual Review of Plant Physiology 31*:545–567.

Vernon, L. P., and G. R. Seely, eds. 1966. *The Chlorophylls.* Academic Press, New York.

CHAPTER 6

Bioenergetics

Throughout this text we have considered various aspects of energy in biological systems. The concept of chemical potential was introduced in Chapter 2 and then applied to the specific case of water. In Chapter 3 we used this thermodynamic approach to discuss the movement of ions. We also considered the use of energy for actively transporting substances toward higher chemical potentials. Chapter 4 dealt with the absorption of light, an event that is followed by various de-excitation reactions for the excited states of the molecules. The photochemistry of photosynthesis (Ch. 5) involves the conversion of such electromagnetic energy into forms that are useful biologically. This last aspect—namely, the production and use of various energy currencies in biological systems—is the topic of the present chapter.

Two energy "currencies" produced in chloroplasts soon after the trapping of radiant energy are ATP and NADPH. These substances represent the two main classes of energy-storage compounds associated with the electron transfer pathways of photosynthesis and respiration. We can appreciate the importance of ATP by noting that about 100 mol of ATP ($\cong$ 50 kg) are hydrolyzed in the synthesis of 1 kg dry weight ($\cong$ 10 kg wet weight) of many microorganisms. Since there is typically only about 10^{-2} mol of ATP per kg dry weight of cells, a great turnover of ATP is necessary both to synthesize new tissue and to maintain mature cells in a state far from equilibrium (see Forrest and Walker, Payne, or Penning de Vries).

In this chapter we will first examine energy storage in terms of the chemical potential changes accompanying the conversion of a set of reactants into their products. This consideration of the Gibbs free energy allows us to determine the amount of chemical energy a given reaction can store or release. We will then evaluate the energy-carrying capacity of ATP in terms of the energetics of its formation and hydrolysis. NADPH can be regarded as possessing electrical energy, the particular amount depending on the oxidation-reduction potential of the system with which it interacts. After considering ATP and NADPH as individual molecules, we will place them in their biological context—namely, as part of the bioenergetic scheme of chloroplasts and mitochondria. We will conclude with a brief comment on the flow of energy from the sun through the biosphere, paying particular attention to the overall photosynthetic efficiency.

GIBBS FREE ENERGY

Using the appropriate thermodynamic relations, we can calculate the energy changes that accompany biological reactions. Two conditions that are often met by physiological processes greatly simplify these calculations. First, most biological reactions take place at *constant temperature*; i.e., they are isothermal. Second, processes in cells or tissues generally take place at *constant pressure*. These two special conditions make the Gibbs free energy, G, a very convenient function for describing energetics in biology, since the decrease in G under these conditions equals the maximum amount of energy available for work. (See Ch. 2 for an introduction to G, and App. VI for a mathematical presentation of the Gibbs free energy.) Biological use of free energy, or "work," takes on a number of different forms, all the way from muscular movement to chemical synthesis and active transport. The difference in Gibbs free energy between two states can be used to predict the spontaneous direction for a reaction and to indicate how much energy the transition makes available for performing work. Biologists are generally interested only in such *changes* in free energy, rather than in the absolute amount, which must be defined relative to some arbitrary level.

The spontaneous direction for a reaction at constant T and P is toward a minimum of the Gibbs free energy of the system; minimum Gibbs free energy is actually achieved at equilibrium (see Fig. 6.1). In principle, a spontaneous process can always be harnessed to do work; the reversal of a spontaneous reaction requires an input of free energy (see p. 62). As we will show below, light can be harnessed to produce a free energy source that causes the phosphorylation of ADP and the reduction of NADP. These

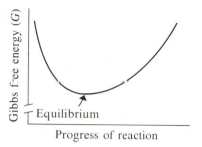

Figure 6.1
Relation between progress of a reaction and the Gibbs free energy of the system (G). When the product concentration is low (abscissa values near the origin), the reaction will spontaneously proceed in its forward direction toward lower values of G. The Gibbs free energy for a reaction attains a minimum at equilibrium; ΔG is then zero for the reaction proceeding a short distance in either direction. For high product concentrations (right side of the abscissa), ΔG to drive the reaction further in the forward direction is positive, indicating that a free energy input is then needed. When ΔG in the forward direction is positive, ΔG for the reverse reaction is negative. The absolute value of G is actually arbitrary $\left(\text{e.g., } G = \sum_j n_j \mu_j, \text{ where each } \mu_j \text{ contains an unknown constant, } \mu_j^*\right)$, and so the ordinate is interrupted in the figure.

two processes are prime examples of energy-requiring reactions at the very heart of chloroplast bioenergetics.

The concept of free energy was introduced in Chapter 2 in presenting the chemical potential μ_j. The chemical potential is actually the partial molal Gibbs free energy with respect to that species, i.e., $\mu_j = (\partial G/\partial n_j)_{T,P,E,h,n_i}$ (Eq. VI.9, App. VI). We must therefore consider the Gibbs free energy of an entire system consisting of many components in order to define the chemical potential of species j. In turn, G depends on each of the species present, an appropriate expression being

$$G = \sum_j n_j \mu_j \tag{6.1}$$

where n_j is the number of moles of species j in some system, μ_j is the chemical potential of species j (Eq. 2.4), and the summation is over all of the species present (see App. VI, where Eq. 6.1 is derived). The Gibbs free energy is expressed relative to some arbitrary zero level; e.g., the arbitrariness in the baseline for G in Equation 6.1 is a consequence of the μ_j^* included in each μ_j (see Fig. 6.1). We will find that the Gibbs free energy as represented by

Equation 6.1 is in a very useful form for our application of free energy relations to bioenergetics.

Chemical Reactions and Equilibrium Constants

From the thermodynamic point of view, we are interested in the overall change in free energy for an individual reaction—or, perhaps, a sequence of reactions. Let us consider a general chemical reaction for which A and B are the reactants and C and D are the products:

$$aA + bB \rightleftharpoons cC + dD \qquad (6.2)$$

where a, b, c, and d are the number of moles of the various species taking part in the reaction.

How much energy is stored (or released) when the reaction proceeds a certain extent in either direction? More specifically, what is the change in Gibbs free energy for the reaction in Equation 6.2 proceeding in the forward direction, with a moles of A and b moles of B reacting to give c moles of C and d moles of D? For most applications, we can consider that the chemical potentials of the species involved are actually constant; in other words, we are concerned with a hypothetical change in the Gibbs free energy were the reaction to take place at certain concentrations and under other fixed conditions. Using Equation 6.1 $\left(G = \sum_{j} n_j \mu_j \right)$, we can express the change in the Gibbs free energy for such a reaction as

$$\Delta G = -a\mu_A - b\mu_B + c\mu_C + d\mu_D \qquad (6.3)$$

where Δn_j is positive for a product and negative for a reactant $\left(\Delta G = \sum_{j} \Delta n_j \mu_j \text{ when all } \mu_j \text{ are constant} \right)$. Equation 6.3 indicates that the free energy change for a chemical reaction is the Gibbs free energy of the products minus that of the reactants.

To transform Equation 6.3 into a more useful form, we need to incorporate expressions for the chemical potentials of the various species involved. The chemical potential of species j has already been presented in Chapter 2, where μ_j was given as a linear combination of its component parts:

$\mu_j = \mu_j^* + RT \ln a_j + \bar{V}_j P + z_j FE + m_j gh$ (Eq. 2.4).* Substituting such chemical potentials of A, B, C, and D into Equation 6.3, and collecting similar terms, we obtain

$$\Delta G = -a\mu_A^* - b\mu_B^* + c\mu_C^* + d\mu_D^*$$

$$+ RT(-a \ln a_A - b \ln a_B + c \ln a_C + d \ln a_D)$$

$$+ P(-a\bar{V}_A - b\bar{V}_B + c\bar{V}_C + d\bar{V}_D)$$

$$+ FE(-az_A - bz_B + cz_C + dz_D)$$

$$+ gh(-am_A - bm_B + cm_C + dm_D) \qquad (6.4)$$

Let us next simplify this equation. The constant terms, $-a\mu_A^* - b\mu_B^* + c\mu_C^* + d\mu_D^*$, can be replaced by ΔG^*, a quantity that we will evaluate below. If the volume of the products, $c\bar{V}_C + d\bar{V}_D$, is the same as that of the reactants, $a\bar{V}_A + b\bar{V}_B$, the factor multiplying P in Equation 6.4 is zero—a situation closely approached for many chemical reactions. The actual value of $P \sum_j n_j \bar{V}_j$ for biochemical reactions is usually relatively small for pressures encountered in plant cells, so we will not retain these terms here.† The algebraic sum of the terms making up the factor multiplying FE is zero because no charge is created or destroyed by the reaction given in Equation 6.2; i.e., the total charge of the products, $cz_C + dz_D$, equals that of the reactants, $az_A + bz_B$. Since no mass is created or destroyed by the reaction, the factor multiplying gh is also zero.

We can now convert Equation 6.4 to a relatively simple form. Using ΔG^*, the constancies of charge and mass, and the generally valid assumption that $P(-a\bar{V}_A - b\bar{V}_B + c\bar{V}_C + d\bar{V}_D)$ is negligible, the change in Gibbs free energy for the reaction in Equation 6.2 proceeding in the forward direction becomes simply

$$\Delta G = \Delta G^* + RT \ln \frac{(a_C)^c (a_D)^d}{(a_A)^a (a_B)^b} \qquad (6.5)$$

---

* μ_j^* is a constant, a_j the activity of species j, $\bar{V}_j$ its partial molal volume, P the pressure in excess of atmospheric, z_j its charge number, F the faraday, E the electrical potential, m_j its mass per mole, and h the vertical position in the gravitational field (see p. 65).

† Ignoring $P \sum_j n_j \bar{V}_j$ may not be valid for reactions occurring under very high pressures (e.g., deep in the ocean), especially if volume changes are suspected, such as those that can occur when two oppositely charged species react to form a neutral one (see Zimmerman).

where various properties of logarithms presented in Appendix IV have been used to obtain the form indicated. Equation 6.5 is a general expression giving the Gibbs free energy stored or released by a chemical reaction.

The quantity ΔG^* in Equation 6.5 has a special value and meaning. At equilibrium, the argument of the logarithm in Equation 6.5, $[(a_C)^c(a_D)^d]/[(a_A)^a(a_B)^b]$, is simply the equilibrium constant K of the reaction given by Equation 6.2. Furthermore, at constant T and P the Gibbs free energy achieves a minimum at equilibrium (Fig. 6.1), and it would not change for a conversion of reactants to products under such conditions; i.e., ΔG equals zero for a chemical reaction proceeding a short distance in either direction at equilibrium under these conditions. Equation 6.5 thus indicates that $\Delta G^* + RT \ln K$ is zero at equilibrium, and so

$$\Delta G^* = -RT \ln K \tag{6.6}$$

where K is the equilibrium constant of the reaction; i.e., K equals the equilibrium value for $[(a_C)^c(a_D)^d]/[a_A)^a(a_B)^b]$. Hence ΔG^* depends rather simply on the equilibrium constant.

Let us next assume that the reactants, A and B, and the products, C and D, initially all have unit activity. Thus $[(a_C)^c(a_D)^d]/[(a_A)^a(a_B)^b]$ is one, and so the logarithm of the activity term is zero. Equation 6.5 thus indicates that, for a moles of A plus b moles of B reacting to form c moles of C plus d moles of D under these conditions, the Gibbs free energy change ΔG is simply ΔG^*. If the equilibrium constant K for the reaction is greater than 1, ΔG^* is negative ($\Delta G^* = -RT \ln K$, Eq. 6.6), and so the reaction in the forward direction is spontaneous for this case of unit activity of reactants and products. For $K > 1$ we would expect the reaction to proceed in the forward direction, since at equilibrium the products are then favored over the reactants—in the sense that $[(a_C)^c(a_D)^d] > [(a_A)^a(a_B)^b]$ at equilibrium. On the other hand, if K is less than one, ΔG^* is positive, and such a reaction would not proceed spontaneously in the forward direction for the given initial condition of unit activity of all reactants and products.

Let us now consider the units of G and ΔG. G is an extensive variable; i.e., it depends on the extent or size of the system and is obtained by summing its values throughout the whole system (see App. VI). Specifically, Equation 6.1 $\left(G = \sum_j n_j \mu_j \right)$ indicates that the Gibbs free energy is the sum, over all species present, of the number of moles of species j (n_j, an extensive variable) times the energy/mole of species j (μ_j, an intensive variable, i.e., a quantity that can be measured at some point in a system such as T, P, $\overline{V}_j$, μ_j). Hence,

G has the dimensions of energy. On the other hand, Equation 6.5 suggests that G has the same units as RT—namely, energy *per mole* (a logarithm is dimensionless). To help us out of this apparent dilemma, let us reconsider the conventions used in Equation 6.2, $aA + bB \rightleftharpoons cC + dD$. We usually write a chemical reaction using the smallest possible integers for a, b, c, and d, not the actual number of moles reacting. In fact, in the equations describing nearly all biochemical reactions, either or both a and c are unity, e.g., ADP + phosphate $\rightleftharpoons$ ATP + H_2O. Another convention is to express the Gibbs free energy change per mole of the species on which attention is being focused, e.g., per mole of a certain reactant $(\Delta G/a)$ or per mole of a certain product $(\Delta G/c)$. So, when a or c represents one mole, ΔG then has the same magnitude, whether as energy or as energy/mole. When we use actual values for ΔG or ΔG^* to describe chemical reactions, we will always indicate on what basis we are using the Gibbs free energy, e.g., "ΔG per mole of ATP formed."

What are the numerical values of ΔG^* per mole of reactant A or product C for K's of 100 and 0.01? Since RT is 2.48 kJ mol^{-1} at 25°C, and since ln equals 2.303 log, an equilibrium constant of 100 corresponds to a $\Delta G^*/a$ or $\Delta G^*/c$ given by Equation 6.6 of $-(2.48$ kJ mol$^{-1})(2.303)$ log (100), or -11.4 kJ mol^{-1} for a moles of reactant A or c moles of product C, while a K of 0.01 leads to a $\Delta G^*/a$ or $\Delta G^*/c$ of $+11.4$ kJ mol^{-1}. In the former case Equation 6.5 indicates that 11.4 kJ of energy per mole of the reactant or product (assuming a or c is 1 mole) would be released, while in the latter case the same amount of energy would be required to drive the reaction in the forward direction, starting with unit activity of all reactants and products.

Interconversion of Chemical and Electrical Energy

To help understand how chemical energy can be converted into electrical energy, and vice versa, we must re-examine the properties of both chemical reactions and the movement of charged species. Let us first consider a chemical reaction such as the dissociation of sodium chloride: NaCl $\rightleftharpoons$ $Na^+ + Cl^-$. Although two charged species are produced upon dissociation of NaCl, no overall change in the electrical components of the chemical potentials of Na^+ and Cl^- occurs. In other words, the electrical term $z_j FE$ for Na^+ $(z_{Na} = +1)$ is balanced by an opposite change in the electrical component of μ_{Cl} $(z_{Cl} = -1)$. Next, let us consider the following type of reaction: $Ag_s \rightleftharpoons Ag^+ + e^-$, i.e., the dissociation of solid silver to an ion plus an electron. Again, no net change in the electrical contributions to the two chemical potentials occurs for the dissociation as written. However, the production of an electron opens up various other possibilities, since electrons can be conducted in special ways to regions where the electrical potential may be different. Such reactions, in which electrons are produced and then conducted to regions of different electrical potential, allow for a very useful interconversion of chemical and electrical energy. The electron-producing or electron-consuming reactions are referred to as *electrode*, or *half-cell*, reactions and occur in batteries as well as in the electron transfer chains located in chloroplast and mitochondrial membranes.

To elaborate on this matter of energy conversion, let us consider mixing ferrous (Fe^{2+}) and cupric (Cu^{2+}) ions in an aqueous solution; we can assume that the common anion is Cl^-. A chemical reaction occurs in which the products are ferric (Fe^{3+}) and cuprous (Cu^+) ions. Since this is a spontaneous process, there is a decrease in Gibbs free energy; also, the reaction is exothermic, since heat is evolved. Next, let us consider the electrode reaction, $Fe^{2+} \rightleftharpoons Fe^{3+} + e^-$. If the electrons produced in such a half-cell initially containing only Fe^{2+} can be *conducted* by a wire to the Cu^{2+} ions (Fig. 6.2), another electrode reaction can occur in a second beaker initially containing only Cu^{2+}—namely, $Cu^{2+} + e^- \rightleftharpoons Cu^+$. Except for heat evolution, the net result in the solutions is the same as that which occurs by mixing Fe^{2+} and Cu^{2+}. When the electrons move in the conductor, however, they can be used to do various types of electrical work, e.g., powering a direct-current electrical motor or a light bulb. Such an arrangement provides a way of converting the change in Gibbs free energy of the two spontaneous half-cell reactions into electrical energy that could then be used for performing work. Indeed, the important thing for obtaining electrical work from the two half-cell reactions is the conducting pathway between them. (See Castellan, MacInnes, Morris, or Weyer for a further discussion of half-cells.)

When electrons are moved to a lower electrical potential ($\Delta E < 0$), e.g., by using the chemical energy in a battery, their electrical energy ($z_j FE$, where $z_j = -1$ for an electron) increases. We can use this increase in the electrical energy of the electrons to power a chemical reaction when the electrons subsequently move spontaneously to higher E. In photosynthesis, light energy is used to move electrons toward lower electrical potentials, which sets up a spontaneous flow of electrons in the other direction using the electron transfer components we introduced in the previous chapter. This latter, energetically downhill, spontaneous electron movement is harnessed to drive the photophosphorylation reaction, ADP + phosphate $\rightleftharpoons$ ATP + H_2O, in the forward direction and thereby to store chemical energy.

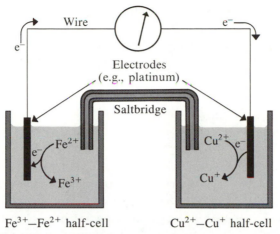

Figure 6.2

Two half-cells, or redox couples, connected by a wire and a saltbridge to complete the electrical circuit. Electrons donated by Fe^{2+} to one electrode are conducted by the wire to the other electrode where they reduce Cu^{2+}. Both electrodes (couples) are necessary before electrons can flow. A saltbridge, which provides a pathway along which ions can move, often contains agar and KCl to minimize the diffusion potentials at the junctions between the saltbridge and the solutions in the beakers (see p. 124).

Now let us consider the interconversion of chemical and electrical energy in more formal terms. Suppose that n moles of electrons ($z_j = -1$) are transferred from one region to another where the electrical potential differs by ΔE, e.g., from one half-cell to another. As we noted on p. 107, the charge carried by a mole of protons is the faraday (F); hence, the total charge moved in the present case is $-nF$. Electrical work is expressed as the charge transported times the electrical potential difference through which it moves (ΔE). The change in the electrical energy of n moles of electrons is therefore $-nF\,\Delta E$. This can be converted to an equal change in Gibbs free energy (ΔG):

$$\Delta G = -nF\,\Delta E \tag{6.7}$$

According to Equation 6.7, the amount of Gibbs free energy which can be stored or released is directly proportional to the difference in electrical potential across which the electrons move. Moreover, this equation indicates that the flow of electrons toward more positive electrical potentials ($\Delta E > 0$) corresponds to a decrease in free energy ($\Delta G < 0$); thus, this movement would be expected to proceed spontaneously, as is indeed the case. We should emphasize that *two* half-cells are necessary to get a ΔE and thus a ΔG for electron transfer (see Fig. 6.2). We will apply these free energy considerations to the energetics of electrons moving from molecule to molecule in the electron transfer chains of chloroplasts and mitochondria.

Redox Potentials

Many organic compounds involved in photosynthesis can accept or donate electrons (see Table 5.2). The electrons spontaneously flow toward higher electrical potentials, which are measured by *redox potentials* in the case of the components involved with electron flow in chloroplast lamellae or the inner membranes of mitochondria. Redox potentials are a measure of the relative chemical potential of electrons accepted or donated by a particular type of molecule. The oxidized and reduced forms of each electron transfer component can be regarded as an electrode, or half-cell. Such a half-cell can interact with the other electron-accepting and electron-donating molecules in the membrane, in which case the electrons spontaneously move toward the component with the higher redox potential.

We will begin by writing a chemical reaction describing a general electrode (half-cell) reaction. We can represent the acceptance or donation of electrons by some species as follows:

$$\text{oxidized form} + q\text{e}^- \;\rightleftharpoons\; \text{reduced form} \tag{6.8}$$

where q is a dimensionless parameter indicating the number of electrons transferred per molecule; and *oxidized* and *reduced* refer to different forms of the same species—e.g., NADP is an oxidized form and $NADPH_2$ (or $NADPH + H^+$) represents the corresponding reduced form (q is 2 in this case; see p. 310). Like any other chemical reaction, an oxidation-reduction reaction such as Equation 6.8 has a change in Gibbs free energy associated with it when the reactants are converted to products. Thus oxidation-reduction, or "redox," reactions can be described by the relative tendency of the redox system, or *couple* (the oxidized plus the reduced forms of the compound), to proceed in the forward direction, which for Equation 6.8 means accepting electrons.

It is more useful to describe redox reactions in terms of relative electrical potentials instead of the equivalent changes in Gibbs free energy. The electrons in Equation 6.8 come from some other redox couple, and whether or not the reaction in Equation 6.8 will proceed in the forward direction depends on the relative electrical potentials of these two couples. It is therefore convenient to assign a particular electrical potential to a system accepting or donating electrons, a value known as its *redox potential*. This oxidation-reduction potential can then be compared with that of another couple to predict the direction for spontaneous electron flow when the two systems are allowed to interact—electrons spontaneously move toward higher redox potentials. The redox potential of species j, E_j, is defined as follows:

$$E_j = E_j^* - \frac{RT}{qF} \ln \frac{(\text{reduced}_j)}{(\text{oxidized}_j)} \qquad (6.9)$$

where E_j^* is an additive constant, q is the number of electrons transferred (the same q as in Eq. 6.8), and (reduced_j) and (oxidized_j) refer to the activities of the two different redox states of species j. Equation 6.9 indicates that the oxidation-reduction potential of a particular redox couple is determined by the ratio of the reduced to the oxidized form plus an additive constant, a quantity that we will consider next.

We can measure electrical potentials when two electrodes (half-cells) are connected and pathways for electron flow are provided (see Fig. 6.2). Since the sum of the electrical potential drops (voltage changes) going completely around such a circuit is zero, we can determine the half-cell potential for a particular electrode if the potential of some standard reference electrode is known. By international agreement, the E_j^* of a hydrogen half-cell ($\frac{1}{2}H_{2\text{gas}} \rightleftharpoons H^+ + e^-$) is arbitrarily set equal to zero for an activity of hydrogen ions of 1 N equilibrated with hydrogen gas at a pressure of one atmosphere, i.e., $E_H^* = 0$. Fixing the zero level of the electrical potential for the hydrogen half-cell removes the arbitrary nature of redox potentials for all half-cells, since the redox potential for any species can then be determined relative to that of the hydrogen electrode. We will replace E_j^* in Equation 6.9 by $E_j^{*,H}$ to emphasize the convention of referring electrode potentials to the standard hydrogen electrode.

According to Equation 6.9, the larger the ratio $(\text{reduced}_j)/(\text{oxidized}_j)$, the more negative the redox potential becomes. Since electrons are negatively charged, a lower E_j corresponds to higher energies for the electrons. Thus, the further Equation 6.8 is driven in the forward direction, the more energy will be required to reduce species j. Likewise, the larger is $(\text{reduced}_j)/(\text{oxidized}_j)$, the higher will be the electrical energy of the electrons which the reduced form of that couple can then donate. When (reduced_j) equals (oxidized_j), E_j equals $E_j^{*,H}$ by Equation 6.9. This $E_j^{*,H}$ is commonly referred to as the *midpoint redox potential* (see the values given in Table 5.2 for the $E_j^{*,H}$'s of some components involved with electron transfer in chloroplasts). For certain purposes, knowledge of the midpoint redox potentials may be sufficient, as we will show later in this chapter (see Lehninger, Morris, or White et al. for more details on redox reactions).

BIOLOGICAL ENERGY CURRENCIES

In photosynthesis photons are captured, initiating an electron flow leading both to the production of NADPH and to a coupled process whereby ATP is formed. Light energy is converted into chemical energy in the formation of a phospho-anhydride (ATP) in an aqueous environment. It is converted into electrical energy by providing a reduced compound (NADPH) under oxidizing conditions. ATP and NADPH are the two energy storage compounds, or "currencies," that we will consider in this section. Both can occur as ions, both can readily diffuse around within a cell or organelle, and both can carry appreciable amounts of energy under biological conditions. In addition to its use in processes like active transport and muscle contraction, the chemical energy stored in ATP is also used in certain biosynthetic reactions involving the formation of anhydrous links, or bonds, in the aqueous milieu of a cell. The relatively high atmospheric levels of O_2 insure that appreciable amounts of this strong oxidizing agent will be present in most biological systems; a reduced compound like NADPH is thus an important currency for energy storage. We will discuss these two compounds in turn, after briefly considering the difference between ATP and NADPH as energy currencies.

Redox couples are assigned a relative electrical energy, whereas chemical reactions have a specific chemical energy. In a chemical reaction certain reactants are transformed into products, and the accompanying change in Gibbs free energy can be calculated. This change in chemical energy need not depend on any other system. For instance, if the concentrations—strictly speaking, the chemical activities—of ADP, phosphate, and ATP as well as certain other conditions (e.g., temperature, pH, Mg^{2+} concentration, and ionic strength) are the same in different parts of a plant, then the Gibbs free energy released upon the hydrolysis of a certain amount of ATP to ADP and phosphate will be the same in each of the different locations. But an oxidation-reduction couple must donate electrons to, or accept electrons from, another redox system, and the change in electrical energy depends on the difference in the redox potential between the two couples. Thus, the amount of electrical energy released when $NADPH_2$ (NADPH + H^+) is oxidized to NADP depends on the redox potential of the particular couple with which $NADPH_2$ interacts.

ATP—Structure and Reactions

To help us understand the bioenergetics of chloroplasts and mitochondria, we need to know how much energy is stored in ATP, that is, the difference between its chemical potential and that of the reactants (ADP and phosphate) used in its formation. We must then look for reactions that have a large enough free energy decrease to drive the ATP production reaction in the energetically uphill direction; this will lead us to a consideration of the energetics of electron flow in organelles—topics we will discuss in the next two sections. Our immediate concern is with the following: (1) the chemical reaction describing ATP formation, (2) the associated change in Gibbs free energy for that reaction, and (3) the implications of the substantial amount of energy storage in ATP.

ADP, ATP, and phosphate can all occur in a number of different charge states in aqueous solutions. Moreover, all three compounds can interact with other species, notably Mg^{2+} and Ca^{2+}. Thus, many different chemical reactions can be used for describing ATP formation. A predominant reaction occurring near neutral pH in the absence of divalent cations is as follows:

$$(6.10)$$

where adenosine is adenine esterified to the 1′ position of the sugar ribose, i.e.,

The attachment of adenosine to the phosphates in ADP and ATP—and in NADP as well as in FAD—is by means of an ester linkage with the hydroxymethyl group on the 5' position of the ribose moiety (see Fig. 6.3).

Equation 6.10 indicates a number of features of ATP production. For instance, the formation of ATP from ADP plus phosphate is a dehydration, while the reversal of Equation 6.10, in which the phospho-anhydride is split with the incorporation of water, is known as ATP *hydrolysis*. Since Equation 6.10 contains H^+, the equilibrium constant depends on pH, $-\log (a_{H^+})$. Moreover, the fractions of ADP, phosphate, and ATP in various states of ionization depend on the pH. Near pH 7 about half of the ADP molecules are doubly charged and half are triply charged, the latter form being indicated in Equation 6.10.* Likewise, ATP at pH 7 is about equally distributed between the forms with charges of -3 and -4. Because of their negative charges in aqueous solutions, both ADP and ATP can readily bind positive ions, especially divalent cations such as Mg^{2+} or Ca^{2+}. A chelate is formed such that Mg^{2+} or Ca^{2+} is electrostatically held between two negatively charged oxygen atoms on the same molecule (consider the many $—O^-$'s occurring on the chemical structures indicated in Eq. 6.10). Also, inorganic phosphate can electrostatically interact with Mg^{2+} and other divalent cations, further increasing the number of complexed forms of ADP, ATP, and phosphate that are possible.

The activities (or concentrations) of a species in all of its ionization states and complexed forms are generally summed to obtain the total activity (or concentration) of that species. The number of relations and equilibrium constants needed to describe a reaction like ATP formation is then reduced to one; i.e., a separate equilibrium constant is not needed for every possible combination of ionization states and complexed forms of all the reactants and products. Using this convention, we can replace Equation 6.10 and many others like it, which also describe ATP formation, by the following general reaction for the phosphorylation of ADP:

$$\text{ADP} + \text{phosphate} \rightleftharpoons \text{ATP} + H_2O \qquad (6.11)$$

We will return to ATP formation as represented by Equation 6.11 after briefly commenting on two important conventions used in biochemistry. First, most equilibrium constants for biochemical reactions are defined at

---

* For simplicity, we are ignoring the charge due to the extra proton bound to an adenine nitrogen, which gives that part of the ADP and ATP molecules a single positive charge at pH 7.

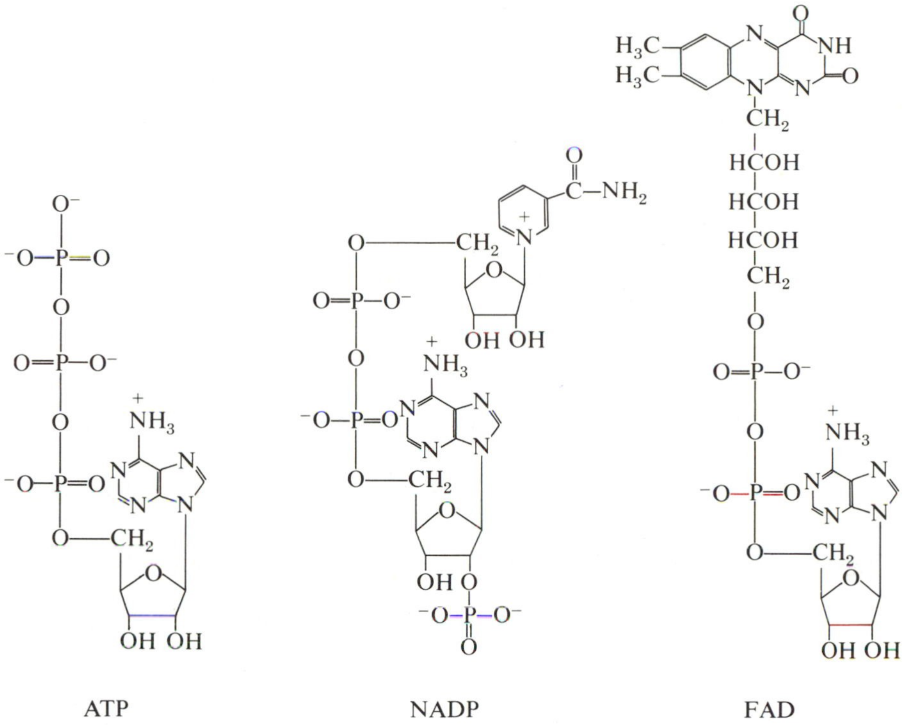

Figure 6.3
Structures of three molecules important in bioenergetics. The dissociations and bindings of protons indicated in the figure are appropriate near pH 7. Note the similarity between the molecules.

a specific pH, usually pH 7 ($a_{H^+} = 10^{-7}$ M). At constant pH the activity of H^+ does not change. Thus, H^+ need not be included as a reactant or product in the expression for the change in Gibbs free energy (Eq. 6.5). In other words, the effect of H^+ in relations such as Equation 6.10 is actually incorporated into the equilibrium constant, which itself generally depends on pH. Second, biological reactions such as ATP formation usually take place in aqueous solutions where the concentration of water does not change appreciably. (The concentrations of other possible reactants and products are much, much less than that of water.) Hence, the (a_{H_2O}) term coming from relations like Equation 6.11 is also usually incorporated into the equilibrium constant. We can illustrate these points concerning a_{H^+} and a_{H_2O} by specifically considering ATP formation as described by Equation 6.10. For the reaction as written, the equilibrium constant K is equal to $[(a_{ATP})(a_{H_2O})]/[(a_{ADP})(a_{phosphate})(a_{H^+})]$; hence, $(a_{ATP})/[(a_{ADP})(a_{phosphate})]$ equals $(a_{H^+}) K/(a_{H_2O})$. For a dilute aqueous solution a_{H_2O} is essentially constant during the reaction; therefore, $(a_{H^+}) K/(a_{H_2O})$—which is conventionally called the equilibrium constant in biochemistry—has a fixed value at a given pH, e.g., at pH 7, $(a_{H^+})K/(a_{H_2O}) = (10^{-7}\ \text{M})K/(a_{H_2O}) = K_{pH\ 7}$.

Next, we will specifically consider the equilibrium constant for ATP formation under biological conditions. Using the above conventions for H^+ and H_2O, an equilibrium constant for Equation 6.11 at pH 7 ($K_{pH\ 7}$) is

$$K_{pH\ 7} = \frac{[ATP]}{[ADP][phosphate]} \cong 5 \times 10^{-6}\ \text{M}^{-1}\ \text{at } 25°C \qquad (6.12)$$

where the total concentration of each species involved is indicated in brackets; i.e., it is experimentally more convenient to measure $K_{pH\ 7}$ using concentrations (indicated by brackets) instead of activities (indicated by parentheses).*

When activities of ions ($a_j = \gamma_j c_j$, Eq. 2.5) are replaced by concentrations (c_j), the resulting equations for equilibrium constants or free energy changes apply only to a particular ionic strength $\left(\frac{1}{2}\sum_j c_j z_j^2;\ \text{see p. 112}\right)$, because the activity coefficients of ions (γ_j) can markedly depend on ionic strength (see Eq. 3.3). The value for $K_{pH\ 7}$ given in Equation 6.12 is suitable only for ionic strengths close to 0.2 M (200 mol m^{-3}), which is an ionic strength that can occur in vivo (a 0.05 M increase or decrease in ionic strength changes K in the opposite direction only about 10%). The magnitude of the equilibrium constant for ATP formation also depends on the concentration of Mg^{2+}. The value of $K_{pH\ 7}$ in Equation 6.12 is appropriate for 10 mM Mg^{2+}, a concentration range occurring in many plant cells (a 5 mM increase or decrease in Mg^{2+} changes K in the opposite direction about 10% to 20%). An equilibrium constant also depends on temperature, $K_{pH\ 7}$ for ATP formation increasing 1% to 5% per °C. A large effect on K is produced by pH, which may not be near 7 and often is unknown in a cell. The equilibrium

---

* Since equilibrium constants for ATP formation are quite small (see Eq. 6.12) and are sensitive to temperature, pH, Mg^{2+}, and ionic strength, measured values of K vary considerably. Actually, instead of K, the standard Gibbs free energy, ΔG^*, is usually determined ($\Delta G^* = -RT \ln K$, Eq. 6.6). Values of ΔG^* for ATP formation range from 28 to 45 kJ mol^{-1}, but a value near 30 kJ mol^{-1} seems most likely at pH 7, 25°C, 10 mM Mg^{2+}, and an ionic strength of 0.2 M (see D. O. Hall in Barber, or Rosing and Slater).

constant for ATP formation increases about 3-fold as the pH is lowered 1 unit from pH 7 and decreases 7-fold as it is raised one unit (see Alberty, D. O. Hall in Barber, or Rosing and Slater).

Gibbs Free Energy Change for ATP Formation

The energetics of a reaction like ATP formation is summarized by its Gibbs free energy change, ΔG. For a general chemical reaction, Equation 6.5 indicates that ΔG equals $\Delta G^* + RT \ln \{[(a_C)^c(a_D)^d]/[(a_A)^a(a_B)^b]\}$, where ΔG^* is $-RT \ln K$ (Eq. 6.6). For the present case, the reactant A is ADP, B is phosphate, the product C is ATP, and the equilibrium constant is given by Equation 6.12. Therefore, our sought-after free energy relationship describing ATP formation is

$$\Delta G = -RT \ln (K_{\mathrm{pH}\,7}) + RT \ln \frac{[\mathrm{ATP}]}{[\mathrm{ADP}][\mathrm{phosphate}]} \qquad (6.13a)$$

which at pH 7 and 25°C becomes

$$\Delta G \cong 30 + 5.71 \log \frac{[\mathrm{ATP}]}{[\mathrm{ADP}][\mathrm{phosphate}]} \ \mathrm{kJ\ (mol\ ATP)^{-1}} \qquad (6.13b)$$

where ln equals 2.303 log, 2.303 RT is 5.71 kJ mol^{-1} at 25°C (App. II), and $-(5.71 \text{ kJ mol}^{-1}) \log (5 \times 10^{-6})$ is 30 kJ mol^{-1} (7.2 kcal mol^{-1}).

It is apparent from Equation 6.13 that ATP usually does not tend to form spontaneously, because the Gibbs free energy change for the reaction is generally quite positive. In fact, the energy required for the phosphorylation of ADP is rather large compared with the free energy changes for most biochemical reactions. Stated another way, much energy can be stored by converting ADP plus phosphate to ATP. Although ATP in an aqueous solution is *thermodynamically* unstable, in that its hydrolysis can release a considerable amount of Gibbs free energy, it still can last for a long enough time in cells to be an important energy currency. In particular, ATP is generally not hydrolyzed very rapidly unless the appropriate enzymes necessary for its use in certain biosynthetic reactions or other energy-requiring processes are present (for long-term energy storage, plants use carbohydrates like the polysaccharide starch).

We will now estimate the changes in Gibbs free energy that might be expected for photophosphorylation under physiological conditions. For purposes of calculation, we will assume that in unilluminated chloroplasts the concentration of ADP is 2.2 mM, phosphate is 10 mM, and ATP is 0.2 mM (1 mM = 1 mol m^{-3}). From Equation 6.13b, the free energy change required to form ATP then is

$$\Delta G = 30 + 5.71 \log \frac{(0.2 \times 10^{-3}\ \text{M})}{(2.2 \times 10^{-3}\ \text{M})(10 \times 10^{-3}\ \text{M})}$$

$$= 35\ \text{kJ (mol ATP)}^{-1}$$

The change in Gibbs free energy required is positive, indicating that energy must be supplied to power photophosphorylation. Moreover, the energy necessary depends in a predictable way on the concentrations of the reactants and the product. After a certain period of time in the light, ATP may increase to 2.2 mM with a concomitant decrease in ADP to 0.2 mM and in phosphate to 8 mM. The free energy required for photophosphorylation under these conditions is 48 kJ (mol ATP)$^{-1}$. Consequently, the further photophosphorylation goes to completion, the greater is the energy required to form more ATP.

The high energy of ATP relative to ADP plus phosphate for usual physiological concentrations is not the property of a single bond but of the local configuration in the ATP molecule, a point we can appreciate by considering the phosphorus atoms in ADP, ATP, and phosphate. Phosphorus is in group V of the third period of the periodic table and has 5 electrons in its outermost shell. It can enter into a total of 5 bonds with 4 oxygen atoms, the bonding to one O being a double bond (consider structures in Eq. 6.10). In inorganic phosphate all 4 bonds are equivalent, so 4 different structures for phosphate exist in resonance with each other. The terminal P of ADP has only 3 resonating forms, since one of the O's is connected to a second phosphorus atom and does not assume a double bond configuration (see Eq. 6.10). When inorganic phosphate is attached to this terminal P of ADP to form ATP, a resonating form is lost both from the ADP and from the phosphate. Since configurations having more resonating structures are in general more probable or stable (lower in energy), energy must be supplied to form ATP from ADP plus phosphate with an accompanying loss of 2 resonating forms.

Let us next examine quantitatively some of the ways in which ATP can be used as a free energy currency. Each of our four examples will relate to a different variable term in the chemical potential ($\mu_j = \mu_j^* + RT \ln a_j + \bar{V}_j P + z_j FE + m_j gh$, Eq. 2.4). To transfer a mole of a neutral compound against a 10-fold increase in activity requires 2.303 RT, or 5.71 kJ (1.364 kcal) of Gibbs free energy at 25°C, while it takes 11.4 kJ for a 100-fold increase and 17.1 kJ for a thousandfold increase in activity. To move a monovalent cation from one side of a membrane to the other, where it has the same concentration but the electrical potential is 0.1 V higher, requires 9.65 kJ mol^{-1} ($F = 96.5$ kJ mol^{-1} V^{-1}, App. II). ATP usually supplies at least 40 kJ mol^{-1} when hydrolyzed and can act as the Gibbs free energy source for the active transport of solutes across membranes toward regions of higher chemical potential. ATP is also the free energy currency for the contraction of muscles. The ATP-driven contraction of the muscles surrounding the left ventricle of the human heart can increase the blood pressure within it by 20 kPa (0.2 bar or 150 mm Hg). This increases the chemical potential of the water in the blood (i.e., the $\bar{V}_w P$ term), which causes it to flow out to the aorta and then to the rest of the circulatory system toward lower hydrostatic pressures. Pressure-driven flow is an efficient way to move fluids; e.g., it takes only 0.02 kJ of Gibbs free energy to increase the pressure of 10^{-3} m^3 (1 litre) of water by 20 kPa. As an example of gravitational work that can be mediated by ATP, the increase in Gibbs free energy as a 50 kg person climbs up 100 m is 49 kJ. Because of its large free energy release upon hydrolysis—about 40 to 50 kJ mol^{-1} (10 to 12 kcal mol^{-1}), and up to 60 kJ mol^{-1} (14 kcal mol^{-1}) in exceptional circumstances—and because of its convenient form as a relatively abundant ion, ATP is an extremely useful cellular energy currency.

NADP-NADPH$_2$ Redox Couple

Another class of energy storage compounds consists of redox couples such as NADP-NADPH$_2$. The reduced form, NADPH$_2$, is produced by noncyclic electron flow in chloroplasts. Photosynthesis in bacteria makes use of a different redox couple, NAD-NADH$_2$. The reduced member of this latter couple causes an electron flow in mitochondria and an associated formation of ATP (see "Mitochondrial Bioenergetics," p. 322). NAD is nicotinamide adenine dinucleotide and differs from NADP in not having a phosphate esterified to the 2′ hydroxy group of the ribose in the adenosine part of the molecule (see Fig. 6.3). Our present discussion will focus on the NADP-NADPH$_2$ couple, but the same arguments and also the same midpoint redox potential apply to the NAD-NADH$_2$ couple (see Lehninger, Morris, or White et al.).

The reduction of a molecule of NADP involves its acceptance of two electrons. Only the nicotinamide portion (illustrated in Eq. 6.14) of NADP is involved in accepting the electrons. The actual half-cell reaction describing this reduction is

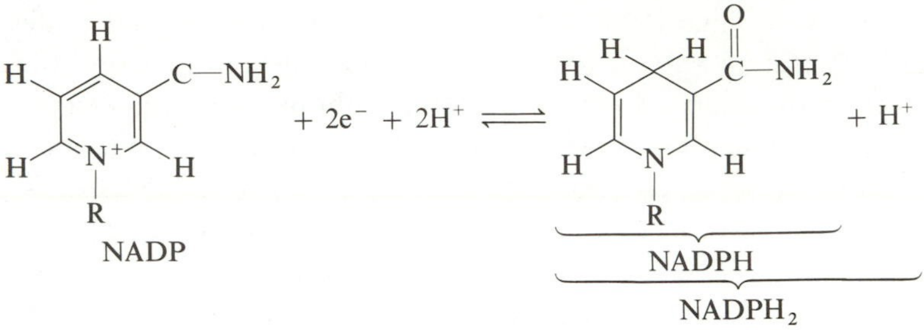

$$+ 2e^- + 2H^+ \rightleftharpoons \qquad + H^+ \qquad (6.14)$$

NADP

NADPH

NADPH$_2$

where R represents a ribose attached at its 1C position to nicotinamide and at its 5C position by a pyrophosphate bridge

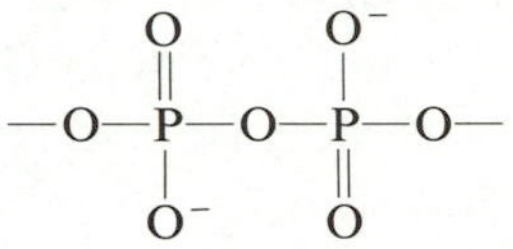

to an adenosine (see structure on p. 303) having the 2′ hydroxy group of its ribose moiety esterified to an additional phosphate (see Fig. 6.3). Adenosine less the ribose is called adenine, which leads to the somewhat inappropriate name of nicotinamide adenine dinucleotide phosphate, or NADP.

The reduction of NADP involves the transfer of two electrons to the nicotinamide ring, plus the attachment of one H^+ to the *para* position (top of the ring for the NADPH indicated in Eq. 6.14). The convention of designating the reduced form as NADPH$_2$ has been adopted in this book when discussing redox couples and in diagrams (e.g., Figs. 5.1, 6.4, and 6.6) to emphasize that *two* electrons are accepted by the NADP molecule during its reduction, although one of the *two* accompanying protons is not attached to the reduced form, as Equation 6.14 indicates. NADP$^+$ is commonly used in the literature for the oxidized form of nicotinamide adenine dinucleotide phosphate—in fact, our NADP could be replaced by NADP$^+$ and our NADPH$_2$ by NADPH + H^+ in the diagrams, "balanced" chemical reactions, and redox couples presented in this book.

A particular half-cell reaction such as Equation 6.14 can accept or donate electrons. We can quantitatively describe this by the oxidation-reduction (redox) potential for that reaction, as expressed by Equation 6.9 {$E_j = E_j^{*,H} - (RT/qF)\ln[(\text{reduced}_j)/(\text{oxidized}_j)]$}. We will use (NADPH$_2$) to represent the activity of all of the various ionization states and complexed forms of the reduced nicotinamide adenine dinucleotide phosphate, while (NADP) has an analogous meaning for the oxidized species of the NADP-NADPH$_2$ couple. For redox reactions of biological interest the midpoint (standard) redox potential is generally determined at pH 7. By using Equation 6.9, in which the number q of electrons transferred per molecule reduced is 2, we can express the oxidation-reduction potential of the NADP-NADPH$_2$ couple (Eq. 6.14) as follows:

$$E_{\text{NADP-NADPH}_2} = E_{\text{pH 7}}^{*,\text{H}} - \frac{RT}{2F} \ln \frac{(\text{NADPH}_2)}{(\text{NADP})} \qquad (6.15a)$$

which at 25°C and pH 7 becomes

$$E_{\text{NADP-NADPH}_2} = -320 - 29.6 \log \frac{(\text{NADPH}_2)}{(\text{NADP})} \text{ mV} \qquad (6.15b)$$

where in the second equation ln has been replaced by 2.303 log, and the numerical value for 2.303 RT/F of 59.2 mV at 25°C (App. II) has been used. The midpoint redox potential, $E_{\text{pH 7}}^{*,\text{H}}$, for the NADP-NADPH$_2$ couple is -320 mV (Table 5.2), a value achieved when (NADPH$_2$) equals (NADP).

To determine whether electrons spontaneously flow toward or away from the NADP-NADPH$_2$ couple, we must compare its redox potential with that of some other redox couple. As indicated above, electrons spontaneously flow toward more positive redox potentials, whereas energy must be supplied to move electrons in the energetically uphill direction of algebraically decreasing redox potential. We will next examine the redox potentials of the various redox couples involved in electron flow in chloroplasts, then in mitochondria.

CHLOROPLAST BIOENERGETICS

In Chapter 5 we introduced various molecules involved with electron transfer in chloroplasts, together with a consideration of the sequence of electron flow between components. Now that the concept of redox potential has been presented, we will resume our discussion of electron transfer in chloroplasts. We will compare the midpoint redox potentials of the various redox systems, not only to help understand the direction of spontaneous electron flow, but also to see how much energy is available between certain pairs of redox couples for use in such reactions as ATP formation (see Hatch and Boardman, or Trebst and Avron, both Ch. 5 references).

Redox Couples

Although the ratio of reduced to oxidized forms of species j affects its redox potential $\{E_j = E_j^{*,\text{H}} - (RT/qF) \ln [(\text{reduced}_j)/(\text{oxidized}_j)]$, Eq. 6.9$\}$, the actual activities of the two forms are generally unknown for biological studies in vivo. Moreover, the value of the local pH (which can affect $E_j^{*,\text{H}}$) is also generally unknown. Consequently, we can usually only compare midpoint (standard) redox potentials determined at pH 7 to predict the direction for spontaneous electron flow in lamellar membranes of chloroplasts. We will follow such an approach here and will assume that free energy is required to transfer electrons to a compound with a more negative *midpoint* redox potential, while the reverse process can go on spontaneously.

The absorption of a photon can markedly affect the redox properties of a pigment molecule. An excited molecule such as trap chl* has an electron in an antibonding orbital (see Ch. 4). It takes less energy to remove such an electron from the excited molecule than when that molecule is in its ground state. Thus, the electronically excited molecule is a better electron donor (reducing agent) than the ground state, inasmuch as it has a considerably more negative redox potential. Once the electron is removed, this oxidized molecule (trap chl$^+$) becomes a very good electron acceptor (oxidizing agent). Such electron acceptance, trap chl$^+$ + D → trap chl + D^+ (Eq. 5.6), involves the ground state of the chlorophyll molecule, which has a rather positive redox potential (see Table 5.2 and Fig. 6.4). The electronic state of trap chl that donates an electron is therefore an excited state with a negative redox potential, while the ground state with its rather positive redox potential can readily accept an electron. In short, the absorption of light energy transforms trap chl's from couples with positive redox potentials to couples with negative redox potentials. In this section we will estimate the redox potential spans at each of the two photosystems in chloroplasts and then diagram the overall pattern of photosynthetic electron flow.

Following light absorption by Chl a or an accessory pigment feeding into Photosystem II, the excitation migrates to P_{680}, where an electron transfer reaction takes place (trap chl* + A ⟶ trap chl$^+$ + A^-, Eq. 5.5). The electron removed from P_{680}^* is replaced by one coming from water, which results in O_2 evolution as described by Equation 5.8. (For energetic considerations, we write Eq. 5.8 as $H_2O \rightleftharpoons \frac{1}{2}O_2 + 2H$.) The water-oxygen system represents a half-cell reaction with a midpoint redox potential of 0.82 V at 25°C, pH 7, and an O_2 pressure of one atmosphere (see Fig. 6.4). Water oxidation and the accompanying O_2 evolution follow spontaneously after the photochemistry at the reaction center of Photosystem II has led to

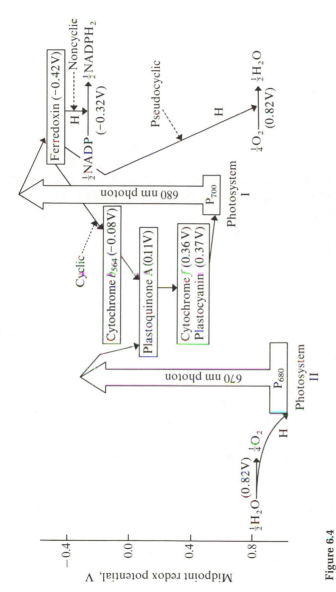

Figure 6.4

Energy aspects of photosynthetic electron flow. The lengths of the arrows emanating from the trap chl's of Photosystems I and II represent the increases in chemical potential of the electrons that occur upon absorption of red light near the λ_{max}'s of the Chl a's. The diagram also shows the various midpoint redox potentials of the couples involved (data from Table 5.2) and the three types of electron flow mediated by ferredoxin.

P_{680}^{+}. Thus, the required oxidant for water—i.e., P_{680}^{+} or some intermediate oxidized by it—must have a redox potential more positive than 0.82 V for the electron to move energetically downhill from water to the trap chl^{+} in the reaction center of Photosystem II. As we indicated in Table 5.2, the redox potential of P_{680} in the ground state is about 0.90 V.

The electron removed from P_{680}^{*} goes to the plastoquinone pool, presumably by way of one or two intermediate plastoquinones (see p. 277). The electron transfer from the first plastoquinone to the plastoquinone pool is spontaneous, and hence the redox potential of this initial plastoquinone must be more negative than the midpoint redox potential of plastoquinone A in the pool (0.113 V; Table 5.2). This initial acceptor in Photosystem II apparently has a redox potential near -0.05 V. Thus, the minimum electrical potential span in Photosystem II is from 0.90 V to -0.05 V, or 0.95 V overall. The energy required to move an electron in the energetically uphill direction toward lower redox potentials in Photosystem II is supplied by a light quantum (Fig. 6.4).

Photosystem II can be excited by 670-nm light (as well as by other wavelengths, this value being near the λ_{max} for the red band of its Chl a). From Equation 4.2a ($E = hc/\lambda_{vacuum}$) and the numerical value of hc (1 240 eV nm, App. II), we can calculate that the energy of 670-nm light is (1 240 eV nm)/ (670 nm), or 1.85 eV per photon. Such a photon could move an electron across 1.85 V, which is more than sufficient energy to cross the redox potential span estimated for Photosystem II. However, the amount of absorbed light energy and the accompanying changes in free energy available for decreasing the redox potential are not the same; i.e., the increase in internal energy U upon light absorption is generally not equal to the change in G (see Eq. VI.4a, App. VI, $G = U + PV - TS$). The magnitude of the increase in Gibbs free energy caused by the absorption of a photon by chlorophyll is only about 70% of the light energy—the exact value depends on the level of illumination as well as on the various pathways competing for the de-excitation of trap chl*. Although an understanding of this situation requires certain entropy considerations beyond the scope of this text, the important aspect for our present discussion is that the absorption of 670-nm light (1.85 eV/photon) leads to an increase in free energy of about (1.85 eV)(0.7), or 1.3 eV (see Knox, as well as Ross and Calvin). This is still sufficient to move an electron across the redox potential span of Photosystem II, which we estimated to be about 0.95 V.

We can make a similar analysis of the energetics for Photosystem I, where the trap chl is P_{700}. The minimum redox potential span across which electrons are moved (see Table 5.2) is from the midpoint redox potential of

0.43 V for the P_{700}-P_{700}^+ couple to the -0.42 V of ferredoxin (the latter is a quite negative midpoint redox potential). Since an intermediate with a midpoint redox potential near -0.55 V apparently reduces ferredoxin, the redox potential span for Photosystem I is about 0.98 V overall. A 680-nm photon, which corresponds to the λ_{max} for the red band of Chl a in Photosystem I, has an energy of 1.82 eV, of which about 1.3 eV is available to increase the free energy. This is ample to move the electron from P_{700} to the intermediate that spontaneously reduces ferredoxin.

From ferredoxin to the next component in the noncyclic electron flow sequence, NADP, electrons go from -0.42 V to -0.32 V (midpoint redox potentials of the couples), for a ΔE of $+0.10$ V (see Fig. 6.4). Moving toward higher redox potentials is energetically downhill for electrons, so this step leading to the reduction of the pyridine nucleotide follows spontaneously from the reduced ferredoxin—the enzyme ferredoxin-NADP reductase catalyzes this downhill process (see Table 5.2).

In noncyclic electron flow, two electrons originating in the water-oxygen couple with a midpoint redox potential of 0.82 V are moved to the redox level of -0.32 V for the reduction of one molecule of the NADP-NADPH$_2$ couple. Since a midpoint redox potential of -0.32 V is more negative than most encountered in biology, NADPH$_2$ can spontaneously reduce most other redox systems; reduced pyridine nucleotides are therefore an important energy currency. Moving electrons from 0.82 V to -0.32 V requires a considerable expenditure of free energy, which helps to explain why light, with its relatively large amount of energy (see Table 4.1), is needed. We can calculate the actual free energy change for the overall process using Equation 6.7 ($\Delta G = -nF \Delta E$):

$$\Delta G = -(2 \text{ mol})(96.5 \text{ kJ mol}^{-1} \text{ V}^{-1})(-0.32 \text{ V} - 0.82 \text{ V})$$

$$= 220 \text{ kJ mol}^{-1}$$

for the overall movement of two moles of electrons along the pathway for noncyclic electron flow—a process that leads to the reduction of one mole of NADP.

The incorporation of CO_2 into a carbohydrate during photosynthesis requires 3 ATP's and 2 NADPH's (see Fig. 5.1). Using these energy currencies, CO_2 fixation could energetically proceed in the absence of light, and so the steps of the reductive pentose cycle are often referred to as the *dark reactions* of photosynthesis (actually, since several of the enzymes are light-activated, not much CO_2 fixation would occur in the dark). The hydrolysis of ATP in chloroplasts can release about 50 kJ mol^{-1} (p. 308); we have just stated that 220 kJ are required to reduce a mole of NADP using electrons coming from water (this is actually the Gibbs free energy change between standard states, since midpoint redox potentials were used in our calculations); and the increase in Gibbs free energy per mole of CO_2 incorporated into a carbohydrate during photosynthesis is 479 kJ (p. 241). Using these numbers, we can estimate the efficiency for free energy storage by the dark reactions. Dividing the energy stored per mole of CO_2 fixed, 479 kJ, by the energy input, (3 mol ATP)(50 kJ mol^{-1} ATP) + (2 mol NADPH)(220 kJ mol^{-1} NADPH), we find that the efficiency is [(479 kJ)/(590 kJ)](100), or 81%! The dark reactions of photosynthesis are indeed extremely efficient.

Figure 6.4 incorporates the midpoint redox potentials of the various components involved with photosynthetic electron transfer discussed both in Chapter 5 (see Table 5.2) and here. The direction for spontaneous electron flow to higher midpoint redox potentials is downward (toward the bottom of Fig. 6.4), while the absorption of light quanta with their relatively large energies corresponds to moving electrons vertically upward to higher energy. The role played by ferredoxin at the crossroads of cyclic, noncyclic, and pseudocyclic electron flow is also illustrated in Figure 6.4.

H^+ Chemical Potential Differences Caused by Electron Flow

We indicated in Chapter 5 that the components involved with electron flow were so situated in the lamellar membranes of chloroplasts that they led to vectorial or unidirectional movement of electrons and protons (see Fig. 5.14). We will now return to this theme and focus on the gradients in H^+ (protons) so created. The difference in the chemical potential of H^+ from the inside to the outside of a thylakoid in the light acts as the energy source to drive photophosphorylation. This was first clearly recognized in the 1960's by Mitchell, who received the 1978 Nobel Prize in chemistry for his enunciation of what has become known as the *chemiosmotic* hypothesis for interpreting the relationship between electron flow, proton movements, and ATP formation (see Clayton, Ch. 5 reference; Mitchell; or Witt).

Figure 6.5 indicates that the O_2-evolution step as well as the electron flow mediated by the plastoquinone pool leads to an accumulation of H^+ inside a thylakoid in the light. This causes the internal H^+ concentration c_H^i or activity a_H^i to increase. These steps are dependent on the light-driven electron flow, which leads to electron movement out across the thylakoid in each of

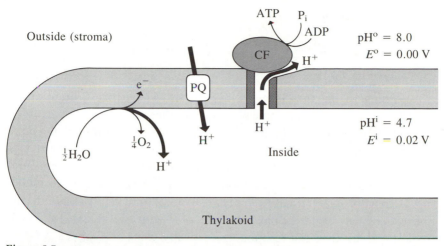

Figure 6.5
Energetics and directionality of the coupling between electron flow and ATP formation in chloroplasts, emphasizing the role played by H^+ (see also Fig. 5.14). The O_2 evolution from H_2O and electron flow via the plastoquinone pool lead to H^+ accumulation inside the thylakoid. This H^+ can move back out through a hydrophobic channel and a coupling factor (CF), leading to ATP formation.

the photosystems (see Fig. 5.14). Such movements of electrons out and protons in can increase the electrical potential inside the thylakoid (E^i) relative to that outside (E^o), allowing an electrical potential difference ($E^i - E^o = E_M$) to develop across a thylakoid. By the definition of chemical potential ($\mu_j = \mu_j^* + RT \ln a_j + z_j FE$, Eq. 2.4 with the pressure and gravitational terms omitted; see p. 113), the difference in chemical potential of H^+ across a membrane is

$$\mu_H^i - \mu_H^o = RT \ln a_H^i + FE^i - RT \ln a_H^o - FE^o$$

$$= RT \ln \frac{a_H^i}{a_H^o} + FE_M \qquad (6.16a)$$

Incorporating the definition of pH (pH $= -\log a_H = -\ln a_H/2.303$) into Equation 6.16a, we obtain

$$\mu_H^i - \mu_H^o = -2.303\ RT(pH^i - pH^o) + FE_M \qquad (6.16b)$$

which at 25°C becomes

$$\mu_H^i - \mu_H^o = 5.71 \, (pH^o - pH^i) + 96.5 \, E_M \text{ kJ mol}^{-1} \qquad (6.16c)$$

where E_M is in volts.

According to the chemiosmotic hypothesis (which might, more appropriately, be termed a "transmembrane hypothesis"), the ATP reaction is driven in the energetically uphill forward direction by protons moving out of the thylakoids in their energetically downhill direction. By measuring the amount of ATP formed in the dark and the accompanying efflux of protons, the stoichiometry is found to be about 3 H^+'s per ATP. Since the formation of ATP in chloroplasts usually requires 40 to 50 kJ (mol ATP)$^{-1}$, as indicated above, the difference in chemical potential of H^+ must be at least 13 to 17 kJ (mol H^+)$^{-1}$ to drive the reaction energetically using 3 H^+'s per ATP. By Equation 6.16c, such an energy difference corresponds to a pH difference of 2.3 to 3.0 pH units, or a difference in electrical potential of 0.14 to 0.17 V. In turn, if the proton chemical potential gradient is established by coupling to electron flow, then an energetically uphill movement of protons of at least 13 to 17 kJ mol^{-1} would be accompanied by an even larger energetically downhill flow of electrons. We will next examine the evidence and thermodynamic requirements for the various steps from electron flow to ATP formation.

Evidence for Chemiosmotic Hypothesis

We will begin by seeing whether the appropriate reactions occur, and whether sufficient energy is available between pairs of electron flow components, to lead to proton accumulation within thylakoids. Let us start with the O_2-evolution step, $\frac{1}{2}H_2O \rightleftharpoons \frac{1}{4}O_2 + H^+ + e^-$ (essentially Eq. 5.8). To obtain H^+ inside thylakoids from the O_2 evolving step, we need: (1) H_2O inside the thylakoids, which can readily diffuse in from the stroma; (2) an oxidant with a redox potential more positive than the 0.82 V of the H_2O-O_2 couple, which is supplied by the P_{680}^+-P_{680} couple in Photosystem II having a midpoint redox potential of 0.90 V (Table 5.2); (3) a pathway for removing electrons, which is provided by Photosystem II, since it moves electrons from the inner to the outer side of the thylakoid membrane (see Fig. 5.14); and (4) removal of O_2 from inside the thylakoid, which readily occurs by outward diffusion of this small neutral molecule. Thus, the asymmetrical nature of the reaction center of Photosystem II, together with the known properties of membranes, allowing small neutral molecules to cross easily while retarding the penetration of charged species, leads to an accumulation of H^+ inside thylakoids owing to the light-dependent, O_2-evolution step.

We argued above that electrical and chemical energies are interconvertible. Thus, we might ask what difference in redox potential would be sufficient to supply the energy necessary to cause ADP plus phosphate to react to form ATP? The minimum change in redox potential necessary to power such ATP formation can be calculated from Equation 6.7, $\Delta G = -nF\,\Delta E$, where F is 96.5 kJ mol^{-1} V^{-1} (App. II). The amount of Gibbs free energy required to phosphorylate a mole of ADP is generally 40 to 50 kJ, so the ΔG for the redox reaction must be at least this negative. When one mole of electrons is transferred, this corresponds to the following electrical potential change:

$$\Delta E = \frac{\Delta G}{-nF} = \frac{(-40 \text{ to } -50 \text{ kJ})}{-(1 \text{ mol})(96.5 \text{ kJ mol}^{-1} \text{ V}^{-1})}$$

$$= 0.41 \text{ to } 0.52 \text{ V}$$

If two electrons were used per ATP, the minimum electrical potential difference would be 0.21 to 0.26 V; if three were used, it would be 0.14 to 0.17 V. We indicated above that 3 H^+'s are apparently needed per ATP. If one H^+ is transported for each electron, then 0.14 to 0.17 V would represent the minimum ΔE needed.

The midpoint redox potentials in Table 5.2 and Figure 6.4 indicate the approximate ΔE's available between the electron flow components. The difference in midpoint redox potential between plastoquinone A (0.113 V) and Cyt f(0.365 V) or plastocyanin (0.360 V) is about 0.25 V, which could supply the 0.14 V to 0.17 V potential drop required for the 3-electron case. Also, plastoquinones can carry protons and electrons (actually, 2 H^+ and 2e^- per molecule). The H^+ is apparently picked up in the stroma and delivered to the inside of the thylakoid, leading to the H^+ accumulation there (Figs. 5.14 and 6.5). The electron, which came from the reaction center of Photosystem II on the outside of the membrane, crosses toward the inside with the H^+, and then is moved back out across the lamellar membrane by the photochemistry in Photosystem I (Fig. 5.14).

A matter related to the coupling of electrons and protons is the relative amounts of ATP produced and NADP reduced in chloroplasts. Three moles of ATP and two of NADPH are needed per mole of CO_2 photosynthetically fixed in the majority of plants, which are referred to as C_3 plants, since CO_2 is incorporated into ribulose-1,5-diphosphate to yield two 3-phosphoglyceric acids, a *three carbon* compound (Fig. 8.5). Four to six ATP's (depending on which of three different decarboxylating enzymes is involved) and two NADPH's are required per CO_2 fixed in C_4 plants, where the first photosynthetic products are *four carbon* organic acids (e.g., oxaloacetic acid). (We will briefly discuss C_3 and C_4 plants in Ch. 8, e.g., pp. 434 and 452.) If the absorption of 8 photons leads to the processing of 4 excitations in each of the two photosystems, then 1 O_2 can be evolved and 4 H^+'s produced inside a thylakoid by Photosystem II, 4 H^+'s would be delivered inside a thylakoid by the plastoquinone pool, and 2 NADPH's would be produced by the overall noncyclic electron flow (see Figs. 5.12 and 6.4). Although this is sufficient NADPH to fix 1 CO_2 in C_3 plants, the 3 ATP's would require 9 H^+'s moving out across the thylakoid instead of the 8 generated. Thus cyclic or perhaps pseudocyclic electron flow could lead to the other H^+ (as we indicated in Ch. 5, no NADP reduction accompanies either of these types of electron flow). For instance, cyclic electron flow takes an electron from ferredoxin to the plastoquinone pool, which moves the electron back to Photosystem I and delivers an H^+ inside the thylakoid. This would suggest that 9 photons are required to produce the 3 ATP's and 2 NADPH's needed per CO_2. So far, the precise quantum requirement and actual electron flow involvement for CO_2 fixation are experimentally unresolved.

One of the most striking pieces of evidence in support of the chemiosmotic hypothesis was obtained in the 1960's by Jagendorf and Uribe (see Jagendorf and Uribe). When chloroplast lamellae were incubated in a solution at pH 4—in which case pH^i presumably attained a value near 4—and then rapidly transferred to a solution with a pH^o of 8 containing ADP and phosphate, the lamellae were capable of leading to ATP formation *in the dark*. Approximately 100 ATP's could be formed per Cyt f. When the difference in pH across the membrane was 2.5 or less, essentially no ATP was formed by the chloroplast lamellae. This is in close agreement with the energetic argument that we presented above, where a minimum ΔpH of 2.3 to 3.0 pH units was predicted as being necessary to lead to ATP formation. Also, if the pH of the external solution is gradually increased from 4 to 8 in the dark (over a period of tens of seconds), the protons "leak" out across the lamellar membranes, ΔpH is relatively small, and no ATP is formed.

The electrical term in the chemical potential of H^+ also can power ATP formation (see Gräber et al. or Witt). For instance, when an E_M of 0.20 V was artificially created across lamellar membranes, ATP formation could be induced in the dark. This is consistent with our prediction that an electrical potential difference of at least 0.14 to 0.17 V would be necessary. In chloroplast thylakoids, E_M in the light appears to be fairly low, e.g., near 0.02 V in the steady state (see Fig. 6.5). However, the electrical term seems to be the main contributor to $\Delta\mu_H$ for the first 1 to 2 seconds after chloroplasts are exposed to a high PAR. Also, the electrical component of the H^+ chemical potential difference can be much larger for the chromatophores of certain photosynthetic bacteria (as indicated on p. 26, chromatophores are small vesicles surrounded by a single membrane; they have 20 to 50 reaction centers). In chromatophores from the purple photosynthetic bacterium *Rhodopseudomonas spheroides*, E_M can be 0.20 V in the light in the steady state.

When chloroplasts are illuminated, electron flow commences, which causes μ_H^i within the thylakoids to increase relative to μ_H^o. We would expect a delay before $\Delta\mu_H$ given by Equation 6.16 is large enough to lead to ATP formation. Indeed, a lag of a fraction of a second to seconds can be observed before photophosphorylation commences at low PAR's, and the lag can be decreased by increasing the PAR (see Avron). We can also reason that a gradient in the chemical potential of H^+ will affect the movement of other ions. For instance, the light-induced uptake of H^+ into the thylakoids is accompanied by a release of Mg^{2+}, which can cause the stromal concentration of this ion to increase by more than 10 mM (see Barber). This released Mg^{2+} can activate various enzymes involved with CO_2 fixation in the stroma, indicating that the ionic readjustments following light-dependent proton movements can act as a cellular control for biochemical reactions.

We can imagine various ways of dissipating the chemical potential gradient of H^+ across the lamellar membranes and thus uncoupling electron flow from ATP formation. Compounds that accomplish this are called *uncouplers*. For instance, neutral weak bases (e.g., methyl amine, CH_3NH_2; or ammonia, NH_3) can readily diffuse into the thylakoids and there combine with H^+. This lowers a_H^i and raises pH^i. Moreover, the protonated base ($CH_3NH_3^+$ or NH_4^+) cannot readily diffuse back out, since it is now charged. The uncoupler nigericin competitively binds H^+ and K^+, and can exchange K^+ outside for H^+ inside, which also tends to lower a_H^i (since nigericin can stoichiometrically lead to H^+ movement one way and K^+ the other, it is referred to as an H^+/K^+ antiport; see p. 154). Detergents can remove certain membrane components, thus making the thylakoids leaky to H^+ and other ions, which also dissipates the $\Delta\mu_H$. Such studies further show the importance of the H^+ chemical potential difference in leading to ATP formation.

The coupling between $\Delta\mu_H$ and ATP is done via an ATPase (more appropriately called an ATP synthetase), although the exact mechanism is not presently understood. As illustrated in Figure 6.5, two components appear to be involved: (1) a coupling factor attached to the outside of the thylakoid, and (2) a hydrophobic part in the membrane to which the coupling factor is bound and through which H^+ can pass (see McCarty, and Shavit).[*] The coupling factor, which is soluble in water and is readily dislodged from the membrane, has a molar mass of about 325 kg mol^{-1} (380 kg mol^{-1} in one study where dissociation of subunits was prevented) and consists of 5 different subunits, most probably occurring in pairs. It makes up about 10% of the thylakoid protein, and appears as spheres about 9 nm in diameter on the outer side of the thylakoid (it is absent where the thylakoids stack together to form grana). Approximately 3 coupling factors occur per pair of reaction centers in the thylakoids. The hydrophobic part of the ATP-synthesizing complex, which occurs embedded in the lamellar membrane, has at least 3 distinct subunits with a total molar mass of about 100 kg mol^{-1} for single copies of each subunit. When the coupling factor is removed, the thylakoid is quite permeable to H^+, suggesting that the hydrophobic part is a channel or transporter for protons. The mechanism of catalysis and the manner in which apparently 3 H^+'s are involved per ATP formed are currently active areas of research.

MITOCHONDRIAL BIOENERGETICS

The activities of chloroplasts and mitochondria are related in a number of ways, as is clear from Figure 6.6. For instance, the O_2 evolved by photosynthesis can be consumed during respiration, while the fate of CO_2 for the two processes is just the opposite. Moreover, ATP formation is coupled to electron flow in both organelles; in mitochondria the electron flow is from a reduced pyridine nucleotide to the oxygen-water half-cell, while in chloroplasts it is in the opposite direction (see Fig. 6.6). From a few to many thousands of mitochondria occur in a given plant cell, their frequency tending to be somewhat less in cells where chloroplasts are abundant. In photosynthetic tissue at night, and at all times in the nongreen parts of a

---

[*] Since the chloroplast coupling factor resembles the analogous coupling factor found in mitochondria, designated F_1, it is usually referred to as CF_1. The hydrophobic part resembles the analogous F_0 in mitochondria, where the subscript o stands for oligomycin, since F_0 made the mitochondrial F_1 ATPase (ATP hydrolyzing activity) sensitive to oligomycin.

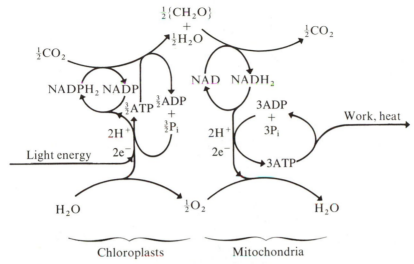

Figure 6.6
Schematic representation of the interrelationship between components involved in chloroplast and mitochondrial bioenergetics.

plant, oxidative phosphorylation in mitochondria is the predominant supplier of ATP for the cells.

Electron Flow Components—Redox Potentials

As with chloroplast membranes, many compounds in mitochondrial membranes can accept or donate electrons. These electrons originate from biochemical cycles occurring in the cytosol as well as in the mitochondrial matrix—most come from the tricarboxylic acid (Krebs) cycle, which leads to the oxidation of pyruvate and the reduction of NAD within mitochondria. The principal components for mitochondrial electron transfer are indicated in Figure 6.7, where the spontaneous electron flow to higher redox potentials is toward the bottom of the figure. As for photosynthetic electron flow, only a relatively few types of compounds are involved in electron transfer in mitochondria—namely, pyridine nucleotides, flavoproteins, quinones, cytochromes, and the water-oxygen couple (some additional iron-plus-sulfur

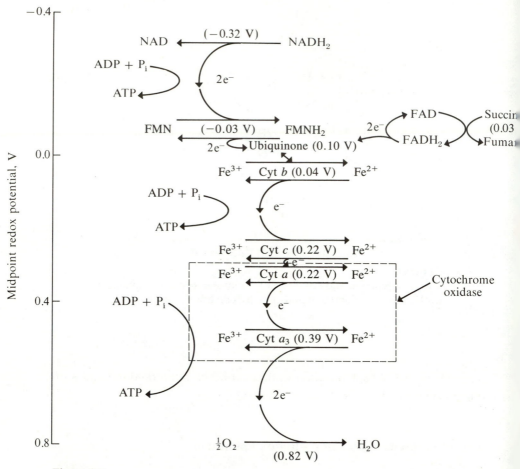

Figure 6.7
Components of the mitochondrial electron transport chain with midpoint redox potentials in parentheses. Also indicated are possible coupling sites where the movement of two moles of electrons toward higher redox potentials can supply sufficient Gibbs free energy to form one mole of ATP. (See Ikuma, Lehninger, Stryer, or White et al.)

containing proteins may be involved, and Fe-S centers can also occur on flavoproteins and cytochromes).

The reduced compounds that introduce electrons directly into the mitochondrial electron transfer chain are NADH[*] and $FADH_2$ (see Fig. 6.7). FAD is flavin adenine dinucleotide, and consists of riboflavin (vitamin B_2)

---

[*] $NADH + H^+$ is referred to as $NADH_2$ when denoting the redox couple in this book; see p. 310 for the analogous use of $NADPH_2$.

bound by a pyrophosphate bridge to adenosine (Fig. 6.3). Upon accepting two electrons and two protons—one H going to each N with a double bond in the riboflavin part of the molecule—FAD is reduced to $FADH_2$. The FAD-$FADH_2$ couple is usually bound to a protein, in which case the combination is referred to as a *flavoprotein* (ferredoxin-NADP reductase, p. 282, is a flavoprotein involved with photosynthetic electron transfer). We note that a flavoprotein containing flavin mononucleotide (FMN) as the prosthetic group also occurs in mitochondria, where FMN is riboflavin phosphate:

Electrons from NADH or $FADH_2$ move by a mitochondrial electron transfer chain to O_2, which is thus reduced to H_2O (see Fig. 6.7). Important members of the mitochondrial electron transfer chain, in order of increasing midpoint redox potential (indicated in parentheses), are NAD-$NADH_2$ (-0.32 V); FMN-$FMNH_2$ (-0.03 V); Cyt b (0.04 V); ubiquinone (0.10 V); Cyt c (0.22 V); Cyt a (0.22 V) plus Cyt a_3 (0.39 V), which are in the cytochrome oxidase complex; and O_2-H_2O (0.82 V). The cytochrome oxidase complex contains 2 Cyt a plus 2 Cyt a_3 molecules and copper on an equimolar basis with the heme (see Fig. 5.13). Both the Cu and the Fe of the heme of Cyt a_3 may be involved with the reduction of O_2 to H_2O (see Wikström et al.). Mitochondria contain coenzyme Q, or ubiquinone, which apparently intervenes in electron transfer between the flavoproteins and just before or after Cyt b (Fig. 6.7). The ubiquinone found in most plant and animal mitochondria differs from plastoquinone A (p. 280) by having 2 methoxy groups in place of the methyl groups on the ring, and 10 instead of 9 isoprene units in the side chain. A c-type cytochrome, referred to as Cyt c_1 in animal mitochondria, may intervene just before Cyt c, and there is evidence for a b-type cytochrome in plant mitochondria which is involved with an electron transfer that bypasses cytochrome oxidase on the way to O_2. Cytochromes a, b, and c are in roughly equal amounts in mitochondria (the ratios vary somewhat with plant species; see Ikuma), while flavoproteins are about 4 times, ubiquinones 7 to 10 times, and pyridine nucleotides 10 to 30 times more abundant than individual cytochromes. Likewise, in chloroplasts the quinones and pyridine nucleotides are much more abundant than the cytochromes (see Table 5.2).

Oxidative Phosphorylation

ATP formation coupled to electron flow in mitochondria is generally called *oxidative phosphorylation*. Since electron flow involves reduction as well as oxidation, more appropriate names would be "respiratory phosphorylation" or "respiratory-chain phosphorylation," terms that are also more consistent with photophosphorylation for the ATP formation occurring in photosynthesis. As with photophosphorylation, the mechanism of oxidative phosphorylation is not yet fully understood in molecular terms. Processes like phosphorylation accompanying electron flow appear to be intimately connected with membrane structure, and thus they are much more difficult to study than the biochemical reactions that take place in solutions. Coupling between electron flow and ATP formation in mitochondria could be via specific chemical components, the electron flow possibly inducing conformational changes facilitating the synthesis of ATP. Alternatively, a chemiosmotic mechanism is possible, and we will discuss some of its characteristics at the end of this section after examining the energetics of electron flow. (See Boyer et al.; Fillingame; Lehninger; B. T. Storey in Davies, Ch. 5 reference; Stryer; or White et al. for further details.)

Experiments with isolated mitochondria have shown that the number of ATP's produced per pair of electrons used to reduce O_2 depends on the particular compound introducing the electrons into the electron transfer pathway. For instance, when a pair of electrons moves from $NADH_2$ to O_2, 3 ATP molecules can be produced (see Fig. 6.7). Certain substrates, such as succinate, lead directly to a reduction of a flavoprotein without first reducing NAD. The oxidation of the $FADH_2$ reduced by succinate leads to the phosphorylation of only 2 ADP's per pair of electrons moving along the mitochondrial electron transfer chain. This suggests that a "coupling site" for ATP formation may exist between $NADH_2$ and FMN, ubiquinone, or Cyt *b* (see Fig. 6.7), where for animal mitochondria a redox potential span of 0.31 V can occur (the actual ΔE, not the difference in midpoint redox potentials; see White et al.). We calculated on p. 319 that a redox potential increase of 0.21 to 0.26 V for a pair of electrons could supply the 40 to 50 kJ required to synthesize a mole of ATP, indicating that the energy may well

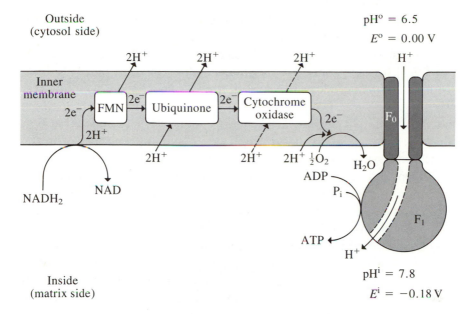

Outside
(cytosol side)

$pH^o = 6.5$

$E^o = 0.00$ V

Inside
(matrix side)

$pH^i = 7.8$

$E^i = -0.18$ V

Figure 6.8

Schematic representation of certain electron flow and ATP synthesis components in mitochondria, emphasizing the directional flows of H^+. The H^+, moved out toward higher μ_H accompanying electron flow along the respiratory chain, can move back through a hydrophobic channel (F_o) and a coupling factor attached to the inside of the inner membrane (F_1), leading to ATP synthesis.

be sufficient. Electrons from Cyt b or ubiquinone move to the subsequent members of the electron transfer chain, where two more coupling sites are expected (see Fig. 6.7). Energetically speaking, coupling sites may exist between Cyt b and Cyt c (0.31 V available for animal mitochondria) and between Cyt c and Cyt a_3 (0.32 V available). Such energetic arguments do not indicate the coupling mechanism, which remains to be clarified.

Let us next examine the coupling between electron flow and ATP formation in mitochondria by the chemiosmotic hypothesis, where the intermediate is the electrochemical gradient of protons. Accompanying electron flow in mitochondria, H^+ is transported from the matrix (see Fig. 1.8) on the inner side of the inner membrane to the space between the limiting membranes but outside the inner membrane (Fig. 6.8). Certain electron flow components may be so situated in the membranes that they can carry out this asymmetrical movement. For instance, FMN could deliver 2 H^+'s to the outside (these protons are probably obtained from $NADH_2$). As in chloroplasts, a pool of quinones exists in the lipid phase of the membranes unbound to proteins. This ubiquinone pool in mitochondria could move 2 H^+'s from the inside to the outside as it carries electrons from a flavoprotein to Cyt c. We also note that 2 H^+'s are necessary for the reduction of $\frac{1}{2}O_2$ to H_2O, and these protons could also be taken up from the internal solution (Fig. 6.8). Since we have indicated that a coupling site may be associated with the cytochrome oxidase complex, we would expect transmembrane proton movement there also, although the molecular details are not obvious. Perhaps the electron flow affects the conformation and dissociation constants in such a way as to bind H^+'s from the matrix and release them on the outside of the membrane (see Wikström et al.). Some experiments suggest that 3 or 4 H^+'s may be moved across the mitochondrial inner membrane per pair of electrons moving across a coupling site (see Boyer et al. or Fillingame). Since the outer mitochondrial membrane is quite permeable to small solutes (mentioned on p. 24), the protons from the matrix can end up in the cytosol. Thus, the pH is higher and the electrical potential lower in the inner region (Fig. 6.8) than it is outside the inner membrane, which is opposite from the polarity found for chloroplasts (Fig. 6.5).

The transport of protons out of the matrix leads to a difference in the H^+ chemical potential across the inner mitochondrial membrane. Using equation 6.16 and the values in Figure 6.8, we can calculate that $\mu_H^i - \mu_H^o = -25$ kJ (mol H^+)$^{-1}$, indicating that the H^+ chemical potential is lower on the inside (about 0.26 eV per H^+ lower). In some cases where ATP formation occurs, E_M is about -0.15 V and $pH^o - pH^i$ is about -0.5, in which case only 17 kJ (mol H^+)$^{-1}$ is available for use in ATP formation from the chemical potential difference of H^+ across the inner mitochondrial membrane. We also note that for chloroplasts most of the $\Delta\mu_H$ is due to the pH term, while for mitochondria the electrical term is apparently more important.

As with chloroplasts, we can uncouple ATP formation from electron flow by adding compounds that dissipate the H^+ chemical potential difference. In addition to the types of compounds mentioned in connection with chloroplast bioenergetics, studies with the ionophore valinomycin have proved useful. Valinomycin acts like an organic ring with a hydrophilic center through which K^+ and NH_4^+ can readily pass but Na^+ and H^+ cannot, and it can thus provide a selective channel when it is embedded in a membrane. If the antiport nigericin, which facilitates K^+-H^+ exchange (p. 321), is added together with the ionophore valinomycin, then protons tend to move back into the matrix through the antiport, thereby diminishing the ΔpH without affecting E_M, while the ionophore causes K^+ entry, thereby collapsing the electrical potential difference. As $\Delta\mu_H$ is thus dissipated, ATP formation ceases.

ATP formation is apparently coupled to the energetically downhill H^+ movement back into the mitochondrial matrix through a hydrophobic protein in the inner membrane (F_o; see footnote, p. 322) and a coupling factor protein (F_1) that protrudes into the matrix from the inner side of the inner membrane (Fig. 6.8). The coupling factor, which looks like a knob about 9 nm in diameter, has a molar mass of approximately 360 kg mol^{-1} and consists of about 5 different polypeptide subunits (1 to 3 copies of each subunit per F_1). It is bound to F_o, which has a molar mass of at least 150 kg mol^{-1}, and consists of at least 4 different polypeptide subunits. When F_1 is removed, the inner membrane becomes leaky to H^+, suggesting that the hydrophobic F_o is a channel or transporter for protons. Apparently, the passage of 2 to 3 H^+'s through the ATP synthetase ($F_o + F_1$) leads to the formation of 1 ATP—energetically speaking, 3 H^+'s would seem to be required under certain conditions leading to ATP formation (see Boyer et al., Cross, or Fillingame). Since ATP is produced in the mitochondrial matrix but often needed in the cytosol, some mechanism for readily moving this fairly heavily charged substance across the mitochondrial inner membrane is likely. In fact, an ADP/ATP antiport exists in the inner membrane, which serves to replenish the internal ADP pool as well. Phosphate is also needed for ATP synthesis, and it enters by a P_i/OH^- antiport (the concentration of OH^- is relatively high in the matrix) or, perhaps, an H^+/P_i symport.

As with chloroplasts, many questions concerning electron flow and the coupled ATP formation in mitochondria remain unanswered. The first part of the mitochondrial electron transfer chain has a number of two-electron carriers (NAD, FMN, and ubiquinone) that must interact with the cytochromes (one-electron carriers). Do the two electrons from NAD move in sequence to the same Cyt b or to two different Cyt b's acting in parallel? Do two electrons come from one Cyt a_3 to reduce an oxygen atom to its state in water? In fact, the reduction of O_2 may involve 1, 2, or 4 electrons coming from each of 4, 2, or 1 cytochrome, respectively. Of perhaps more interest is deciding on the actual mechanism of ATP formation—whether it is some variation of a chemical coupling hypothesis or a chemiosmotic one. This involves examining the ratio of electrons to protons and the ratio of protons to ATP, where there are many similarities between chloroplasts and mitochondria, but also some perplexing differences.

ENERGY FLOW IN THE BIOSPHERE

The foregoing discussion of the way organisms utilize energy on an organelle level sets the stage for the consideration of bioenergetics in a broader context. We will begin with certain biochemical aspects and then discuss the overall flow of energy from the sun through the biosphere. We will consider photosynthetic efficiency as well as transfer of energy from plants to animals. This material will serve as a transition between the molecular and cellular levels, considered up to now, and the organ and organism levels of the succeeding three chapters.

In Chapter 4 (p. 195) we indicated that the radiation input of the sun into the earth's atmosphere is $1\,360$ W m^{-2}. Some of the radiant energy is incorporated into the free energy increases needed to form ATP and NADPH in chloroplasts (pp. 311–322). In turn, these energy currencies lead to the reductive fixation of CO_2 into a carbohydrate in photosynthesis (Fig. 5.1). In the same photosynthetic cells, in other plant cells, and in animal cells, the carbohydrates formed during photosynthesis can serve as the free energy source for respiration, which leads to the generation of ATP by oxidative phosphorylation (p. 326).

When used as a fuel in respiration, the carbohydrate glucose is first broken down into two molecules of pyruvate in the cytosol. Pyruvate enters the mitochondria and is eventually oxidized to CO_2 and H_2O by the tricarboxylic acid cycle. One mole of glucose thus leads to the formation of 36 moles of ATP.* As we noted on p. 241, the Gibbs free energy released by the complete oxidation of a mole of glucose is $2\,872$ kJ, while about 50 kJ may be required for the phosphorylation of one mole of ADP in mitochondria. Hence, the efficiency of the many-faceted conversion of Gibbs free energy from glucose to ATP is [(36 mol ATP/mol glucose)(50 kJ/mol ATP)/($2\,872$ kJ/mol glucose)](100), or 63%, indicating that the Gibbs free energy in glucose can be rather efficiently mobilized to produce ATP. Such ATP can then be used by the cells to actively transport ions, to synthesize proteins, and to provide for growth and maintenance in other ways. We can readily appreciate that, if free energy were not constantly supplied to their cells, plants and animals would drift toward equilibrium and thus die.

One of the consequences of the flux of energy through the biosphere is the formation of rather complex and improbable molecules such as proteins

---

* Two NADH's are produced in the cytosol during the oxidation of glucose to pyruvate. Depending on the carrier used to transport these NADH's into the mitochondria, 2 or 3 ATP's can be produced per NADH. The figure of 36 ATP per glucose assumes 2 ATP's for each such NADH. See Lehninger, or White et al.

and nucleic acids. Such compounds represent a considerably greater amount of Gibbs free energy than an equilibrium mixture containing the same relative amounts of the various atoms. (Equilibrium corresponds to a minimum in Gibbs free energy; see Fig. 6.1.) For instance, the atoms in the nonaqueous components of cells have an average of about 26 kJ mol^{-1} more Gibbs free energy than the same atoms at equilibrium (see Morowitz). The Boltzmann energy distribution (Eq. 3.21b) predicts that at equilibrium the fraction of atoms with kinetic energy in excess of E is equal to $e^{-E/RT}$, which for 26 kJ mol^{-1} is only 0.000028 at 25°C. Thus, only a very small fraction of atoms would have a kinetic energy equal to the average enrichment in Gibbs free energy per atom of the nonaqueous components in cells. It is the flux of energy from the sun through plants and animals (see Fig. 6.6) that leads to such an energy enrichment in the molecules and, furthermore, ensures that biological systems will be maintained in a state far from equilibrium, an essential condition for life.

Incident Light—Stefan–Boltzmann Law

To help understand the energy available to the biosphere, we need to reconsider some properties of radiating bodies. In Chapter 4 we indicated that the distribution of radiant energy per unit wavelength interval was proportional to $\lambda^{-5}/(e^{hc/\lambda kT} - 1)$, where T is the temperature of the radiation source (p. 197). This form of Planck's radiation distribution formula applies to an object that radiates maximally, a so-called blackbody. When blackbody radiation is integrated over all wavelengths, we can determine the maximum amount of energy radiated by an object. Appropriate integration of Planck's radiation distribution formula leads to the following expression:

$$\text{maximum radiant energy flux density} = \sigma T^4 \qquad (6.17)$$

where σ is a constant and T is in Kelvin degrees (°C + 273). Although Equation 6.17 can be derived from quantum-physical considerations developed by Planck in 1900, it was first proposed in the latter part of the 19th century. In 1879 Stefan empirically determined that the maximum radiation was proportional to the fourth power of the absolute temperature; in 1884 Boltzmann interpreted this in terms of classical physics. The coefficient of proportionality σ was deduced from measurements then available. It has become known as the Stefan–Boltzmann constant, and equals 5.670×10^{-8} W m^{-2} K^{-4} (App. II). In the case where the object does not radiate as a blackbody, the radiant energy flux density at the surface of the radiator equals $e\sigma T^4$, where e is the emissivity (see p. 348). Emissivity depends on the surface material of the radiating body and achieves its maximum value of unity for a blackbody (see Gates for further details).

We will now estimate the amount of energy radiated from the sun's surface and how much of this is annually incident on the earth's atmosphere. Using Equation 6.17, which is known as the Stefan–Boltzmann law, and the effective surface temperature of the sun, about 5 800 K (p. 197), the rate of energy radiation per unit area of the sun's surface is

$$J_{energy} = (5.670 \times 10^{-8} \text{ W m}^{-2} \text{ K}^{-4})(5\,800 \text{ K})^4 = 6.4 \times 10^7 \text{ W m}^{-2}$$

The entire output of the sun is about 3.84×10^{26} W, which leads to 1.21×10^{34} J year^{-1}. Since the amount incident on the earth's atmosphere is 1 360 W m^{-2} and the projected area of the earth is 1.276×10^{14} m^2, the annual energy input into the earth's atmosphere from the sun is (1 360 J m^{-2} s^{-1})(1.276 $\times$ 10^{14} m^2)(3.16 $\times$ 10^7 s year^{-1}), or 5.48×10^{24} J year^{-1}.

Absorbed Light and Photosynthetic Efficiency

Only a small fraction of the sun's energy incident on the earth's atmosphere each year is actually absorbed by photosynthetic pigments, and only a small fraction of the absorbed energy is stored as chemical energy of the photosynthetic products (see Gates; Leith and Whittaker, Ch. 5 reference; and Nobel). Specifically, approximately 5% of the 5.48×10^{24} J annually incident on the earth's atmosphere is absorbed by chlorophyll or other photosynthetic pigments, leading to a radiant energy input into this part of the biosphere of about 2.7×10^{23} J year^{-1}. How much of this energy is annually stored in photosynthetic products? As we indicated in Chapter 5, a net of approximately 7.2×10^{13} kg of carbon are annually fixed by photosynthesis. For each mole of CO_2 (12 g carbon) incorporated into carbohydrate, approximately 479 kJ of Gibbs free energy are stored. The total amount of energy stored each year by photosynthesis is thus (7.2 $\times$ 10^{16} g year^{-1})(1 mol/12 g)(4.79 $\times$ 10^5 J mol^{-1}), or 2.9×10^{21} J year^{-1}. Hence, only about 1% of the radiant energy absorbed by photosynthetic pigments ends up stored by plant cells.

There are many different ways to represent the efficiency of photosynthesis. If we express it on the basis of the total solar irradiation incident on the earth's atmosphere (5.48×10^{24} J year^{-1}), it is only 0.053%. This low figure averages in many places of low productivity, e.g., open oceans, polar icecaps, winter landscapes, and arid regions. Furthermore, not all solar radiation reaches the earth's surface, and much that does is in the infrared (see Fig. 4.2); the efficiency would be 0.093% if we considered only the solar irradiation reaching the ground. Nevertheless, even with the apparently very low overall energy conversion, the trapping of solar energy by photosynthesis is the only source of free energy used to sustain life.

What is the highest possible efficiency for photosynthesis? For low levels of *red* light, the conversion of radiant energy into the Gibbs free energy of photosynthetic products can be up to 34% in the laboratory (p. 241). Solar irradiation incident on plants spans many wavelengths, slightly less than half of this radiant energy being in the region that can be absorbed by photosynthetic pigments, 400 to 700 nm (see p. 195 and Fig. 7.2). If all the incident wavelengths from 400 to 700 nm (about half of the radiant energy) were absorbed by photosynthetic pigments, and 8 photons were required per CO_2 fixed, the maximum photosynthetic efficiency for the use of low levels of incident solar irradiation would be just under half of 34%, say 15%. But some sunlight is reflected or transmitted by plants (see Fig. 7.3), and some is absorbed by nonphotosynthetic pigments in the plant cells. Thus, the maximum photosynthetic efficiency for using incident solar energy under ideal conditions of temperature, water supply, and physiological status of plants in the field would be closer to 10%.

Actual measurements of photosynthetic efficiency in the field have indicated that up to about 7% of the incident solar energy can be stored in photosynthetic products for a rapidly growing crop under ideal conditions. Usually, the PAR (photosynthetically active radiation) level of the upper leaves of vegetation is too high for all excitations of the photosynthetic pigments to be used for the photochemistry of photosynthesis (see p. 268). The energy of many absorbed photons is therefore simply wasted as heat, especially when leaves of C_3 plants are exposed to a PAR of more than 600 μmol m^{-2} s^{-1} (p. 452). Hence, the maximum sustained efficiency for the conversion of solar energy into Gibbs free energy stored in photosynthetic products is often near 3% for a crop when averaged over a day in the growing season (see Gates or Loomis et al., Ch. 9 reference). For all vegetation averaged over a year, the efficiency is about 0.5% of the incident solar irradiation. This is consistent with our previous statement that about 1% of the radiant energy absorbed by photosynthetic pigments is stored in the products of photosynthesis, since only about half of the solar irradiation incident on plants is absorbed by chlorophylls, carotenoids, and phycobilins.

Food Chains and Material Cycles

We will now consider the fate of the Gibbs free energy stored in photosynthetic products when animals enter the picture. We begin by noting that across each step in a food chain the free energy decreases, as required by the second law of thermodynamics. For instance, growing herbivores generally retain only 10% to 20% of the free energy of the ingested plant material, while a mature animal uses essentially all of its Gibbs free energy consumption just to remain in a state far from equilibrium. Growing carnivores will store about 10% to 20% of the free energy content of herbivores or other animals they eat. Thus, there is a sizable loss in Gibbs free energy for each link in a food chain, and as a consequence there are seldom more than four links, or steps, in a chain.

Although modern agriculture tends to reduce the length of our food chain for meat, humans still make a rather large demand on the Gibbs free energy available in the biosphere. The global average intake of free energy is about 10 MJ person^{-1} day^{-1} (2 400 kcal person^{-1} day^{-1}; see Gates). For a world population of 5 billion, the annual consumption of Gibbs free energy for food would be 1.8×10^{19} J year^{-1}. (Humans also consume plants and animals for clothing, shelter, firewood, papermaking, and in many other ways.) Thus, our food consumption alone amounts to almost 1% of the 2.9×10^{21} J year^{-1} stored in photosynthetic products. If we were to eat only carnivores that ate herbivores with a 10% retention in Gibbs free energy across each step in the food chain, we would indirectly be responsible for the consumption of the entire storage of energy by present-day photosynthesis. Fortunately, we obtain most of our free energy requirements directly from plants. The average daily consumption in the United States is 13 MJ person^{-1}, of which just over 9 MJ comes from plants and just under 4 MJ from animals. In turn, the animals consumed in this case require an average Gibbs free energy input of 26 MJ of plant material, which corresponds to 15% efficiency in free energy retention across this link in the food chain.

The harnessing of solar radiation by photosynthesis starts the flow of Gibbs free energy through the biosphere. In addition to maintaining individual chemical reactions as well as entire plants and animals in a state far from equilibrium, the annual degradation of chemical energy to heat sets up various cycles (see Morowitz). We have already indicated some of these, e.g., O_2 is evolved in photosynthesis and then consumed by respiration, while CO_2 cycles in the reverse direction between these two processes (Fig. 6.6). There is also a cycling between ATP and ADP + phosphate at the cellular level, as well as the inevitable birth-death cycle of organisms. In addition, we can recognize a cycling of nitrogen, phosphorus, and sulfur in the biosphere. All of these material cycles can be regarded as a consequence of the unidirectional flow of Gibbs free energy, which becomes less after each step along the way.

Problems

6.1. A reaction $A + B \rightleftharpoons C$ has a ΔG^* of -17.1 kJ mol^{-1} of reactant or product at 25°C. Assume that activity coefficients are unity. In which direction will the reaction proceed under the following conditions? (a) The concentrations of A, B, and C are all 1 molal. (b) The concentrations of A, B, and C are all 1 millimolal. (c) The concentrations of A, B, and C are all 1 micromolal. (d) What is the equilibrium constant for the reaction?

6.2. Consider the following two half-cell reactions at 25°C:

$$A \rightleftharpoons A^+ + e^- \qquad \Delta G^* = 8.37 \text{ kJ mol}^{-1}$$

$$B \rightleftharpoons B^+ + e^- \qquad \Delta G^* = 2.93 \text{ kJ mol}^{-1}$$

Assume that the midpoint redox potential of the second reaction is $+0.118$ V and that all activity coefficients are unity. (a) If the redox potential of the B–B^+ couple is 0.000 V, what is the ratio of B^+ to B? (b) What is the midpoint redox potential of the A–A^+ couple? (c) Suppose that all reactants and products are initially 1 molal but that the couples are in separate solutions of equal volume. If the half-cells are electrically connected with a metal wire, what is the initial electrical potential difference between them, and in which direction do electrons flow? (d) If all reactants and products are initially 1 molal, what is the concentration of each at equilibrium in a single solution? (e) Qualitatively, how would the answer to (d) change if the initial conditions were identical to (c), but as the electrons flow through the wire they do electrical work?

6.3. Suppose that in the dark 2 mM ADP and 5 mM phosphate occur inside chloroplasts, which are initially devoid of ATP. Assume that the temperature is 25°C and that the pH is 7. (a) What is the ATP concentration at equilibrium? (b) When the chloroplasts are illuminated, the ADP concentration decreases to 1 mM. What is the new concentration of ATP, and what is the change in Gibbs free energy for continued photophosphorylation? (c) If ferredoxin has a redox potential of -0.580 V and the activity of NADPH$_2$ is 3% of that of NADP, what is the difference in redox potential between the two couples? (d) How much Gibbs free energy is available between the two couples in (c)? Is this enough for the continued formation of ATP under the conditions of (b)?

6.4. Consider the following two mitochondrial cytochromes:

$$\text{Cyt } b \text{ (Fe}^{2+}) \rightleftharpoons \text{Cyt } b \text{ (Fe}^{3+}) + e^- \qquad E_b^{*,\text{H}} = 0.040 \text{ V}$$

$$\text{Cyt } c \text{ (Fe}^{3+}) + e^- \rightleftharpoons \text{Cyt } c \text{ (Fe}^{2+}) \qquad E_c^{*,\text{H}} = 0.220 \text{ V}$$

Assume that the temperature is 25°C, the chemical activity of Cyt b (Fe^{2+}) is 20% of that of the oxidized form, and activity coefficients are equal to unity. (a) What is the redox potential of Cyt b? (b) If the concentration of Cyt c (Fe^{2+}) is 1 mM, what is the concentration of ferricytochrome c such that the Cyt c couple can just transfer electrons back to Cyt b? (c) If the free energy to form 1 mole of ATP is 32 kJ in mitochondria, what must be the redox potential of Cyt c such that a coupling site could use *one* electron going from Cyt b to Cyt c per ATP? (d) If the 32 kJ gained upon the hydrolysis of one mole of ATP were used to transfer two moles of electrons in the energetically uphill direction from Cyt c to Cyt b, what would be the maximum redox potential of Cyt c? (e) Assume that ATP is 0.2 mM, ADP is 0.5 mM, and phosphate is 4.0 mM in mitochondria. What is the answer to (d) using the Gibbs free energy from ATP hydrolysis under these conditions?

References

Alberty, R. A. 1968. Effect of pH and metal ion concentration on the equilibrium hydrolysis of adenosine triphosphate to adenosine diphosphate. *Journal of Biological Chemistry 243*:1337–1343.

Avron, M. 1978. Energy transduction in photophosphorylation. *FEBS Letters 96*: 225–232.

Barber, J., ed. 1976. *The Intact Chloroplast. Topics in Photosynthesis*, Vol. 1. Elsevier, Amsterdam.

Boyer, P. D., B. Chance, L. Ernster, P. Mitchell, E. Racker, and E. C. Slater. 1977. Oxidative phosphorylation and photophosphorylation. *Annual Review of Biochemistry 46*:955–1026.

Castellan, G. W. 1971. *Physical Chemistry*, 2nd ed. Addison-Wesley, Reading, Massachusetts.

Cross, R. L. 1981. The mechanism and regulation of ATP synthesis by F$_1$-ATPases. *Annual Review of Biochemistry 50*:681–714.

Fillingame, R. H. 1980. The proton-translocating pumps of oxidative phosphorylation. *Annual Review of Biochemistry 49*:1079–1113.

Forrest, W. W., and D. J. Walker. 1971. The generation and utilization of energy during growth. *Advances in Microbial Physiology 5*:213–274.

Gates, D. M. 1980. *Biophysical Ecology*. Springer-Verlag, Berlin.

Gräber, P., E. Schlodder, and H. T. Witt. 1977. Conformational change of the chloroplast ATPase induced by a transmembrane electric field and its correlation to phosphorylation. *Biochimica et Biophysica Acta 461*:426–440.

Ikuma, H. 1972. Electron transport in plant respiration. *Annual Review of Plant Physiology 23*:419–436.

Jagendorf, A. T., and E. Uribe. 1966. ATP formation caused by acid-base transition of spinach chloroplasts. *Proceedings of the National Academy of Sciences of the United States 55*:170–177.

Knox, R. S. 1977. Photosynthetic efficiency and exciton transfer and trapping. In *Primary Processes of Photosynthesis*, J. Barber, ed. Elsevier, Amsterdam. Pp. 55–97.

Lehninger, A. L. 1975. *Biochemistry*, 2nd ed. Worth, New York.

MacInnes, D. A. 1961. *The Principles of Electrochemistry*. Dover, New York.

McCarty, R. E. 1979. Roles of a coupling factor for photophosphorylation in chloroplasts. *Annual Review of Plant Physiology 30*:79–104.

Mitchell, P. 1979. Compartmentation and communication in living systems. Ligand conduction: a general catalytic principle in chemical, osmotic and chemiosmotic reaction systems. *European Journal of Biochemistry 95*:1–20.

Morowitz, H. J. 1979. *Energy Flow in Biology*. Ox Bow Press, Woodbridge, Connecticut.

Morris, J. G. 1968. *A Biologist's Physical Chemistry*. Addison-Wesley, Reading, Massachusetts.

Nobel, P. S. 1974. Free energy: the currency of life. In *Nature in the Round, A Guide to Environmental Science*, N. Calder, ed. Viking Press, New York. Pp. 157–167.

Payne, W. J. 1970. Energy yields and growth of heterotrophs. *Annual Review of Microbiology 24*:17–52.

Penning de Vries, F. W. T. 1972. Respiration and growth. In *Crop Processes in Controlled Environments*, A. R. Rees, K. E. Cockshull, D. W. Hand, and R. G. Hurd, eds. Academic Press, London. Pp. 327–347.

Rosing, J., and E. C. Slater. 1972. The value of $\Delta G°$ for the hydrolysis of ATP. *Biochemica et Biophysica Acta 267*:275–290.

Ross, R. T., and M. Calvin. 1967. Thermodynamics of light emission and free-energy storage in photosynthesis. *Biophysical Journal 7*:595–614.

Shavit, N. 1980. Energy transduction in chloroplasts: structure and function of the ATPase complex. *Annual Review of Biochemistry 49*:111–138.

Stryer, L. 1981. *Biochemistry*, 2nd ed. W. H. Freeman, San Francisco.

Weyer, E. M., ed. 1968. Bioelectrodes. *Annals of the New York Academy of Sciences 148*, Article 1: 1–287.

White, A., P. Handler, E. L. Smith, R. L. Hill, and I. R. Lehman. 1978. *Principles of Biochemistry*, 6th ed. McGraw-Hill, New York.

Wikström, M., K. Krab, and M. Saraste. 1981. Proton-translocating cytochrome complexes. *Annual Review of Biochemistry 50*:623–655.

Witt, H. T. 1979. Energy conversion in the functional membrane of photosynthesis. Analysis by light pulse and electric pulse methods. The central role of the electric field. *Biochimica et Biophysica Acta 505*:355–427.

Zimmerman, A. M., ed. 1970. *High Pressure Effects on Cellular Processes*. Academic Press, New York.

CHAPTER 7

Temperature—Energy Budgets

We have already encountered many aspects of temperature. For instance, when introducing the special properties of water in Chapter 2, we noted that physiological processes generally take part within a fairly narrow temperature range. Such a range often dictates the boundaries of plant distributions, which can be quite apparent with respect to freezing limits. The water vapor content of air at saturation, which is nearly the situation occurring inside a leaf, is very temperature dependent. Biochemical reactions usually exhibit a temperature optimum, although specific reactions in many plants can acclimate to different temperature regimes. In Chapter 3 we discussed the Boltzmann energy distribution, Arrhenius plots, and the temperature coefficient, all of which involve the thermal energy of molecular motion. Light absorption (Ch. 4) causes molecules to attain states that are simply too improbable to be reached by collisions based on thermal energy. Transitions from some excited state to another one at lower energy or to the ground state can occur by radiationless transitions, where the energy is released as heat that is eventually shared by the surrounding molecules. The surface temperature of an object indicates both the wavelengths where radiation from it will be maximal (Wien's displacement law, Ch. 4) and also the total energy radiated (Stefan–Boltzmann law, Ch. 6). The temperature of an object is the net result of *all* the ways that energy can enter or leave it. In this chapter we will examine these various ways, especially for leaves. We will then be

339

able to predict the temperatures of leaves, as well as more massive plant parts, based on the ambient environmental conditions.

We should emphasize at the outset that individual plants as well as environmental conditions vary tremendously. Thus, in this and the next two chapters we will be indicating an approach to the study of plant physiology and physiological ecology, rather than a compendium of facts suitable for all situations. However, certain basic features should become obvious. For instance, CO_2 uptake during photosynthesis is accompanied by a water efflux through the stomata. This water loss during transpiration serves to cool the leaf. Also, energy influxes are balanced against effluxes by changing leaf temperature, which affects the amount of radiation emitted by the leaf as well as the heat conducted to the surrounding air. Another generality is that the temperatures of small leaves tend to be closer to those of the air than do the temperatures of large leaves. Since it difficult to appreciate the relative contributions of the various factors without first making calculations, we will use representative values for leaf and environmental parameters to describe the gas fluxes and the energy balance of leaves. We can then begin to understand the consequences of certain adaptations of plants to their environment as well as to recognize the experimental data needed for future refinements of our calculations.

ENERGY BUDGET—RADIATION

The principle of the conservation of energy (the first law of thermodynamics) states that energy cannot be created or destroyed, but only changed from one form to another. We will apply this principle to the energy balance of a leaf, which exists in an environment with many energy fluxes. We can summarize the various contributors to the energy balance of a leaf as follows:

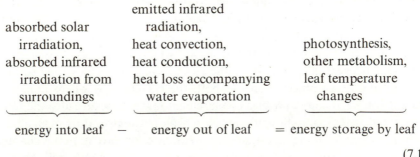

$$
\underbrace{\begin{array}{c} \text{absorbed solar} \\ \text{irradiation,} \\ \text{absorbed infrared} \\ \text{irradiation from} \\ \text{surroundings} \end{array}}_{\text{energy into leaf}} - \underbrace{\begin{array}{c} \text{emitted infrared} \\ \text{radiation,} \\ \text{heat convection,} \\ \text{heat conduction,} \\ \text{heat loss accompanying} \\ \text{water evaporation} \end{array}}_{\text{energy out of leaf}} = \underbrace{\begin{array}{c} \text{photosynthesis,} \\ \text{other metabolism,} \\ \text{leaf temperature} \\ \text{changes} \end{array}}_{\text{energy storage by leaf}}
$$

(7.1)

Since heat moves *into* a leaf when its temperature is less than that of the surrounding turbulent air, Equation 7.1 actually describes the case where the leaf temperature is greater than the temperature of the surroundings. Also, when water condenses onto a leaf, the leaf will gain heat. In such cases, the appropriate energy terms in equation 7.1 change sign.

The various terms in Equation 7.1 differ greatly in magnitude. For instance, the energy storage terms generally are relatively small. As a basis for comparison, we will consider the average amount of solar irradiation incident on the earth's atmosphere (the term "irradiation" refers to incident radiation; see p. 193). This radiant flux density, the solar constant, is about $1\ 360$ W m^{-2} (p. 195). The solar irradiation absorbed by an exposed leaf is often about half this during the daytime. Hence, any process under 7 W m^{-2} would correspond to less than 1% of the absorbed solar irradiation for full sunlight. What is the amount of energy stored by photosynthesis? A typical net rate of CO_2 fixation by a photosynthetically active leaf is 10 μmol m^{-2} s^{-1} (p. 440). As indicated in Chapter 5 (p. 241), about 479 kJ of energy are stored per mole of CO_2 fixed into photosynthetic products. Hence, photosynthesis by leaves might store $(10 \times 10^{-6}$ mol m^{-2} s$^{-1})(479 \times 10^3$ J mol$^{-1})$, or 5 W m^{-2}, which is only about 1% of the rate of absorption of solar irradiation under the same conditions. In some cases the rate of photosynthesis can be higher (see p. 441). But we can generally ignore the contribution of photosynthesis to the energy balance of a leaf. Other metabolic processes in a leaf, such as respiration and photorespiration, are usually even less important on an energy basis than photosynthesis, so that they too can generally be ignored.

We will now consider the amount of energy that can be stored because of changes in leaf temperature. For purposes of calculation, we will assume that a leaf has the high specific heat of water (4.19 kJ kg^{-1} °C^{-1} = 1.00 cal g^{-1} °C^{-1} at 20°C), is 400 μm thick, and has an overall density of 500 kg m^{-3} (0.5 g cm^{-3})—a leaf is often 50% air by volume. Hence, the mass per unit leaf area in this case is $(400 \times 10^{-6}$ m)(500 kg m$^{-3})$, or 0.20 kg m^{-2}. If 7 W m^{-2} were stored by temperature changes in such a leaf, its temperature would rise at the rate of $(7$ J m^{-2} s$^{-1})/[(4\ 190$ J kg^{-1} °C$^{-1})(0.20$ kg m$^{-2})]$, or 0.008°C s^{-1} (i.e., 0.5°C min^{-1}). Since this is a faster temperature change than that actually sustained for very long periods by leaves, we can assume that very little energy is stored (or released) in the form of leaf temperature changes. Hence, all three energy storage terms indicated in Equation 7.1 usually are relatively small for a leaf.

When the energy storage terms in Equation 7.1 are ignored, the remaining contributors to the energy balance of a leaf are either radiation or heat terms.

We can then simplify our energy balance relation for a leaf as follows:

$$
\underbrace{
\begin{array}{c}
\text{absorbed solar irradiation,} \\
\text{absorbed infrared irradiation} \\
\text{from surroundings}
\end{array}
}_{\text{energy into leaf}}
\cong
\underbrace{
\begin{array}{c}
\text{emitted infrared radiation,} \\
\text{heat convection,} \\
\text{heat conduction,} \\
\text{heat loss accompanying} \\
\text{water evaporation}
\end{array}
}_{\text{energy out of leaf}}
\tag{7.2}
$$

The heat conducted and convected from leaves is sometimes referred to as *sensible* heat, while that associated with the evaporation or the condensation of water is known as *latent* heat. In this section we will consider each of the terms in Equation 7.2 in turn.

Solar Irradiation

Solar irradiation can reach a leaf in many different ways, the most obvious being as *direct* sunlight (see p. 193 for certain radiation terminology). Alternatively, sunlight can be *scattered* by molecules and particles in the atmosphere before striking a leaf. Finally, both the direct and the scattered solar irradiation can be *reflected* by the surroundings toward a leaf.[*] In Figure 7.1 we summarize these various possibilities, which lead to six different ways (all containing the letter S in the figure) by which solar irradiation can impinge on a leaf. The individual energy fluxes may involve the upper surface of a leaf, its lower surface, or perhaps both surfaces—in Figure 7.1 the direct solar irradiation (S^{direct}) is incident only on the upper surface of the leaf. To proceed with the analysis in a reasonable fashion, we will obviously need to make many simplifying assumptions and approximations (see Gates 1980, Idso et al., Robinson, and van Wijk).

Some of the solar irradiation can be scattered or reflected from clouds before being incident on a leaf. On a cloudy day, the diffuse sunlight emanating from the clouds—S^{cloud}, or *cloudlight*—is substantial, while S^{direct} may be greatly reduced. For instance, a sky overcast by a fairly thin cloud layer 100 m thick might absorb or reflect away from the earth about 50% of

---

[*] Generally, the term *scattering* denotes the irregular changes in the direction of light caused by small particles or molecules; *reflection* refers to the change in direction of irradiation at a surface.

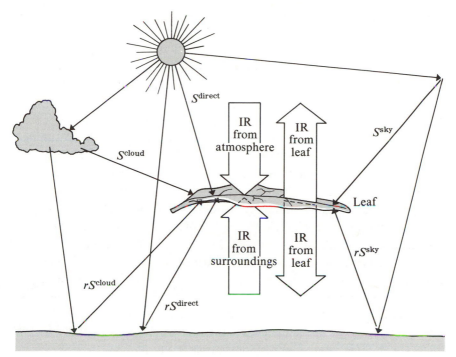

Figure 7.1

Schematic illustration of eight forms of radiant energy incident on an exposed leaf, and the infrared radiation emitted from its two surfaces.

the incident solar irradiation and diffusely scatter most of the rest toward the earth. A cloud 1 km thick will usually scatter somewhat less than 10% of the S^{direct} incident on it toward the earth (the actual numbers depend on the type and density of the clouds). Because the relative amounts of absorption, scattering, reflection, and transmission by clouds all depend on wavelength, cloudlight has a somewhat different wavelength distribution than direct solar irradiation.

Solar irradiation scattered by air molecules leads to S^{sky}, or *skylight*. Actually, scattering is generally divided into two categories—Rayleigh scattering, due to molecules; and Mie scattering, due to particles such as dust. Skylight differs considerably from the wavelength distribution for S^{direct}, since S^{sky} is enriched in the shorter wavelengths (see Fig. 7.2). Rayleigh scattering of solar radiation by the molecules in the atmosphere is approximately proportional to $1/\lambda^4$—light scattering from air-borne dust particles is approximately proportional to $1/\lambda$—and so the shorter wavelengths are scattered more. Because of the Rayleigh scattering, a preponderance of the shorter wavelengths, e.g., those below 500 nm, are scattered out of the direct solar beam (see Fig. 7.2) and become skylight. This explains the blue color of the sky (and its skylight) during the daytime—and the red color of the sun at sunset. In terms of energy, S^{sky} can be up to 10% of S^{direct} for a horizontal leaf (Fig. 7.1) on a cloudless day when the sun is overhead (Fig. 7.2) and can exceed S^{direct} when the sun is near the horizon.

For convenience, we will refer to the direct sunlight plus the cloudlight and skylight as the *global irradiation*, S. Generally, $S^{cloud} + S^{sky}$ is referred to as diffuse shortwave irradiation, a readily measured quantity, whereas S^{cloud} and S^{sky} are often hard to measure separately. Thus, the global irradiation equals the direct plus the diffuse shortwave irradiation. The value of the global irradiation varies widely with the time of day, the time of year, latitude, altitude, and atmospheric conditions. Consequently, it is an important parameter to specify when considering the energy balance of a leaf. The maximum value of S is about $1\,360$ W m^{-2}—the solar irradiation incident on the earth's atmosphere (p. 195). But, because of scattering and absorption of solar irradiation by atmospheric gases (Fig. 4.2), S on a cloudless day with the sun directly overhead in a dust-free sky is about $1\,000$ W m^{-2} at $1\,000$ m above sea level and often about 5% less at sea level.

Sunlight may impinge on a leaf as direct solar irradiation, cloudlight, or skylight. These three components of global irradiation may first be reflected from the surroundings before being incident on a leaf (see Fig. 7.1). Although the reflected global irradiation can be incident on the leaf from all angles, for a horizontal exposed leaf it falls primarily on the lower surface (Fig. 7.1). The reflected sunlight, cloudlight, and skylight often amount to 10% to 30% of the global irradiation. A related quantity is the fraction of the incident shortwave irradiation reflected from the earth's surface, termed the *albedo*, which is about 0.35 for dry sandy soil, 0.25 for dry clay, 0.10 for peat soil, 0.25 for most crops, and 0.15 for forests (see Rosenberg).[*]

Each of these six forms of solar irradiation (reflected as well as direct sunlight, cloudlight, and skylight) can have a different wavelength distribution. Since absorption depends on wavelength, the fraction of each one absorbed can in principle be different. Moreover, the fraction reflected also depends on wavelength. For simplicity, we will assume that the same absorptance (defined shortly) applies to S^{direct}, S^{cloud}, and S^{sky}, as well as to the reflected forms of these irradiations. We will also assume that the same re-

---

[*] The albedo can vary with the angle of incidence of the direct solar beam, being greater at smaller angles of incidence.

flectance applies to each component of the global irradiation. We can then represent the absorption of all forms of solar irradiation by a leaf as follows:

absorbed direct, scattered, and reflected solar irradiation

$$\cong a(S^{direct} + S^{cloud} + S^{sky}) + ar(S^{direct} + S^{cloud} + S^{sky})$$

$$= a(1 + r)S \tag{7.3}$$

where the *absorptance a* is the fraction of the global radiant energy flux density S absorbed by the leaf, and the *reflectance r* is the fraction of S reflected from the surroundings onto the leaf.[*]

Absorbed Infrared Irradiation

Besides the absorption of the various components of solar irradiation, additional *infrared*, or *thermal*, radiation is also absorbed by a leaf (see Eq. 7.2). Any object with a temperature above 0 K ("absolute zero") emits such thermal radiation, including a leaf's surroundings as well as the sky. The peak in the spectral distribution of thermal radiation can be described by Wien's displacement law, which states that the wavelength for maximum emission of energy, λ_{max}, times the temperature of the emitting body, T, equals 2.9 × 10^6 nm K (p. 198). Since the temperature of the surroundings is generally near 300 K, λ_{max} for radiation from them is close to (2.9 × 10^6 nm K)/ (300 K), or 10 000 nm, which is 10 μm. Thus, the emission of thermal radiation from the surroundings occurs predominantly at wavelengths far into the infrared. Because of its wavelength distribution, we will also refer to thermal radiation as both infrared radiation and *longwave* radiation (over 99% of the radiant energy from the surroundings occurs at wavelengths longer than 4 μm, and over 98% of the solar or *shortwave* irradiation is for wavelengths shorter than this).

Most of the thermal radiation from the sky comes from H_2O, CO_2, and other molecules in the atmosphere that emit considerable radiation from 5 to 8 μm and above 13 μm. Moreover, the concentration of these gases varies, so that the effective temperature of the sky, T^{sky}, as judged from its radiation, also varies (to be considered shortly). For instance, clouds contain much water in the form of vapor, droplets, or crystals, and this leads to a

---

[*] Absorptance is often called *absorptivity*, and reflectance is called *reflectivity*, especially when dealing with smooth surfaces of uniform composition.

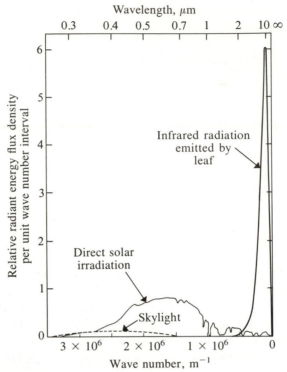

Figure 7.2

Wave number and wavelength distributions for direct solar
irradiation, skylight, and the radiation emitted by a leaf at 25°C.
The areas under the curves indicate the total energy radiated;
e.g., S^{direct} here is 840 W m^{-2}, and the IR emitted is 900 W m^{-2}.

substantial emission of infrared radiation, so that T^{sky} can be as high as
280 K on a cloudy day or night. On the other hand, a dry cloudless dust-free
atmosphere might have a T^{sky} as low as about 220 K (see Munn or Robinson).
Since effective sky temperatures are below 300 K, thermal radiation from
the sky generally occurs at longer wavelengths in the IR than for thermal
radiation from the surroundings, i.e., the soil, rocks, other leaves, and plants.

We will now consider the amount of thermal irradiation absorbed by a
leaf. We will suppose that the infrared irradiation from the surroundings,
acting like a planar source at an effective temperature of T^{surr}, is incident on
the lower surface of the leaf, and that the upper surface of the leaf is exposed
to the sky, which acts like a planar source with an effective temperature of
T^{sky} (Fig. 7.1). In Chapter 6 we introduced the Stefan–Boltzmann law, which
indicates that the amount of radiation emitted by a body depends markedly

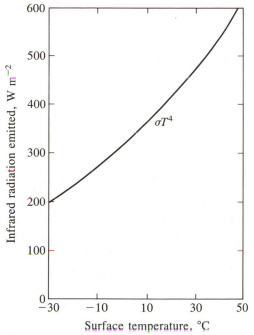

Figure 7.3

Rate of emission of infrared (longwave) radiation per unit area by a blackbody ($e_{IR} = 1.00$) as a function of its surface temperature.

on its temperature (Eq. 6.17, maximum radiant energy flux density $= \sigma T^4$). The Stefan–Boltzmann law predicts the maximum rate of energy radiation, i.e., the radiation emitted by a perfect radiator, the so-called blackbody (pp. 197, 331; see also Fig. 7.3). Here we will use *effective temperature* in the sense that the *actual* radiant energy flux density equals $\sigma(T_{effective})^4$. T^{sky}, for instance, is not the temperature we would measure at some particular location in the sky; but $\sigma(T^{sky})^4$ does indicate the actual amount of radiant energy from the sky. By the Stefan–Boltzmann law, with effective temperatures to give the radiation emitted by the surroundings and the sky, the IR absorbed by a leaf is

$$\text{infrared irradiation absorbed} = a_{IR}\sigma[(T^{surr})^4 + (T^{sky})^4] \qquad (7.4)$$

where the absorptance a_{IR} is the fraction of the energy of the incident infrared irradiation absorbed by the leaf.

Emitted Infrared Radiation

Thermal radiation is also *emitted* by a leaf. Such radiation occurs at wavelengths far into the IR, since leaf temperatures, like those of its surroundings, are near 300 K. This is illustrated in Figure 7.2, where the emission of radiant energy from a leaf at 25°C is plotted both in terms of wavelength and wave number.* Using the wave-number scale makes it easier to illustrate the spectral distribution of radiation from both sun and leaf in the same figure; moreover, the area under the curves is proportional to the total radiant energy. The λ_{max} for sunlight is in the visible region; for the leaf it is in the infrared near 10 μm. Figure 7.2 also indicates that essentially all of the thermal radiation emitted by a leaf has wave numbers less than 5×10^5 m^{-1}, which corresponds to IR wavelengths greater than 2 μm.

We will express the IR emitted by a leaf at a temperature T^{leaf} using the Stefan–Boltzmann law (Eq. 6.17), which describes the maximum rate of radiation per unit area. For the general emission case we can incorporate a coefficient known as the *emissivity*, or *emittance*, e, which takes on its maximum value of unity for a perfect, or blackbody, radiator.[†] Since infrared radiation is emitted by both sides of a leaf (see Fig. 7.1), the factor 2 is necessary to describe its energy loss by thermal radiation:

$$\text{infrared radiation emitted} = 2e_{IR}\sigma(T^{leaf})^4 \tag{7.5}$$

Like our other flux density relations, Equation 7.5 is expressed on the basis of unit area for one side of a leaf. The marked temperature dependency of emitted infrared radiation is indicated in Figure 7.3.

Values for a, a_{IR}, and e_{IR}

The parameters a, a_{IR}, and e_{IR} obviously help determine the energy balance of a leaf. We will first consider how the absorptance of a leaf depends on

---

* Wave number, which we introduced in Problems 4.2 and 4.6, is the frequency of the radiation divided by the speed of light, i.e., v/c, which is $1/\lambda_{vacuum}$ ($\lambda_{vacuum}v = c$ by Eq. 4.1). Hence, wave number is proportional to energy (see Eq. 4.2a, $E_\lambda = hv = hc/\lambda_{vacuum}$).

† The actual radiant energy flux density equals $e\sigma(T_{actual})^4$, which is the same as $\sigma(T_{effective})^4$. We will use emissivities and actual temperatures to describe the energy radiated by leaves; effective temperatures are usually employed for thermal radiation from the surroundings and the sky, since their temperatures are difficult to measure or, indeed, hypothetical. Empirical equations incorporating the influence of air temperature, water vapor content, and clouds can also be used to predict the longwave radiation from the sky (see Campbell, and Monteith 1973).

Wavelength, μm

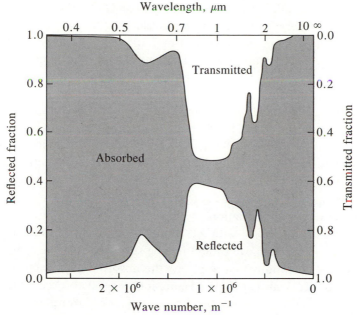

Figure 7.4
Representative fractions of irradiation absorbed, transmitted, and reflected by a leaf, as a function of wave number and wavelength. The sum $a_\lambda + r_\lambda + t_\lambda$ is unity. (See Gates 1965, Gates 1970, and Woolley.)

wavelength and then discuss its emittance for infrared radiation.

Figure 7.4 indicates that leaf absorptance at a particular wavelength, a_λ, varies considerably with the spectral region. For example, a_λ averages about 0.8 in the visible (0.4 to 0.7 μm). Such relatively high fractional absorption by a leaf is due mainly to the photosynthetic pigments; the local minimum in a_λ near 0.55 μm is where chlorophyll absorption is relatively low and thus a larger fraction of the incident light is reflected or transmitted (Fig. 5.3), leading to the green color (see Table 4.1) of leaves. Figure 7.4 indicates that a_λ is rather small from just past 0.7 μm up to nearly 1.2 μm. This is quite important for minimizing the energy input into a leaf, since much global irradiation occurs in this interval of the infrared. The fraction of irradiation absorbed becomes essentially unity for infrared irradiation beyond 2 μm. This does not lead to excessive heating of the leaves from absorption of global irradiation, since very little radiant emission from the sun occurs beyond 2 μm (see Fig. 7.2).

In our equations we have used two absorptances, a (in Eq. 7.3) and a_{IR} (in Eq. 7.4). In contrast to a_λ, these coefficients represent absorptances for a particular wavelength *range*. For example, *a* refers to the fraction of the incident solar energy absorbed (the wavelength distributions for direct sunlight and skylight are presented in Fig. 7.2). For most leaves, *a* is between 0.4 and 0.6, with 0.5 being typical. The shortwave absorptance can differ between the upper and the lower surfaces of a leaf, and also *a* tends to be lower for lower sun angles in the sky, since there is then more scattering of the shorter wavelengths (Fig. 7.2) and a_λ for the relatively enriched longer wavelengths is lower (Fig. 7.4; see Gates 1980). For the region 0.4 to 0.7 μm (designated "photosynthetically active radiation", or PAR, in Ch. 4), the overall leaf absorptance is generally 0.75 to 0.90, as indicated above. Figure 7.4 shows that nearly all the IR irradiation beyond 2 μm is absorbed by a leaf. In fact, a_{IR} for leaves is usually from 0.95 to 0.98, and we will use a value of 0.96 for purposes of calculation.

Next, let us compare a_{IR} with e_{IR}. Since the emission of radiation is the reverse of its absorption, we should not be surprised that the same sort of electronic considerations that apply to the absorption of electromagnetic radiation (see Ch. 4) also apply to its emission. A good absorber of radiation is also a good emitter. In more precise language, the absorptance a_λ equals the emittance e_λ when they refer to the same wavelength (referred to as Kirchhoff's radiation law)—for a blackbody, $a_\lambda = e_\lambda = 1.00$ at all wavelengths (see Kreith or Monteith 1973). Since radiation from the surroundings and the sky occurs in essentially the same region of the infrared as that emitted by a leaf, e_{IR} is about the same as a_{IR}, e.g., 0.96.

Let us now relate e_λ and a_λ to radiation quantities that we introduced in Chapter 4. The amount of radiant energy emitted by a blackbody per unit wavelength interval is proportional to $\lambda^{-5}/(e^{hc/\lambda kT} - 1)$, as predicted by Planck's radiation formula (p. 198). When we multiply this maximum radiation by e_λ at each of the wavelengths involved, we obtain the actual spectral distribution of the emitted thermal radiation. The absorptance a_λ is related to the absorption coefficient ε_λ introduced in Chapter 4. Equation 4.15 indicates that $\log J_0/J_b$ equals $\varepsilon_\lambda cb$, where c is the concentration of the absorbing species, b is the optical path length, J_0 is the incident flux density, and J_b is the flux density of the emergent beam when only absorption takes place (i.e., in the absence of reflection and scattering). The fraction of irradiation absorbed at a particular wavelength, $(J_0 - J_b)/J_0$, is the absorptance, a_λ. Thus, a_λ equals $1 - J_b/J_0$, which is $1 - 10^{-\varepsilon_\lambda cb}$ by Equation 4.15. Hence, the absorptance tends to be greater for wavelengths where the pigments absorb more (higher ε_λ), for thicker leaves, and for leaves with higher pigment concentrations.

Net Radiation

We have now considered each of the terms that involve radiation in the energy balance of a leaf (Eq. 7.2). It is often convenient to single out these these quantities and to refer to them as the *net radiation* balance for the leaf:

$$\begin{array}{c} \text{absorbed solar irradiation} \\ + \\ \text{absorbed infrared irradiation} \\ \text{from surroundings} \end{array} - \begin{array}{c} \text{emitted infrared} \\ \text{radiation} \end{array} = \text{net radiation} \quad (7.6a)$$

We can also use Equations 7.3 through 7.5 to express the net radiation balance:

$$\begin{array}{c} a(1+r)S \\ + \\ a_{IR}\sigma[(T^{surr})^4 + (T^{sky})^4] \end{array} - 2e_{IR}\sigma(T^{leaf})^4 = \text{net radiation} \quad (7.6b)$$

Before continuing with our analysis of the energy balance of a leaf, we will examine representative values for each of the terms in the net radiation.

Examples for Radiation Terms

Let us consider a horizontal leaf exposed to full sunlight at sea level where the global irradiation S is 840 W m^{-2}. We will assume that the absorptance of the leaf for global irradiation a equals 0.60, and that the reflectance of the surroundings r is 0.20. By Equation 7.3, the direct plus the reflected sunlight, cloudlight, and skylight absorbed by the leaf is

$$a(1+r)S = (0.60)(1.00 + 0.20)(840 \text{ W m}^{-2})$$

$$= 605 \text{ W m}^{-2}$$

To calculate the infrared irradiation absorbed by the leaf, we will let a_{IR} be 0.96, the temperature of the surroundings be 20°C, and the sky temperature be $-20°C$. Using Equation 7.4, with a Stefan–Boltzmann constant of 5.67 $\times$

10^{-8} W m^{-2} K^{-4} (App. II), the absorbed IR is

$$a_{IR}\sigma[(T^{surr})^4 + (T^{sky})^4]$$

$$= (0.96)(5.67 \times 10^{-8} \text{ W m}^{-2} \text{ K}^{-4})[(293 \text{ K})^4 + (253 \text{ K})^4]$$

$$= 624 \text{ W m}^{-2}$$

Hence, the total irradiation load on the leaf is (605 W m^{-2}) + (624 W m^{-2}), or 1 229 W m^{-2}.

In the present case the rate of energy input per unit leaf area (1 229 W m^{-2}) is nearly the size of the solar constant (1 360 W m^{-2}). About half the total heat load on the leaf is contributed by the various forms of irradiation from the sun (605 W m^{-2}), and half by IR irradiation from the surroundings plus the sky (624 W m^{-2}). Since the sky generally has a much lower effective temperature for radiation than the surroundings, the upper surface of an exposed leaf usually receives less IR than the lower one: $a_{IR}\sigma(T^{sky})^4$ here is 223 W m^{-2}, while the IR absorbed by the lower surface of the leaf, $a_{IR}\sigma$ $(T^{surr})^4$, is 401 W m^{-2} (see Fig. 7.3). Changes in the angle of an exposed leaf generally have a small influence on the total irradiation load (less than 300 W m^{-2}), since, the IR, the scattered, and the reflected irradiation received by a leaf come from all angles. But leaf angle can still have important implications for the interaction of certain plants with their environment, as we will see near the end of this chapter.

To estimate the infrared radiation emitted by a leaf, we will let e_{IR} be 0.96 and the leaf temperature be 25°C. By Equation 7.5, the energy loss by thermal radiation is

$$2e_{IR}\sigma(T^{leaf})^4 = (2)(0.96)(5.67 \times 10^{-8} \text{ W m}^{-2} \text{ K}^{-4})(298 \text{ K})^4$$

$$= 859 \text{ W m}^{-2}$$

In this case, the IR emitted by both sides of the exposed leaf is about 50% greater than the leaf's absorption of either solar or infrared irradiation. (Leaves shaded by other leaves are not fully exposed to T^{sky}, and for them IR emitted can be approximately equal to IR absorbed, a matter to which we will return on p. 378).

The net radiation balance (see Eq. 7.6) for our exposed leaf is (1 229 W m^{-2}) − (859 W m^{-2}), or 370 W m^{-2}. As we will discuss below, such excess energy is dissipated by conduction, convection, and the evaporation of water accompanying transpiration. However, most of the energy input into a leaf is actually balanced by the emission of infrared radiation. The IR emitted in the present case, for example, amounts to $(859 \text{ W m}^{-2})/(1\,229 \text{ W m}^{-2})$, or 70% of the energy input from all sources of incident irradiation. The radiation terms for this exposed leaf are summarized in Table 7.1, top line.

Since many conditions affect the net radiation balance for leaves, let us now consider some other examples. At an elevation of 2 000 m, S at noon on a cloudless day might be $1\,050 \text{ W m}^{-2}$. Because of the higher incident global irradiation at 2 000 m, the leaf there will absorb 141 W m^{-2} more direct, scattered, and reflected solar irradiation than our leaf at sea level. The effective temperature of the sky is generally slightly lower at the higher elevation ($-25°C$ vs. $-20°C$ in Table 7.1). We will assume that the surroundings are at the same temperature as for our example at sea level. The total irradiation input is then $1\,363 \text{ W m}^{-2}$, which is 134 W m^{-2} greater than at sea level (see Table 7.1). For the net radiation to be the same in the two cases (370 W m^{-2}), the leaf at the higher altitude must emit 134 W m^{-2} more thermal radiation than the leaf at sea level. As Table 7.1 indicates, this could be accomplished if T^{leaf} for the leaf at 2 000 m were 11°C higher than the one at sea level.

Certain plants have silvery or shiny leaves, which increases the amount of solar irradiation reflected. The fraction of S reflected by the leaf may increase from typical values of 0.1 or 0.2 (see Fig. 7.4) to 0.3 for silvery leaves, with an accompanying reduction in the absorptance from 0.6 to 0.5 or lower (see Gates 1970). This reduction in absorptance can have a marked influence on T^{leaf}. Other conditions remaining the same, a reduction of the absorptance a by only 0.1 can cause the leaf temperature to go from 36°C to 26°C (Table 7.1). Such a reduction in leaf temperature can have substantial effects on transpiration and photosynthesis. This is particularly apparent for desert plants in hot environments. For instance, seasonal variations in pubescence (epidermal hairs) can increase the shortwave reflectance of leaves of the desert shrub *Encelia farinosa* produced in dry, hot periods by about 0.2 over the reflectance for leaves from cool, wet periods. The higher reflectance decreases T^{leaf} by about 5°C for this species; this leads to photosynthetic rates that are 10% to 60% higher during the dry, hot periods than they would otherwise be and at the same time conserves water, since transpiration rates averaged 23% less (see Ehleringer and Mooney).

Let us next consider the effect of clouds on leaf temperature at night. On a clear night at sea level the effective sky temperature might be $-20°C$, while for a heavy cloud cover it could be 1°C because of the IR emitted by the clouds (Table 7.1). If the temperature of the surroundings were 1°C in both cases, the infrared absorbed by the leaf, $a_{\text{IR}}\sigma[(T^{\text{surr}})^4 + (T^{\text{sky}})^4]$, would

Table 7.1

Representatives values for the various terms in the net radiation balance of an exposed leaf. Equation 7.6 is used, taking a as 0.60 (except where indicated), r as 0.20, and both a_{IR} and e_{IR} as 0.96. (See text for interpretations.)

Condition	Global irradiation, S (W m^{-2})	Absorbed solar irradiation, $a(1+r)S$ (W m^{-2})	Temperature of surroundings (°C)	Sky temperature (°C)	Absorbed infrared $a_{IR}\sigma[(T^{surr})^4 + (T^{sky})^4]$ (W m^{-2})	Leaf temperature (°C)	Emitted infrared $2e_{IR}\sigma(T^{leaf})^4$ (W m^{-2})	Net radiation (W m^{-2})
Sea level on cloudless day	840	605	20	−20	624	25	859	370
2 000 m on cloudless day	1 050	756	20	−25	607	36	993	370
Silvery leaf ($a = 0.50$) at 2 000 m on cloudless day	1 050	630	20	−25	607	26	867	370
Sea level on cloudy night	0	0	1	1	614	1	614	0
Sea level on cloudless night	0	0	1	−20	530	−9	530	0
	0	0	1	−20	530	−1	597	−67

be 84 W m^{-2} lower on the clear night because of less IR from the cloudless sky (Table 7.1). For there to be no net gain or loss of energy by a leaf from all forms of radiation, i.e., $a_{IR}\sigma[(T^{surr})^4 + (T^{sky})^4] = 2e_{IR}\sigma(T^{leaf})^4$, the leaf temperature would have to be 1°C on the cloudy night and -9°C on the clear one (Table 7.1). Thus, on the clear night with the surroundings at 1°C, T^{leaf} could be considerably below freezing, and so heat would be conducted from the surroundings to the leaf, raising its temperature to perhaps -1°C. Using this leaf temperature, we can calculate that there would be a net energy *loss* by radiation of 67 W m^{-2} on the clear night (see Table 7.1). A plant on a cloudless night may therefore freeze even though the temperatures of both air and surroundings are above 0°C, since the excess of IR emitted over that absorbed can lower the leaf temperature below the freezing point. Such freezing of leaves on clear nights is a severe problem for certain agricultural crops (see Geiger or Schwintzer). A method for avoiding freezing damage employed for certain citrus and other orchards is to spray water on the trees. The release of the heat of fusion as this water freezes maintains the plant tissues at 0°C, which is above the freezing point of their cell sap.

WIND—HEAT CONDUCTION AND CONVECTION

Now that we have considered the net radiation balance of a leaf, we will examine other ways that energy may be exchanged with the environment. For instance, heat can be conducted from one body to a cooler one in contact with it by molecular and/or electronic collisions. We will assume that the leaf and its petiole (the leaf stem) have the same temperature, and that in general a temperature difference exists across the air boundary layer adjacent to the leaf surface.* Heat can be *conducted* across this air layer by the random thermal collisions of the gas molecules. Heat convection, on the other hand, involves turbulent movement of a fluid, brought about for example by differences in pressure. There are two types of convection, *free* and *forced*. Free (natural) convection occurs when the heat transferred from a leaf causes the air outside the unstirred layer to expand and thus to decrease in density; this more buoyant warmer air then moves upward and thereby moves heat away from the leaf. Forced convection, caused by wind, can also

---

* Frictional interactions between a fluid and a solid phase moving with respect to each other lead to boundary layers of fluid adhering to the solid phase and across which heat and mass are exchanged. In Chapter 1 we considered unstirred layers of water adjacent to membranes (see p. 28), and in this section we will consider boundary layers of air adjacent to leaves.

remove the heated air outside the boundary layer. As the wind speed increases, more and more heat is dissipated by forced relative to free convection. However, even at a wind speed of only 0.1 m s^{-1}, which is essentially the lowest occurring naturally, forced convection dominates free convection as a means of heat loss from most leaves (0.10 m s^{-1} = 0.36 km hour^{-1} = 0.22 mile hour^{-1}, a speed that is often called "still air"). We can therefore generally assume that heat is *conducted* across the boundary layer adjacent to the leaf and then removed by forced *convection* in the surrounding turbulent air. We will shortly introduce certain dimensionless numbers that can help indicate whether forced or free convection should prevail. However, before discussing conduction and convection, we will examine some general characteristics of wind, paying particular attention to the air boundary layers adjacent to plant parts.

Wind—General Comments

Wind can influence plant growth, reproduction, distribution, and even death (see Nobel 1981). It can mechanically deform plants and also can disperse pollen, plant propagules, disease organisms, as well as gas molecules like CO_2 and pollutants. Many effects of wind depend on the air boundary layers next to the aerial surfaces of a plant across which mass and heat exchanges occur with the environment.

Wind is caused by differences in air pressure that result from differential absorption of shortwave irradiation by the earth's surface as well as by clouds. On a macro scale, wind speed is affected by land surface features such as mountains and canyons, while about 10 m above a plant canopy wind can be mostly influenced by plants, becoming completely arrested at their surfaces. In coastal regions wind speed can annually average as high as 7 m s^{-1} at 10 m above the ground (see Grace), while in topographically flat inland areas 1 m s^{-1} would be a typical mean wind speed. Wind speeds are generally lower at night.

Near smooth surfaces air can flow in sheets, or lamina, nearly parallel to the surface. But turbulent flow, where air movement is not parallel and orderly, characterizes the wind regime near vegetation. In fact, the standard deviation of wind speed divided by the mean wind speed, the *turbulence intensity*, is often about 0.4 near vegetation, indicating that wind has considerable temporal variation (see Cionco). In such turbulent regimes air can be described as moving in packets or "eddies" (to be considered in Ch. 9).

Sites experiencing greater wind speeds often tend to have shorter vegetation, as for alpine tundra or the procumbent forms on coastal dunes (see Jaffe 1980). Also, stem elongation can be reduced 2- to 3-fold by high winds. On the other hand, environmental growth chambers tend to have low wind speeds (generally below 0.5 m s^{-1}), which can lead to rather spindly plants that are unlike their field-grown counterparts. The retardation of stem elongation and the increase in girth caused by wind are mainly due to the development of shorter cells with thicker cell walls. Agronomic implications of wind-induced sway have been recognized; e.g., increasing the spacing between nursery stock leads to sturdier trees of larger trunk diameter. Buttresses at the base of tree trunks and roots are more common on the windward side—such a location is more effective in resisting wind-caused upsetting moments than if the buttresses or enhanced root growth were on the leeward side, since the tensile strength of wood is greater than its compressional strength. A consequence of a prevailing wind direction are "flag trees," where branches occur mainly to the leeward. Many of these effects of wind on plant morphology have been shown to be hormonal responses (see Jaffe 1982).

Air Boundary Layers

Wind speed affects the thickness of the air boundary layer next to a leaf or other plant part. Since boundary layers affect the heat exchange and hence temperature of the plant part, any plant process depending on temperature can be affected by wind speed. The boundary layer is a region dominated by the shearing stresses originating at some surface; such layers arise for any solid immersed in a fluid, e.g., a leaf in air or under water (see Grace or Schlichting). Adjacent to the leaf is a laminar sublayer of air (Fig. 7.5), where movement is parallel to the leaf surface; air movement is arrested at the surface and has increasing speed at increasing distances. Diffusion perpendicularly away from the leaf surface is by molecular motion in the laminar sublayer. Farther from the surface, especially on the downwind part of a leaf, the boundary layer becomes turbulent (Fig. 7.5). Here, heat and gas movements are eddy-assisted; i.e., the air swirls around in vortices and behaves as if it were moving in little units or packets. Instead of trying to describe the transfer processes in the laminar and the turbulent portions, both of which change in thickness across a leaf's surface, we use an effective or equivalent boundary layer thickness, δ^{bl}, averaged over the whole leaf

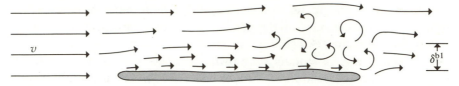

Figure 7.5
Schematic illustration of originally nonturbulent air flowing over a flat leaf, indicating the laminar sublayer (shorter straight arrows), the turbulent region, and the effective boundary layer thickness δ^{bl}. The arrows indicate the relative speed and direction of the air movement.

surface (this is often referred to as the displacement air boundary layer). Actually, the boundary layer is somewhat thinner at the upwind or leading edge of a leaf than at its center (Fig. 7.5); as a consequence, the temperature there can differ from that in the center of the leaf. Since there is no sharp discontinuity of wind speed between the air adjacent to the leaf and the free airstream, the definition of boundary layer thickness is somewhat arbitrary, and so it is generally defined operationally, e.g., in terms of the flux of heat across it, which we will consider shortly.

As we have indicated, δ^{bl} represents an *average* thickness of the unstirred air layer adjacent to a leaf (or to leaflets for a compound leaf). The main factors affecting δ^{bl} are the ambient wind speed and leaf size, with leaf shape exerting a secondary influence. Partly for convenience, but mainly because it has proved experimentally justifiable, we will handle the effect of leaf size on boundary layer thickness by the characteristic dimension l, which is the mean length of a leaf in the direction of the wind. Based on hydrodynamic theory for laminar flow adjacent to a flat surface as modified by actual observations under field conditions (see Pearman et al.), there is an approximate, but useful, expression for the average thickness of the boundary layer next to a flat leaf:

$$\delta^{bl}_{(mm)} = 4.0\sqrt{\frac{l_{(m)}}{v_{(m\,s^{-1})}}} \tag{7.7}$$

where $l_{(m)}$ is the mean length of the leaf in the downwind direction in m, $v_{(m\,s^{-1})}$ is the ambient wind speed in m s^{-1}, and $\delta^{bl}_{(mm)}$ is the average thickness of the boundary layer in mm (the factor 4.0 has dimensions of m s$^{-1/2}$, as we can deduce by using the indicated units for the three variables).

Instead of the numerical factor 4.0 in Equation 7.7, hydrodynamic theory actually predicts a factor closer to 6.0 for the "displacement," or effective, boundary layer thickness adjacent to a flat plate (see Cowan, Monteith 1973, and especially Schlichting). However, wind tunnel measurements under high turbulence intensity, as well as field measurements, have indicated that 4.0 is actually a more suitable factor for leaves (or 3.7; see Gates 1980). This divergence from theory relates to the relatively small size of leaves, their irregular shape, leaf curl, leaf flutter, and, most importantly, to the high turbulence intensity occurring under field conditions. Moreover, the dependency of δ^{bl} on $l^{0.5}$, which applies to large flat surfaces, does not always give the best fit to the data; wind tunnel measurements for various leaf shapes and sizes give values for the exponent of from 0.3 to 0.5 (see Gates and Papian). Thus, various shape considerations, including the leaf dimension perpendicular to the wind direction, can affect δ^{bl}. In addition, instead of the dependence on $v^{-0.5}$ indicated in Equation 7.7, actual measurements with leaves are best described by exponents of from -0.5 to -0.7. Consequently, Equation 7.7 must be regarded as only a useful approximation for indicating how the average boundary layer thickness varies with leaf size and wind speed.

Average wind speeds generally range from "still air" values of approximately 0.1 m s^{-1} up to about 10 m s^{-1} (36 km hour^{-1}, or 22 miles hour^{-1}), while exposed leaves commonly experience wind speeds near 1 m s^{-1}. Since the thickness of the air boundary layer enters into many calculations of heat and gas fluxes for leaves, the magnitudes of δ^{bl} for a wide variety of wind speeds and leaf dimensions are presented in Table 7.2.

Table 7.2

Thickness in mm of the boundary layer adjacent to a leaf, $\delta^{bl}_{(mm)}$, for a variety of leaf sizes and wind speeds. Calculations were made assuming that $\delta^{bl}_{(mm)} = 4.0\sqrt{l_{(m)}/v_{(m\,s^{-1})}}$ (Eq. 7.7, where $l_{(m)}$ is the mean leaf length in the wind direction in m and $v_{(m\,s^{-1})}$ is the ambient wind speed in m s^{-1}. (Note that 1 km hour^{-1} = 0.278 m s^{-1}, and 1 mile hour^{-1} = 0.447 m s^{-1}; see App. III.)

		\multicolumn{7}{c}{$v_{(m\,s^{-1})}$}						
		0.10	0.28	0.45	1.00	2.78	4.47	10.00
	0.002	0.57	0.34	0.27	0.179	0.107	0.085	0.057
	0.01	1.26	0.76	0.60	0.40	0.24	0.189	0.126
$l_{(m)}$	0.05	2.8	1.69	1.33	0.89	0.54	0.42	0.28
	0.25	6.3	3.8	3.0	2.0	1.20	0.95	0.63
	0.50	8.9	5.3	4.2	2.8	1.70	1.34	0.89

Boundary Layers for Bluff Bodies

Although relatively flat leaves can be described by the boundary layer considerations presented above (Fig. 7.5 and Eq. 7.7), many plant parts such as stems, branches, inflorescences, fruits, and even certain leaves (e.g., those of *Allium cepa*, the onion) represent three-dimensional objects. Air flow is intercepted by such *bluff bodies* and forced to move around them. Here, we will consider two particular shapes, cylinders and spheres. In the next subsection we will present heat flux equations for cases of cylindrical and spherical symmetry as well as for flat leaves.

On approximately the upwind half of a cylinder, a laminar boundary layer develops (Fig. 7.6). It is analogous to the laminar sublayer for flat plates (Fig. 7.5), and its air movements can be theoretically analyzed (see Kreith or Schlichting). On the downwind portion, the air flow becomes turbulent, can be opposite to the wind in direction, and in general is quite complicated. Nevertheless, an effective boundary layer thickness can be estimated (see Nobel 1974). For turbulence intensities appropriate to field conditions, $\delta^{bl}_{(mm)}$ in mm can be expressed as follows:

$$\delta^{bl}_{(mm)} = 5.8 \sqrt{\frac{d_{(m)}}{v_{(m\,s^{-1})}}} \qquad \text{cylinder} \qquad (7.8)$$

where $d_{(m)}$ is the cylinder diameter in m. As is the case for flat leaves, the boundary layer is thinner for smaller objects or at higher wind speeds.

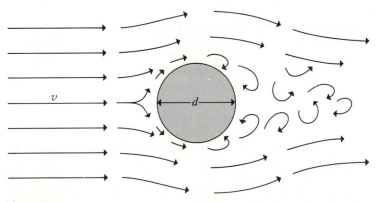

Figure 7.6
Schematic illustration of air flow around a cylinder. Flow can be laminar on the upwind half, but turbulence develops on the downwind side.

A similar analysis for the effective average boundary layer around a sphere under turbulent intensities appropriate to field conditions is

$$\delta^{bl}_{(mm)} = 2.8 \sqrt{\frac{d_{(m)}}{v_{(m\,s^{-1})}} + \frac{0.25}{v_{(m\,s^{-1})}}} \qquad \text{sphere} \qquad (7.9)$$

where $d_{(m)}$ is here the diameter of the sphere in m (see Nobel 1975). Equation 7.9 has been successfully used to predict boundary layer thicknesses across which heat transfer occurs for the approximately spherical fruit of *Vitis vinifera*, the grape (see Smart and Sinclair).

Heat Conduction/Convection Equations

Now that we have considered the average boundary layer thickness for objects of various shapes, let us return to a consideration of convective heat exchange, where the heat is first conducted across the boundary layer and then convected away in the moving airstream. For the one-dimensional case, heat flow by conduction equals $- K \, \partial T / \partial x$, where K is the thermal conductivity coefficient and $\partial T / \partial x$ is the temperature gradient (this relation is sometimes referred to as Fourier's heat-transfer law). Since heat can be conducted across the boundary layers on either side of a leaf, we need to incorporate the factor 2 to describe the total rate of heat flux by conduction per unit area of one side of a leaf. (For convenience, we will assume that the boundary layers on the two sides are of equal thickness, δ^{bl}.) The amount of heat conducted across the boundary layers and convected away from a leaf per unit time and area therefore is

$$J^C_H = -2K^{air} \frac{\partial T}{\partial x}$$

$$= 2K^{air} \frac{(T^{leaf} - T^{ta})}{\delta^{bl}} \qquad (7.10)$$

where J^C_H is the rate of heat conduction per unit area (e.g., W m^{-2}), K^{air} is the thermal conductivity coefficient of air, T^{leaf} is the leaf temperature, and T^{ta} is the temperature of the turbulent air outside a boundary layer of thickness δ^{bl}. Since heat is conducted from the solid surface of the leaf across the adjacent unstirred air, J^C_H does not depend on whether the stomata are opened or closed—we have a planar source at a specific temperature, T^{leaf}, from which heat is conducted across the boundary layer to the outside turbulent air, at T^{ta}. The heat flux density in Equation 7.10 is considered positive when heat goes from the leaf to the surrounding air. Of course, heat is conducted into the leaf when T^{leaf} is less than T^{ta}, in which case J_H^C is negative.

All of our flux equations used so far in this book have been for one-dimensional cases. Since we have introduced the average thickness of the boundary layer for cylinders (Eq. 7.8) and spheres (Eq. 7.9), let us also con-sider the appropriate fluxes for such cases, which can have many biological applications. For the case of cylindrical symmetry, thermal properties could change in the radial direction perpendicularly away from the cylinder axis, but not with angle around the cylinder or length along the axis. The heat flux density for cylindrical symmetry is

$$J_H^C = \frac{K^{air}(T^{surf} - T^{ta})}{r \ln\left(\dfrac{r + \delta^{bl}}{r}\right)} \qquad \text{cylinder} \qquad (7.11)$$

where r is the cylinder radius, T^{surf} is its surface temperature, δ^{bl} is calcu-lated by Equation 7.8, and the other quantities have the same meaning as for the one-dimensional case, Equation 7.10 (note that J_H^C is the heat flux density at the cylinder surface). For the case of spherical symmetry, where properties vary only in the radial direction and not with any angle, the heat flux density at the sphere's surface for conduction across the boundary layer and then convection in the surrounding turbulent air is

$$J_H^C = \frac{(r + \delta^{bl})K^{air}(T^{surf} - T^{ta})}{r\delta^{bl}} \qquad \text{sphere} \qquad (7.12)$$

where r is the radius of the sphere.

Values of the thermal conductivity coefficient for dry air at various tem-peratures are given in W m^{-1} °C^{-1} in Appendix II. Since the conduction of heat in a gas phase is based on the random thermal motion of the molecules, the composition of air can influence the value of K^{air}. Our main concern here is with the effect of water vapor content on K^{air}, an influence that is relatively small at the temperatures usually encountered by leaves. For instance, K^{air} at 20°C is only 1% less for 100% relative humidity than it is for dry air. Air can hold considerably more water vapor at 40°C, where K^{air} is 2% lower for water-saturated air than that of 0% relative humidity.

Dimensionless Numbers

In many studies on heat and gas fluxes, relationships between parameters are expressed in terms of dimensionless groups or numbers. This facilitates comparisons between objects of different sizes and for different wind speeds — i.e., dimensionless numbers allow application of data to different, but geometrically similar, situations (see Campbell, Gates 1980, Monteith 1973, Schlichting, and Vogel 1981). For instance, dimensionless numbers were used to determine the effects of wind speed on δ^{bl} for the various shapes presented above (see Nobel 1974 and Nobel 1975). Dimensionless numbers can also be used to study boundary layer phenomena and flow characteristics for water (see Wheeler and Neushul), where water speeds can exceed 10 m s^{-1} in intertidal regions owing to wave action, although water speeds are usually from 0.01 to 0.2 m s^{-1} in coastal regions of lakes and oceans. We will here consider three dimensionless numbers that are particularly important for discussing heat fluxes.

Before presenting these dimensionless numbers, we will indicate one common convention for describing heat conduction across boundary layers, a convention that is invariably used for objects of irregular shape and is often used for geometrically regular objects. Instead of expressions involving δ^{bl} (e.g., Eq. 7.10 for flat plates, Eq. 7.11 for cylinders, and Eq. 7.12 for spheres), the following relation is used to describe the heat flux density across the air boundary layer:

$$J_H^C = h_c(T^{surf} - T^{ta}) \tag{7.13}$$

where h_c is called the heat convection coefficient (or the convective heat-transfer coefficient; Eq. 7.13 is known as Newton's law of cooling). Upon comparing Equation 7.13 with 7.10 and noting that h_c generally refers to unit surface area of one side of a leaf [hence, the total heat flux density is $2h_c(T^{surf} - T^{ta})$ for a flat leaf], we find that h_c equals K^{air}/δ^{bl} for flat leaves (slightly more complex relations hold for cylinders and spheres). Even when the boundary layer thickness cannot be determined analytically, e.g., for the irregular shapes of cacti with their surface ribbing and projecting spines, Equation 7.13 can still be used to relate the convective heat exchange to the temperature difference between the air and the plant part. Appropriate units of h_c are W m^{-2} °C^{-1}.

Now that we have introduced the heat convection coefficient, we will proceed to define our first dimensionless number, the *Nusselt number*. We will represent the size of a particular plant part by a characteristic dimension d, which for a flat plate is the quantity l in Equation 7.7 and for a cylinder or sphere is the diameter. This leads us to

$$\text{Nusselt number} = \text{Nu} = \frac{h_c d}{K^{\text{air}}} \tag{7.14}$$

Using our above comment on h_c for flat leaves ($h_c = K^{\text{air}}/\delta^{\text{bl}}$), we note that their Nusselt number equals d/δ^{bl}; i.e., Nu relates the characteristic dimension to the boundary layer thickness, and thus Nusselt numbers are useful for describing heat transfer occurring across boundary layers (as indicated above, boundary layers occur for any solid in a fluid).

Next, we will introduce a dimensionless number that describes flow characteristics—e.g., it can help indicate whether the flow will be laminar or turbulent. This quantity indicates the ratio of inertial forces (due to momentum, which tends to keep things moving) to viscous forces (due to frictional interactions, which tend to slow things down), and it is known as the *Reynolds number*:

$$\text{Reynolds number} = \text{Re} = \frac{vd}{v} \tag{7.15}$$

where v is the kinematic viscosity (ordinary viscosity divided by density) for the fluid ($1.53 \times 10^{-5} \text{ m}^2 \text{ s}^{-1}$ for dry air at 20°C). At low Reynolds numbers viscous forces dominate inertial forces and the flow tends to be laminar, while at high Reynolds numbers (above about 10^4 for plant parts; see Gates 1980) the flow becomes turbulent owing to the dominance of inertial forces. For a cylindrical plant part such as a branch 0.1 m in diameter, this means that turbulence sets in for wind speeds above about 1.5 m s^{-1}. Although air movement is different below compared to above such critical Reynolds numbers, Equations 7.7 through 7.12 are satisfactory for most applications to plants in either flow regime.[*]

At very low wind speeds and large values of $T^{\text{surf}} - T^{\text{ta}}$, free convection can dominate forced convection for large objects. In such cases, the Reynolds number should be replaced in heat flow studies by the *Grashof number*,

---

[*] For flat leaves under field conditions, analysis by dimensionless numbers indicates that Nu = 0.97 Re$^{0.5}$ (see Pearman et al. and Gates 1980). Hence, d/δ^{bl} equals $0.97 \, (vd/v)^{1/2}$ by Equations 7.14 and 7.15, and so δ^{bl} is $(1/0.97)(v/vd)^{1/2}d = (1/0.97)(1.53 \times 10^{-5} \text{ m}^2 \text{ s}^{-1})^{1/2}(d/v)^{1/2} = 4.0 \times 10^{-3} \sqrt{d/v} \text{ m s}^{-1/2}$, which is essentially Equation 7.7.

which takes into account buoyant forces. Specifically, the Grashof number Gr indicates the tendency of a parcel of air to rise or fall, and thus it describes the tendency for free convection. In fact, Gr represents the ratio of buoyant times inertial forces to the square of the viscous forces:

$$\text{Grashof number} = \text{Gr} = \frac{g\beta\,\Delta T\,d^3}{v^2} \tag{7.16}$$

where g is the gravitational acceleration, β is the coefficient of volumetric thermal expansion (i.e., the fractional change in volume with temperature at constant pressure, which equals $1/T$ for an essentially perfect gas such as air, e.g., $\beta = 3.4 \times 10^{-3}\,°\text{C}^{-1}$ at 20°C), and ΔT is the temperature difference from the object's surface to the ambient air ($T^{\text{surf}} - T^{\text{ta}}$).

Using the Grashof and the Reynolds numbers, we can indicate whether forced or free convection dominates in a particular case. Since Re equals inertial forces/viscous forces and Gr equals buoyant × inertial forces/(viscous forces)^2, Re^2/Gr equals inertial forces/buoyant forces. Thus, Re^2/Gr reflects forced convection/free convection. Experiments show that forced convection accounts for nearly all heat transfer when Re^2/Gr > 10, and free convection accounts for nearly all heat transfer when Re^2/Gr < 0.1, while the intervening region has mixed convection; i.e., both forced and free convection should then be considered, especially for Re^2/Gr near 1. Using Equations 7.15 and 7.16, we obtain:

$$\frac{\text{Re}^2}{\text{Gr}} = \frac{\left(\dfrac{vd}{v}\right)^2}{\left(\dfrac{g\beta\,\Delta T\,d^3}{v^2}\right)} = \frac{v^2}{g\beta\,\Delta T\,d}$$

$$= \frac{v^2}{(9.8\text{ m s}^{-2})(3.4\times10^{-3}\,°\text{C}^{-1})\Delta T\,d}$$

$$= (30\text{ s}^2\,°\text{C m}^{-1})\frac{v^2}{\Delta T\,d} \tag{7.17}$$

For a ΔT of 5°C and a d of 0.1 m, Equation 7.17 indicates that Re^2/Gr equals 1 when the wind speed is

$$v = \left[\frac{(5°\text{C})(0.1\text{ m})}{(30\text{ s}^2\,°\text{C m}^{-1})}\right]^{1/2} = 0.1\text{ m s}^{-1}$$

Thus, for wind speeds greater than 0.1 m s^{-1}, which as we indicated above is often referred to as still air, forced convection will dominate free convection when ΔT is 5°C and d is 0.1 m. This domination of forced convection over free convection will occur for most of our applications.

Examples of Heat Conduction/Convection

Let us now calculate the heat conduction across the air boundary layer for a leaf at 25°C when the surrounding turbulent air is at 20°C. We will consider a leaf with a mean length in the wind direction of 10 cm and a wind speed of 0.8 m s^{-1}. From Equation 7.7, the boundary layer thickness is

$$\delta^{bl}_{(mm)} = 4.0 \sqrt{\frac{(0.10 \text{ m})}{(0.8 \text{ m s}^{-1})}} = 1.4 \text{ mm}$$

which is a rather typical value for a leaf. Using Equation 7.12 and K^{air} of 0.0259 W m^{-1} °C^{-1} (appropriate for 20°C to 25°C; App. II), we can calculate the heat flux density for conduction across the boundary layer:

$$J^C_H = \frac{(2)(0.0259 \text{ W m}^{-1} \text{ °C}^{-1})(25°C - 20°C)}{(1.4 \times 10^{-3} \text{ m})}$$

$$= 190 \text{ W m}^{-2}$$

A leaf at 25°C at sea level on a sunny day could have a net radiation balance of 370 W m^{-2} (see Table 7.1, top line). In the present case, just over half this energy input by radiation is dissipated by conduction of heat across the boundary layers on each side of the leaf (190 W m^{-2} total), followed by forced convection in the surrounding turbulent air outside the boundary layers.

A leaf with a thick boundary layer can have a temperature quite different from that of the surrounding air, since air is a relatively poor conductor of heat. Specifically, K^{air} is relatively low compared with the thermal conductivity coefficients for liquids and most solids. A large leaf in a low wind might have a boundary layer 4 mm thick (see Table 7.2 or Eq. 7.7 for the wind speeds and leaf sizes implied by this). If J^C_H and K^{air} are the same as in the previous paragraph, where the difference between leaf and turbulent air temperatures was 5°C when the boundary layer was 1.4 mm thick, then $T^{leaf} - T^{ta}$ will be (4 mm/1.4 mm)(5°C), or 14°C, for a δ^{bl} of 4 mm. Thus, the combination of large leaves and low wind speeds favors a large drop in temperature across the boundary layers. On the other hand, a small leaf in a moderate wind could have a δ^{bl} of 0.2 mm. For the same J_H^C and K^{air} as above, $T^{leaf} - T^{ta}$ would be (0.2 mm/1.4 mm)(5°C), or 0.7°C, for this thin boundary layer. Hence, small leaves tend to have temperatures quite close to that of the air, especially at moderate to high wind speeds. This close coupling between leaf and air temperatures for small leaves can keep the leaf temperature low enough for optimal photosynthesis (often 30°C to 35°C) in hot, sunny climates. Also, the lower the leaf temperature, the lower is the concentration of water vapor in the pores of the cell walls of mesophyll cells, and consequently less water would then tend to be lost in transpiration (Ch. 8), an important consideration in arid and semi-arid regions.

We can apply the same heat conduction analysis to massive structures such as the stems of cacti. Let us consider a cylindrical stem 0.25 m in diameter exposed to the sun on a cloudless day at a high elevation of 2 000 m and calculate the air temperature expected if we use the radiation terms presented in Table 7.1, line 2 (net radiation balance of 370 W m^{-2} for a surface temperature of 36°C). Cacti are Crassulacean acid metabolism (CAM) plants (to be discussed on p. 445), so the stomata tend to be closed during the daytime, making transpirational cooling then minimal. Consequently, if we ignore heat conduction or storage in the stem, the net radiation will be balanced by heat conduction across the boundary layer. For an ambient wind speed of 1.0 m s^{-1}, the boundary layer thickness (Eq. 7.8) is 2.9 mm. Assuming cylindrical symmetry (Eq. 7.11), we can then calculate the temperature of the surrounding turbulent air such that a net radiation input of 370 W m^{-2} is just balanced by heat conduction across the boundary layer:

$$T^{ta} = T^{surf} - \frac{J_H^C\, r \ln\left(\frac{r + \delta^{bl}}{r}\right)}{K^{air}}$$

$$= 36°C - \frac{(370 \text{ W m}^{-2})(0.125 \text{ m}) \ln\left(\frac{0.125 \text{ m} + 0.0029 \text{ m}}{0.125 \text{ m}}\right)}{(0.0257 \text{ W m}^{-1}\,°C^{-1})}$$

$$= -5°C$$

Actually, however, we would not expect the stem to be 41°C above T^{ta}—in fact, our assumption that no heat is stored or conducted in the stem is incorrect, an aspect we will reconsider toward the end of this chapter.

LATENT HEAT—TRANSPIRATION

Evaporation of water is clearly a cooling process. Water evaporates at the air-liquid interfaces along the pores in the cell walls of mesophyll, epidermal, and guard cells (see Fig. 1.2), and then diffuses out of a leaf. Thus, transpiration represents a mode for heat loss by a leaf (Eq. 7.2). A leaf can have a latent heat gain if dew or frost condenses onto it, as we will discuss shortly.

Heat Flux Density Accompanying Transpiration

We will represent the flux density of water vapor diffusing out of a leaf during transpiration by J_{wv}. If we multiply the amount of water leaving per unit time and per unit leaf area, J_{wv}, by the energy necessary to evaporate a unit amount of water at the temperature of the leaf, H_{vap}, we obtain the heat flux density accompanying transpiration, J_H^T:

$$J_H^T = J_{wv}H_{vap} = \frac{H_{vap}D_{wv}\Delta c_{wv}^{total}}{\Delta x^{total}} = \frac{H_{vap}D_{wv}(c_{wv}^e - c_{wv}^{ta})}{\Delta x^{total}} \qquad (7.18)$$

where Fick's first law (Eqs. 1.1 and 1.8) has been used to express J_{wv} in terms of the diffusion coefficient for water vapor, D_{wv}, and the total drop in water vapor concentration, Δc_{wv}^{total}, over some effective total distance, Δx^{total}. In turn, Δc_{wv}^{total} equals the water vapor concentration at the sites of evaporation within a leaf, c_{wv}^e, minus the value in the turbulent air just outside the boundary layer, c_{wv}^{ta}. In the next chapter we will discuss in detail the effective distance, resistance, and conductance for water vapor flux.

How much of the heat load on a leaf is dissipated by the evaporation of water during transpiration? For an exposed leaf of a typical mesophyte during the daytime, J_{wv} is about 4 mmol m^{-2} s^{-1} (p. 415). In Chapter 2 we noted that water has a high heat of vaporization, e.g., 44.0 kJ mol^{-1} at 25°C (values at other temperatures are given in App. II). By Equation 7.18, the heat flux density out of the leaf by transpiration then is

$$J_H^T = (4 \times 10^{-3} \text{ mol m}^{-2} \text{ s}^{-1})(44.0 \times 10^3 \text{ J mol}^{-1}) = 180 \text{ W m}^{-2}$$

For the leaf described in the top line of Table 7.1, a heat loss of 180 W m^{-2} by evaporation would dissipate slightly under half of the net radiation balance (370 W m^{-2}), the rest of the energy input being removed by heat conduction across the boundary layer followed by forced convection (see Fig. 7.7a). A heat loss accompanying transpiration reduces leaf temperatures

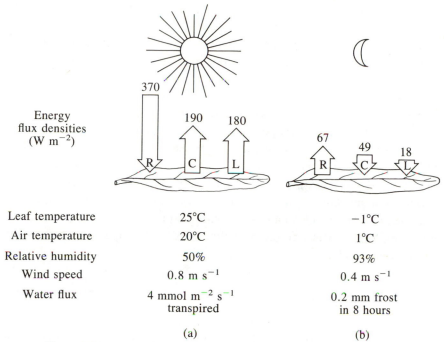

	(a)	(b)
Leaf temperature	25°C	−1°C
Air temperature	20°C	1°C
Relative humidity	50%	93%
Wind speed	0.8 m s^{-1}	0.4 m s^{-1}
Water flux	4 mmol m^{-2} s^{-1} transpired	0.2 mm frost in 8 hours

Figure 7.7
Energy budget for an exposed leaf: (a) at midday, and (b) at night with frost. The flux densities in W m^{-2} are indicated for net radiation (R), conduction/convection (C)—also referred to as sensible heat—and latent heat (L).

during the daytime. Although such latent heat losses can be adaptive, evaporation and its associated cooling are an inevitable consequence of gas exchange by leaves, where opening of the stomata is necessary for substantial rates of CO_2 uptake.

Heat Flux Density for Dew or Frost Formation

So far, we have regarded c_{wv}^e as greater than c_{wv}^{ta}, in which case there is a net loss of water from the leaf and a consequent heat dissipation. When the turbulent air is warmer than the leaf, however, the water vapor concentration in the turbulent air may be greater than that in the leaf (in Ch. 2 we noted that the water vapor concentration and partial pressure at saturation increase rapidly with temperature; also, see values for P_{wv}^* and c_{wv}^* in App. II). If c_{wv}^{ta} is greater than the water vapor concentration in the leaf, then there will be a net diffusion of water vapor toward the leaf. This can cause c_{wv} at the leaf surface to increase, and it may reach c_{wv}^*, the saturation value. If c_{wv}^{ta} is greater than this c_{wv}^*, dew—or frost, if the leaf temperature is below freezing—can form as water vapor diffuses toward the leaf and then condenses onto its surface, which is cooler than the turbulent air.[*]

Condensation of water vapor leads to a heat *gain* by a leaf. Since water condensation is the reverse of the energy-dissipating process evaporation, the heat gain per unit amount of water condensed is the heat of vaporization of water at the temperature of the leaf, H_{vap}. If the condensation is on the leaf surface, the diffusion step is across the air boundary layers of thickness δ^{bl} that occur on either side of a leaf. We can then modify Equation 7.18 to the following form to describe the heat flux density accompanying such water vapor condensation:

$$\text{heat gain by dew formation} = \frac{2H_{vap}D_{wv}(c_{wv}^{ta} - c_{wv}^{leaf*})}{\delta^{bl}} \qquad (7.19)$$

where c_{wv}^{leaf*} is the saturation concentration of water vapor at the temperature of the leaf. The factor 2 is necessary because water vapor diffuses toward the leaf across the boundary layer on each side (as usual, we will assume that these layers are of equal average thickness δ^{bl}).

The temperature to which the turbulent air must be reduced at constant pressure for it to become saturated with water vapor is known as the *dew point*, or dew temperature, $T_{dew\ point}^{ta}$. When T^{leaf} is below $T_{dew\ point}^{ta}$, the turbulent air contains more water vapor (c_{wv}^{ta}) than the air at the leaf surface can hold (c_{wv}^{leaf*}). Water vapor then diffuses toward the leaf, and this can lead to dew formation. If T^{leaf} is below $0°C$ and less than $T_{dew\ point}^{ta}$, the water that condenses onto the leaf surface will freeze. Under such conditions we should replace H_{vap} in Equation 7.19 by the heat of sublimation, H_{sub}, at that particular leaf temperature to describe the heat gain by frost formation.

Examples of Frost and Dew Formation

As an example of nighttime frost formation, we will consider a leaf with boundary layers 2.0 mm thick and a temperature of $-1°C$ when the sur-

---

[*] Condensation resulting from water emanating from the soil is sometimes called "distillation", the term "dew" then being reserved for water coming from the air above.

rounding air is at 1°C and 93% relative humidity. The leaf can be the same leaf considered above (p. 366), where we estimated a boundary layer thickness of 1.4 mm during the daytime; at night, the wind speed is generally lower—if v were halved, then the boundary layer would be 41% thicker, or 2.0 mm, by Equation 7.7. D_{wv} is 2.13×10^{-5} m^2 s^{-1} at 0°C, H_{sub} is 51.0 kJ mol^{-1} at -1°C, and c_{wv}^* is 0.251 mol m^{-3} at -1°C and 0.288 mol m^{-3} at 1°C (App. II). By Equation 7.19, the heat gain by frost formation is:

$$\text{heat gain by frost formation} = (2)(51.0 \times 10^3 \text{ J mol}^{-1})(2.13 \times 10^{-5} \text{ m}^2 \text{ s}^{-1})$$

$$\times \frac{[(0.93)(0.288 \text{ mol m}^{-3}) - (0.251 \text{ mol m}^{-3})]}{(2.0 \times 10^{-3} \text{ m})}$$

$$= 18 \text{ W m}^{-2}$$

Since T^{leaf} is here less than T^{ta}, heat is conducted *into* the leaf from the air. By Equation 7.10 [$J_H^C = 2K^{\text{air}}(T^{\text{leaf}} - T^{\text{ta}})/\delta^{\text{bl}}$, where K^{air} is 0.0243 W m^{-1} °C^{-1} at 0°C; App. II], the heat conduction across the boundary layer is -49 W m^{-2}, where the minus sign indicates that heat is conducted into the leaf. The bottom line of Table 7.1 indicates that this leaf can lose 67 W m^{-2} by net radiation. Thus, the heat inputs from frost formation (18 W m^{-2}) and heat conduction (49 W m^{-2}) balance the energy loss by net radiation (see Fig. 7.7b).

How long would it take to form a layer of frost 0.1 mm thick on each side of a leaf for the above conditions? The amount of water deposited in kg m^{-2} s^{-1}, i.e., J_{wv}, would be the heat gain by frost formation divided by H_{sub} (see Eq. 7.18, $J_H^T = J_{wv}H_{vap}$). This J_{wv}, divided by the density of ice (917 kg m^{-3}), would give the rate of frost accumulation in m s^{-1}. To accumulate 0.1 mm worth of ice on each side of a leaf would therefore take

$$t = \frac{0.2 \text{ mm}}{J_{wv}/\rho_{ice}} = \frac{(917 \text{ kg m}^{-3})(0.2 \times 10^{-3} \text{ m})}{J_H^T/H_{sub}}$$

$$= \frac{(0.18 \text{ kg m}^{-2})}{(18 \text{ J s}^{-1} \text{ m}^{-2})/(2.83 \times 10^6 \text{ J kg}^{-1})}$$

$$= 2.8 \times 10^4 \text{ s} \qquad (8 \text{ hours})$$

Dew formation can be as high as 0.5 mm per night, which can be an important source of water for leaves in certain regions for part of the year (see Levitt, Oke, Rosenberg, or Rutter). For the lichen *Ramalina maciformis* in the Negev Desert, dew sufficient to lead to photosynthesis the next day occurred on over half of the nights, and the annual dewfall was about 30 mm (see Kappen et al.).

Dew or frost formation is favored on cloudless nights and when the relative humidity of the surrounding turbulent air is close to 100%. As we mentioned above, T^{leaf} tends to be further from T^{ta} for large leaves than for small ones. On a cloudless night when T^{leaf} is less than T^{surr} and T^{ta} (see Table 7.1), the larger exposed leaves will generally dip below $T^{ta}_{dew\ point}$ sooner than the small ones, and hence dew or frost tends to form on the larger leaves first. For convenience, we have assumed that the leaf has a uniform temperature. In fact, however, the boundary layer tends to be thinner at the edges of a leaf (Fig. 7.5), so T^{leaf} is somewhat closer to T^{ta} at the edges than at the center. Thus, dew or frost generally forms first at the center of a leaf, where δ^{bl} is larger and the temperature slightly lower than at the leaf edges.

SOIL

Soil acts as an extremely important component in the energy balance of plants. For instance, shortwave irradiation can be reflected from its surface, it is the source of longwave radiation that can correspond to a temperature considerably different from that of the surrounding air, and heat can be conducted to or from stems in their region of contact with the soil. Also, considerable energy can be stored by the soil, contrary to the case for most leaves (see Eqs. 7.1 and 7.2). Although the soil surface can have large diurnal oscillations in temperature, the soil temperature at moderate depths of 1 m or so can be fairly steady, even from season to season. The soil surface temperature for the whole earth averages about 13°C over a year.

Thermal Properties

Soil has a relatively high heat capacity. To raise the temperature of 1 kg of dry sand by 1°C takes about 0.8 kJ, with similar values for dry clay or loam. Since the density of soil solids is about 2 600 kg m^{-3} and soil is about half pores by volume, the overall density of dry soil is about 1 300 kg m^{-3}— values generally range from 1 200 kg m^{-3} for dry loam, slightly higher for clay, and up to 1 500 kg m^{-3} for dry sand. Thus, the volumetric heat capacity of dry soil is about 1.2 MJ m^{-3} °C^{-1}. For a moist loam containing 20% water by volume (water has a volumetric heat capacity of 4.18 MJ m^{-3} °C^{-1} at 20°C), it is about 2.0 MJ m^{-3} °C^{-1} (see van Wijk).

The relatively high heat capacity of soil means that a considerable amount of energy can be involved with its temperature changes. For instance, if the soil temperature for the upper 0.4 m of a moist soil with a volumetric heat capacity of 2.0 MJ m^{-3} °C^{-1} increases by an average of 2°C during the daytime, then (0.4 m)(2°C)(2.0 MJ m^{-3} °C^{-1}), or 1.6 MJ, can be stored per m^2 of ground. Nearly all the heat stored by the soil during a day is released that night, since the temperature near the top of the soil changes diurnally. Generally, the soil temperature at depths below 0.4 m changes less than 0.5°C during a day or night, although annually it will vary considerably more, as we will indicate below.

Although it has a substantial volumetric heat capacity, soil does not have a very high thermal conductivity coefficient, K^{soil}. Heat is therefore not conducted very readily in a soil, where the heat flux density by conduction is

$$J_H^C = -K^{soil} \frac{\partial T}{\partial z} \tag{7.20}$$

Heat can of course be conducted in all directions in the soil, instead of only vertically, as we will consider here (see Eq. 7.20). For Equation 7.20, J_H^C can be expressed in W m^{-2}, the temperature gradient in °C m^{-1}, and therefore the soil thermal conductivity coefficient in W m^{-1} °C^{-1}, just as for K^{air}. K^{soil} depends on the soil water content; replacement of soil air (a relatively poor heat conductor) by water will increase K^{soil}. For instance, K^{soil} can vary from 0.2 W m^{-1} °C^{-1} for a dry soil to 2 W m^{-1} °C^{-1} for a wet one. By comparison, K^{water} is 0.60 W m^{-1} °C^{-1} and K^{air} is 0.026 W m^{-1} °C^{-1} near 20°C (see App. II, Rose, and van Wijk).

During the daytime the surface of the ground can be considerably warmer than the underlying layers of the soil. Since soil exposed to the turbulent air tends to be drier than the underlying layers, the thermal conductivity coefficient can be lower near the surface. For the upper part of a fairly moist sandy loam, K^{soil} may therefore be 0.6 W m^{-1} °C^{-1} and $\partial T/\partial z$ may be -100°C m^{-1} (at least for the upper 0.05 m or so; z is here considered positive into the soil). Using Equation 7.20, the heat flux density by conduction into the soil is

$$J_H^C = -(0.6 \text{ W m}^{-1}\,°C^{-1})(-100°C \text{ m}^{-1}) = 60 \text{ W m}^{-2}$$

This heat conducted into the soil could lead to the daytime heat storage calculated above, 1.6 MJ m^{-2}, in (1.6 × 10^6 J m^{-2})/(60 J m^{-2} s^{-1}), or 2.67 × 10^4 s, which is about 7 hours.

Soil Energy Balance

The components of the energy balance for the soil surface are similar to those for leaves (see Eqs. 7.1 and 7.2). However, we must also take into consideration the heat conducted into the soil itself (Eq. 7.20), which leads to a gradual temperature change of its upper layers.

The absorption and emission of radiation usually takes place in the upper mm of the soil. Using Equations 7.3 through 7.6, we can generally represent the net radiation balance for the soil surface by $aS + a_{IR}\sigma(T^{surr})^4 - e_{IR}\sigma(T^{soil})^4$, where the values of all parameters are those at the soil surface. (If the soil is exposed directly to the sky, T^{surr} should be replaced by T^{sky} for the incident IR.) For a soil exposed to direct sunlight, the net energy input by radiation can be quite large—the temperature of the uppermost part of the crust can exceed 70°C in a desert.

Let us now consider the flux for heat conduction between the overlying air and the soil, as well as the heat flux accompanying water evaporation in the soil. The heat conducted across the relatively still air next to the soil surface is equal to $-K^{air}\partial T/\partial z$, where K^{air} is the thermal conductivity coefficient of air at the local temperature (see Eq. 7.10, which indicates that the heat conducted from both sides of a leaf is $-2K^{air}\partial T/\partial x$). The heat conducted *within* the soil can be calculated using Equation 7.20 ($J_H^C = -K^{soil}\partial T/\partial z$). For a given layer of soil where the water vapor flux density changes by ΔJ_{wv}, the heat loss accompanying water evaporation (or heat gain accompanying condensation) equals $\Delta J_{wv}H_{vap}$, where H_{vap} is the heat necessary to evaporate a unit amount of water at the local soil temperature (see Eq. 7.18, $J_H^T = J_{wv}H_{vap}$). Except in the upper mm or so, the main energy flux in the soil is for heat conduction, not for radiation or for phase changes of water. Because of the large heat capacity of water, its movement in the soil can also represent an important means of heat movement.

Variations in Soil Temperature

Since the energy flux in the soil is often mainly by heat conduction, we can fairly readily estimate the soil temperature at various depths. But complications arise because of the heterogeneous nature of soil as well as the many types of plant cover. Thus, to obtain some idea of diurnal and annual temperature variations, we will assume that the volumetric heat capacity (C_P) and the thermal conductivity coefficient (K^{soil}) are both constant with depth, and we will ignore water movement in the soil. Moreover, we will assume that the soil surface temperature varies sinusoidally around an average value $\bar{T}^{surf}$, with a daily or annual amplitude of ΔT^{surf}, a useful approximation that can be tested. We then obtain the following relation for the temperature T at time t and depth z (see Campbell, de Vries, Rosenberg, and van Wijk):

$$T = \bar{T}^{surf} + \Delta T^{surf} e^{-z/d} \cos\left(\frac{2\pi t}{p} - \frac{z}{d}\right) \tag{7.21}$$

where the damping depth d, which is the depth in the soil where the variation in temperature has been damped down to $1/e$ of the value at the soil surface, equals $(pK^{soil}/\pi C_P)^{1/2}$; and p is the period (24 hours, or 8.64×10^4 s, for a diurnal variation and 365 times longer for an annual variation). We note that z equals zero at the surface and is considered positive downward, and that the soil surface has its maximum temperature, $\bar{T}^{surf} + \Delta T^{surf}$, when t equals 0.*

Let us next estimate the depths where the variation in soil temperature would be only $\pm 1°C$, in one case diurnally and in another annually. As above, we will use a C_P of 2 MJ m^{-3} °C^{-1} and a K^{soil} of 0.6 W m^{-1} °C^{-1}. The damping depth for the diurnal case then is

$$d = \left[\frac{(8.64 \times 10^4 \text{ s})(0.6 \text{ J s}^{-1} \text{ m}^{-1} \text{ °C}^{-1})}{(\pi)(2 \times 10^6 \text{ J m}^{-3} \text{ °C}^{-1})}\right]^{1/2}$$

$$= 0.091 \text{ m}$$

and it is 1.7 m for the annual case. From Equation 7.21 the diurnal variation in temperature at depth z is $\pm \Delta T^{surf} e^{-z/d}$. The amplitude of the diurnal variation in soil surface temperature for bare soil is often 15°C (i.e., $\bar{T}^{surf}_{max} - T^{surf}_{min} = 30°C$), and the annual amplitude for variations in average daily surface temperatures is generally somewhat less (e.g., 10°C). The depth where the diurnal variation in temperature is $\pm 1°C$ is then

$$z = -d \ln\left(\frac{1°C}{\Delta T^{surf}}\right) = -(0.091 \text{ m}) \ln\left(\frac{1°C}{15°C}\right)$$

$$= 0.25 \text{ m}$$

---

* Equation 7.21 is of the form $y = A + B \cos \alpha$, where A is the average value of y and B is the amplitude of the variation.

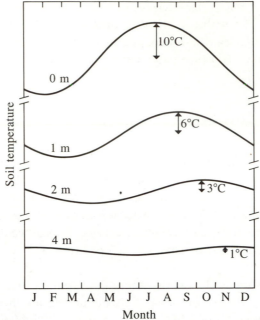

Figure 7.8
Simulated annual variation in soil temperatures at the indicated
depths. The average daily temperature at the soil surface was
assumed to vary sinusoidally, with a maximum on August 1
and an annual amplitude of 10°C. See text for other conditions.

which is consistent with our previous statement that soil temperatures
generally change less than 0.5°C diurnally at about 0.4 m. On an annual
basis, the ± 1°C variation would occur at a depth of -3.9 m (Fig. 7.8).

The factor z/d occurring in the cosine in Equation 7.21 indicates that the
peak of the temperature "wave" arrives later at greater depths in the soil.
This peak occurs when the cosine equals 1, which can correspond to $2\pi t/p =
z/d$ (since $\cos 0 = 1$). The speed of movement of the wave $\Delta z/\Delta t$ is thus
$2\pi d/p$. Figure 7.8 shows this effect at various depths on an annual basis.
For instance, at 4 m the peak temperature occurs $4\frac{1}{2}$ months later than at
the surface.

FURTHER EXAMPLES OF ENERGY BUDGETS

So far in this chapter our discussion of the energy fluxes of leaves has been
for those in exposed locations. Although their temperatures depend on

the net effect of all the energy fluxes we have discussed, many observations and calculations indicate that exposed leaves when sunlit tend to be above air temperature for T^{ta} up to about 30°C and below T^{ta} for air temperatures above about 35°C (see Gates 1980). This primarily reflects the increasing importance of infrared radiation emission as leaf temperature rises [see Eq. 7.5, $J_{\text{IR}} = 2e_{\text{IR}}\sigma(T^{\text{leaf}})^4$] and the increase with temperature of the water vapor concentration in the leaves, which affects transpiration (to be discussed in the next chapter). These energy budget interactions often lead to temperatures for exposed leaves that are more favorable for photosynthesis than is the ambient air temperature. We can easily extend our analysis to include leaves shaded by overlying ones. Also, our brief description of the thermal properties of soil serves as an introduction to the energy balance of certain plant parts where heat storage and conduction within the tissue are important. We will conclude this section with some comments on the time constants for changes in leaf temperature.

Leaf Shape and Orientation

Leaf sizes and shapes vary tremendously, which can have important consequences for leaf temperature (see Gates 1980, Parkhurst and Loucks, and Taylor). Leaves developing in full sunlight tend to have smaller areas when mature than leaves on the same plant that develop in the shade—"sun" leaves range from about 70% to less than 10% of the surface area of "shade" leaves. When shade leaves are placed in exposed locations, their larger size leads to thicker boundary layers (Eq. 7.7), less convective heat loss (Eq. 7.10), and consequently greater differences from air temperature than for sun leaves. For leaves above air temperature, this can lead to high transpiration rates, since the water vapor concentration at saturation depends more or less exponentially on temperature, as indicated in Chapter 2 (also see Fig. 8.4 and App. II). Moreover, the amount of CO_2 photosynthetically fixed per unit of water transpired (the *water use efficiency*, a quantity that we will consider at the end of the next chapter) was considerably higher for sun leaves than for shade leaves in exposed sunlit locations, but was higher for shade leaves than for sun leaves in shaded locations for model calculations (see Parkhurst and Loucks) and for a desert shrub, *Hyptis emoryi* (see Smith and Nobel).

Lobing and dissection tend to decrease the effective length across a leaf in the direction of the wind and hence to reduce δ^{bl}, with a consequent enhancement of convective exchange. The greater lobing often observed for sun leaves compared with shade leaves thus reduces the heating of sun leaves above air temperature (see Vogel 1970). Heat convection can be approximately twice as great for a pinnate leaf than for a simple leaf of the same area (see Balding and Cunningham).

Certain plants, especially those exposed to intense shortwave irradiation, have vertically oriented leaves, e.g., willow, many species of *Eucalyptus*, and certain chaparral and desert shrubs. Over the course of a day, vertical leaves often intercept nearly as much light as do horizontal leaves, but they intercept less at midday, when air temperatures tend to be high. Higher leaf temperatures at midday would lead to extra transpiration for a given stomatal opening and, perhaps, to temperatures above those photosynthetically optimal. Also, leaves generally become more nearly vertical upon wilting, which would reduce interception of shortwave irradiation for sun angles 45° or more above the horizontal. As we indicated on p. 352, shortwave plus longwave irradiation absorbed by a leaf at midday could decrease by about 300 W m^{-2} if a leaf were rotated from horizontal to vertical. For the exposed horizontal leaf we have considered (radiation terms in the top line of Table 7.1, convection on p. 366, and transpiration on p. 368; see Fig. 7.7a), such rotation would decrease the leaf temperature from 25°C to 19°C, or by 6°C. Consistent with this are observations that rotating the leaves of *Cercis canadensis* (redbud) and *Erythrina berteroana* from vertical to horizontal at midday increased their temperatures by 2°C to 6°C (see Gates 1980). Also, leaf temperature can be influenced by seasonal differences in leaf orientation and by diurnal solar "tracking" movements, as occurs for cotton, many clovers, and certain desert annuals (e.g., see Ehleringer and Forseth). *Malvastrum rotundifolium*, a winter annual inhabiting warm deserts but growing during the cool part of the year, tracked the sun so well that the direct solar beam was within 20° of being perpendicular to the leaf surface throughout the day. This tracking raised the leaf temperature to photosynthetically more optimal values as well as led to better light interception (see Mooney and Ehleringer).

Shaded Leaves Within Plant Communities

Let us next consider a shaded leaf at the same temperature as its surroundings. As we indicated on p. 352, the infrared radiation absorbed by such a leaf would be the same as the IR emitted by it (when $a_{IR} = e_{IR}$ and $T^{leaf} = T^{surr}$). The net radiation would then be due solely to the various forms of solar radiation that reach the leaf. Since the transmission by leaves is fairly high from 0.7 to 2 μm (Fig. 7.4), much of the solar radiation reaching a shaded leaf would be in a region not useful for photosynthesis. The leaf absorptance in this range is lower than for the solar irradiation incident on an exposed leaf (Fig. 7.4). For purposes of calculation, we will assume that S on the shaded leaf is 70 W m^{-2}, that its absorptance a is 0.30, and that the reflectance r of the surroundings has the rather high value of 0.4 because of considerable reflection of radiation by the other leaves within the plant community. By Equations 7.3 and 7.6, the net radiation balance is then $(0.30)(1.00 + 0.40)(70 \text{ W m}^{-2})$, or 30 W m^{-2}.

Our shaded leaf has a rather low rate of photosynthesis because the amount of radiation in the visible region reaching it is fairly small. Moreover, the stomata generally are partially closed at lower illumination levels, which increases the stomatal resistance and further decreases photosynthesis (Ch. 8). Three factors tend to reduce the flux of water vapor out of a shaded leaf: the increase in stomatal resistance, which reduces transpiration; a lower wind speed for a protected than for an exposed leaf, which leads to a thicker air boundary layer; and a concentration of water vapor in the turbulent air that is generally higher than at the top of the plant canopy. Instead of the water vapor flux density of 4 mmol m^{-2} s^{-1} that can occur for an exposed leaf of a typical mesophyte (p. 415), J_{wv} for the shaded leaf might be only 0.7 mmol m^{-2} s^{-1}. By Equation 7.18 ($J_H^T = J_{wv}H_{vap}$), the heat dissipation by the latent heat term would be 30 W m^{-2}. No heat is conducted across the boundary layers if the leaf is at the same temperature as the surrounding air. Thus, the heat loss by water evaporation (30 W m^{-2}) here balances the energy gain from the absorption of shortwave irradiation for this shaded leaf.

Heat Storage

We indicated at the beginning of this chapter that very little energy could be stored in the temperature changes of leaves. However, massive plant parts, such as tree trunks, can store considerable energy, as can the soil. We will represent the heat storage rate (e.g., in J s^{-1}, or W) as follows:

$$\text{heat storage rate} = C_P V \frac{\Delta T}{\Delta t} \tag{7.22}$$

where C_P is the volumetric heat capacity (e.g., in J m^{-3} °C^{-1}; see values introduced above for soil), and V is the volume that undergoes a change in temperature ΔT in the time interval Δt.

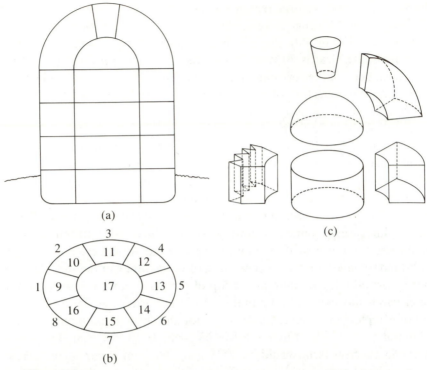

Figure 7.9
System of nodes or subvolumes used in energy balance studies on the barrel
cactus, *Ferocactus acanthodes*: (a) vertical section indicating the division of
the stem into various levels, (b) horizontal section indicating the surface
nodes (1 through 8) and interior nodes (9 through 17), and (c) three-dimensional representation of certain nodes. (Modified from Lewis and Nobel;
used by permission.)

As indicated above, heat storage can be important for any massive plant
part. To model its energy balance, such a part can be divided into a number
of isothermal subvolumes, which are generally referred to as *nodes* in heat
transfer studies (see Kreith). This has been applied to the energy balance
of massive stems of cacti, which have been divided into about 100 nodes
for the barrel-shaped *Ferocactus acanthodes* (Fig. 7.9) and over 200 nodes
for the tall, columnar *Carnegiea gigantea*, commonly known as saguaro or
the giant cactus (see Lewis and Nobel, or Nobel 1980). The stem was divided
into surface nodes, which had no volume and hence no heat storage, plus
interior nodes, which had volume but no radiation, boundary-layer, or

latent-heat terms (Fig. 7.9). These interior nodes were thus involved with changes in temperature leading to heat storage (Eq. 7.22) and heat conduction to or from surface nodes as well as to or from other interior nodes (describable by equations of the form of Eqs. 7.10 and 7.20).

Once the stem has been divided into nodes, an energy balance for each node can be calculated using the various environmental parameters we described above—greater precision requires a greater number of nodes. The analysis is complicated, in part because the various contributors depend on temperature in different ways. Specifically, shortwave and longwave absorption is independent of temperature, longwave emission depends on T^4, conduction depends on a temperature gradient or difference (e.g., $T^{surf} - T^{ta}$), saturation water vapor content—which can affect the latent heat term—varies approximately exponentially with temperature, and heat storage depends on $\Delta T / \Delta t$. When these various energy terms for all the nodes are incorporated into a simulation model, or when cactus temperatures are directly measured in the field, parts of the cactus stem facing the sun are found to range up to 15°C above air temperature for stems approximately 0.25 m in diameter, not the 41°C calculated when heat storage and conduction in the tissue are ignored (p. 367). Moreover, parts facing away from the sun can be below air temperature, and considerable lags (of a few hours) are observed in the heating of the center of the stem.

We indicated near the beginning of this chapter that consumption or production of energy by metabolic processes can generally be ignored in the energy budget of a leaf. An interesting situation where metabolic heat production must be reckoned with occurs for the inflorescences of many members of the Araceae (*Arum* family), where high respiratory rates can substantially raise the inflorescence temperature and lead to considerable heat storage. An extreme example is presented by the 2- to 9-g inflorescence or spadix of *Symplocarpus foetidus* (eastern skunk cabbage). By consuming O_2 at the same rate as an active mammal of the same size (heat production of about 0.10 W g^{-1}), the tissue temperature can be 15°C to 35°C above ambient air temperatures of -15°C to 15°C for at least two weeks (see Knutson).

Time Constants

A matter related to heat storage is the time constant for temperature changes. Analogous to our use of lifetimes in Chapter 4, we will define a time constant τ as the time required for the change of surface temperature from some initial value T_0^{surf} to within $1/e$ (37%) of the overall change to a final value approached asymptotically (T_∞^{surf}):

$$T^{surf} = (T_0^{surf} - T_\infty^{surf})e^{-t/\tau} + T_\infty^{surf} \tag{7.23}$$

If we ignore transpiration and assume uniform tissue temperatures, then the time constant is

$$\tau = \frac{(V/A)\, C_P}{4e_{IR}\sigma(T^{surf} + 273)^3 + K^{air}/\delta^{bl}} \tag{7.24}$$

where V is the volume of a plant part having total surface area A (V/A indicates the mean depth for heat storage, which would be half its thickness for a leaf), C_P is the heat capacity per unit volume, and T^{surf} is expressed in °C. For a rapidly transpiring leaf, the time constant would be about 50% less than indicated by Equation 7.24 (see Gates 1980, and especially Monteith 1981).

Let us next estimate the time constants for a leaf and a cactus stem. If we consider the leaf discussed on p. 341 and assume it has a temperature of 25°C and a boundary layer thickness of 1.4 mm (p. 366), by Equation 7.24 the time constant is then

$$\tau = \frac{(200 \times 10^{-6}\ m)(2.1 \times 10^6\ J\ m^{-3}\ °C^{-1})}{(4)(0.96)(5.67 \times 10^{-8}\ J\ s^{-1}\ m^{-2}\ °C^{-4})(298°C)^3 + \dfrac{(0.0259\ J\ s^{-1}\ m^{-1}\ °C^{-1})}{(1.4 \times 10^{-3}\ m)}}$$

$$= 17\ s$$

Thus, 63% of the overall change in leaf temperature would occur in only 17 s, indicating that such leaves respond fairly rapidly to variations in environmental conditions, consistent with our statement on p. 341 that very little heat can be stored by means of temperature changes of such leaves. On the other hand, stems of cacti can store appreciable amounts of heat. For the stem portrayed in Figure 7.9, V/A is about 0.05 m, and C_P is nearly twice as high as for a leaf, since cacti generally have a much smaller volume fraction of intercellular air spaces. Assuming the other factors are the same as in the above calculation for a leaf, τ is then 8.7×10^3 s (2.4 hours) for the cactus. Indeed, massive stems do have very long time constants for thermal changes.

Among other effects, long thermal time lags help avoid overheating of tree trunks as well as seeds and roots in the soil during rapidly moving fires.

Hence, they are crucial for understanding certain aspects of fire ecology (see Rundel). For instance, temperatures near the soil surface average about 300°C during fires in different ecosystems, while at 0.1 m below the soil surface they rarely exceed 50°C. Energy budget analyses can provide information on the multitude of physiological and ecological processes involving temperature—from frost to fire to photosynthesis.

Problems

7.1. An exposed leaf at 10°C has an a_{IR} and e_{IR} of 0.96, while a is 0.60 and r is 0.10. Suppose that a cloudy sky has an effective temperature for radiation of 2°C, while for a clear sky it is -40°C. (a) If the absorbed IR equals the emitted IR for the leaf, what are the temperatures of the surroundings for a clear sky and for a cloudy one? (b) What are the λ_{max}'s for emission of radiant energy by the leaf and the surroundings for (a)? (c) If the global irradiation is 700 W m^{-2} and the temperature of the surroundings is 9°C on a clear day, how much radiation is absorbed by the leaf? (d) Under the conditions of (c), what percentage of the energy input by absorbed irradiation is dissipated by the emission of thermal radiation? (e) Assume that the clouds block out the sunlight and the skylight. Let S^{cloud} be 250 W m^{-2} on the upper surface of the leaf, while 15% as much is reflected onto the lower surface. If the temperature of the surroundings is 9°C, what is the net radiation for the leaf?

7.2. Consider a circular leaf 0.12 m in diameter at 25°C. The ambient wind speed is 0.8 m s^{-1}, and the ambient air temperature is 20°C. (a) What is the mean distance across the leaf in the direction of the wind? (b) What is the boundary layer thickness? What would δ^{bl} be, assuming that the mean distance is the diameter? (c) What is the heat flux density conducted across the boundary layer? (d) If the net radiation balance for the leaf is 300 W m^{-2}, what is the transpiration rate such that the leaf temperature remains constant?

7.3. Suppose that the global irradiation absorbed by the ground below some vegetation averages 100 W m^{-2}. We will assume that the bulk of the vegetation is at 22°C, the top of the soil is at 20°C, and that both emit like ideal blackbodies. (a) What is the net radiation balance for the soil? (b) Suppose that there are 4 plants/m^2 of ground and that their average stem diameters are 3 cm. If the thermal conductivity coefficient of the stem is the same as that of water, and the temperature drops from that of the ground to that of the bulk of the vegetation in 0.8 m, what is the rate of heat conduction in W down each stem? What is the average value of such J_H^C per m^2 of the ground? (c) Suppose that the 4 mm of air immediately above the ground acts like a boundary layer and that the air temperature at 4 mm is 21°C. What is the rate of heat conduction from the soil into the air? (d) What is J_H^C into the soil in the steady state if 0.3 mmol m^{-2} s^{-1} of water evaporates from the upper part of the soil where the radiation is absorbed? (e) If K^{soil} is the same as K^{water}, what is $\partial T/\partial z$ in the upper part of the soil?

7.4. Let us consider a spherical cactus 0.2 m in diameter with essentially no stem mass below ground. The surface of the cactus averages 25.0°C in late afternoon and is 50% shaded by spines. Assume that the ambient wind speed is 1.0 m s^{-1}, the ambient air temperature is 20.0°C, the global irradiation perpendicular to the solar beam is 1 000 W m^{-2}, the effective temperature of the surroundings (including the sky) is -20°C, $a_{IR} = e_{IR} = 0.97$, and the spines have no transpiration or heat storage. (a) What is the mean boundary layer thickness for the stem and the heat conduction across it? (b) What is the stem heat convection coefficient? (c) Assuming that the spines can be represented by cylinders 1.2 mm in diameter, what is their boundary layer thickness? (d) If the net radiation balance averaged over the spine surface is due entirely to shortwave irradiation (i.e., $IR_{absorbed} = IR_{emitted}$), and if the maximum shortwave measured perpendicular to the cylinder surface is 100 W m^{-2}, what is T^{spine} to within 0.1°C? Ignore spine heat conduction to the stem. (e) What is the absorbed-minus-emitted longwave radiation at the stem surface in the presence and absence of spines? (f) Assume that 30% of the incident shortwave is absorbed by the stem surface for the plant with spines. What is the net energy balance averaged over the stem surface? What is the hourly change in mean tissue temperature? Assume that the volumetric heat capacity is 80% of that of water, and ignore transpiration as well as diffuse shortwave irradiation.

References

Balding, F. R., and G. L. Cunningham. 1976. A comparison of heat transfer characteristics of simple and pinnate leaf models. *Botanical Gazette 137*:65–74.

Campbell, G. S. 1977. *An Introduction to Environmental Biophysics.* Springer-Verlag, Heidelberg.

Cionco, R. M. 1972. Intensity of turbulence within canopies with simple and complex roughness elements. *Boundary-Layer Meteorology 2*:453–465.

Cowan, I. R. 1968. Mass, heat and momentum exchange between stands of plants and their atmospheric environment. *Quarterly Journal of the Royal Meteorological Society 94*:523–544.

de Vries, D. A. 1975. Heat transfer in soils. In *Heat and Mass Transfer in the Biosphere,* Part 1, *Heat and Mass Transfer in Processes in Plant Environment,* D. A. de Vries and N. H. Afgan, eds. Halsted Press, Wiley, New York. Pp. 5–28.

Ehleringer, J., and I. Forseth. 1980. Solar tracking by plants. *Science 210*:1094–1098.

Ehleringer, J. R., and H. A. Mooney. 1978. Leaf hairs: effects on physiological activity and adaptive value to a desert shrub. *Oecologia 37*:183–200.

Gates, D. M. 1965. Energy, plants, and ecology. *Ecology 46*:1–13.

Gates, D. M. 1970. Physical and physiological properties of plants. In *Remote Sensing.* National Academy of Sciences, Washington, D.C. Pp. 224–252.

Gates, D. M. 1980. *Biophysical Ecology.* Springer-Verlag, Heidelberg.

Gates, D. M., and L. E. Papian. 1971. *Atlas of Energy Budgets of Plant Leaves*. Academic Press, New York.

Geiger, R. 1966. *The Climate Near the Ground*. Harvard University Press, Cambridge.

Grace, J. 1977. *Plant Responses to Wind*. Academic Press, London.

Idso, S. B., D. G. Baker, and D. M. Gates. 1966. The energy environment of plants. *Advances in Agronomy 18*:171–218.

Jaffe, M. J. 1980. Morphogenetic responses of plants to mechanical stimuli or stress. *BioScience 30*:239–243.

Jaffe, M. J. 1982. Wind and other mechanical effects in the development and behavior of plants, with special emphasis on the role of hormones. In *Hormonal Regulation of Development III*, J. Macmillan, ed. *Encyclopedia of Plant Physiology, New Series*, Vol. 9C. Springer-Verlag, Berlin. In press.

Kappen, L., O. L. Lange, E.-D. Schulze, M. Evenari, and U. Buschbom. 1979. Ecophysiological investigations on lichens of the Negev Desert. VI. Annual course of the photosynthetic production of *Ramalina maciformis* (Del.) Bory. *Flora 168*: 85–108.

Knutson, R. M. 1974. Heat production and temperature regulation in eastern skunk cabbage. *Science 186*:746–747.

Kreith, F. 1973. *Principles of Heat Transfer*, 3rd ed. Intext Educational Publishers, New York.

Levitt, J. 1980. *Responses of Plants to Environmental Stresses*, 2nd ed., Vol. II, *Water, Radiation, Salt, and Other Stresses*. Academic Press, New York.

Lewis, D. A., and P. S. Nobel. 1977. Thermal energy exchange model and water loss of a barrel cactus, *Ferocactus acanthodes*. *Plant Physiology 60*:609–616.

Monteith, J. L. 1973. *Principles of Environmental Physics*. American Elsevier, New York.

Monteith, J. L. 1981. Coupling of plants to the atmosphere. In *Plants and their Atmospheric Environment*, J. Grace, E. D. Ford, and P. G. Jarvis, eds. Blackwell, Oxford. Pp. 1–29.

Mooney, H. A., and J. R. Ehleringer. 1978. The carbon gain benefits of solar tracking in a desert annual. *Plant, Cell, and Environment 1*:307–311.

Munn, R. E. 1966. *Descriptive Micrometeorology*. Academic Press, New York.

Nobel, P. S. 1974. Boundary layers of air adjacent to cylinders. Estimation of effective thickness and measurements on plant material. *Plant Physiology 54*:177–181.

Nobel, P. S. 1975. Effective thickness and resistance of the air boundary layer adjacent to spherical plant parts. *Journal of Experimental Botany 26*:120–130.

Nobel, P. S. 1980. Morphology, surface temperatures, and northern limits of columnar cacti in the Sonoran desert. *Ecology 61*:1–7.

Nobel, P. S. 1981. Wind as an ecological factor. In *Physiological Plant Ecology*, O. L. Lange, P. S. Nobel, C. B. Osmond, and H. Ziegler, eds. *Encyclopedia of Plant Physiology, New Series*, Vol. 12A. Springer-Verlag, Berlin. Pp. 475–500.

Oke, T. R. 1978. *Boundary Layer Climates*. Wiley, New York.

Parkhurst, D. F., and O. L. Loucks. 1972. Optimal leaf size in relation to environment. *Journal of Ecology 60*:505–537.

Pearman, G. I., H. L. Weaver, and C. B. Tanner. 1972. Boundary layer heat transfer coefficients under field conditions. *Agricultural Meteorology 10*:83–92.

Robinson, N. ed. 1966. *Solar Radiation*. Elsevier, Amsterdam.

Rose, C. W. 1966. *Agricultural Physics*. Pergamon, Oxford.

Rosenberg, N. J. 1974. *Microclimate: The Biological Environment*. Wiley, New York.

Rundel, P. W. 1981. Fire as an ecological factor. In *Physiological Plant Ecology*, O. L. Lange, P. S. Nobel, C. B. Osmond, and H. Ziegler, eds. *Encyclopedia of Plant Physiology, New Series*, Vol. 12A. Springer-Verlag, Berlin. Pp. 501–538.

Rutter, A. J. 1975. The hydrological cycle in vegetation. In *Vegetation and the Atmosphere*, Vol. 1, *Principles*, J. L. Monteith, ed. Academic Press, New York. Pp. 111–154.

Schlichting, H. 1968. *Boundary-Layer Theory*, 6th ed. McGraw-Hill, New York.

Schwintzer, C. R. 1971. Energy budgets and temperatures of nyctinastic leaves on freezing nights. *Plant Physiology 48*:203–207.

Smart, R. E., and T. R. Sinclair. 1976. Solar heating of grape berries and other spherical fruits. *Agricultural Meteorology 17*:241–259.

Smith, W. K., and P. S. Nobel. 1977. Temperature and water relations for sun and shade leaves of a desert broadleaf, *Hyptis emoryi*. *Journal of Experimental Botany 28*:169–183.

Taylor, S. E. 1975. Optimal leaf form. In *Perspectives of Biophysical Ecology*, D. M. Gates and R. B. Schmerl, eds. Springer-Verlag, Heidelberg. Pp. 73–86.

van Wijk, W. R., ed. 1966. *Physics of Plant Environment*, 2nd ed. North-Holland, Amsterdam.

Vogel, S. 1970. Convective cooling at low airspeeds and the shapes of broad leaves. *Journal of Experimental Botany 21*:91–101.

Vogel, S. 1981. *Life in Moving Fluids; the Physical Biology of Flow*. Willard Grant Press, Boston.

Wheeler, W. W. and M. Neushul. 1981. The Aquatic Environment. In *Physiological Plant Ecology*, O. L. Lange, P. S. Nobel, C. B. Osmond, and H. Zeigler, eds. *Encyclopedia of Plant Physiology, New Series*, Vol. 12A. Springer-Verlag, Berlin. Pp. 229–247.

Woolley, J. T. 1971. Reflectance and transmittance of light by leaves. *Plant Physiology 47*:656–662.

CHAPTER 8

Leaves and Fluxes

In this chapter we will draw upon a number of previously introduced topics to discuss gas fluxes for leaves. Specifically, we will reconsider the transpiration of water and the photosynthetic fixation of carbon dioxide. The driving forces for such fluxes are differences in CO_2 and H_2O concentrations or mole fractions. We will quantitatively discuss the movement of gases into and out of leaves in terms of the various resistances or conductances involved. The use of resistances to describe gas fluxes quantitatively was first applied to leaves by Brown and Escombe in 1900. Over the years, consideration of resistance networks has become very convenient for specifying which parts of the pathway are most limiting for photosynthesis or transpiration. Recently, use of two distinctly different forms of conductance has become more popular, especially for describing transpiration, and we will consider both forms. We will see that greater stomatal opening may be an advantage for photosynthesis, but can result in excessive transpiration. Consequently, a benefit/cost index, such as the amount of CO_2 fixed per unit of water lost, can be important for evaluating ecological aspects of gas exchange. Figure 8.1 indicates how the flux densities of water vapor and CO_2 can be measured for a leaf. Although we will usually be referring to leaves, the discussion of gas fluxes is also applicable to stems, flower petals, and other plant parts.

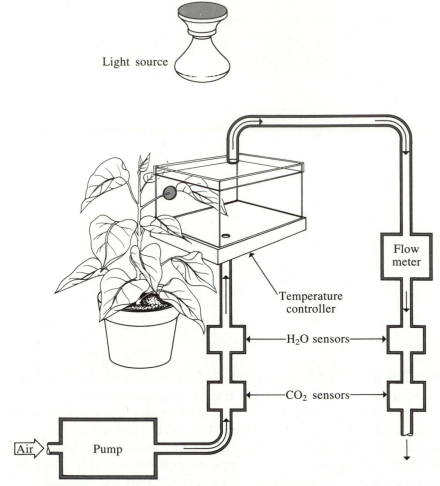

Figure 8.1
Experimental arrangement for the measurement of leaf transpiration
and photosynthesis. The water vapor and CO_2 content of the gas
entering a transparent chamber enclosing a leaf is compared with that
leaving. If the water vapor concentration goes from 0.6 mol m^{-3}
entering to 1.0 mol m^{-3} leaving for a gas flow rate of 10^{-5} m^3 s^{-1}
(10 cm^3 s^{-1}), then J_{wv} for a leaf of area 10^{-3} m^2 (10 cm^2) would be
(1.0 mol m^{-3} − 0.6 mol m^{-3})(10^{-5} m^3 s^{-1})/(10^{-3} m^2), or 0.004 mol
m^{-2} s^{-1}.

RESISTANCES AND CONDUCTANCES—TRANSPIRATION

The resistances and conductances we will discuss in this section are those encountered by water vapor as it diffuses from the pores in the cell walls of mesophyll cells or other sites of water evaporation out into the turbulent air surrounding a leaf. We will define these quantities for the intercellular air spaces, the stomata, the cuticle, and the boundary layer next to a leaf. As explained later in this chapter, CO_2 diffuses across the same gaseous phase resistances or conductances as water vapor, and also across a number of other components in the liquid phases of mesophyll cells.

Throughout this book we have used equations of the following general form: flux density = proportionality coefficient × force (e.g., pp. 11, 117). The proportionality coefficients in such flux density expressions are measures of conductivity. To represent force, instead of using the gradient in chemical potential—which, as we noted in Chapter 3, is a very general force—we often use quantities that are more convenient experimentally, such as differences in concentration (consider Fick's first law, Eq. 1.1; see pp. 12 and 120), partial pressure, or mole fraction. Such "forces," however, do not have the proper units for force; and so, to be correct, the proportionality coefficient is not the conductivity, but rather the *conductance*:

$$\text{flux density} = \text{conductivity} \times \text{force} \qquad (8.1a)$$

$$\text{flux density} = \text{conductance} \times \text{"force"} \qquad (8.1b)$$

We might equally well choose an alternative form for the relation between forces and fluxes: flux density = force/resistivity, where *resistivity* is the reciprocal of conductivity. In turn, we can define a *resistance*, which is the reciprocal of conductance:

$$\text{flux density} - \frac{\text{force}}{\text{resistivity}} \qquad (8.1c)$$

$$\text{flux density} = \frac{\text{"force"}}{\text{resistance}} \qquad (8.1d)$$

To help understand the difference between resistivity and resistance, we will consider the electrical usage of these terms. Electrical resistivity is a fundamental physical property of a material; e.g., such resistivities are tabulated in suitable handbooks. Electrical resistance, on the other hand, describes a particular component in an electrical circuit, i.e., a particular piece of material. We can readily measure the magnitude of resistance in the laboratory or purchase a resistor of known resistance in an electronics store.

Electrical resistivity, ρ, generally has the units of ohm m, while electrical resistance, $R = \rho \Delta x / A$ (see p. 117), is expressed in ohms. Thus, besides differing conceptually (compare Eq. 8.1c and d), resistance and resistivity differ in their units. In fact, much of our attention in this chapter will be devoted to the units for parameters such as resistance, since many different systems are in use in plant physiology, ecology, meteorology, and other related disciplines.

We will initially consider both the resistance and the conductance of a particular component, and we will present the expressions for both resistance and conductance for series versus parallel arrangements of components. Since the flux density is directly proportional to conductance (Eq. 8.1b), but inversely proportional to resistance (Eq. 8.1d), conductance terminology is often more convenient for discussing plant responses to environmental factors. Also, resistance has no upper limit—i.e., it varies from some minimal value to infinity—whereas conductances varies between zero and some maximum value. However, resistance terminology can be easier to use when a substance must cross a series of components in sequence, such as CO_2 diffusing across the cell wall, plasmalemma, cytosol, and chloroplast membranes. To help familiarize the reader with the various conventions found in the literature, we will use both systems throughout this chapter, emphasizing conductances for transpiration and resistances for photosynthesis.

Boundary Layer Adjacent to Leaf

As a starting point for our discussion of gas fluxes across air boundary layers, let us consider the one-dimensional form of Fick's first law of diffusion, $J_j = -D_j \partial c_j / \partial x$ (Eq. 1.1). As in Chapter 1, we will replace the concentration gradient by the difference in concentration across some distance. In effect, we are considering cases not too far from equilibrium, so the flux density will depend linearly on the force, and the force can be represented by the difference in concentration. The distance that interests us here is the one occurring across the air boundary layer adjacent to the surface of a leaf, δ^{bl} (see Ch. 7 for a discussion of boundary layers and equations for their thickness). Consequently, Fick's first law assumes the following form for the diffusion of species j:

$$J_j = -D_j \frac{\partial c_j}{\partial x} = D_j \frac{\Delta c_j}{\Delta x} = D_j \frac{\Delta c_j^{bl}}{\delta^{bl}}$$

$$= g_j^{bl} \Delta c_j^{bl} = \frac{\Delta c_j^{bl}}{r_j^{bl}} \tag{8.2}$$

Equation 8.2 is a very general relation showing how the net flux density of species j depends on its diffusion coefficient D_j and on the difference in its concentration Δc_j^{bl} across a distance δ^{bl} of the air. The net flux density J_j is toward regions of lower c_j, as we will specifically consider for water vapor and CO_2. Outside the boundary layer of thickness δ^{bl} is a region of air turbulence, where we will assume that the concentrations of gases are the same as in the bulk atmosphere (an assumption that we will remove in the next chapter). Equation 8.2 indicates that J_j equals Δc_j^{bl} multiplied by a conductance, g_j^{bl}, or divided by a resistance, r_j^{bl}, an aspect that we will consider next.

The air boundary layers on both sides of a leaf represent important barriers in the pathways for the entry of CO_2 and the exit of H_2O, as was clearly shown by Raschke in the 1950's. Movement across these layers is by diffusion of the gas molecules in response to differences in concentration. Using Equation 8.2, we can represent the conductance and the resistance of a boundary layer of air as follows:

$$g_j^{bl} = \frac{J_j}{\Delta c_j^{bl}} = \frac{D_j}{\delta^{bl}} = \frac{1}{r_j^{bl}} \tag{8.3}$$

Since the SI unit for the diffusion coefficient D_j is $m^2 \, s^{-1}$ (see Ch. 1), and since the thickness of the boundary layer δ^{bl} is in m, g_j^{bl} is therefore in $(m^2 \, s^{-1})/(m)$, or $m \, s^{-1}$ (values are generally expressed in $mm \, s^{-1}$), and r_j^{bl} is in $s \, m^{-1}$ (values are often expressed in $s \, cm^{-1}$). J_j is expressed per unit leaf area, and so g_j^{bl} and r_j^{bl} also relate to unit area of a leaf.

We note that D_j is a fundamental measure of conductivity (values available in suitable handbooks) describing the diffusion of species j in a given medium or material. On the other hand, δ^{bl} characterizes a particular situation, since the thickness of the boundary layer depends on wind speed and leaf size (see Ch. 7, e.g., Eq. 7.7). Thus, r_j^{bl} as defined by Equation 8.3 describes a particular component of the path, analogous to the resistance (R) used in Ohm's law. Recalling Equation 1.9, $P_j = D_j K_j / \Delta x$, we recognize that D_j / δ^{bl} is the permeability coefficient for species j, P_j, as it diffuses across an air boundary layer of thickness δ^{bl}. Thus, g_j^{bl} as defined by Equation 8.3 is equal to P_j. When something readily crosses a region of air by diffusion, P_j and g_j^{bl} are large and r_j^{bl} is small. Using resistances and conductances, we can describe gas fluxes into and out of leaves employing a number of relations originally developed for the analysis of electrical circuits (see Gaastra, J. L. Monteith in Evans, or Lake).

We will next estimate values that might be expected for g_{wv}^{bl} and r_{wv}^{bl}—the conductance and the resistance, respectively—for water vapor diffusing across the boundary layer of air next to a leaf. In Chapter 7 we indicated that the boundary layer thickness in mm ($\delta^{bl}_{(mm)}$) for a flat leaf under field conditions is $4.0\sqrt{l_{(m)}/v_{(m\,s^{-1})}}$ (Eq. 7.9), where $l_{(m)}$ is the mean length of the leaf in the direction of the wind in m and $v_{(m\,s^{-1})}$ is the ambient wind speed in m s^{-1}. Let us consider a relatively thin boundary layer of 0.3 mm and a thick one of 3 mm (see Table 7.2 for the values of $l_{(m)}$ and $v_{(m\,s^{-1})}$ that this implies). For water vapor diffusing in air at 20°C, D_{wv} is 2.4×10^{-5} m^2 s^{-1} (App. II). Using Equation 8.3, we obtain

$$g^{bl}_{wv} = \frac{(2.4 \times 10^{-5} \text{ m}^2 \text{ s}^{-1})}{(0.3 \times 10^{-3} \text{ m})} = 8 \times 10^{-2} \text{ m s}^{-1} = 80 \text{ mm s}^{-1}$$

$$r^{bl}_{wv} = \frac{(0.3 \times 10^{-3} \text{ m})}{(2.4 \times 10^{-5} \text{ m}^2 \text{ s}^{-1})} = 13 \text{ s m}^{-1}$$

for the thin boundary layer, and g^{bl}_{wv} equals 8 mm s^{-1} and r^{bl}_{wv} equals 130 s m^{-1} for the thick one (Table 8.1).* Usually, boundary layer conductances are larger and resistances are smaller than their respective values for diffusion along the stomatal pores, which we will now examine.

Stomata

As we indicated in Chapter 1 (p. 4), stomata help control the exit of water vapor from leaves and the entry of CO_2 into them. Although the epidermal cells occupy a much greater fraction of the leaf surface area than do the stomatal pores, the waxy cuticle covering them greatly reduces the water loss from their cell walls to the turbulent air surrounding a leaf, the *cuticular transpiration*. The usual pathway for water vapor leaving a leaf during transpiration is therefore through the stomata.

The stomatal aperture is controlled by the conformation of the two guard cells surrounding a pore (see Figs. 1.2 and 8.2). These cells are generally kidney-shaped (dumbbell-shaped in grasses), may be 40 μm long, and, unlike ordinary epidermal cells, contain chloroplasts. When the guard cells are relatively flaccid, the stomatal pore is nearly closed, as is the case for most

---

* Instead of estimating the boundary layer conductance or resistance based on δ^{bl} calculated by Equation 7.7, which cannot account for all the intricacies of different leaf shapes, it is often more expedient to construct a paper replica of the leaf; if this "leaf" is then moistened, the observed J_{wv} from it for a certain Δc^{bl}_{wv} will indicate g^{bl}_{wv} or r^{bl}_{wv} (cf. Eq. 8.2), since water vapor will cross only the boundary layer in this case.

Table 8.1
Summary of representative values of conductances and resistances for water vapor diffusing out of leaves. See text for specific calculations (conductances in mmol m^{-2} s^{-1} are from Eq. 8.8), as well as Cowan and Milthorpe, Körner et al., and Kramer.

Component condition	Conductance		Resistance
	mm s^{-1}	mmol m^{-2} s^{-1}	s m^{-1}
Boundary layer			
thin	80	3 000	13
thick	8	300	130
Stomata			
large area—open	32	1 300	31
small area—open	1.7	70	600
closed	0	0	∞
mesophytes—open	2 to 20	80 to 800	50 to 500
certain xerophytes and trees—open	1	40	1 000
Cuticle			
crops	0.1 to 0.4	4 to 16	2 500 to 10 000
many trees	0.05 to 0.2	2 to 8	5 000 to 20 000
many xerophytes	0.01 to 0.1	0.4 to 4	10 000 to 100 000
Intercellular air spaces			
calculation	24 to 240	1 000 to 10 000	4.2 to 42
waxy layer			
typical	50 to 200	2 000 to 8 000	5 to 20
certain xerophytes	10	400	100
typical	20 to 100	800 to 4 000	10 to 50
Leaf (lower surface)			
crops—open stomates	2 to 10	80 to 400	100 to 500
trees—open stomates	0.5 to 3	20 to 120	300 to 2 000

plants at night. Upon illumination, guard cells take up K$^+$, whose concentration may increase by 0.3 M or more in a cell (see Fischer, Jarvis and Mansfield, or Raschke). The K$^+$ uptake raises the internal osmotic pressure and thus lowers the internal water potential; water then spontaneously flows from the epidermal cells into the guard cells. This water entry leads to an increase in the internal hydrostatic pressure of a pair of guard cells, which in turn causes them to expand and their opposed cell walls (those on either side of the pore) to become concave. As the kidney-shaped guard cells thus bow outward, an elliptical pore develops between the two cells. The formation of this pore is a consequence of the anisotropic properties of the cell wall surrounding each guard cell (see Aylor et al.). The distance between the guard cells across the open pore (the pore "width") is generally 5 to 10 μm, and the major axis of the elliptical pore may be about 20 μm.

The water relations of the special epidermal cells immediately surrounding the guard cells, which are referred to as subsidiary cells, are also crucial for stomatal opening (see Cooke et al., Cowan 1972 and 1977, and Tyree and Yianoulis). For instance, solutes like K^+ can move from the subsidiary cells to the guard cells, causing water to leave the subsidiary cells and their internal hydrostatic pressure to decrease while P^i of the guard cells increases. Once the stomatal pore has opened, water evaporates from the inner side of the guard cells and subsidiary cells and then diffuses out of the leaf. This lowers the water potential in the cell walls, which in turn can cause water to leave the protoplasts of the guard cells and subsidiary cells, thereby lowering their P^i and causing some stomatal closure (local water flow depends on the local Ψ and leads to different kinetics for the changes of P^i in guard cells versus subsidiary cells). The ensuing partial stomatal closure reduces the water exit from the leaf and hence reduces evaporation, thereby allowing P^i to build back up in the guard cells and subsidiary cells, leading to re-opening of the stomata. The resulting oscillation of stomatal pore aperture can have a period of about 30 to 60 minutes, although it is often damped out by changes in environmental factors such as wind and irradiation.

What regulates the opening of stomatal pores? This question has proved difficult to answer, in part because a number of factors are involved. An initial event may be active H^+ extrusion from the guard cells, followed by passive K^+ uptake, the latter perhaps via facilitated diffusion (p. 158). Or K^+ may be actively transported into guard cells during opening and actively transported out during closing. After stomatal opening is initiated, the added K^+ within the guard cells is electrically balanced—partly by a Cl^- influx (perhaps via an OH^--antiport or an H^+-symport; see p. 154), but usually to a greater degree by the production of organic anions like malate in the guard cells. For some plants stomatal opening increases in the light only up to a PAR of about 200 μmol m^{-2} s^{-1}, but for others it may increase all the way up to full sunlight (2 000 μmol m^{-2} s^{-1}).

The degree of stomatal opening often depends on the CO_2 concentration in the guard cells, which reflects their own carbohydrate metabolism as well as the CO_2 level in the air within the leaf (see Cowan 1977, Hall, and Raschke). For instance, upon illumination, the CO_2 concentration in the leaf mesophyll is decreased by photosynthesis, resulting in decreased CO_2 levels in the guard cells, which somehow triggers stomatal opening. CO_2 can then enter the leaf and photosynthesis can continue (to be discussed again near the end of this chapter). In the dark, respiration generally leads to relatively high CO_2 levels in the leaves, and stomata are then triggered to close. Lowering the CO_2 concentration in the ambient air can induce stomatal opening in the *dark* (again, as a response to the low CO_2 level in the guard cells), which indicates that the energy for opening can be supplied by mitochondrial respiration. Stomata may tend to close somewhat as the relative humidity of the ambient air decreases, another means of regulating water loss. If a constant water vapor concentration difference from inside the leaf to the surrounding air is maintained, stomata generally tend to open with increasing temperature up to temperatures optimal for photosynthesis, and often somewhat above. Stomatal movements can also be controlled hormonally. In particular, abscisic acid is formed in illuminated leaves during periods of water stress. This leads to stomatal closure, which conserves water, although at the expense of a decrease in photosynthesis. The halftimes for stomatal movements are generally 5 to 20 minutes, with closing usually more rapid than opening.

For leaves of dicots, stomata are usually more numerous on the lower surface than on the upper one. In some such species, stomata may even be nearly absent from the upper epidermis. On the other hand, many grasses and certain plants with vertically oriented leaves have approximately the same number of stomata per unit area on each side. A frequency of 40 to 300 stomata per mm^2 is representative for the lower surface of most leaves. The pores of the open stomata generally occupy 0.2% to 2% of the leaf surface area. Thus, the area for diffusion of gases through the stomatal pores in the upper or lower epidermis of a leaf, A^{st}, is much less than the leaf surface area, A (see Meidner and Mansfield).

Stomatal Conductance and Resistance

We can apply Fick's first law in the form $J_j = D_j \Delta c_j / \Delta x$ (Eq. 8.2) to describe the diffusion of gases along stomatal pores. Let the depth of the stomatal pore be δ^{st} (see Fig. 8.2); the concentration of species j changes by Δc_j^{st} along the distance δ^{st}. In the steady state, the amount of species j moving per unit time toward or away from the leaf (J_j times the leaf area A) must equal the amount of species j moving per unit time through the stomata (the flux density within the stomata, $D_j \Delta c_j^{st}/\delta^{st}$, times the stomatal area A^{st} that occurs for leaf area A). This leads us to the obvious but extremely important conclusion that the stomata constrict the area available for the diffusion of

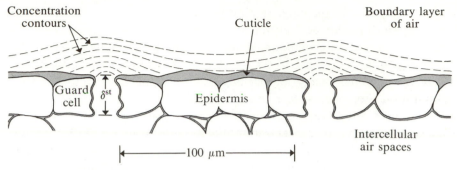

Figure 8.2
Concentration contours of water vapor outside open stomata.

gas molecules. Recognition of this constricting effect that the stomata have on the flux density of species j gives us the following relations:

$$J_j = D_j \frac{\Delta c_j^{st}}{\delta^{st}} \frac{A^{st}}{A}$$

$$= \frac{D_j n a^{st}}{\delta^{st}} \Delta c_j^{st} \tag{8.4}$$

where n is the number of stomata per unit area of the leaf, and a^{st} is the average area per stomatal pore; i.e., na^{st} equals the fraction of the leaf surface area occupied by stomatal pores, A^{st}/A. The flux density J_j in Equation 8.4 refers to the rate of movement of species j *per unit area of the leaf*, a quantity considerably easier to measure than the flux density within a stomatal pore.

The area available for water vapor diffusion abruptly changes from A^{st} to A at the leaf surface. On the other hand, the three-dimensional surfaces of equal concentration must fan out from each stomatal pore (Fig. 8.2). This geometrical aspect could introduce considerable complications, but fortunately we can still use a one-dimensional form for Fick's first law to describe gases moving across the boundary layer next to a leaf, although we need to make an "end correction" to allow for the diffusion pattern at the end of the stomatal pores. The distance between stomata is often about 100 μm, which is considerably less than the thicknesses of nearly all boundary layers (see Table 7.2). Thus, the three-dimensional concentration contours from adjacent stomata tend to overlap in the boundary layer (Fig. 8.2). When this occurs, the concentration of water vapor varies only slightly in planes parallel to the leaf surface, but changes substantially in the direction perpendicular to the leaf surface, and we can therefore generally use a one-dimensional form of Fick's first law (see Cooke and Rand; Meidner and Mansfield; and Penman and Schofield). However, the concentration patterns on both ends of a stomatal pore cause the pore to have an effective depth greater than δ^{st} by about the mean, or effective, "radius" of the pore, r^{st}.[*] Since stomatal pores are approximately elliptical, not circular, we will define r^{st} as follows: $\pi(r^{st})^2 = a^{st}$, where a^{st} is the effective area of the pore.

For our subsequent applications to transpiration and photosynthesis, we will define a stomatal conductance, g_j^{st}, and resistance, r_j^{st}, for the diffusion of species j. We have previously described analogous quantities for the boundary layer (Eq. 8.3) based on a form of Fick's first law for the diffusion of species j. Here we will use Equation 8.4 to define the conductance and resistance for the diffusion of gases along the length of the stomata:

$$g_j^{st} = \frac{J_j}{\Delta c_j^{st}} = \frac{D_j n a^{st}}{\delta^{st} + r^{st}} = \frac{1}{r_j^{st}} \tag{8.5}$$

Equation 8.5 incorporates the effective depth of the stomatal pore, $\delta^{st} + r^{st}$, where r^{st} is the mean pore radius just introduced. When the width of a stomatal pore is about 0.1 to 0.3 μm, as occurs when the pores are nearly closed, the mean free path for molecules diffusing in air is about the same as the dimensions of the opening.[†] In this case, the molecular interactions with the sides of the stomatal pore are important, and this affects the value of D_j. Even though we generally ignore this interaction and make certain geometrical approximations for δ^{st}, r^{st}, and a^{st}, Equation 8.5 is still quite useful for estimating stomatal conductances and resistances.

Let us now calculate some values for the stomatal conductance for the diffusion of water vapor. We will consider air at 20°C for which D_{wv} is 2.4×10^{-5} m^2 s^{-1}. The stomatal conductance tends to be high when a large portion of the leaf surface area (e.g., $n a^{st} = 0.02$) is occupied by open stomata of relatively short pore depth (e.g., $\delta^{st} = 10$ μm). If we assume that the mean radius r^{st} is 5 μm, Equation 8.5 predicts that the conductance

[*] Formulas for the end correction due to the three-dimensional nature of the concentration gradients at each end of a stomatal pore vary and are more complicated than that used here, r^{st} (see Cook and Rand; Meidner and Mansfield; Monteith 1973, Ch. 7 reference; and Parlange and Waggoner). However, the differences among the various correction formulas are generally relatively small compared to δ^{st}.

[†] The mean free path of a gas molecule, which is the average distance such a molecule travels before colliding with another gas molecule, is about 0.1 μm at pressures and temperatures normally experienced by plants.

would then be

$$g_{wv}^{st} = \frac{(2.4 \times 10^{-5} \text{ m}^2 \text{ s}^{-1})(0.02)}{(10 \times 10^{-6} \text{ m} + 5 \times 10^{-6} \text{ m})}$$

$$= 3.2 \times 10^{-2} \text{ m s}^{-1} = 32 \text{ mm s}^{-1}$$

At the other extreme, the open stomata may occupy only 0.4% of the lower surface of a leaf ($na^{st} = 0.004$), and the pore depth may be rather large, e.g., 50 μm. Again assuming that r^{st} is 5 μm, the stomatal conductance calculated using Equation 8.5 is 1.7 mm s^{-1}, a rather small value for open stomata (Table 8.1).

The stomatal conductance is usually 2 to 20 mm s^{-1} for water vapor diffusing out through the open stomata of most mesophytes (r_{wv}^{st} of 50 to 500 s m^{-1} or 0.5 to 5 s cm^{-1}). In general, agricultaral crops tend to have high values of g_{wv}^{st}, while for certain xerophytes and many trees with open stomata it may be only 1 mm s^{-1} (Table 8.1). Some xerophytes have sunken stomata leading to another conductance in series with g_{wv}^{st}, which would slightly decrease the overall conductance for water loss by transpiration. Of course, as the stomatal pores close, the conductance decreases accordingly, since g_{wv}^{st} is proportional to the stomatal area by Equation 8.5. Since g_{wv}^{st} is essentially the only conductance in the whole diffusion pathway that is variable over a wide range, changes of the openings of the stomatal pores provide a plant with an important control mechanism for regulating the movement of gases into or out of the leaves.

Cuticle

Some water molecules diffuse across the waxy cuticle of the epidermal cells, which is termed the *cuticular transpiration*. This movement of H_2O across the cuticle probably involves liquid water as well as water vapor. We will identify a cuticular conductance, g_j^c, and resistance, r_j^c, for the diffusion of species j across the cuticle. These quantities are in parallel with the analogous quantities for the stomata—i.e., species j can leave the leaf either by crossing the cuticle or by going through the stomatal pores (see Fig. 8.1). The conductance for cuticular transpiration, g_{wv}^c, generally ranges from 0.05 to 0.3 mm s^{-1} for different species, although it can be considerably lower for xerophytes with thick cuticles that greatly restrict water loss (Table 8.1). Thus, g_{wv}^c is usually much smaller than g_{wv}^{st} for open stomata. When the stomata are nearly closed (low g_{wv}^{st}), the cuticular transpiration can exceed the loss of water through the stomata. If the cuticle is mechanically damaged or develops cracks, as can occur for older leaves, g^c_{wv} can be raised, and more water will then move out of the leaf by this pathway.

Intercellular Air Spaces

Another conductance encountered by the diffusion of species j in plant leaves is that of the intercellular air spaces, g^{ias}_j. Because of the irregular shape of these air spaces, which generally account for about 50% of the leaf volume, it is difficult to estimate g^{ias}_j accurately from geometrical considerations alone. However, the intercellular air spaces do act like an unstirred air layer across which substances must diffuse; hence we will describe g^{ias}_j by a relation similar to Equation 8.3 ($g^{bl}_j = D_j/\delta^{bl}$). As a first approximation, we can regard the thickness of the unstirred air layer in the present case as an average distance from the surfaces of the mesophyll cells within a leaf to the inner side of a stomatal pore, δ^{ias}. By analogy with Equations 8.3 and 8.5, we can express g^{ias}_j as

$$g^{ias}_j = \frac{D_j}{\delta^{ias}} = \frac{1}{r^{ias}_j} \tag{8.6}$$

For convenience, we will combine a number of factors to get an effective distance, δ^{ias}. For instance, we will let δ^{ias} include the effective length of the air-filled part of the cell wall pores from the cell wall surface to the sites where the evaporation of water or the dissolving of CO_2 takes place. This length is greater than the actual distance along the pores, since we must correct for the decrease in the cross-sectional area available for diffusion, a decrease caused by the presence of the nongaseous parts of the cell wall. Also, δ^{ias} includes the effective length of the thin waxy layer that generally occurs on the surfaces of mesophyll cells. This thin waxy layer has a conductance of 50 to 200 mm s^{-1} for mesophytes, but may be much lower for certain xerophytes where cutinization of the mesophyll cells is appreciable (Table 8.1). In addition, δ^{ias} incorporates the fact that the entire cross section of the mesophyll region is not available for diffusion of gases, but that the flow is constricted to the intercellular air spaces, which have a smaller cross-sectional area than the corresponding leaf area, i.e., $A^{ias}/A < 1$. Thus, that part of the δ^{ias} referring to the intercellular air spaces per se equals the actual distance involved times the ratio A/A^{ias}. If the mesophyll region were about half air by volume (a rather typical value for a leaf), A/A^{ias} would be approximately two.

The effective length δ^{ias}, including all of the factors just enumerated, might range from 100 μm to 1 mm for most leaves. Equation 8.6 indicates that the water vapor conductance across the intercellular air spaces might then range from an upper limit of

$$g_{wv}^{ias} = \frac{(2.4 \times 10^{-5} \text{ m}^2 \text{ s}^{-1})}{(100 \times 10^{-6} \text{ m})} = 0.24 \text{ m s}^{-1} = 240 \text{ mm s}^{-1}$$

for a δ^{ias} of 100 μm down to about 24 mm s^{-1} for a δ^{ias} of 1 mm ($D_{wv} = 2.4 \times 10^{-5}$ m^2 s^{-1} at 20°C; see App. II). Thus, the conductance of the intercellular air spaces is relatively large (the resistance is relatively small) compared with the other conductances encountered by gases diffusing into or out of leaves (Table 8.1).*

Fick's First Law and Conductances or Resistances

The form of Fick's first law that we are using to describe the flux density of water vapor out of a leaf or of CO_2 into one is appropriate for a one-dimensional situation (e.g., $J_j = D_j \Delta c_j / \Delta x = g_j \Delta c_j = \Delta c_j / r_j$). The H_2O lost from a leaf during transpiration can emanate from the cell walls of its mesophyll cells—actually, much if not most of the water evaporates from the inner sides of the guard cells and adjacent subsidiary cells (see Cooke and Rand; Cowan 1977; Tanton and Crowdy; and Tyree and Yianoulis). We can imagine that the water vapor moves in the intercellular air spaces toward the leaf surface by diffusing down planar fronts of sucessively lower concentration (see Figs. 8.2 and 8.5). Our imaginary planar fronts are parallel to the leaf surface, so the direction for the fluxes is perpendicular to the leaf surface. When we reach the inner side of a stomatal pore, the area for diffusion is reduced from A^{ias} to A^{st}. In other words, we are still discussing the advance of planar fronts in one dimension, but the flux is now constricted to the stomatal pores. (A small amount of water constituting the cuticular transpiration diffuses across the cuticle in parallel with the stomatal fluxes.) Finally, the movement of water vapor across the boundary layer at the leaf

---

* Since most water evaporation occurs in the immediate vicinity of the guard cells, whereas most CO_2 must enter the mesophyll cells, δ^{ias} is somewhat shorter for water vapor than for CO_2. Also, CO_2 and water vapor can enter or leave cells along the length of the intercellular air spaces, which complicates the analysis (see Cooke and Rand; and Parkhurst). In any case, g_j^{ias} generally is relatively large compared to the other conductances.

surface is again a one-dimensional diffusion process—in this case across a distance δ^{bl}. Thus, all fluxes of gases that we will consider here are moving perpendicular to the leaf surface, and thus a one-dimensional form of Fick's first law is usually appropriate. We generally express gas fluxes on the basis of unit leaf area. All conductances and resistances are also given per unit leaf area, as we mentioned above.

We should next ask whether Δc_j is an accurate representation of the force for diffusion. Also, do the coefficients g_j and r_j relating flux densities to Δc_j change greatly as other quantities vary over physiological ranges? For instance, the coefficients could depend on concentration of species j or on temperature. Finally, are other quantities—such as changes in chemical potential $\Delta \mu_j$, changes in water potential $\Delta \Psi_{wv}$, or changes in partial pressure ΔP_j—more appropriate than Δc_j for flux calculations?

We will begin by writing the equation for the chemical potential of species j in a gas phase, μ_j^{vapor}. By Equations 2.20 and VI.10 (in App. VI), μ_j^{vapor} equals $\mu_j^* + RT \ln (P_j/P_j^*)$, where P_j is the partial pressure of species j in the gas phase and P_j^* is its maximum possible partial pressure at that temperature and one atmosphere of total pressure (we are ignoring the gravitational term of μ_j^{vapor}, since it does not influence the fluxes of gases into or out of leaves). At constant T the most general way to represent the force promoting the movement of species j in the x-direction is $-\partial \mu_j^{vapor}/\partial x$ (see p. 118), which becomes $-RT \, \partial[\ln (P_j/P_j^*)]/\partial x$ for our above representation of the chemical potential, and in turn equals $-(RT/P_j) \, \partial P_j/\partial x$. For an ideal gas, $P_j V$ equals $n_j RT$. The total number of moles of species j divided by the volume (n_j/V) is the concentration of species j, c_j; hence P_j $(= n_j RT/V)$ can be replaced by $c_j RT$. Making this substitution for P_j, and using the very general flux relation given by Equation 3.6 $[J_j = u_j c_j(-\partial \mu_j/\partial x)]$, we can express the flux density of gaseous species j as follows:

$$J_j = u_j c_j \left(\frac{-\partial \mu_j^{vapor}}{\partial x} \right) = -u_j c_j \frac{RT}{P_j} \frac{\partial P_j}{\partial x}$$

$$= -u_j \frac{\partial P_j}{\partial x} = -u_j RT \frac{\partial c_j}{\partial x} \qquad (8.7)$$

$$= \frac{u_j RT}{\Delta x} \Delta c_j$$

where in the last line of the equation we have replaced the negative concentration gradient, $-\partial c_j/\partial x$, by an average concentration gradient, $\Delta c_j/\Delta x$. In essence, Equation 8.7 represents a thermodynamic derivation of Fick's first law for a gas phase (see p. 120).

Upon comparing Equation 8.7 with Equation 8.2, we see that the conductance g_j equals $u_j RT/\Delta x$. Thus, g_j is essentially independent of concentration—the very slight dependence of u_j or D_j (which equals $u_j RT$; see p. 120) on concentration can be ignored for gases. On the other hand, if we replace $-\partial \mu_j^{\text{vapor}}/\partial x$ by $\Delta \mu_j^{\text{vapor}}/\Delta x$, Equation 8.7 becomes $J_j = u_j c_j \Delta \mu_j^{\text{vapor}}/\Delta x$. The factor multiplying $\Delta \mu_j^{\text{vapor}}$ depends directly on concentration, and so in addition to knowing $\Delta \mu_j^{\text{vapor}}$, we would also need to specify c_j before calculating J_j. Moreover, c_j can vary near a leaf. Thus, $\Delta \mu_j^{\text{vapor}}$ is not used to represent the force in the relations for transpiration or photosynthesis. Since $\Delta \Psi_{wv}$ equals $\Delta \mu_{wv}/\bar{V}_w$ (Eq. 2.21), Equation 8.7 would become $J_{wv} = (u_{wv} c_{wv} \bar{V}_w/\Delta x) \Delta \Psi_{wv}$ if differences in water potential were used to represent the force. Again, the coefficient depends on concentration, and so using $\Delta \Psi_{wv}$ is not appropriate for describing transpiration. But Equation 8.7 also indicates that J_j could equal $(u_j/\Delta x) \Delta P_j$. This can lead us to an alternative formulation for flux relations in which conductance has less dependence on temperature and ambient pressure.

Instead of partial pressures, this alternative formulation uses mole fractions (see p. 73) to represent the driving force. We have already encountered a difference in mole fraction representing a driving force when we discussed water flow ($J_{V_w} = P_w \Delta N_w$, p. 97). For the case of perfect gases, which approximates situations of biological interest, Dalton's law of partial pressures indicates that the mole fraction of species j, N_j, equals P_j/P, where P is the total pressure. We can then modify Equation 8.7 as follows:

$$J_j = -u_j \frac{\partial P_j}{\partial x} = u_j \frac{\Delta P_j}{\Delta x} = \frac{u_j RT}{RT} \frac{P}{\Delta x} \frac{\Delta P_j}{P}$$

$$= \frac{D_j P}{RT \Delta x} \Delta N_j = g'_j \Delta N_j$$

(8.8)

where we have incorporated the relationship between diffusion coefficients and mobility, $D_j = u_j RT$, and the perfect gas law, $PV = nRT$, to lead to the conductance g'_j.

Let us next see how the conductance introduced in Equation 8.2 ($g_j = D_j/\Delta x$) and that introduced in Equation 8.8 [$g'_j = D_j P/(RT \Delta x) = g_j P/(RT)$] depend on environmental parameters like pressure and temperature. We have noted that, for gases, D_j depends inversely on pressure (p. 20) and approximately on temperature raised to the power 1.8 (p. 120); i.e.,

$$D_j \cong D_j^0 \left(\frac{T}{273} \right)^{1.8} \frac{P^0}{P} \tag{8.9}$$

where D_j^0 is the diffusion coefficient of species j at 273 K (0°C) and an ambient pressure of P^0 (often taken as one atmosphere).[*] As the pressure decreases, D_j increases proportionally (Eq. 8.9), and so does g_j ($g_j = D_j/\Delta x$, Eq. 8.2). On the other hand, as pressure changes, g_j' is unchanged [$g_j' = D_j P/(RT \Delta x)$, Eq. 8.8]. As temperature increases, D_j and hence g_j increase approximately as $T^{1.8}$ (Eq. 8.9), whereas g_j' only increases approximately as $T^{0.8}$. Thus, g_j' has no dependence on pressure and much less dependence on temperature than does g_j, and thus the former conductance is being increasingly used to describe gas fluxes (see Cowan 1977, Farquhar et al. 1978, or Hall).

We can illustrate the difference between the two types of conductance by comparing a leaf with *constant* anatomical properties under various environmental conditions. If a leaf at sea level and 10°C were heated to 40°C, g_j would increase 20% even though the leaf's anatomical properties are unchanged. If it were then transferred to 2 000 m, g_j would increase 26% more. On the other hand, heating from 10°C to 40°C would increase g_j' only 8%, and transferring to 2 000 m would not affect g_j'.

Since ΔN_j is dimensionless, g_j' has the same units as J_j, e.g., mol m^{-2} s^{-1}. At 1 atmosphere and 20°C, P/RT equals 41.6 mol m^{-3}. Thus, a conductance g_j of 1 mm s^{-1} then corresponds to a conductance g_j' of (1 mm s^{-1})(41.6 mol m^{-3}), or 41.6 mmol m^{-2} s^{-1}. In our ensuing discussion of transpiration and photosynthesis, we will give examples of the use of resistances and both forms of conductance (see Table 8.1).

WATER VAPOR FLUXES ACCOMPANYING TRANSPIRATION

Here we will describe the flux of water vapor out of a leaf during transpiration in terms of the conductances and resistances involved. We will represent the conductances and resistances using symbols (namely, —∿∿∿—) borrowed from electrical circuit diagrams. Typical values for the components will be given

---

[*] Experimental errors occur in the measurement of D_j and its temperature dependence (see Fuller et al., Ch. 3 reference; Marrero and Mason; and Reid et al.). We will use $D_{wv} = 2.13 \times 10^{-5}$ m^2 s^{-1} and $D_{CO_2} = 1.33 \times 10^{-5}$ m^2 s^{-1} at 0°C and one atmosphere (0.1013 MPa); values at other temperatures are given in Appendix II.

along with the differences in water vapor concentration and mole fraction across them. Our analysis of water vapor fluxes will indicate the important control of transpiration exercised by the stomata.

Conductance and Resistance Network

Water vapor that evaporates from cell walls of mesophyll cells or the inner side of leaf epidermal cells diffuses through the intercellular air spaces to the stomata and then into the outside air. We have already introduced the four components involved—two are strictly anatomical (intercellular air spaces and cuticle), one depends on anatomy and yet responds to metabolic as well as environmental factors (stomata), and one depends on leaf morphology and wind speed (boundary layer). Figure 8.3 summarizes the symbols and uses an electrical circuit analogy to illustrate how the four components can be arranged.

The conductance g_{wv}^{ias} and the resistance r_{wv}^{ias} include all parts of the pathway from the site of water evaporation to the leaf epidermis. The water can evaporate at the air-water interface for mesophyll cells, the inner side of epidermal cells (including guard cells), and even from cells of the vascular tissue in the leaf before diffusing in the generally tortuous pathways of the intercellular air spaces. However, most of the water appears to come from the guard cells or the immediately adjoining subsidiary cells, as we indicated above. Before evaporation, the water generally has to cross a thin waxy layer on the cell walls of the cells within the leaf, this layer often being as thick as 0.1 μm (see Esau). After crossing the waxy layer, the water vapor diffuses through the intercellular air spaces and then through the stomata (conductance $= g_{wv}^{st}$, resistance $= r_{wv}^{st}$) to reach the boundary layer adjacent to the leaf surface. Alternatively, water in the cell walls of leaf cells might move as a liquid to the cell walls on the cuticle side of epidermal cells, where it could evaporate and the vapor then diffuse across the cuticle (or water may move across the cuticle as a liquid) before reaching the boundary layer adjacent to the leaf. The pathway for such cuticular transpiration (conductance $= g_{wv}^{c}$, resistance $= r_{wv}^{c}$) is in parallel with the transpiration that occurs through the stomatal pores. For simplicity, we have not included in Figure 8.3 the intermediate case of water evaporating from the cell walls of mesophyll cells and then moving through the intercellular air spaces before crossing the waxy cuticle (the resistance of this pathway is indistinguishable from r_{wv}^{c}, since $r_{wv}^{c} \gg r_{wv}^{ias}$). The final component encountered by the diffusing water vapor is that of the boundary layer just outside the leaf (conductance $= g_{wv}^{bl}$, resistance $= r_{wv}^{bl}$).

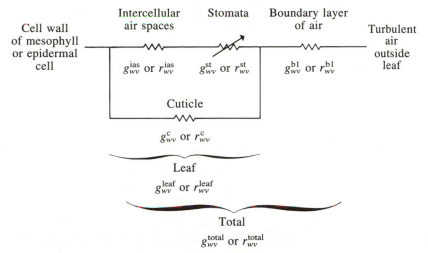

Figure 8.3
Conductances and resistances involved in water vapor flow accompanying transpiration, as arranged into an electrical circuit. The symbol for the stomatal component indicates that it is variable.

We will first analyze the electrical circuit presented in Figure 8.3 in terms of resistances and then in terms of conductances. We begin by noting that resistances r_{wv}^{ias} and r_{wv}^{st} occur in series. The total resistance of a group of resistors in series, r_{series}, is simply the sum of the individual resistances $\left(r_{series} = \sum_i r_i, \text{ where } r_i \text{ is the resistance of series resistor } i \right)$. Thus, the resistance of the pathway from the site of water evaporation across the intercellular air spaces and through the stomata is $r_{wv}^{ias} + r_{wv}^{st}$. This resistance is in parallel with r_{wv}^c, the cuticular resistance. The reciprocal of the total resistance of a group of resistors in parallel is the sum of the reciprocals of the individual resistances $\left(1/r_{parallel} = \sum_i 1/r_i, \text{ where } r_i \text{ is the resistance of parallel resistor } i \right)$. For two resistors in parallel, $r_{parallel}$ is equal to $r_1 r_2/(r_1 + r_2)$. Since $(r_{wv}^{ias} + r_{wv}^{st})$ and r_{wv}^c are in parallel, it is often convenient to express their combined resistance as the leaf resistance for water vapor, r_{wv}^{leaf}:

$$r_{wv}^{leaf} = \frac{(r_{wv}^{ias} + r_{wv}^{st})(r_{wv}^c)}{r_{wv}^{ias} + r_{wv}^{st} + r_{wv}^c} \qquad (8.10)$$

When the stomata are open, the cuticular resistance (r_{wv}^c) is generally much larger than the resistance in parallel with it, $r_{wv}^{ias} + r_{wv}^{st}$. In that case, the leaf resistance as given by Equation 8.10 is approximately equal to $r_{wv}^{ias} + r_{wv}^{st}$.

The leaf resistance is in series with that of the boundary layer, r_{wv}^{bl}. Thus the total resistance for the flow of water vapor from the site of evaporation to the turbulent air surrounding the leaf (r_{wv}^{total}) is $r_{wv}^{leaf} + r_{wv}^{bl}$ (see Fig. 8.3). We now must face the complication created by the presence of two leaf surfaces representing parallel pathways for the diffusion of water vapor from the interior of the leaf to the surrounding turbulent air. We will represent the leaf resistance for the pathway through the upper surface by $r_{wv}^{leaf_u}$ and that for the lower surface by $r_{wv}^{leaf_l}$. Each of these resistances is in series with that of an air boundary layer—$r_{wv}^{bl_u}$ and $r_{wv}^{bl_l}$ for the upper and the lower surfaces, respectively. The parallel arrangement of the pathways through the upper and lower surfaces of a leaf leads to the following comprehensive expression for the total resistance to the diffusion of water vapor from a leaf:

$$r_{wv}^{total} = \frac{(r_{wv}^{leaf_u} + r_{wv}^{bl_u})(r_{wv}^{leaf_l} + r_{wv}^{bl_l})}{r_{wv}^{leaf_u} + r_{wv}^{bl_u} + r_{wv}^{leaf_l} + r_{wv}^{bl_l}} \tag{8.11}$$

Now let us reconsider the analysis of the electrical circuit using conductances. For a group of conductances in series, the reciprocal of the total conductance ($1/g_{series}$) is the sum of the reciprocals of the individual conductances $\left(1/g_{series} = \sum_i 1/g_i\right)$, whereas the conductance of a group of conductances in parallel is the sum of the individual conductances $\left(g_{parallel} = \sum_i g_i\right)$. Thus, the water vapor conductance of the intercellular air spaces and stomata in series is $g_{wv}^{ias} g_{wv}^{st}/(g_{wv}^{ias} + g_{wv}^{st})$, and the water vapor conductance of the leaf (see Eq. 8.10 for resistance) is

$$g_{wv}^{leaf} = \frac{g_{wv}^{ias} g_{wv}^{st}}{g_{wv}^{ias} + g_{wv}^{st}} + g_{wv}^c \tag{8.12}$$

Next, we note that g_{wv}^{leaf} is in series with a boundary layer conductance g_{wv}^{bl} and that the two sides of a leaf act as parallel conductances for moving water vapor from the interior of a leaf. We therefore obtain the following expression for the total conductance of a leaf with air boundary layers on each side:

$$g_{wv}^{total} = \frac{g_{wv}^{leaf_u} g_{wv}^{bl_u}}{g_{wv}^{leaf_u} + g_{wv}^{bl_u}} + \frac{g_{wv}^{leaf_l} g_{wv}^{bl_l}}{g_{wv}^{leaf_l} + g_{wv}^{bl_l}} \tag{8.13}$$

Values of Conductances

We generally assume that the average thickness of the boundary layer is the same on the two sides of a leaf, in which case $g_{wv}^{bl_u}$ equals $g_{wv}^{bl_l}$. However, this does not simplify our analysis of gas fluxes very much, except in special cases. Usually the boundary layer conductance is much greater than that of the leaf. If, in addition, $g_{wv}^{leaf_l}$ is much greater than $g_{wv}^{leaf_u}$, as can occur for leaves with stomata primarily on the lower (abaxial) surface, then the diffusion of water vapor is mainly out through the lower surface of the leaf. This is often the case for deciduous trees. The total conductance for water vapor diffusion from the sites of evaporation inside the leaf to the turbulent air surrounding the leaf (g_{wv}^{total}) is then approximately $g_{wv}^{leaf_l} g_{wv}^{bl_l}/(g_{wv}^{leaf_l} + g_{wv}^{bl_l})$ (see Eq. 8.13). For the symmetrical case where $g_{wv}^{leaf_l}$ equals $g_{wv}^{leaf_u}$ and the boundary layers are of equal thickness (hence, $g_{wv}^{bl_l} = g_{wv}^{bl_u}$), the conductance is the same through either surface; g_{wv}^{total} is then double that through either surface, e.g., $2g_{wv}^{leaf_l} g_{wv}^{bl_l}/(g_{wv}^{leaf_l} + g_{wv}^{bl_l})$, by Equation 8.13. This can occur for certain monocots—e.g., certain grasses, oats, barley, wheat, corn—which can have a fairly equal stomatal frequency on the two leaf surfaces. For many plants, two to four times more stomata occur on the lower surface of the leaves than on the upper one, and so A^{st}/A is considerably larger for the lower surface. In such cases it may be necessary to use the rather cumbersome expression for g_{wv}^{total} given by Equation 8.13.

Let us next examine some representative values for the conductances encountered by water vapor as it diffuses out of leaves. For crop plants such as beet, spinach, tomato, and pea, g_{wv}^{ias} is generally 20 to 100 mm s^{-1} (800 to 4 000 mmol m^{-2} s^{-1}), and for open stomata g_{wv}^{st} usually ranges from 5 to 20 mm s^{-1} (200 to 800 mmol m^{-2} s^{-1}). Both of these conductances tend to be somewhat lower for many other agricultural crops, but the maximum $g_{wv}^{leaf_l}$ is still 2 to 10 mm s^{-1} (80 to 400 mmol m^{-2} s^{-1}) for most cultivated plants (equivalent to resistances of 100 to 500 s m^{-1}, or 1 to 5 s cm^{-1}). Compared to crop plants, the maximum g_{wv}^{leaf} is usually somewhat lower for leaves of deciduous trees and conifer needles (see Table 8.1). Of course, as the stomata close, g_{wv}^{st} and hence g_{wv}^{leaf} decrease accordingly. The minimum value for leaf water vapor conductance occurs for fully closed stomata and essentially equals the cuticular conductance (see Eq. 8.12). Although measured values of g_{wv}^{c} vary considerably, in part because of the difficulty in telling when stomata are fully closed, it may be as low as 0.1 mm s^{-1} (4 mmol m^{-2} s^{-1}) for cultivated plants, 2-fold lower for trees, and 10-fold lower for certain xerophytes (Table 8.1).

Effective Lengths and Resistance

Let us next examine a simplified expression that is often adequate for describing the total resistance to diffusion of water vapor from the sites of evaporation in cell walls to the turbulent air surrounding a leaf. We will restrict our attention to when nearly all the water vapor diffusing from the leaf moves out across the lower epidermis and when cuticular transpiration is negligible. By Equations 8.10 and 8.11, the total resistance then is

$$r_{wv}^{total} \cong r_{wv}^{leaf_1} + r_{wv}^{bl_1}$$

$$= r_{wv}^{ias} + r_{wv}^{st_1} + r_{wv}^{bl_1}$$

$$= \frac{1}{D_{wv}} \left(\delta^{ias} + \frac{\delta^{st} + r^{st}}{na^{st}} + \delta^{bl} \right) \tag{8.14}$$

where Equations 8.3, 8.5, and 8.6 have been used to obtain the bottom line (see Penman and Schofield).

When diffusion is described by a one-dimensional form of Fick's first law, Equation 8.14 indicates that each part of the pathway contributes its own effective length influencing the movement of water vapor out of the leaf. For instance, the effective length of the boundary layer is its thickness, δ^{bl}. The distance δ^{ias} includes the effective depth of the cell wall pores, the effective thickness of the waxy layer on the mesophyll cells, and the constriction on the region available for diffusion of gases within a leaf caused by the presence of mesophyll cells (see p. 399). A similar constricting effect greatly increases the effective length of the stomatal pores over the value of $\delta^{st} + r^{st}$. Specifically, Equation 8.14 indicates that the effective length of the stomatal pores is $(\delta^{st} + r^{st})/(na^{st})$, which is considerably greater than $\delta^{st} + r^{st}$ since na^{st} is much less than one (recall that $na^{st} = A^{st}/A$, p. 396, so na^{st} is a dimensionless number indicating the *fraction* of the surface area occupied by stomatal pores). The large effective length of the stomatal pores caused by the factor $1/na^{st}$ is an alternative way of viewing the constricting effect of the stomata.

If we need to incorporate yet another series resistance for gaseous diffusion, we can include the effective length of the additional part of the pathway within the parentheses in Equation 8.14. For instance, the stomata in some xerophytes and conifers occur sunken beneath the leaf surface in cavities or crypts. We can estimate the effective length of this additional part of the pathway by using appropriate geometrical approximations. The effective length divided by the diffusion coefficient gives us the resistance directly.

Also, many leaves are covered by epidermal "hairs," which can be unicellular projections of the epidermal cells or multicellular appendages with the cells often occurring in a row. Such *trichomes* cause an additional air layer to be held next to the leaf surface. The thickness of this layer may equal the average distance that the hairs project perpendicularly away from the leaf surface (more than 1 mm for some leaves). This length can be added within the parenthesis in Equation 8.14 to calculate the overall resistance. Also, the extra resistance for water vapor diffusion caused by the hairs can be separately calculated as the thickness of the additional air layer divided by D_{wv}. Some water may also evaporate from the surfaces of the epidermal hairs; this evaporation can lead to a substantial complication in the analysis of water vapor fluxes for leaves with numerous trichomes.

Leaf Water Vapor Concentrations and Mole Fractions

As we can see from relations like Equation 8.2 ($J_j = g_j \Delta c_j = \Delta c_j / r_j$), the conductances or resistances of the various parts of the pathway determine the drops in concentration across each component when the flux density is constant. Here we will apply this condition to a specific consideration of water vapor concentrations and mole fractions in a leaf. We will also discuss the important effect of temperature on the water vapor content of air.

Let us represent the difference in water vapor concentration across the intercellular air spaces by Δc_{wv}^{ias}, that across the stomatal pores by Δc_{wv}^{st}, and that across the boundary layer adjacent to the leaf surface by Δc_{wv}^{bl}. Then the overall drop in water vapor concentration from the cell walls where the water evaporates to the turbulent air surrounding a leaf, Δc_{wv}^{total}, is

$$\Delta c_{wv}^{total} = c_{wv}^{e} - c_{wv}^{ta}$$

$$= \Delta c_{wv}^{ias} + \Delta c_{wv}^{st} + \Delta c_{wv}^{bl} \qquad (8.15)$$

where c_{wv}^{e} is the concentration of water vapor at the evaporation sites, presumably in the cell wall pores, and c_{wv}^{ta} is its value in the turbulent air surrounding the leaf. We also note that $\Delta c_{wv}^{ias} + \Delta c_{wv}^{st}$ is equal to Δc_{wv}^{c}, the difference in water vapor concentration from the cell walls of epidermal cells across the cuticle to the leaf surface, since the endpoints of the pathway are the same in each case (Fig. 8.3).

The rate of water vapor diffusion per unit leaf area, J_{wv}, is equal to the difference in water vapor concentration multiplied by the conductance across which Δc_{wv} occurs ($J_j = g_j \Delta c_j$, Eq. 8.2). In the steady state, when the flux density of water vapor and the conductance of each component are constant, this relation holds both for the overall pathway and any individual segment of it. Some water evaporates from the cell walls of mesophyll cells along the pathway within the leaf. Also, much of the water lost from a leaf evaporates from the inner sides of epidermal and guard cells. Consequently, J_{wv} is actually not spatially constant in the intercellular air spaces. For simplicity, however, we generally assume that J_{wv} is unchanging from the mesophyll cell walls all the way out to the turbulent air outside the leaf. When water vapor moves out mainly across the lower epidermis of the leaf, and when cuticular transpiration is negligible, we obtain the following relations in the steady state:

$$J_{wv} = g_{wv}^{total} \Delta c_{wv}^{total} = g_{wv}^{ias} \Delta c_{wv}^{ias}$$

$$= g_{wv}^{st} \Delta c_{wv}^{st} = g_{wv}^{bl} \Delta c_{wv}^{bl} \tag{8.16}$$

Equation 8.16 indicates that the drop in water vapor concentration across a particular component is inversely proportional to its conductance when J_{wv} is constant (Δc_{wv} is then directly proportional to the resistance, i.e., $J_{wv} = \Delta c_{wv}/r_{wv}$). Since g_{wv}^{ias} and g_{wv}^{bl} are generally larger than g_{wv}^{st} when the stomata are open (Table 8.1), the largest Δc_{wv} then occurs across the stomata by Equation 8.16.

The quantity c_{wv} is sometimes referred to as the *absolute* humidity. Its value in the turbulent air generally does not change very much during the day unless there is precipitation or other marked changes in the weather, while the *relative* humidity can vary greatly as the air temperature changes. For instance, air that is saturated with water vapor at 20°C (100% relative humidity) drops to 50% relative humidity when heated at constant pressure to 32°C. Most data for c_{wv} are expressed as mass of water/unit volume of air, and therefore we will use $g\,m^{-3}$ as one of our units for water vapor concentration. A unit for c_{wv} that is often more appropriate is $mol\,m^{-3}$, and we will use it as well (see Fig. 8.4).

Equations 8.15 and 8.16 can be adapted to water vapor expressed as a mole fraction, N_{wv}, simply by replacing c with N throughout. For instance, N_{wv}^{total} equals $N_{wv}^{e} - N_{wv}^{ta}$, and J_{wv} equals $g_{wv}^{total} \Delta N_{wv}^{total}$ (we will here use the same symbol for conductance in either system, the units being clear by whether changes in concentration or mole fraction represent the driving force). Water vapor is generally the third most prevalent gas in air, although its mole fraction is relatively low compared to levels for O_2 or N_2. Its saturation value is quite temperature dependent; e.g., N_{wv}^* equals 0.0231 at 20°C and 0.0420

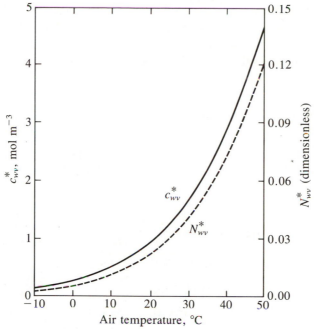

Figure 8.4

Variation in saturation values of water vapor concentration (c_{wv}^*) and mole fraction (N_{wv}^*) with temperature. Ambient pressure (needed to determine N_{wv}^*) was assumed to be one atmosphere (0.1013 MPa). Values are summarized in Appendix II.

at 30°C (Fig. 8.4). In fact, the water vapor content of air at saturation (represented by c_{wv}^*, N_{wv}^*, or P_{wv}^*) increases nearly exponentially with temperature.[*] Appendix II lists c_{wv}^*, N_{wv}^*, and P_{wv}^* at various temperatures.

Examples of Water Vapor Levels in a Leaf

We will now consider the actual relative humidities, concentrations, and mole fractions of water vapor in different parts of the transpiration pathway associated with a leaf. We will suppose that the leaf temperature is 25°C

---

[*] It may be helpful for understanding this temperature dependence to note that the fraction of molecules in the upper part of the Boltzmann distribution (p. 140) which have a high enough energy to escape from the liquid increases exponentially with temperature.

and that the turbulent air surrounding the leaf is at 20°C with a relative humidity of 50%, which corresponds to a c_{wv}^{ta} of 0.48 mol m^{-3} and an N_{wv}^{ta} of 0.0115 (Fig. 8.5).

Let us begin by assuming that the water vapor in the pores of the cell walls of mesophyll cells is in equilibrium with the water in the cell wall and that $\Psi^{cell\ wall}$ is -1.0 MPa. Using Equation 2.21 and omitting the gravity term, we can calculate the corresponding % relative humidity (RH) at 25°C as follows:

$$\Psi^{cell\ wall} = \Psi_{wv}^{e} = \frac{RT}{\bar{V}_w} \ln\left(\frac{RH}{100}\right)$$

or

$$RH = 100e^{\bar{V}_w \Psi^{cell\ wall}/RT}$$

$$= 100e^{(-1.0\ MPa/137.3\ MPa)} = 99.3\%$$

Thus, the air in the pores of the mesophyll cell walls is nearly saturated with water vapor. From Appendix II, c_{wv}^{*} equals 1.28 mol m^{-3}, and N_{wv}^{*} equals 0.0313 at 25°C; therefore 99.3% relative humidity corresponds to a c_{wv}^{e} of (0.993)(1.28 mol m^{-3}), or 1.27 mol m^{-3}, and an N_{wv}^{e} of 0.0311. Hence, the difference in water vapor concentration from the mesophyll cell walls to the turbulent air is 1.27 mol m^{-3} $-$ 0.48 mol m^{-3}, or 0.79 mol m^{-3} ($\Delta c_{wv}^{total} = c_{wv}^{e} - c_{wv}^{ta}$, Eq. 8.15), and $N_{wv}^{e} - N_{wv}^{ta}$ is 0.0311 $-$ 0.0115, or 0.0196 (see Fig. 8.5).*

Equation 8.15 indicates that Δc_{wv}^{total} is the sum of three components, Δc_{wv}^{ias}, Δc_{wv}^{st}, and Δc_{wv}^{bl}. Likewise, ΔN_{wv}^{total} equals $\Delta N_{wv}^{ias} + \Delta N_{wv}^{st} + \Delta N_{wv}^{bl}$. The magnitude of each of these differences in water vapor level is inversely proportional to the conductance across which the drop occurs—e.g., $g_{wv}^{x}\Delta N_{wv}^{x} = g_{wv}^{total}\Delta N_{wv}^{total}$—and so ΔN_{wv}^{x} is larger where g_{wv}^{x} is smaller, where x refers to any series component in the pathway (see Eq. 8.16). Values of

---

* The condition $N_{wv}^{e} > N_{wv}^{ta}$ necessary for transpirational efflux of water vapor from a leaf leads to $N_{air}^{ta} > N_{air}^{e}$, where N_{air} represents the mole fraction of everything but water vapor—i.e., $N_{wv} + N_{air} = 1$ (there is a relatively small ΔN_{CO_2} resulting from CO_2 uptake during photosynthesis, which is mostly compensated for by a small ΔN_{O_2} acting in the opposite direction). Thus we expect a diffusion of air (mainly N_2) into a leaf, which can lead to slightly higher pressures in leaves (generally 0.1 to 1 kPa higher) compared to the ambient pressure outside. This internally elevated pressure can lead to mass flow within a plant, such as along the stems of the yellow waterlily, *Nuphar luteum*, where a ΔN_{wv} of 0.01 is accompanied by an air pressure 0.2 kPa higher inside a young leaf (see Dacey).

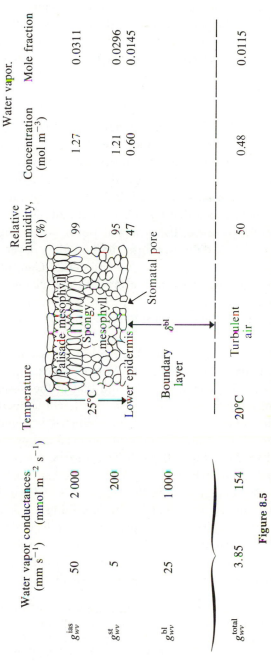

Figure 8.5
Representative values of quantities influencing the diffusion of water vapor out of a leaf. Conductances are given for the indicated parts of the pathway assuming water moves out only through the lower surface and ignoring cuticular transpiration.

c_{wv}, N_{wv}, and relative humidity are given in Figure 8.5, which also indicates specific values of the conductances as well as the overall series conductance for the diffusion of water vapor from the sites of evaporation to the turbulent air ($1/g_{\text{series}} = \sum 1/g_i$). The largest drop in water vapor level occurs across the stomatal pores, since they have the smallest conductance in the present case. For instance, $\Delta N_{wv}^{\text{st}}$ equals $g_{wv}^{\text{total}} \, \Delta N_{wv}^{\text{total}}/g_{wv}^{\text{st}}$, which is (154 mmol m^{-2} s^{-1})(0.0196)/(200 mmol m^{-2} s^{-1}), or 0.0151 (see Fig. 8.5).

We note that a small drop in relative humidity, here from 99% in the cell walls of mesophyll cells to 95% at the inner side of the stomata, is necessary for the diffusion of water vapor across the intercellular air spaces (the greater the fraction of water evaporating from near the guard cells, the smaller the drop across the intercellular air spaces). There is a fairly large humidity drop across the stomatal pores, such that the relative humidity is 47% at the leaf surface (Fig. 8.5). After crossing the stomatal pores, water vapor moves energetically *downhill* as it diffuses across the boundary layer from 47% relative humidity at 25°C to 50% relative humidity at 20°C in the turbulent air surrounding the leaf (Fig. 8.5). Since a temperature change is involved in this part of the pathway, we must express the driving force for the diffusion of water vapor in terms of the difference in concentration or mole fraction, not the change in relative humidity.

In this text we will generally assume that the leaf has a uniform temperature.* For instance, the leaf temperature is 25°C in the example presented, including the air in the intercellular spaces and that in the stomatal pores. The turbulent air outside the leaf generally has a different temperature, here 20°C. Fick's first law (e.g., $J_j = D_j \Delta c_j/\Delta x$, Eq. 8.2) strictly applies only to isothermal situations. For instance, D_j depends on the absolute temperature (D_j is proportional to $T^{1.8}$; Eq. 8.9), so $\Delta(D_j c_j)$ may be nonzero and lead to a flux even when Δc_j is zero. Fortunately, even when there are temperature differences between leaves and the turbulent air, Fick's first law in the form of Equation 8.2 generally proves adequate for describing the fluxes of H_2O and CO_2, our primary concern in this chapter. However, flux relations based on differences in mole fraction (e.g., Eq. 8.8), which have a much lower dependence on temperature for their conductance, are preferred when there is a temperature difference from leaf to air, as is usually the case for plants.

---

* If most of the water evaporation occurs in the immediate vicinity of a guard cell, temperature differences of 0.1°C to 0.3°C could develop locally (see Cowan 1977, and Tyree and Yianoulis). Differences in boundary layer thickness near the edge of a leaf (see p. 358) can create even larger temperature differences. Indeed, temperature differences of a few degrees C can develop across the width of a moderate-sized leaf at a wind speed of 1 m s^{-1}, in part reflecting spatial stomatal variation (see Clark and Wigney).

Fluxes of Water Vapor

Based on the quantities indicated in Figure 8.5, we can readily calculate the flux density of water vapor moving out of the leaf. Specifically, J_{wv} equals $g_{wv}^{\text{total}} \, \Delta c_{wv}^{\text{total}}$ or $g_{wv}^{\text{total}} \, \Delta N_{wv}^{\text{total}}$, e.g.,

$$J_{wv} = (3.85 \times 10^{-3} \text{ m s}^{-1})(1.27 \text{ mol m}^{-3} - 0.48 \text{ mol m}^{-3})$$

$$= 3.0 \times 10^{-3} \text{ mol m}^{-2} \text{ s}^{-1} = 3.0 \text{ mmol m}^{-2} \text{ s}^{-1}$$

If we express $\Delta c_{wv}^{\text{total}}$ in mass per unit volume, then J_{wv} is the rate of movement of mass per unit leaf area; e.g., $\Delta c_{wv}^{\text{total}}$ is here 14 g m^{-3}, and so J_{wv} is 0.054 g m^{-2} s^{-1}. Based on the decrease in mole fraction of water vapor, J_{wv} is $(154 \text{ mmol m}^{-2} \text{ s}^{-1})(0.0196)$, or 3.0 mmol m^{-2} s^{-1}.

For simplicity, we have so far been considering the movement of water vapor across only the lower surface of the leaf, and we have ignored cuticular transpiration. Cuticular transpiration is generally small compared with the transpiration through open stomata that is in parallel with it; i.e., the drop in water vapor level across the cuticle is essentially the same as the drop across the stomata, while the stomatal conductance for open stomata is much greater than the cuticular conductance (see Table 8.1). We can add the cuticular transpiration to that through the stomata to get the total transpiration through one side of a leaf. To obtain the overall rate of water vapor diffusing out of both sides of a leaf, we could scale up the J_{wv} calculated for the lower surface by an appropriate factor—the reciprocal of the fraction of transpiration through the lower surface. For instance, about 70% of the water loss in transpiration might be through the lower surface for a representative mesophyte. The J_{wv} of 3.0 mmol m^{-2} s^{-1} which we calculated in the previous paragraph would then become $(1/0.70)(3.0 \text{ mmol m}^{-2} \text{ s}^{-1})$, or 4.3 mmol m^{-2} s^{-1}, when both leaf surfaces are considered; i.e., $J_{wv}^{l} = 0.70 \times$ the overall J_{wv} implies that the overall $J_{wv} = J_{wv}^{l}/0.70$. Alternatively, we could use the actual g_{wv}^{total} as given by Equation 8.13, which considers the two leaf surfaces in parallel, or we could measure g_{wv}^{total} experimentally. This conductance times Δc_{wv} or ΔN_{wv}, as appropriate, would give J_{wv} through both surfaces, but expressed per unit area of one side of the leaf; i.e., the flux density would still be expressed on the same basis as the other fluxes in this chapter. For many cultivated plants and other mesophytes under the above conditions ($T^{\text{leaf}} = 25°C$, $T^{\text{ta}} = 20°C$, relative humidity$^{\text{ta}} = 50\%$), J_{wv} for open stomata is from 1 to 5 mmol m^{-2} s^{-1}. Therefore our previous example corresponds to a slightly above-averge transpiration rate. Unfortunately, many systems of units are used for the rates of water vapor leaving unit area of leaves. The conversion factors for the more common systems of expressing transpiration are summarized in Table 8.2.

Control of Transpiration

Let us now reconsider the values of the various conductances affecting the diffusion of water vapor through the intercellular air spaces, out the stomata, and across the boundary layer at the leaf surface. Usually g_{wv}^{ias} is relatively large, and g_{wv}^{bl} is rarely less than 10 mm s^{-1}, while g_{wv}^{st} is generally less than 10 mm s^{-1} and decreases as the stomata close. Consequently, control for limiting transpiration usually rests with the stomata, not with the boundary layer or the intercellular air spaces. When g_{wv}^{st} is at least a few times smaller than g_{wv}^{bl}, as generally occurs under field conditions, moderate changes in the ambient wind speed have relatively little effect on g_{wv}^{total}. However, the boundary layer conductance can be the main determinant of J_{wv} for the fruiting bodies of Basidiomycetes (fungi), which have no stomata. For instance, near the ground, where the wind speed is relatively low (generally below 0.2 m s^{-1}) and the fruiting bodies occur, g_{wv}^{bl} exerts the main control on transpiration for *Lycoperdon perlatum* and *Scleroderma australe* (see Nobel 1975, Ch. 7 reference).

Stomata tend to close as a leaf wilts. Assuming that g_{wv}^{st} decreases 20-fold, and using values presented in Figure 8.5, we note that g_{wv}^{total} would decrease about 15-fold, e.g., from 154 mmol m^{-2} s^{-1} for a g_{wv}^{st} of 200 mmol m^{-2} s^{-1} to 9.9 mmol m^{-2} s^{-1} for a g_{wv}^{st} of 10 mmol m^{-2} s^{-1} (this latter conductance is 100 times lower than g_{wv}^{bl} and 200 times lower than g_{wv}^{ias} in the present case). The decrease in g_{wv}^{total} causes transpiration to decrease to only about 6% of its former value—such stomatal closure is a common response of plants to water stress. When the stomata close tightly, only cuticular transpiration remains, which can have a conductance of 1 to 10 mmol m^{-2} s^{-1} (Table 8.1). In that case, cuticular transpiration accounts for all of the loss of water vapor from the leaves.

CO_2 CONDUCTANCES AND RESISTANCES

We will next consider the main function of a leaf, photosynthesis, in terms of the conductances and resistances encountered by CO_2 as it diffuses from the turbulent air, across the boundary layers next to a leaf, through the

Table 8.2

Conversion factors for some of the more common units used in expressing transpiration, CO$_2$ concentrations, and photosynthesis. Values were determined using quantities in Appendices II and III. To convert from one set to another, a quantity expressed in the units in the left column should be multiplied by the factor in the column of the desired units.

Transpiration units	mmol m^{-2} s^{-1}	mg m^{-2} s^{-1}
μmol H$_2$O cm^{-2} s^{-1}	10	180.2
mol H$_2$O m^{-2} hour^{-1}	0.278	5.01
μg H$_2$O cm^{-2} s^{-1}	0.555	10
μg H$_2$O cm^{-2} min^{-1}	9.25×10^{-3}	0.1667
mg H$_2$O dm^{-2} min^{-1}	9.25×10^{-2}	1.667
g H$_2$O dm^{-2} hour^{-1}	1.542	27.8
kg H$_2$O m^{-2} hour^{-1}	15.42	278

CO$_2$ concentrations	mmol m^{-3}	mg m^{-3}	Pa	mole fraction $\times 10^6$ (ppm by volume) (μlitre litre^{-1}) (μbar bar^{-1})
nmol CO$_2$ cm^{-3} (mmol CO$_2$ m^{-3}, μM)	1	44.0	2.44	24.4
ng CO$_2$ cm^{-3} (mg CO$_2$ m^{-3})	0.0227	1	0.0554	0.554
μlitre CO$_2$ litre^{-1}* (ppm CO$_2$ by volume)	0.0410	1.806	0.1	1
μbar*	0.0410	1.806	0.1	1
Pa*	0.410	18.06	1	10

Photosynthesis units	μmol CO$_2$ m^{-2} s^{-1}
ng CO$_2$ cm^{-2} s^{-1}	0.227
nmol cm^{-2} s^{-1}	10
mg CO$_2$ m^{-2} s^{-1}	22.7
mg CO$_2$ dm^{-2} hour^{-1} (kg CO$_2$ hectare^{-1} hour^{-1})	0.631
kg carbohydrate hectare^{-1} hour^{-1}	0.92
mm^3 CO$_2$ cm^{-2} hour^{-1}	0.114

* To convert a volume/volume number like ppm to a mole/volume or a pressure unit, or vice versa, we need to know the temperature and the pressure; an air temperature of 20°C and a pressure of 0.1 MPa (1 bar) were used here. To adjust for other temperatures and pressures, the perfect gas law ($PV = nRT$) must be employed; e.g., 1 ppm of CO$_2$ at temperature T_x in K and pressure P_x in MPa is equal to 0.410 $(293.15/T_x)(P_x)$ mmol m^{-3}.

stomata, across the intercellular air spaces, into the mesophyll cells, and eventually into the chloroplasts. The situation is obviously more complicated than the analogous one of the movement of water vapor during transpiration, since CO_2 must not only diffuse across the same components encountered by water vapor moving in the opposite direction, but must also cross the cell wall of a mesophyll cell, the plasmalemma, part of the cytosol, the membranes surrounding a chloroplast, and some of the chloroplast stroma. We will specifically indicate the resistance of each component, resistances being somewhat easier to present than conductances for the series of components involved in the pathway for CO_2 movement. We will summarize these resistances, as well as both forms of conductances, in a table.

Resistance and Conductance Network

Figure 8.6 illustrates the various conductances and resistances affecting CO_2 as it diffuses from the turbulent air surrounding a leaf up to the sites in the chloroplasts where it is incorporated into photosynthetic products. For simplicity, we will initially restrict our attention to the diffusion of CO_2 into the leaf across the lower epidermis only. When the parallel pathways through the upper and lower surfaces of a leaf are both important, we can readily modify our equations to handle the reduction in resistance, or increase in conductance, encountered between the turbulent air surrounding the leaf and the cell walls of its mesophyll cells (see Eqs. 8.11 and 8.13). We will also ignore the cuticular path for CO_2 entry into the leaf for the same sort of reason we neglected this part of the pathway when discussing transpiration—namely, $r^c_{CO_2}$ is generally considerably greater than $r^{ias}_{CO_2} + r^{st}_{CO_2}$, the resistance in parallel with it.

The first three resistances encountered by CO_2 entering a leaf through the lower epidermis ($r^{a_l}_{CO_2}$, $r^{st_l}_{CO_2}$, and $r^{ias}_{CO_2}$) have analogs in the case of transpiration. We can thus transfer Equation 8.14 to our present discussion, changing only the subscripts:

$$r^{leaf_l}_{CO_2} + r^{bl_l}_{CO_2} = \frac{1}{D_{CO_2}}\left(\delta^{ias} + \frac{\delta^{st} + r^{st}}{na^{st}} + \delta^{bl}\right) \tag{8.17}$$

where $r^{leaf_l}_{CO_2}$ is the resistance of the intercellular air spaces plus the stomatal pores of the lower leaf surface to the diffusion of CO_2. We also note that $r^{leaf_l}_{CO_2} + r^{bl_l}_{CO_2}$ is analogous to r^{total}_{wv}, the *total* resistance encountered by water vapor (see Fig. 8.6).

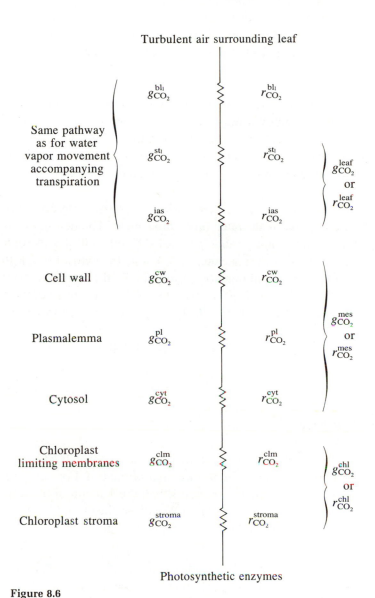

Figure 8.6

Principal conductances and resistances involved in the movement of CO_2 from the turbulent air surrounding a leaf, across the lower epidermis, up to the enzymes involved in the fixation of CO_2 into photosynthetic products in the chloroplasts of mesophyll cells.

The diffusion coefficient for CO_2, D_{CO_2}, is 1.51×10^{-5} m^2 s^{-1} in air at 20°C (App. II). This is smaller than D_{wv} (2.42×10^{-5} m^2 s^{-1}), because CO_2 molecules are heavier and thus diffuse more slowly than do H_2O molecules (see p. 18). By Equations 8.14 and 8.17, $(r_{CO_2}^{leaf_1} + r_{CO_2}^{bl_1})/r_{wv}^{total}$ equals D_{wv}/D_{CO_2}, which is $(2.42 \times 10^{-5}$ m^2 s$^{-1})/(1.51 \times 10^{-5}$ m^2 s$^{-1})$, or 1.60. Consequently, CO_2 diffusing from the turbulent air up to the cell walls of mesophyll cells encounters 60% more resistance than water vapor diffusing in the opposite direction over the same pathway. Likewise, the gas phase conductance would be $(100\%)/(1.60)$, or only 62%, as great for CO_2 as for water vapor.[*] Representative values for CO_2 conductances and resistances are indicated in Table 8.4 for leaves of crops and trees with open stomata (see Table 8.1 for analogous values for water vapor).

As indicated in Figure 8.6, five additional resistances are involved in CO_2 flow compared with water vapor movement. The new components of the pathway are the nongaseous parts of the cell wall of a mesophyll cell (resistance $= r_{CO_2}^{cw}$), a plasmalemma ($r_{CO_2}^{pl}$), and the cytosol ($r_{CO_2}^{cyt}$), then the chloroplast limiting membranes ($r_{CO_2}^{clm}$), and finally the interior of the chloroplasts ($r_{CO_2}^{stroma}$). For convenience we will divide these five resistances into two parts, the mesophyll resistance to CO_2, $r_{CO_2}^{mes}$, and the chloroplast resistance, $r_{CO_2}^{chl}$:

$$r_{CO_2}^{mes} = r_{CO_2}^{cw} + r_{CO_2}^{pl} + r_{CO_2}^{cyt} \tag{8.18a}$$

$$r_{CO_2}^{chl} = r_{CO_2}^{clm} + r_{CO_2}^{stroma} \tag{8.18b}$$

Mesophyll Area

The area of the cell walls of mesophyll cells across which CO_2 can diffuse is considerably larger than the surface area of the leaf. For the constricting effect caused by the stomata, we used A^{st}/A, the fraction of the leaf surface area occupied by stomatal pores. Here we will use the ratio A^{mes}/A to indicate the *increase* in area available for CO_2 diffusion, where A^{mes} is the total area

---

[*] Actually, movement across the boundary layer is partly by diffusion, where the ratio D_{wv}/D_{CO_2} applies, and partly by turbulent mixing (see Fig. 7.7), where molecular differences are obliterated. Thus, $r_{CO_2}^{bl}/r_{wv}^{bl}$ would be expected to be intermediate between the extremes of 1.60 and 1.00, and indeed it is found to be $(D_{wv}/D_{CO_2})^{2/3}$, which is 1.37 (see Monteith or Schlichting, both Ch. 7 references). Although Equation 8.17 is generally satisfactory, certain situations may warrant replacing δ^{bl}/D_{CO_2} by $1.37 \, \delta^{bl}/D_{wv}$, which equals $0.86 \, \delta^{bl}/D_{CO_2}$. We also note that the effective boundary layer thicknesses for water vapor and heat transfer are quite similar, being within 7% for flat plates (see Monteith, Ch. 7 reference) and within 5% for cylinders and spheres (see Nobel 1974 and Nobel 1975, both Ch. 7 references).

of the cell walls of mesophyll cells that is exposed to the intercellular air spaces, and A is the area of one side of the same leaf.

Although A^{mes}/A varies with plant species as well as with leaf development, it is generally between 10 and 40 for mesophytes (see Björkman; El-Sharkway and Hesketh; Esau; and Nobel 1980). We can appreciate the large value of A^{mes}/A by examining Figures 1.2 and 8.5, which indicate that there is a considerable amount of air space and hence exposed cell walls within a leaf; e.g., the palisade mesophyll is generally 10% to 40% air by volume, while the spongy mesophyll is 50% to 80% air. Although there is a greater fraction of air in the spongy mesophyll region, the palisade cells usually have a greater total cell wall area exposed to the intercellular air spaces than do spongy mesophyll cells. Xerophytes tend to have a somewhat more highly developed palisade region than do mesophytes (in some cases, the spongy mesophyll cells are even absent in xerophytes), which leads to values of 20 to 50 for A^{mes}/A of many xerophytes.

To help appreciate the magnitude of A^{mes}/A, let us consider some geometrical idealizations. For a single layer of uniform spheres in an orthogonal array (i.e., a cubic lattice), A^{mes}/A would be $4\pi r^2/(2r \times 2r)$, or π. Hence, 3 layers of spherical cells would have an A^{mes}/A of 3π, or 9.4 (Fig. 8.7a). If the radius of the spheres were halved but the thickness of the array remained unchanged, then A^{mes}/A would double (Fig. 8.7a versus 8.7b). In a more realistic representation of a leaf with a single palisade layer having cells four times as long as wide and two layers of spherical spongy cells, A^{mes}/A

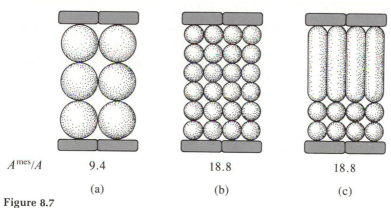

A^{mes}/A	9.4	18.8	18.8
	(a)	(b)	(c)

Figure 8.7

Representations of mesophyll cells showing how geometry affects A^{mes}/A. Spheres or cylinders with hemispherical ends in an orthogonal array lead to the indicated A^{mes}/A. The length of the lateral walls of the "palisade" cells in (c) is 6 times the radius.

would be 18.8, two-thirds of which is contributed by the palisade cells (Fig. 8.7c). Moreover, two-thirds of the exposed cell wall area of the palisade cells actually occurs on their lateral walls. To show that the relatively large contribution of the lateral walls is quite typical, we will represent the palisade cells as cylinders of radius r and length l with hemispheres on each end. The total area of the lateral surface of the cylinder ($2\pi rl$) is generally greater than that of the two hemispherical ends ($4\pi r^2$), since l is usually considerably greater than r for such palisade cells—e.g., r may be 10 to 20 μm, while l is 30 to 100 μm for representative palisade cells. When such cylinders are packed together to form a layer of palisade cells, nearly the entire surface area of the lateral walls is exposed to the intercellular air spaces. This surface area and most of that of both ends of a palisade cell are available for the inward diffusion of CO_2.

The anatomy of the mesophyll region can be greatly influenced by the illumination condition under which the leaf develops. Development in a dark or shaded environment can lead to what is known as a *shade* leaf, while differentiation under moderate to high illumination can lead to a *sun* leaf (see Esau; also see App. VIII). Besides being smaller in area, sun leaves usually are thicker and have a higher proportion of palisade cells than do shade leaves on the same plant. Moreover, their palisade cells are generally longer (larger l for the cylinders). Consequently, A^{mes}/A can be two to four times larger in sun leaves than in shade leaves on the same plant. For example, growing *Plectranthus parviflorus* at a PAR (photosynthetically active radiation; see p. 193) of 17 μmol m^{-2} s^{-1} for 12-hour days led to thin leaves with an A^{mes}/A of 11, whereas a PAR of 810 μmol m^{-2} s^{-1} led to thick leaves with an A^{mes}/A of 50 (see Nobel et al.).

Although PAR generally has the greatest influence, A^{mes}/A can also be influenced by changes in other environmental factors during leaf development (see Nobel 1980, and Patterson). For instance, higher temperatures generally lead to smaller cells and up to a 40% higher A^{mes}/A. Reduced cell size generally accompanies water stress, but the influences on A^{mes}/A vary with species, ranging from no change to a 50% increase in A^{mes}/A. Higher salinities during leaf development generally lead to thicker leaves, which can be accompanied by a corresponding increase in cell dimensions with no change in A^{mes}/A, or sometimes by an increase in A^{mes}/A.

Resistance Formulation for Cell Components

The resistance to diffusion of some molecular species across a barrier equals the reciprocal of its permeability coefficient (see p. 391). We will let $P^j_{CO_2}$ be the permeability coefficient for CO_2 diffusion across barrier j. To express the resistance of a particular mesophyll or chloroplast component on a leaf area basis, we must also incorporate the factor A^{mes}/A to allow for the actual area available for diffusion. Since the area of the plasmalemma is about the same as that of the cell wall, and the chloroplasts generally occupy a single layer around the periphery of the cytosol (Fig. 1.1), the factor A^{mes}/A will be applied to all the diffusion steps of CO$_2$ with which we are concerned in mesophyll cells (all 5 individual resistances in Eq. 8.18). In other words, we are imagining for simplicity that the cell wall, the plasmalemma, the cytosol, and the chloroplasts are all in layers having equal areas. Thus, the resistance of any of the mesophyll or chloroplast components for CO$_2$ diffusion, $r^j_{CO_2}$, is reduced from $1/P^j_{CO_2}$ by the reciprocal of the same factor, A^{mes}/A:

$$
r^j_{CO_2} = \frac{1}{A^{mes}/A} \frac{1}{P^j_{CO_2}} = \frac{A}{A^{mes} P^j_{CO_2}}
$$

$$
= \frac{A \Delta x^j}{A^{mes} D^j_{CO_2} K^j_{CO_2}} = \frac{1}{g^j_{CO_2}}
$$

(8.19)

where the second line of this equation follows from the definition of a permeability coefficient, $P_j = D_j K_j / \Delta x$ (Eq. 1.9). In particular, Δx^j is the thickness of the jth barrier, $D^j_{CO_2}$ is the diffusion coefficient of CO$_2$ in it, and $K^j_{CO_2}$ is a suitably defined partition coefficient.

Cell Wall Resistance

Let us begin our discussion of the newly introduced resistances by evaluating the components of $r^{cw}_{CO_2}$, the resistance encountered by CO$_2$ as it diffuses through the water-filled interstices of the cell wall from the interface with the intercellular air spaces on one side to the plasmalemma on the other side. We will use Equation 8.19 to describe this resistance: $r^{cw}_{CO_2} = A \Delta x^{cw}/(A^{mes} D^{cw}_{CO_2} K^{cw}_{CO_2})$.

The distance across the barrier Δx^{cw} is the average thickness of the cell walls of the mesophyll cells. The diffusion coefficient for the gas CO$_2$ dissolved in water is 1.7×10^{-9} m^2 s^{-1} at 20°C (Table 1.1). However, the effective $D^{cw}_{CO_2}$ is probably lower by a factor of three or four, because the water-filled interstices represent slightly less than half the cell wall and their course through the cell wall is rather tortuous (see p. 37). Thus, $D^{cw}_{CO_2}$ may be about 5×10^{-10} m^2 s^{-1}. In addition to moving as the dissolved gas, "CO$_2$" may also diffuse across the cell wall as H$_2$CO$_3$ or HCO$_3^-$. The diffusion coefficients of these two species in the cell wall are also most likely about 5×10^{-10} m^2 s^{-1}. However, the possible presence of H_2CO_3 and HCO_3^- makes the effective *concentration* of "CO_2" in the cell wall uncertain.

In Chapter 1 we introduced a partition coefficient (p. 29) to describe the ratio of the concentrations of some species in two adjacent phases. For instance, $K_{CO_2}c_{CO_2}$ could be the actual concentration of CO_2 in some region where concentrations are difficult to measure, c_{CO_2} is the equilibrium concentration of CO_2 in an adjacent region where it is readily measured, and K_{CO_2} is the partition coefficient. Similarly, we will express every $K_{CO_2}^j$ in the mesophyll cells as the actual concentration of all forms of CO_2 in component j divided by the concentration of CO_2 in an adjacent air phase at equilibrium, $c_{CO_2}^j$. A concentration referred to as $c_{CO_2}^j$ will thus equal the actual concentration in component j divided by $K_{CO_2}^j$. (This convention allows us to discuss fluxes in a straightforward manner, since CO_2 then diffuses toward regions of lower $c_{CO_2}^j$ regardless of the actual concentrations and partition coefficients involved.) Here we are interested in the partitioning of CO_2 between the air in the cell wall pores and the various species of CO_2 in the adjacent water within the cell wall interstices. $K_{CO_2}^{cw}$ is thus the actual concentration of CO_2 plus H_2CO_3 and HCO_3^- in the cell wall wall water divided by the concentration of CO_2 in air in equilibrium with the cell wall water.

The concentrations of the various forms of "CO_2" present in an aqueous phase are temperature dependent and extremely sensitive to pH.[*] For instance, the equilibrium concentration of CO_2 dissolved in water divided by that of CO_2 in an adjacent gas phase, $c_{CO_2}/c_{CO_2}^{air}$, decreases just over 2-fold from 10°C to 40°C (Table 8.3). This partition coefficient is itself not very pH dependent, but the equilibrium concentration of HCO_3^- in water relative to that of dissolved CO_2 is markedly affected by pH. In particular, CO_2 dissolved in an aqueous solution may interact with OH^- to form bicarbonate, which then associates with H^+ to form H_2CO_3: $CO_2 + OH^- + H^+ \rightleftharpoons HCO_3^- + H^+ \rightleftharpoons H_2CO_3$. Alternatively, CO_2 in water might form H_2CO_3, which then dissociates to HCO_3^- and H^+. The interconversions of CO_2 and H_2CO_3 are actually relatively slow unless a suitable catalyst, such as the enzyme carbonic anhydrase, is present (see Forster et al., as well as Maren). Since H^+ is involved in these reactions, its concentration, i.e., the pH, will affect the amount of HCO_3^- in solution, which in turn depends on the CO_2 concentration—the equilibrium concentration

---

[*] The concentrations also depend on the presence of other solutes, but this effect is presumably small for the cell wall water.

Table 8.3
Influence of temperature and pH on partitioning of "CO$_2$"
between an aqueous solution and an adjacent air phase. The
partition coefficient for all three forms of "CO$_2$" at various
pH's is for 20°C.

Temperature (°C)	$\dfrac{c_{CO_2}}{c^{air}_{CO_2}}$	pH	$\dfrac{c_{CO_2} + c_{H_2CO_3} + c_{HCO_3^-}}{c^{air}_{CO_2}}$
0	1.71	4	0.88
10	1.19	5	0.91
20	0.88	6	1.23
30	0.67	7	4.4
40	0.53	8	35
50	0.44		

of H$_2$CO$_3$ is only about 1/400 of that of the dissolved CO$_2$, and so our main concern will be with CO$_2$ and HCO$_3^-$.

At 20°C, $(c_{CO_2} + c_{H_2CO_3} + c_{HCO_3^-})/c^{air}_{CO_2}$ (our definition for $K^{cw}_{CO_2}$) is about 1 from pH 4 to pH 6, but increases markedly above pH 7 (Table 8.3). These equilibrium values are affected by temperature in approximately the same way as the partition coefficient $c_{CO_2}/c^{air}_{CO_2}$ cited above; e.g., going from 20°C to 30°C they change by about (0.67)/(0.88), or 0.76, which is a decrease of 24%. We note that this appreciable temperature dependence of K_{CO_2} results in a similar temperature dependence of $r^j_{CO_2}$—see Equation 8.19. Although the pH in the cell walls of mesophyll cells within a leaf is not known with certainty, it is probably under 6. Thus, at usual leaf temperatures, $K^{cw}_{CO_2}$ will be close to unity, the value we will use for calculation.

Let us now estimate a value for $r^{cw}_{CO_2}$. We will assume that the thickness of the cell walls Δx^{cw} is 1.0 μm, that the diffusion coefficient in the cell walls for solutes like CO$_2$ or HCO$_3^-$, $D^{cw}_{CO_2}$, is 5×10^{-10} m^2 s^{-1}, and that $K^{cw}_{CO_2}$ is 1.0. The magnitude of $r^{cw}_{CO_2}$ also depends on the relative surface area of the mesophyll cells compared with the leaf, a quantity that varies considerably. We will let A^{mes}/A be 20, a representative value for mesophytes. Using Equation 8.19, we can then calculate the resistance of the cell walls to the diffusion of CO$_2$:

$$r^{cw}_{CO_2} = \frac{(1 \times 10^{-6} \text{ m})}{(20)(5 \times 10^{-10} \text{ m}^2 \text{ s}^{-1})(1)} = 100 \text{ s m}^{-1}$$

Table 8.4
Summary of representative values of conductances and resistances for CO_2 diffusing into leaves. Certain values are calculated in the text. See Longstreth et al., Šetlik, Waggoner, Zelitch, and references for Table 8.1.

	Conductance		Resistance
Component	mm s^{-1}	mmol m^2 s^{-1}	s m^{-1}
Leaf (lower surface)—gas phase			
crops—open stomata	1.2 to 6	50 to 250	170 to 830
trees—open stomata	0.4 to 2	16 to 80	500 to 2 500
Cell wall	10	400	100
Plasmalemma	2	80	500
Cytosol	100	4 000	10
Mesophyll			
estimation	1.6	66	600
measurements—mesophytes	1.2 to 5	50 to 200	200 to 800
Chloroplast			
estimation	>2	>80	<500
measurements	>2.5	>100	<400

This is a rather small value for a CO_2 resistance (see Table 8.4), and indicates that the cell walls of the mesophyll cells generally do not represent a major barrier to the diffusion of the various species of CO_2.

Plasmalemma Resistance

Let us now examine $r_{CO_2}^{pl}$, the resistance of the plasmalemmas of mesophyll cells to diffusion of the various species of CO_2. Although we do not know the actual permeability coefficients of the plasmalemmas of mesophyll cells for CO_2 and HCO_3^-, we expect them to be much lower for a charged species like HCO_3^- (see p. 33). For instance, $P_{HCO_3^-}^{pl}$ might be about 10^{-8} m s^{-1}. On the other hand, CO_2 is a small neutral linear molecule that enters cells extremely easily (see Forster et al.). The permeability coefficient of CO_2 entering plant cells is probably at least 10^{-4} m s^{-1}, which is in the same range as $P_{H_2O}^{pl}$ (p. 30). Using Equation 8.19 and a value of 20 for A^{mes}/A, the plasmalemma CO_2 resistance would be

$$r_{CO_2}^{pl} = \frac{1}{(20)(10^{-4} \text{ m s}^{-1})} = 500 \text{ s m}^{-1}$$

Based on the relative values for the two permeability coefficients, the resistance to the diffusion of HCO$_3^-$ across the plasmalemma could be about 10^4 times higher, namely 5×10^6 s m^{-1}. Because of the extremely high resistance for HCO$_3^-$, we conclude that bicarbonate does not *diffuse* across the plasmalemma at the rates necessary to sustain photosynthesis. But the resistance calculated for CO$_2$ (500 s m^{-1}) is only moderate, which suggests that diffusion of CO$_2$ can be adequate for moving this substrate of photosynthesis across the plasmalemma.

HCO$_3^-$ or even CO$_2$ could be actively transported across the plasmalemma, or perhaps could cross by facilitated diffusion (p. 158). Facilitated diffusion would act like a low resistance pathway in parallel with the ordinary diffusion pathway, and thus would reduce the effective resistance of the plasmalemma to either of these species. Unfortunately, the actual mechanism for CO$_2$ or HCO$_3^-$ movement across the plasmalemmas of mesophyll cells is not known with certainty, although $r_{CO_2}^{pl}$ for diffusion of CO$_2$ is probably low enough to account for the CO$_2$ fluxes actually observed.

Cytosol Resistance

The resistance of the cytosol of the mesophyll cells to the diffusion of CO$_2$ is small, since the chloroplasts are located around the periphery of most mesophyll cells. The average distance from the plasmalemma to the chloroplasts, Δx^{cyt}, is only 0.1 to 0.3 μm. We will use a value of 0.2 μm for Δx^{cyt} when estimating $r_{CO_2}^{cyt}$ using Equation 8.19.

Apparently because of the presence of fibrous proteins, chloroplasts generally tend to remain stationary and quite close to the plasmalemma of a mesophyll cell. The diffusion coefficient of CO$_2$ in this region of the cytosol is somewhat less than its value in water at 20°C, 1.7×10^{-9} m^2 s^{-1}. For purposes of calculation, we will let $D_{CO_2}^{cyt}$ be 1×10^{-9} m^2 s^{-1} for the various species of CO$_2$ diffusing from the plasmalemma to the chloroplasts. The value of $K_{CO_2}^{cyt}$ of mesophyll cells is not known, primarily because of the lack of knowledge about cytosolic pH. In any case, $K_{CO_2}^{cyt}$ cannot be much less than unity, and values greater than 1.0—the magnitude we will assume here—will not change our conclusions about the relative value of $r_{CO_2}^{cyt}$. As before, we will let A^{mes}/A be 20. Using Equation 8.19, we calculate that the cytosolic resistance for CO$_2$ is only 10 s m^{-1}, a very small value for a resistance to CO$_2$ diffusion (Table 8.4). Thus, the location of the chloroplasts around the periphery of a mesophyll cell is a "good" design in that it causes the resistance of this part of the pathway for CO_2 diffusion to be quite low.

Mesophyll Resistance

Let us now compare the measured values of the mesophyll resistance with the magnitudes just calculated for its three components. Most of the measurements of $r_{CO_2}^{mes}$ have been rather indirect, but they do indicate that this resistance is generally 200 to 800 s m^{-1} for mesophytes (Table 8.4). Based on our estimates of 100 s m^{-1} for $r_{CO_2}^{cw}$, 500 s m^{-1} for $r_{CO_2}^{pl}$, and 10 s m^{-1} for $r_{CO_2}^{cyt}$, we would predict a value near 600 s m^{-1} for $r_{CO_2}^{mes}$ (by Eq. 8.18a, $r_{CO_2}^{mes} = r_{CO_2}^{cw} + r_{CO_2}^{pl} + r_{CO_2}^{cyt}$). Thus, our estimate based on the *diffusion* of CO_2 across each barrier is consistent with the measured values of mesophyll resistance.

There are many assumptions and specific choices for parameter values involved in the calculation of $r_{CO_2}^{mes}$ here. For instance, we let A^{mes}/A be 20, while many leaves have values from 30 to 40; the latter ratios would reduce our $r_{CO_2}^{mes}$ to 300 to 400 s m^{-1}. On the other hand, the cell walls of some mesophyll cells are 2 μm thick, which would increase our calculated $r_{CO_2}^{cw}$ to about 200 s m^{-1}. Permeability coefficients of the plasmalemmas of mesophyll cells for CO_2 have so far not been adequately measured. Our assumed value of 10^{-4} m s^{-1} is probably a lower limit. In particular, $K_{CO_2}^{pl}$ is most likely at least ten times higher than $K_{H_2O}^{pl}$, while the diffusion coefficients of both species are about the same (within a factor of two of each other). Thus, $P_{CO_2}^{pl}$ may be larger than 10^{-4} m s^{-1}, a possible value for $P_{H_2O}^{pl}$ (by Eq. 1.9, $P_j = D_j K_j/\Delta x$, where Δx^{pl} is the same in each case). By comparison, we noted in Chapter 1 (p. 31) that P_j for another small molecule, O_2, crossing erythrocyte membranes has an extremely high value of 0.3 m s^{-1}. A higher value for $P_{CO_2}^{pl}$ would decrease our estimate for $r_{CO_2}^{pl}$, and thus for $r_{CO_2}^{mes}$.

Chloroplast Resistance

The last two structural resistances encountered by the CO_2 used in photosynthesis are due to the chloroplasts ($r_{CO_2}^{chl} = r_{CO_2}^{clm} + r_{CO_2}^{stroma}$, Eq. 8.18b). As is the case for the plasmalemma, the resistance of the chloroplast limiting membranes, $r_{CO_2}^{clm}$, would be extremely large for the diffusion of HCO_3^- and moderate to low for the diffusion of CO_2. Numerous experiments indicate that the chloroplast limiting membranes are actually rather permeable to small solutes, so that $r_{CO_2}^{clm}$ may be lower than $r_{CO_2}^{pl}$, which we estimated to be at most 500 s m^{-1}. Since Δx^{stroma} might be 0.3 to 1 μm, while Δx^{cyt} is about 0.2 μm and the other pertinent parameters (A^{mes}/A, D_{CO_2}, K_{CO_2}) are approximately the same in the two cases, $r_{CO_2}^{stroma}$ could be a few times larger than $r_{CO_2}^{cyt}$, which we calculated to be only 10 s m^{-1} (if the pH in the chloroplast stroma were near or above 7, this would increase $K_{CO_2}^{chl}$ above unity and reduce $r_{CO_2}^{stroma}$ accordingly). In any case, $r_{CO_2}^{stroma}$ is a relatively small resistance.

At present, all we can really do is estimate an upper limit of 500 s m^{-1} for the resistance to the diffusion of CO_2 into chloroplasts and across their stroma. Measurement of $r_{CO_2}^{chl}$ in vivo is also quite difficult—analysis of available data indicates that most likely it is less than 400 s m^{-1}. Although active transport or facilitated diffusion of CO_2 or HCO_3^- into chloroplasts may occur and thus lower the effective resistance, the experimental values for $r_{CO_2}^{chl}$ are compatible with diffusion of CO_2 across the chloroplast limiting membranes. All the resistances we have just discussed and their corresponding conductances, expressed in both mm s^{-1} and mmol m^{-2} s^{-1}, are summarized in Table 8.4.

CO$_2$ FLUXES ACCOMPANYING PHOTOSYNTHESIS

Now that we have discussed the CO_2 resistances and conductances, we are ready to examine CO_2 fluxes. We will do this for photosynthesis, a process that consumes CO_2, as well as for respiration and photorespiration, two processes that evolve CO_2. Our analysis will use an electrical circuit analogy so that we can represent the interrelationships between the various factors influencing net CO_2 uptake by a leaf.

Photosynthesis

The flux density of CO_2 into the chloroplasts represents the gross rate of photosynthesis per unit leaf area, $J_{CO_2}^{ps}$. In the steady state, $J_{CO_2}^{ps}$ is the same as the flux density entering the leaf, J_{CO_2}, corrected for any other reactions evolving or consuming CO_2, as we will consider later (most evidence indicates that CO_2, not HCO_3^-, is the substrate for photosynthesis in chloroplasts, and we will therefore focus our attention on CO_2). We will represent the average rate of photosynthesis per unit *volume* of chloroplasts by v_{CO_2}, which can have units of mol of CO_2 fixed m^{-3} s^{-1}. If chloroplasts have an average or effective thickness Δx^{chl} in a direction perpendicular to the plasmalemma, the rate of photosynthesis per unit leaf area $J^{ps}_{CO_2}$ can be represented as

$$J^{ps}_{CO_2} = A^{mes} v_{CO_2} \Delta x^{chl} / A \qquad (8.20)$$

where the factor A^{mes}/A relates the chloroplast area to the leaf area. We can also derive Equation 8.20 by regarding the chloroplasts as a flat layer of average thickness Δx^{chl} occupying some area, e.g., A^{mes}. The rate of photosynthesis in this volume ($v_{CO_2} A^{mes} \Delta x^{chl}$) is equal to the gross CO_2 flux per unit leaf area times the leaf area corresponding to A^{mes}, i.e., $J^{ps}_{CO_2} A$.

Observations indicate that the gross rates of CO_2 fixation by leaves and by isolated chloroplasts are proportional to the CO_2 level over the lower range of its concentration, and eventually reach an upper limit when the CO_2 concentration is raised sufficiently high. One way to describe such behavior is with a Michaelis–Menten type of expression:

$$v_{CO_2} = \frac{V_{maxCO_2} c^{chl}_{CO_2}}{K_{MCO_2} + c^{chl}_{CO_2}} \qquad (8.21)$$

where V_{maxCO_2} is the maximum rate of CO_2 fixation per unit volume and K_{MCO_2} is essentially a Michaelis constant for CO_2 fixation, i.e., the value of $c^{chl}_{CO_2}$ at which v_{CO_2} equals $\frac{1}{2}V_{maxCO_2}$ [see Eq. 3.27a, $J^{in}_j = J^{in}_{jmax} c^o_j/(K_j + c^o_j)$, and the discussion that follows it]. Although convenient, using a Michaelis–Menten type of expression for the photosynthetic rate per unit volume may not always be justified for such a complicated series of reactions as photosynthesis (see P. Gaastra in Evans, Lommen et al., Tenhunen et al., Waggoner, and Zelitch).

In Equation 8.21, V_{maxCO_2} and, to some extent, K_{MCO_2} depend on PAR and temperature. For instance, V_{maxCO_2} is zero in the dark, since photosynthesis then ceases, and is directly proportional to PAR up to about 50 μmol m^{-2} s^{-1}. If we continually increase the PAR, V_{maxCO_2} can reach an upper limit, its value for light saturation. This generally occurs by about 600 μmol m^{-2} s^{-1} for most C_3 plants, whereas photosynthesis in C_4 plants is usually not light-saturated even at full sunlight, 2 000 μmol m^{-2} s^{-1} (see pp. 319 and 452 for comments on C_3 and C_4 plants; also see Fig. 8.12 for PAR responses of leaves of C_3 plants and a C_4 plant). Photosynthesis is maximal at certain temperatures, often from 30°C to 40°C. V_{maxCO_2} increases as the leaf temperature is raised to the optimum, and then decreases with a further increase in temperature.

The values of V_{maxCO_2} at light saturation and at the optimal temperatures for photosynthesis vary with the particular plant species, but they are generally from 2 to 10 mol m^{-3} s^{-1}. We can also estimate V_{maxCO_2} from measurements of the maximum rates of CO$_2$ fixation by isolated chloroplasts. These maximum rates—which are sustained for only rather short periods and are for optimal conditions—can be 100 mmol of CO$_2$ fixed (kg chlorophyll)$^{-1}$ s^{-1} [360 μmol (mg chlorophyll)$^{-1}$ hour^{-1} in the customary units], which is equal to approximately 3 mol m^{-3} s^{-1} (1 kg chlorophyll is contained in about 0.035 m^3 of chloroplasts in vivo). The activity of a key enzyme for CO$_2$ fixation, ribulose-1,5-bisphosphate carboxylase/oxygenase, in vitro can have rates equivalent to 200 mmol (kg chlorophyll)$^{-1}$ s^{-1}. Usually, the estimates of V_{maxCO_2} using isolated chloroplasts or enzymes are somewhat lower than its values determined for a leaf. Measurements using leaves generally indicate that K_{MCO_2} is 5 to 15 μM.

In Chapter 5 we noted that the processing time for CO$_2$ fixation was about 5 ms (p. 268). Eight photons are required, which are absorbed by approximately 2 400 chlorophyll molecules in chloroplasts where the chlorophyll concentration is about 30 mol m^{-3} (p. 226). Hence, the photosynthetic rate per unit volume of chloroplasts would be

$$v_{\text{CO}_2} = \frac{(30 \text{ mol chlorophyll m}^{-3})}{(5 \times 10^{-3} \text{ s/CO}_2)(2.4 \times 10^3 \text{ chlorophylls})}$$

$$= 2.5 \text{ mol CO}_2 \text{ m}^{-3} \text{ s}^{-1}$$

Under high PAR, $c_{\text{CO}_2}^{\text{chl}}$ is generally limiting for photosynthesis, and so this v_{CO_2} may be only 40% to 70% of the V_{maxCO_2} at a particular temperature. Conversely, the 5 ms length of the processing time may in large measure reflect the suboptimal CO$_2$ levels in the chloroplasts (these comments apply to C$_3$ plants, whereas $c_{\text{CO}_2}^{\text{chl}}$ is generally near the saturation value for photosynthesis in the mesophyll cells of C$_4$ plants, a topic to which we will return at the end of this chapter—see also Fig. 8.11). A related factor is the increase in atmospheric CO$_2$, primarily due to the burning of fossil fuels, from about 300 ppm by volume in 1900 to 340 ppm in 1980, when the annual rate of increase were nearly 2 ppm. Other things being equal, this increase in $c_{\text{CO}_2}^{\text{ta}}$ will lead to rise in $c_{\text{CO}_2}^{\text{chl}}$ and hence in v_{CO_2} (see Eq. 8.21).

Respiration and Photorespiration

So far, the only process involving CO$_2$ that we have considered in this chapter is photosynthesis. But we cannot neglect the CO$_2$ produced within mesophyll cells by respiration and photorespiration. If mitochondrial respiration in leaf cells were the same in the light as in the dark, when it can be readily measured, respiration would produce about 5% as much CO_2 as is consumed by photosynthesis at a moderate PAR. In those plants that photorespire, the rate of light-stimulated production of CO_2 by photorespiration is often about 30% (range, 20% to 60%) of the rate of CO_2 fixation into photosynthetic products.

Since we have already considered respiration in Chapter 6, we will next briefly comment on photorespiration (see Edwards and Huber, Lorimer, Osmond et al., and Tolbert). Photorespiration is the uptake of O_2 and the evolution of CO_2 in the light resulting from glycolate synthesis in chloroplasts and subsequent glycolate and glycine metabolism in peroxisomes and mitochondria (Fig. 8.8). A crucial role is played by the enzyme ribulose-1,5-bisphosphate carboxylase/oxygenase, which has a molar mass of about 550 kg mol^{-1} and consists of eight large subunits (55 kg mol^{-1}) and 8 small subunits (14 kg mol^{-1}). It constitutes about half of the soluble protein in the leaves of C_3 plants and perhaps one-sixth of the soluble protein in C_4 plants, making it the most abundant protein in the world. This carboxylase/oxygenase can interact with CO_2, leading to photosynthesis, or with O_2, leading to photorespiration (Fig. 8.8).* In photorespiration, ribulose-1,5-bisphosphate is split into 3-phosphoglycerate and 2-phosphoglycolate, the latter undergoing dephosphorylation in the chloroplasts and then entering the peroxisomes. Although CO_2 can be released by a decarboxylation of glycolate in the peroxisomes, the main product of glycolate metabolism is glycine, which then moves to the mitochondria where the CO_2 is released (Fig. 8.8). Since the generation of ribulose-1,5-bisphosphate depends on the C_3 photosynthetic pathway, photorespiration is influenced by PAR and by temperature, although not quite in the same manner as photosynthesis. For instance, photosynthesis usually doubles in going from 20°C to 30°C, while photorespiration often triples over this interval of leaf temperature—i.e., the oxygenase activity is apparently favored over the carboxylase activity with increasing temperature, and thus photorespiration increases at the expense of photosynthesis at higher temperatures. As far as CO_2 fixation is concerned, photorespiration apparently undoes what photosynthesis has done. We might then ask whether photorespiration benefits a plant—a question that so far has no entirely convincing answer, although it may reduce photoinhibition (damage to the light-harvesting reactions that can occur under high irradiation).

---

* The competition for the same active site on the enzyme ribulose-1,5-bisphosphate carboxylase/oxygenase by the two gases O_2 and CO_2, which thus act as alternative substrates, can be modeled by a modification of Equation 8.21 (see Farquhar et al. 1980b, Farquhar and von Caemmerer, and Tenhunen et al.).

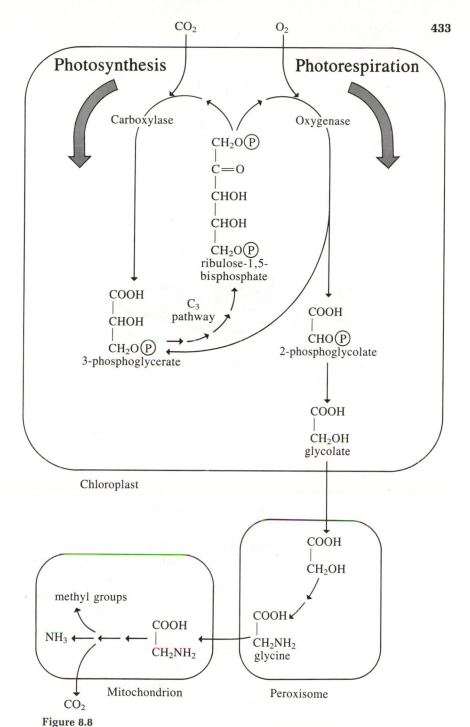

Figure 8.8
Schematic illustration of ribulose-1,5-bisphosphate carboxylase/oxygenase acting as the branch point for photosynthesis and photorespiration. All three of the organelles involved, but only a few of the biochemical steps, are indicated (Ⓟ represents phosphate).

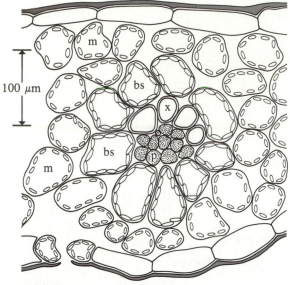

Figure 8.9
Schematic transverse section through a leaf of a C_4 plant, indicating a
vascular bundle containing xylem (x) and phloem (p) cells, a concentric
layer of bundle sheath cells (bs), and the surrounding mesophyll cells
(m). Bundle sheath cells of C_4 plants appear more conspicuously green
than mesophyll cells, since the former generally contain more and/or
larger chloroplasts (which are granaless in some types of C_4 plants).

Not all plants photorespire significantly. Many of these "nonphotorespiring" plants are tropical monocots, and all tend to have a leaf anatomy
different from that of plants with high rates of photorespiration (Fig. 1.2
versus Fig. 8.9). Nonphotorespirers have a conspicuous group of chloroplast-
containing cells surrounding the vascular bundles known as the *bundle
sheaths* (the bundle sheaths in photorespirers tend to have smaller cells with
few or no chloroplasts). Outside a bundle sheath in nonphotorespirers are
mesophyll cells where CO_2 is fixed into 4-carbon dicarboxylic acids via a
C_4 pathway elaborated by Hatch and Slack in the 1960's (see Hatch and
Slack, and O'Leary). In this C_4 pathway, CO_2 in the form of HCO_3^- reacts
with phosphoenolpyruvate via the enzyme phosphoenolpyruvate carboxylase located in the cytosol of the mesophyll cells. The initial product is
oxaloacetate, which is rapidly converted to malate and aspartate. For all
chloroplasts in photorespiring (C_3) plants, and for the chloroplasts of the
bundle sheath cells of C_4 plants, photosynthesis uses the ordinary C_3 pathway
of the Calvin cycle—known since the 1940's—where CO_2 is incorporated

into ribulose-1,5-bisphosphate, yielding two molecules of the 3-carbon compound, 3-phosphoglycerate (Fig. 8.8). Biochemical shuttles in C$_4$ plants move the 4-carbon compounds initially produced in the light, such as malate, from the mesophyll cells into the bundle sheath cells (Fig. 8.9). Decarboxylation of these compounds raises the CO$_2$ level in the bundle sheath cells to much higher levels than would be expected based on diffusion of CO$_2$ in from the atmosphere (to over 1 500 ppm CO$_2$ in one estimate; see Ehleringer and Björkman). Because of the high CO$_2$ level, the carboxylase activity of ribulose-1,5-bisphosphate carboxylase/oxygenase is dominant over (i.e., outcompetes) the oxygenase activity, and so photosynthesis takes place in the bundle sheath cells with very little photorespiration.*

Comprehensive CO$_2$ Resistance Network

We can now develop an analytical framework within which to represent CO$_2$ fixation in photosynthesis and its evolution in respiration and photorespiration. The net flux of CO$_2$ into a leaf, J_{CO_2}, is a measure of the apparent (net) CO$_2$ assimilation rate by photosynthesis. The gross or "true" rate of photosynthesis, $J_{CO_2}^{ps}$, minus the rate of CO$_2$ evolution by respiration and photorespiration per unit leaf area, $J_{CO_2}^{r+pr}$, is related to J_{CO_2} as follows:

$$J_{CO_2} = J_{CO_2}^{ps} - J_{CO_2}^{r+pr} \qquad (8.22)$$

Equation 8.22 summarizes the overall steady state balance of CO$_2$ fluxes for leaves.

We will find it convenient to consider CO$_2$ fluxes and resistances for photosynthesis, respiration, and photorespiration using an electrical circuit analogy (Fig. 8.10). The sources of CO$_2$ for photosynthesis are the turbulent air surrounding a leaf (represented by the E battery in Fig. 8.10) and respiration plus photorespiration (the e battery). The E battery actually corresponds to the drop in CO$_2$ concentration from the turbulent air surrounding a leaf to the enzymes of photosynthesis inside chloroplasts, $c_{CO_2}^{ta} - c_{CO_2}^{chl}$, which represents the driving force for CO$_2$ diffusion. The batteries lead to currents that correspond to fluxes of CO$_2$; e.g., I, the current from the E battery, corresponds to J_{CO_2}, while i represents the flux density of CO$_2$ emanating from respiration and photorespiration, $J_{CO_2}^{r+pr}$. The current I crosses the resistances $r_{CO_2}^{al}$, $r_{CO_2}^{leaf_1}$, and $r_{CO_2}^{mes}$ before being joined by i (see Fig. 8.10). The

---

* Raising the external CO$_2$ level to 1 500 ppm or so similarly would virtually eliminate photorespiration in C$_3$ plants (see Fig. 8.11); such CO$_2$ enrichment is common in greenhouses containing commercially valuable C$_3$ crops.

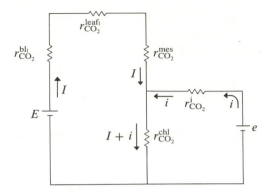

Figure 8.10
Electrical circuit indicating the resistances affecting photosynthesis, respiration, and photorespiration. The sources of CO_2 are the turbulent air surrounding the leaf (represented by the battery of electromotive force E) and respiration plus photorespiration (the e battery). The current I corresponds to the net CO_2 influx into the leaf (J_{CO_2}), i represents CO_2 evolution by respiration and photorespiration ($J_{CO_2}^{r+pr}$), and $I + i$ corresponds to gross photosynthesis ($J_{CO_2}^{ps}$).

current i encounters $r_{CO_2}^{i}$, which is the resistance to the movement of CO_2 out of mitochondria and then across a short distance in the cytosol. Both I and i cross the resistance $r_{CO_2}^{chl}$, since CO_2 coming from the surrounding air, as well as that evolved by respiration and photorespiration, can be used for photosynthesis in the chloroplasts (this is a way of paraphrasing Eq. 8.22).

To analyze the electrical circuit in Figure 8.10, we will make use of Ohm's law ($\Delta E = IR$) and Kirchhoff's second law.* The latter rule—which is also known as the *loop theorem* and is a consequence of the conservation of energy—states that the overall change in electrical potential in going completely around a closed loop is zero. By considering a complete pathway around the left part of the electrical circuit in Figure 8.10, we obtain the following relationship:

$$E - I(r_{CO_2}^{a_1} + r_{CO_2}^{leaf_1} + r_{CO_2}^{mes}) - (I + i)(r_{CO_2}^{chl}) = 0 \qquad (8.23)$$

Let us now insert more familiar quantities into Equation 8.23. E can be replaced by $c_{CO_2}^{ta} - c_{CO_2}^{chl}$, I by J_{CO_2}, and i by $J_{CO_2}^{r+pr}$. Upon moving the concentration term to the opposite side of the equation and changing signs,

---

* Kirchhoff's first law for electrical circuits states that the algebraic sum of the currents at any junction equals zero. For instance, at the junction in Figure 8.10 where current I meets current i, the current leaving that point equals $I + i$.

we obtain

$$J_{CO_2}(r_{CO_2}^{bl_1} + r_{CO_2}^{leaf_1} + r_{CO_2}^{mes} + r_{CO_2}^{chl}) + J_{CO_2}^{r+pr} r_{CO_2}^{chl} = c_{CO_2}^{ta} - c_{CO_2}^{chl} \quad (8.24)$$

Equation 8.24 indicates that, if respiration or photorespiration increase, then the net photosynthetic rate will decrease when other factors are unchanged.

Compensation Points

The atmospheric CO_2 concentration at which the CO_2 evolved by respiration and photorespiration is exactly compensated for by a CO_2 consumption in photosynthesis is known as the *CO_2 compensation point*. We can use Figure 8.10 and Equation 8.24 to demonstrate the CO_2 compensation point for photosynthesis in terms of forces and fluxes. If we continuously decrease $c_{CO_2}^{ta}$ for a leaf initially having a net uptake of CO_2, J_{CO_2} will decrease and eventually become zero when $c_{CO_2}^{ta} - c_{CO_2}^{chl}$ equals $J_{CO_2}^{r+pr} r_{CO_2}^{chl}$ (see Eq. 8.24). Thus, reducing the concentration of CO_2 in the turbulent air surrounding an illuminated leaf will cause the cessation of net CO_2 fixation when we reach the CO_2 compensation point.

In a C_3 plant, the CO_2 compensation point is considerably higher than for a C_4 plant. At the compensation point, $c_{CO_2}^{ta} - c_{CO_2}^{chl}$ equals $J_{CO_2}^{r+pr} r_{CO_2}^{chl}$, and $J_{CO_2}^{r+pr}$ is larger when photorespiration is appreciable. Most C_4 plants— e.g., sugar cane, sorghum, maize, bermuda grass, Sudan grass, and *Amaranthus*—have CO_2 compensation points of 3 to 10 ppm CO_2 (by volume) in the turbulent air (Fig. 8.11). Most dicotyledons and temperate monocots are C_3 plants—e.g., cotton, tobacco, tomato, lettuce, oaks, maples, roses, wheat, and orchard grass—and have CO_2 compensation points of 40 to 100 ppm CO_2 at 25°C (Fig. 8.11). A few species, e.g., some species of *Mollugo*, *Moricandia*, and *Panicum*, appear to have intermediate CO_2 compensation points near 25 ppm CO_2, and shifts between C_3 and C_4 patterns can even occur during leaf development. The CO_2 compensation points generally increase with increasing temperature and decreasing PAR, the values given being appropriate at 25°C when light is not limiting for photosynthesis (see Zelitch).[*]

If we continuously reduce the amount of light incident on a leaf from the value for direct sunlight, we eventually reach a PAR level where there

---

[*] CO_2 levels expressed in ppm (by volume) are equivalent to mole fractions, since by the perfect gas law $V_{CO_2}/V_{total} = n_{CO_2}/n_{total} = N_{CO_2}$. Also, N_{CO_2} equals P_{CO_2}/P_{total}, and so ppm numerically indicates (Pa CO_2 partial pressure)/(MPa total ambient pressure), or (μbar/bar). See Table 8.2 for various conversions.

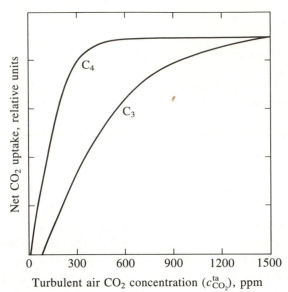

Figure 8.11
Dependence of net CO_2 uptake on external CO_2 level for leaves of representative C_3 and C_4 plants. C_3 plants require a higher $c_{CO_2}^{ta}$ at the CO_2 compensation point ($J_{CO_2} = 0$) and for CO_2 saturation than do C_4 plants.

is no net photosynthesis. This PAR level, where J_{CO_2} is zero, is known as the *light compensation point* for photosynthesis. Since photorespiration depends on photosynthetic products, both photorespiration and gross photosynthesis decrease as the PAR is lowered. Hence, the light compensation point for leaves is approximately the same for C_3 and C_4 plants—at 20°C and 340 ppm CO_2, light compensation usually occurs at about 6 to 10 μmol m^{-2} s^{-1} for C_3 plants and 4 to 8 μmol m^{-2} s^{-1} for C_4 plants. The light compensation point will be quite important in our consideration of the leaf canopy in the next chapter. For example, the uppermost leaves reach the light compensation point on dark days or at sunset, while leaves shaded by many overlying leaves can actually be at (or below) the light compensation point when the exposed leaves have appreciable net rates of photosynthesis.

At either compensation point, J_{CO_2} is zero when $c_{CO_2}^{ta} - c_{CO_2}^{chl}$ equals $J_{CO_2}^{r+pr} r_{CO_2}^{chl}$ (see Eq. 8.24). For the light compensation point, $c_{CO_2}^{ta}$ is unchanged but $c_{CO_2}^{chl}$ increases owing to a decrease in CO_2 fixation by photosynthesis at the low PAR. If we lower the PAR below the light compensation point, J_{CO_2} reverses its direction, which means a net flux of CO_2 out of the leaf. At night, J_{CO_2} becomes equal to $-J_{CO_2}^{r+pr}$, because $J_{CO_2}^{ps}$ is zero since gross photosynthesis stops upon cessation of illumination (see Eq. 8.22). The respiratory flow of CO_2 out of the leaf is then driven by the higher CO_2 concentration in the mitochondria than in the turbulent air, encountering the resistances $r_{CO_2}^i$, $r_{CO_2}^{mes}$, $r_{CO_2}^{leaf_1}$, and $r_{CO_2}^{bl_1}$, in that order. Thus, our electrical circuit in Figure 8.10, and Equation 8.24 derived from it, are able to portray the CO_2 compensation point, the light compensation point, as well as the general interrelations of the fluxes of CO_2 for photosynthesis, photorespiration, and respiration in the light and dark.

Fluxes of CO$_2$

Using Equation 8.24, we can relate the apparent or net rate of photosynthesis, J_{CO_2}, to the various resistances, respiration plus photorespiration, and the overall drop in CO_2 concentration, $\Delta c_{CO_2}^{total} = c_{CO_2}^{ta} - c_{CO_2}^{chl}$. Let us first rearrange Equation 8.24 into the following form:

$$J_{CO_2} = \frac{c_{CO_2}^{ta} - c_{CO_2}^{chl}}{r_{CO_2}^{bl_1} + r_{CO_2}^{leaf_1} + r_{CO_2}^{mes} + \left(1 + \dfrac{J_{CO_2}^{r+pr}}{J_{CO_2}}\right) r_{CO_2}^{chl}}$$

$$= \frac{\Delta c_{CO_2}^{total}}{r_{CO_2}^{total}} \tag{8.25}$$

where we have used our customary definition of resistance to obtain $r_{CO_2}^{total}$, the total effective resistance for CO_2 fixation—namely, resistance equals concentration drop divided by flux density (see Eq. 8.1d). We note that $r_{CO_2}^{total}$ depends on $J_{CO_2}^{r+pr}$, which is a consequence of the rather complicated electrical circuit (Fig. 8.10) needed to represent the various CO_2 components. Sometimes it may be convenient to rearrange Equation 8.24 in other ways, e.g., J_{CO_2} equals $(\Delta c_{CO_2}^{total} - J_{CO_2}^{r+pr} r_{CO_2}^{chl})/(r_{CO_2}^{bl_1} + r_{CO_2}^{leaf_1} + r_{CO_2}^{mes} + r_{CO_2}^{chl})$. This form clearly shows that J_{CO_2} is zero at the compensation points ($\Delta c_{CO_2}^{total} = J_{CO_2}^{r+pr} r_{CO_2}^{chl}$). Also, using Equation 8.22, we can manipulate the factor in Equation 8.25 containing $J_{CO_2}^{r+pr}$ as follows:

$$1 + \frac{J_{CO_2}^{r+pr}}{J_{CO_2}} = \frac{J_{CO_2}^{ps}}{J_{CO_2}} = \frac{J_{CO_2}^{ps}}{J_{CO_2}^{ps} - J_{CO_2}^{r+pr}} = \frac{1}{1 - \dfrac{J_{CO_2}^{r+pr}}{J_{CO_2}^{ps}}}$$

The appropriate form of this factor to use in Equation 8.25 depends on which ratio of fluxes we know.

Let us next consider specific values for the various parameters affecting net CO_2 uptake. We will use a CO_2 concentration of 340 ppm in the turbulent air, which corresponds to (340)(0.0410), or 14 mmol CO_2 m^{-3} at 20°C and a pressure of 0.1 MPa (conversion factor in Table 8.2). Although we do not have reliable measurements of $c_{CO_2}^{chl}$, it may be about 3 mmol m^{-3} for a photorespiring plant at saturating PAR. At 20°C, respiration plus photorespiration might be 30% as large as net photosynthesis. For purposes of calculation, we will let the gas phase resistance $r_{CO_2}^{bl_1} + r_{CO_2}^{leaf_1}$ be 400 s m^{-1}, let $r_{CO_2}^{mes}$ be 600 s m^{-1}, and let $r_{CO_2}^{chl}$ be 300 s m^{-1} (see Table 8.4). From Equation 8.25, we then calculate the net photosynthesis to be

$$J_{CO_2} = \frac{(14 \times 10^{-3} \text{ mol m}^{-3} - 3 \times 10^{-3} \text{ mol m}^{-3})}{400 \text{ s m}^{-1} + 600 \text{ s m}^{-1} + (1.0 + 0.3)(300 \text{ s m}^{-1})}$$

$$= \frac{(11 \times 10^{-3} \text{ mol m}^{-3})}{(1\,390 \text{ s m}^{-1})} = 7.9 \ \mu\text{mol m}^{-2} \text{ s}^{-1}$$

So far, we have considered CO_2 diffusing into a leaf only across its lower surface. In the general case, CO_2 can move in across its upper surface as well, which we can incorporate into our considerations by appropriately reducing the resistance $r_{CO_2}^{bl_1} + r_{CO_2}^{leaf_1}$ ($r_{CO_2}^{mes}$ and $r_{CO_2}^{chl}$ are unaffected when CO_2 diffuses in through both sides of a leaf). If 30% of the CO_2 diffused in through the upper side of a leaf, the effective resistance between the turbulent air and the surfaces of the mesophyll cells would be only 70% as great as $r_{CO_2}^{bl_1} + r_{CO_2}^{leaf_1}$.* For instance, $r_{CO_2}^{bl_1} + r_{CO_2}^{leaf_1}$ is here 400 s m^{-1}, so the resistance of this part of the pathway is reduced to (0.70)(400 s m^{-1}), or 280 s m^{-1}, if 30% of the CO_2 enters through the upper surface. This reduces $r_{CO_2}^{total}$ from 1 390 s m^{-1} to 1 270 s m^{-1}; the relatively small decrease reflects the substantial resistance of the unaffected nongaseous parts of the pathway for CO_2 movement ($r_{CO_2}^{mes}$ and $r_{CO_2}^{chl}$) compared with the gaseous parts. For the above values, the reduction in $r_{CO_2}^{total}$ raises J_{CO_2} to 8.6 μmol m^{-2} s^{-1}.

Let us next consider what would happen to CO_2 uptake if the stomata provided no resistance to CO_2 entry whatsoever. Instead of a gas phase resistance for CO_2 of 280 s m^{-1}, it might then be only 50 s m^{-1} for the two

---

* See p. 415 for our discussion of the analogous situation in transpiration. Instead of knowing the relative fluxes through the two sides, we might know the respective resistances. We could then use Equation 8.11 (with CO_2 replacing wv as subscripts) to determine the resistance of the two leaf surfaces in parallel. Quite often the gas phase resistance for water vapor is measured for the two leaf surfaces, in which case the gas phase CO_2 resistance is obtained by multiplying r_{wv}^{total} by D_{CO_2}/D_{wv}.

leaf surfaces acting in parallel. The $r_{CO_2}^{total}$ would thus be lowered to 1 040 s m^{-1}, which would raise J_{CO_2} to 10.6 μmol m^{-2} s^{-1}, an increase of 23%. In other words, removing both epidermises completely would enhance CO_2 uptake by only 23%, indicating that the stomata do not greatly restrict the photosynthetic rate in the present case, although their pores occupy only a very small fraction of the leaf surface area.

Range in Photosynthetic Rates

As we might expect, the net rate of photosynthesis varies considerably with plant species, temperature, PAR, and other conditions (Fig. 8.12; see also Šetlík or Zelitch). For instance, the maximum J_{CO_2} is often near 5 μmol m^{-2} s^{-1} for the leaves of some trees. Certain C_3 crop plants like sugar beet, soybean, and tobacco can have a J_{CO_2} of 20 μmol m^{-2} s^{-1} at saturating PAR and temperatures near 30°C. For C_4 plants, J_{CO_2} tends to be larger, since $J_{CO_2}^{r+pr}$ is small and also the liquid phase resistances ($r_{CO_2}^{mes}$ and $r_{CO_2}^{chl}$) are often relatively small (see Körner et al., and Longstreth et al.). Under optimal conditions of high PAR and a leaf temperature of 35°C, J_{CO_2} can exceed 40 μmol m^{-2} s^{-1} for bermuda grass, maize, sorghum, sugar cane, and certain other C_4 plants (as well as a few C_3 species). An extremely high value of 67 μmol m^{-2} s^{-1} can occur for the C_4 plant *Hilaria rigida* at full sunlight (Fig. 8.12).

The influence of A^{mes}/A on J_{CO_2} deserves special comment. A^{mes}/A can be two or more times larger for sun leaves than for shade leaves on the same plant; this reduces $r_{CO_2}^{total}$ and thus enhances the maximal photosynthetic rates of sun leaves compared with shade leaves. Maximal rates of CO_2 uptake for C_3 plants can thus vary with A^{mes}/A for leaves on a single plant, within a species, and sometimes even between species (Fig. 8.12). The large area of the cell walls of mesophyll cells that is available for CO_2 diffusion keeps $r_{CO_2}^{cw}$ and the other CO_2 resistances in the mesophyll cells at reasonable values. For instance, if the mesophyll cells were tightly packed into a layer with no intervening air spaces, A^{mes}/A could equal 2.0. This occurs for the moss *Mnium ciliare*, whose leaves are one cell thick with the lateral walls completely touching; thus, the only area available for CO_2 to diffuse from the gas phase into the cells is their end walls, which have a total area twice that of one side of the leaf (see Nobel 1980). Instead of the cell wall resistance of 100 s m^{-1} that we calculated for an A^{mes}/A of 20 (Table 8.4), $r_{CO_2}^{cw}$ alone would be 1 000 s m^{-1} for an A^{mes}/A of 2. Thus, the evolution of a leaf anatomy with a profusion of mesophyll cell surface area leading to a large value for

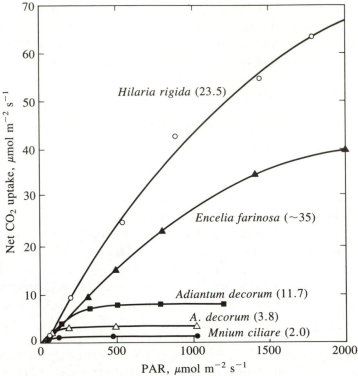

Figure 8.12
Photosynthetic responses of various species to photosynthetically active radiation. Curves were obtained at ambient CO_2 and O_2 concentrations, optimal temperatures, and the A^{mes}/A indicated in parentheses. [Sources: for the C_4 desert grass *Hilaria rigida*, P. S. Nobel, *Ecology 61*:252–258 (1980); for the C_3 desert composite *Encelia farinosa*, J. Ehleringer, O. Björkman, and H. A. Mooney, *Science 192*: 376–377 (1976); and for the C_3 maidenhair fern *Adiantum decorum* and the C_3 moss *Mnium ciliare*, P. S. Nobel, *Physiologia Plantarum 40*:137–144 (1977).]

A^{mes}/A allows $r_{CO_2}^{mes}$ and $r_{CO_2}^{chl}$ to be fairly low, with a correspondingly high value for J_{CO_2} (see Fig. 8.12).

CO_2 Conductances

Our analysis for CO_2 fluxes could be carried out using conductances. Also, we could divide the CO_2 pathway into a gas phase component from the turbulent air up to the mesophyll cells and a liquid phase component representing the mesophyll cells. For instance, the CO_2 concentration in the intercellular air spaces next to the mesophyll cells is

$$c_{CO_2}^{ias} = c_{CO_2}^{ta} - \Delta c_{CO_2}^{gas}$$

$$= c_{CO_2}^{ta} - \frac{J_{CO_2}}{g_{CO_2}^{gas}} \tag{8.26}$$

where $g_{CO_2}^{gas}$ is the CO_2 conductance for the gas phase across which the CO_2 concentration drops by $\Delta c_{CO_2}^{gas}$. In turn, $g_{CO_2}^{gas}$ equals $g_{wv}^{total}/1.60$, where g_{wv}^{total} can be defined by Equation 8.13 and 1.60 represents D_{wv}/D_{CO_2}. Under optimal photosynthetic conditions, $c_{CO_2}^{ias}$ is generally about twice as high for leaves of C_3 compared with C_4 plants (discussed below). Indeed, $c_{CO_2}^{ias}$ (or $N_{CO_2}^{ias}$) has proved to be an important parameter for evaluating photosynthesis at the mesophyll or liquid phase level as well as for studying the regulation of stomatal opening.

We can identify a liquid phase CO_2 conductance for the part of the pathway from the mesophyll cell walls up to the CO_2-fixation enzymes:

$$J_{CO_2} = g_{CO_2}^{liquid} (c_{CO_2}^{ias} - c_{CO_2}^{chl}) \tag{8.27}$$

If respiration and photorespiration can be ignored, then $1/g_{CO_2}^{liquid}$ equals $1/g_{CO_2}^{mes} + 1/g_{CO_2}^{chl}$ (equivalently, $r_{CO_2}^{liquid} = r_{CO_2}^{mes} + r_{CO_2}^{chl}$). Otherwise, we could return to Figure 8.10 and note that $c_{CO_2}^{ias} - c_{CO_2}^{chl} = J_{CO_2}/g_{CO_2}^{mes} + J_{CO_2}^{ps}/g_{CO_2}^{chl} = J_{CO_2}[1/g_{CO_2}^{mes} + (J_{CO_2}^{ps}/J_{CO_2})/g_{CO_2}^{chl}]$, which in turn equals $J_{CO_2}/g_{CO_2}^{liquid}$ by Equation 8.27. This intermingling of conductances and fluxes again reflects the complication of having more than one source of CO_2 that can be fixed photosynthetically (see Eq. 8.24). If we are interested in the photosynthetic properties of the mesophyll cells themselves, we might wish to express the CO_2 conductance on the basis of mesophyll cell surface area:

$$g_{CO_2}^{cell} = \frac{g_{CO_2}^{liquid}}{A^{mes}/A} \tag{8.28}$$

where $g_{CO_2}^{cell}$ is the cellular conductance for CO_2. For optimal temperatures, ambient CO_2, and full sunlight, $g_{CO_2}^{cell}$ can be about 4 times higher for C_4 plants, which reflects their superiority over C_3 plants in processing CO_2 under these conditions (see Longstreth et al.).

Many units are used to express photosynthesis and CO_2 fluxes for leaves. Conversion factors for some of the more common units are summarized in

Table 8.2. For example, 16 mg CO_2 dm^{-2} hour^{-1} corresponds to (16)(0.631), or 10 μmol m^{-2} s^{-1}. The amount of chlorophyll per unit leaf area generally ranges from 0.2 to 0.8 g m^{-2}, with 0.4 to 0.5 g m^{-2} being fairly typical (see Björkman). Thus, 10 μmol m^{-2} s^{-1} might correspond to (10 μmol m^{-2} s^{-1})/ (0.4 g chlorophyll m^{-2}), or 25 μmol (g chlorophyll)$^{-1}$ s^{-1}, which equals 90 μmol CO_2 fixed (mg chlorophyll)$^{-1}$ hour^{-1}.

WATER USE EFFICIENCY

As we have indicated, stomatal opening leading to the CO_2 uptake necessary for photosynthesis results in an inevitable loss of water. A useful parameter relating the two fluxes involved and showing the total CO_2 fixed (benefit) per unit water lost (cost) is the *water use efficiency*, WUE:

$$\text{WUE} = \frac{\text{mass } CO_2 \text{ fixed}}{\text{mass } H_2O \text{ transpired}} \qquad \text{mass basis} \qquad (8.29a)$$

$$\text{WUE} = \frac{\text{mol } CO_2 \text{ fixed}}{\text{mol } H_2O \text{ transpired}} \qquad \text{mole basis} \qquad (8.29b)$$

A convenient unit for WUE on a mass basis, which is the more commonly used index, is g CO_2 (kg H_2O)$^{-1}$. A related quantity is the *transpiration ratio*, which is the reciprocal of the water use efficiency and hence represents the water lost per CO_2 fixed.

Values of WUE

From the values of J_{CO_2} and J_{wv} calculated in this chapter, we can determine a WUE for the leaf of a representative C_3 mesophyte. Specifically, we obtained a J_{CO_2} of 8.6 μmol CO_2 fixed m^{-2} s^{-1} (p. 440) and a J_{wv} of 4.3 mmol H_2O transpired m^{-2} s^{-1} (p. 415). By Equation 8.29b, the water use efficiency is

$$\text{WUE} = \frac{(8.6 \times 10^{-6} \text{ mol } CO_2 \text{ m}^{-2} \text{ s}^{-1})}{(4.3 \times 10^{-3} \text{ mol } H_2O \text{ m}^{-2} \text{ s}^{-1})} = 0.0020 \ CO_2/H_2O$$

On a mass basis, this corresponds to a WUE of 4.9 g CO_2 (kg H_2O)$^{-1}$ (the molar masses of CO_2 and H_2O are 44.0 g mol^{-1} and 18.0 g mol^{-1}, respectively). We also note that the transpiration ratio in the above case is 500

H_2O/CO_2. This substantial water loss per CO_2 fixed is generally not a problem when plenty of water is available for transpiration. Plants in such environments often have a high g_{wv}^{total}, which leads to a somewhat higher $g_{CO_2}^{total}$ and somewhat higher rates of photosynthesis than for plants with a moderate g_{wv}^{total}.

Any loss of water can be potentially harmful for plants growing in arid regions, many of which have evolved a novel way of fixing CO_2 in a manner leading to a high water use efficiency. For example, many species in the family Crassulaceae, as well as other desert succulents, have their stomata closed during the daytime. This greatly reduces transpiration, but also essentially eliminates the net influx of CO_2 at that time. The stomata open at night, CO_2 then diffuses in, and it is fixed into malate (e.g., by carboxylation of phosphoenolpyruvate) and other organic acids. During the next day these organic acids are decarboxylated, and the released CO_2 is retained within the plant because of the closed stomata. This CO_2 is then fixed into photosynthetic products by means of the C_3 pathway. Plants with this CO_2 fixation mechanism are referred to as CAM (Crassulacean acid metabolism) plants, since such reactions were initially studied extensively in the Crassulaceae, although apparently first observed in the Cactaceae in 1804.

As just indicated, stomata for CAM plants tend to open at night, when leaf and air temperatures are lower than daytime values. The concentration of water vapor in the pores of the cell walls of chlorenchyma cells (c_{wv}^e) is then much lower, markedly reducing the rate of transpiration. For example, leaf temperatures of *Agave deserti* can be 25°C in the afternoon and 5°C at night (Fig. 8.13), leading to saturation water vapor concentrations of 23.0 g m^{-3} and 6.8 g m^{-3}, respectively (App. II). For air with a water vapor content of 4.0 g m^{-3}, which is fairly typical during the wintertime in the native habitat of *Agave deserti*, Δc_{wv}^{total} would be 7 times higher at 25°C than at 5°C, and therefore so would J_{wv} for the same degree of stomatal opening. Clearly, nocturnal stomatal opening can result in water conservation and hence a higher water use efficiency. For the CAM plant *Agave deserti* on the day depicted in Figure 8.13, the WUE was 56 g CO_2 (kg H_2O)$^{-1}$, and it was 40 g CO_2 (kg H_2O)$^{-1}$ when averaged over a whole year (see Nobel 1976)—both very high values.

Changes in the thickness of the air boundary layers adjacent to a leaf have a greater influence on the flux of water vapor than on the flux of CO_2. For instance, the total resistance for water vapor diffusion can equal $r_{wv}^{bl} + r_{wv}^{st} + r_{wv}^{ias}$ (Eq. 8.14), whereas $r_{CO_2}^{bl} + r_{CO_2}^{st} + r_{CO_2}^{ias}$ (Eq. 8.17) is generally less than half of the total resistance for CO_2 diffusion. Thus, changes in wind speed have a smaller fractional effect on $r_{CO_2}^{total}$ than on r_{wv}^{total}. Similarly, partial

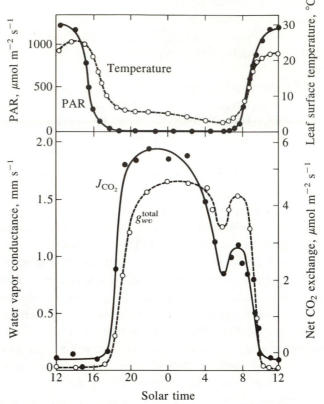

Figure 8.13

Photosynthetically active radiation (on a horizontal surface), leaf surface temperature, water vapor conductance, and net CO_2 exchange for *Agave deserti* on clear winter days in the northwestern Sonoran Desert. (Modified from Nobel 1976; used by permission.)

stomatal closure can appreciably reduce g_{wv}^{total}, but lead to a smaller fractional reduction in $g_{CO_2}^{total}$. For instance, stomata often begin to close when the PAR drops below about 200 μmol m^{-2} s^{-1}, which reduces J_{wv} more than it reduces J_{CO_2}, resulting in water conservation. Certain xerophytes have fairly low maximal values for g_{wv}^{st}—the maximal stomatal conductance is generally less than 2 mm s^{-1} for *Agave deserti* (see Fig. 8.13) compared to over 10 mm s^{-1} for many mesophytes (Table 8.1). Such reduced maximal stomatal conductance reduces transpiration to a greater degree than CO_2 uptake, with a consequent enhancement in water use efficiency (see Eq. 8.26).

Even though photosynthesis and transpiration depend on environmental conditions (e.g., J_{wv} depends on c_{wv}^e and c_{wv}^{ta}), we can still make some generalizations about WUE for different types of plants. Specifically, WUE's averaged over a day for mature leaves are usually 1 to 3 g CO_2 (kg H_2O)$^{-1}$ for C_3 plants, 2 to 5 g CO_2 (kg H_2O)$^{-1}$ for C_4 plants, and 10 to 40 g CO_2 (kg H_2O)$^{-1}$ for CAM plants (see Osmond et al., and Szarek and Ting). C_4 plants have approximately double the WUE of C_3 plants, since C_4 plants tend to have lower gas phase conductances (which conserves water with a relatively small negative effect on photosynthesis) and higher liquid phase conductances (which affects photosynthesis positively) than C_3 plants (also see Körner et al., and Longstreth et al.). Maximizing WUE may not always be adaptive—e.g., water may not be limiting for an aquatic plant.

Elevational Effects on WUE

We would expect both transpiration and photosynthesis to be affected by elevation, since diffusion coefficients depend inversely on ambient (barometric) pressure $[D_j = D_j^0 (T/273)^{1.8} (P^0/P)$, Eq. 8.9], and since the partial pressures of water vapor and CO_2 generally decrease with elevation. At sea level, barometric pressure averages 0.101 MPa, while it averages 0.080 MPa at 2 000 m and about 0.056 MPa at 5 000 m. Thus, diffusion coefficients would be nearly twice as large at 5 000 m as at sea level owing to this pressure change, which correspondingly increases the gas phase conductances based on Δc (e.g., Eq. 8.2), although those based on ΔN (Eq. 8.8) are unchanged (see p. 403). For a typical lapse rate of $-5°C$ per km of elevation,[*] temperatures would decrease from 30°C at sea level to 5°C at 5 000 m, which by itself would decrease diffusion coefficients 14% by Equation 8.9. The partial pressure of CO_2 is reduced more or less in concert with the reduced barometric pressure; i.e., the mole fraction of CO_2 is approximately constant with elevation. The partial pressure of water vapor in the air also tends to decrease with elevation, and under isothermal conditions the driving force for water loss (both Δc_{wv} and ΔN_{wv}) increases, as does transpiration.

Because of the interaction of many factors, especially the numerous temperature effects on both transpiration and photosynthesis, it is difficult to predict the effect of elevation on water use efficiency. When the turbulent mixing aspect in the boundary layer is ignored (see footnote, p. 420), the

---

[*] At the dry adiabatic lapse rate (9.8°C decrease in temperature per km increase in altitude), a rising parcel of dry air will cool by expansion due to the decrease in air pressure and will achieve the same temperature as the surrounding air—a case of neutral stability. That is, air movement is then neither favored nor retarded by buoyancy (see Monteith 1973 or Oke, both Ch. 7 references). Observed lapse rates are often about $-5°C$ to $-6°C$ km^{-1}.

higher D_{CO_2} and lower P_{CO_2} essentially compensate in the gas phase (compensation is exact under isothermal conditions). Diffusion coefficients in the liquid phase are unaffected by barometric pressure (hence, liquid phase conductances expressed in mm s^{-1} are unaffected by the pressure changes with elevation, but those expressed in mmol m^{-2} s^{-1} must be adjusted by the fractional changes in P; see p. 403). Thus, if we ignore changes in stomatal aperture and temperature, increases in elevation decrease $P_{CO_2}^{ias}$ and $c_{CO_2}^{ias}$, which will reduce J_{CO_2}. The greater the liquid phase conductance relative to that of the gas phase, the less is the elevational effect—in the limit of infinite liquid phase conductance, photosynthesis would be essentially unaffected by changes in ambient pressure (see Gale 1972a). However, the lowering of temperature with elevation could have a large effect on photosynthesis, although the optimal temperature for photosynthesis can acclimate somewhat (usually by 2°C to 15°C) to match the average ambient temperature of the environment (see Berry and Björkman, Osmond et al., and Patterson).

As we discussed in Chapter 7, many factors affect leaf temperature, and an energy budget analysis is necessary to calculate T^{leaf} and thus to indicate the effects of elevation on transpiration. Leaves at higher elevations generally experience a higher net radiation balance (more incident shortwave irradiation although somewhat less incident longwave irradiation; see Eq. 7.26) and lower air temperatures. For small leaves, which tend to be close to air temperature, transpiration generally decreases with elevation. For a 5 cm × 5 cm leaf in full sunlight, transpiration was unaffected at a lapse rate of -1°C km^{-1}, but decreased 25% at 1 000 m compared to sea level at a lapse rate of -5°C km^{-1} (see Gale 1972b). For large leaves in full sunlight, transpiration at typical lapse rates may decrease only slightly with elevation, since large leaves are further above air temperature at higher elevations (see Smith and Geller). For temperature inversion conditions (increasing temperature with elevation), transpiration can actually increase with elevation (see Gale 1972b). Thus, since photosynthesis and transpiration can vary in so many ways with elevation, effects of elevation on WUE must be judged case by case.

Stomatal Control of WUE

To maximize water use efficiency, stomatal opening must be synchronized with the capability for CO_2 fixation (see Cowan 1977 and 1981, Cowan and Farquhar, Farquhar et al. 1978 and 1980, and Raschke). Stomatal opening could be regulated by the CO_2 level in the intercellular air spaces, a depletion of $N_{CO_2}^{ias}$ by photosynthesis leading to an increase in $g_{CO_2}^{st}$, which would then let more CO_2 into the leaf under PAR levels and other conditions favorable for photosynthesis. This is an example of a *feedback* system, as $N_{CO_2}^{ias}$ feeds a signal back to the stomata, which in turn leads to a change in $N_{CO_2}^{ias}$. Also, PAR may directly affect the metabolism of guard cells, which indeed contain chloroplasts and hence could utilize PAR. This is an example of a *feedforward* system, as charges in stomatal aperture due to photosynthetic responses of chloroplasts then feed forward (or anticipate) and adjust CO_2 entry into the leaf, thereby matching photosynthesis by the mesophyll region to environmental conditions.

Stomatal opening is also affected by the leaf water status. For instance, stomata tend to close as a leaf begins to wilt, especially after the leaf water potential drops below some threshold level and abscisic acid (ABA) is produced (as we mentioned on p. 395). In fact, ABA can induce stomatal closure even when $N_{CO_2}^{ias}$ favors opening. The water status thus affects stomatal opening and hence transpiration, which in turn feeds back onto the leaf water status. Stomatal opening is usually increased by higher N_{wv}^{ta}. This is another example of a feedforward system, as it anticipates the effect of the ambient water vapor concentration on transpiration—e.g., higher N_{wv}^{ta} means less "force" on water exit from a leaf, so the stomata can open wider without it leading to excessive transpiration.[*] These various processes regulating stomatal movements interact with each other—we will examine the consequences of this for gas exchange by leaves.

As a refinement on our consideration of water use efficiency, we should consider what the optimal behavior of stomata might be over the course of a day. For instance, WUE would be maximized by minimal stomatal opening, since transpiration is decreased more than photosynthesis by partial stomatal closure (see Eq. 8.29); i.e., J_{wv} changes proportionally more than does J_{CO_2} as g_{wv}^{st} changes (Fig. 8.14). However, this might lead to very little CO_2 uptake. Thus, a more pertinent consideration might be what the maximum amount of CO_2 is that could be taken up for a certain amount of water available for transpiration. The amount of water that could be lost would depend on plant condition and environmental factors, and should be considered over the course of a whole day. To help analyze the relationship between gas fluxes, curves showing J_{CO_2} versus J_{wv} can be drawn for any PAR, temperature, wind speed, or relative humidity occurring for a given leaf during the

---

[*] When N_{wv}^{ta} is low, the feedforward system can cause stomatal closure to such an extent that transpiration can actually be *less* under such conditions than at high humidity (see Schulze et al.).

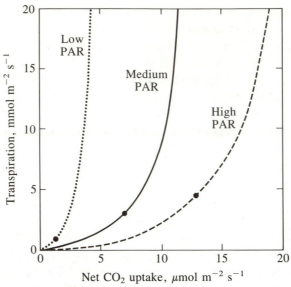

Figure 8.14

Relation between net photosynthesis (J_{CO_2}) and transpiration (J_{wv}) as stomatal conductance is varied. The three curves depict various PAR levels, indicated as "low," "medium," and "high." The circles indicate where the slope $\partial J_{wv}/\partial J_{CO_2}$ is 1 000 H_2O/CO_2 ($\partial J_{CO_2}/\partial J_{wv} = 0.0010$ CO_2/H_2O). Cuticular transpiration is ignored.

day, the location of the curves varying but still conforming to the general shape illustrated in Figure 8.14. In addition to environmental factors, the location of the curves is influenced by leaf properties such as size, age, A^{mes}/A, and shortwave absorptance. Nevertheless, it has been hypothesized (see Cowan and Farquhar) that the stomata will open or close depending on the various feedback and feedforward processes in such a way that the following relation is obeyed:

$$\frac{\partial J_{wv}/\partial g_{wv}^{st}}{\partial J_{CO_2}/\partial g_{wv}^{st}} = \frac{\partial J_{wv}}{\partial J_{CO_2}} = \lambda = \text{constant} \tag{8.30}$$

where $\partial J_{wv}/\partial g_{wv}^{st}$ and $\partial J_{CO_2}/\partial g_{wv}^{st}$ represent the sensitivity of transpiration and photosynthesis, respectively, to changes in stomatal conductance.

The solid circles in Figure 8.14 indicate the CO_2 and H_2O fluxes that can occur at different times of the day for a leaf with a λ of 1 000 H_2O/CO_2.

At low PAR, there is little stomatal opening, so little water would be used under conditions where the rate of photosynthesis inherently cannot be very high. Stomatal opening is much greater at high PAR, and so both transpiration and photosynthesis then are greater, but the local slope of the J_{CO_2} versus J_{wv} curve is still the same. Let us next consider what would happen to the fluxes if we moved along a curve away from one of the solid circles, e.g., the one for medium PAR (Fig. 8.14). If the stomatal opening changed so that J_{wv} increased by 1.0 mmol m^{-2} s^{-1}, then J_{CO_2} would increase by 0.8 μmol m^{-2} s^{-1}. To lead to the same total transpiration for the day, we would have to decrease J_{wv} by the same amount at another time, which for simplicity we can also consider for the medium PAR curve (note that all the solid circles occur for the same slope, 1 000 H$_2$O/CO$_2$). A decrease in J_{wv} of 1.0 mmol m^{-2} s^{-1} would then be accompanied by a decrease in J_{CO_2} of 1.4 μmol m^{-2} s^{-1}. When the effect of both changes is considered, we get the same total transpiration, but a lower net CO$_2$ uptake. In fact, the criterion expressed in Equation 8.30 leads to the maximum amount of CO$_2$ fixed for a particular amount of water transpired as well as to the minimum amount of water transpired for a particular amount of CO$_2$ fixed in a day (see Cowan and Farquhar). Thus, the stomatal conductance would be varying in such a way that the water use efficiency for the entire day is maximized.

The value of λ can change during the growth of a plant. For instance, λ can be small (e.g., 300 H$_2$O/CO$_2$) when water is in short supply. In such cases, constancy of λ requires that stomata should close partially near midday, when temperatures are the highest and transpiration is potentially the greatest. Also, water stress conditions in general lead to higher ABA levels in the leaves and a lower λ. On the other hand, a large λ (e.g., 1 300 H$_2$O/CO$_2$) occurs when the plant is far from water stress and no midday stomatal closure then takes place.

The proposed constancy of $\partial J_{wv}/\partial J_{CO_2}$ helps interpret the partial stomatal closure that occurs at midday when water is limiting, as well as the nocturnal stomatal closure, when PAR is limiting. We can also use the constancy of $\partial J_{wv}/\partial J_{CO_2}$ to help interpret experiments in which a single environmental factor is varied, such as the driving force for water vapor loss, ΔN_{wv}. If the stomata maintained a constant J_{wv}, then changes in g_{wv}^{st} would be the inverse of changes in ΔN_{wv} (when cuticular transpiration is ignored). On the other hand, maintenance of constant J_{CO_2} as ΔN_{wv} is varied would require constancy of $g_{CO_2}^{st}$, which equals $g_{wv}^{st}/1.60$. In fact, however, varying ΔN_{wv} over a 4-fold range for *Nicotiana glauca* (tobacco), *Corylus avellana* (hazel; see Farquhar et al. 1980a), and *Vigna unguiculata* (cowpea; see Hall and Schulze) led to stomatal behavior resulting in variation of both J_{wv} and J_{CO_2}, but $\partial J_{wv}/\partial J_{CO_2}$ was approximately constant. Consideration of other implications and determining the range of validity of the hypothesis represented by Equation 8.30 await future research.

C_3 Versus C_4 Plants

As a final topic we will recapitulate some of the characteristics of C_3 and C_4 plants that we have introduced and also examine the influence of stomata on maximum photosynthetic rates under optimal conditions.

The ecological advantages of the C_4 pathway are most apparent for environments having high PAR, high temperature, and limited water supply. C_4 plants are effective at high PAR, since, while photosynthesis for leaves of many C_3 plants saturates below 600 μmol m^{-2} s^{-1}, most C_4 plants have an increasing J_{CO_2} as the PAR is raised up to 2 000 μmol m^{-2} s^{-1} (see Fig. 8.12). Optimal temperatures for net CO_2 uptake are usually 20°C to 35°C for C_3 plants but 30°C to 45°C for C_4 plants; this can be interpreted by considering the enzyme ribulose-1,5-bisphosphate carboxylase/oxygenase, where the CO_2-evolving photorespiration, which has very low rates in C_4 plants, becomes proportionally more important at higher temperatures and thus reduces the net CO_2 uptake at the higher temperatures for C_3 plants. The optimal temperature for photosynthesis is actually fairly variable, since it can change by 10°C in a matter of days even for mature leaves, allowing for seasonal acclimation of photosynthetic performance. C_4 plants can more readily cope with limited water supply, since the gas phase conductance for CO_2 is relatively lower and the liquid phase CO_2 conductance is relatively higher for C_4 than for C_3 plants. Thus, C_4 plants tend to become dominant in deserts, grasslands, and certain subtropical regions, i.e., areas of high PAR, high temperature, and limited water supply, as indicated above (CAM plants have an even higher WUE than C_4 plants and achieve their greatest relative importance in regions of high PAR and very limited water supply). Low rates of photorespiration and the associated high WUE allow C_4 plants to become very successful weeds—in fact, 8 out of the 10 agriculturally most noxious weeds have the C_4 pathway (see Holm et al.).

At 30°C and for an absorbed PAR up to about 100 μmol m^{-2} s^{-1}, leaves of C_3 and C_4 plants can have a very similar quantum yield (approximately 0.053 mol CO_2/mol photons at an $N_{CO_2}^{ta}$ of 325 ppm; see Ehleringer and Björkman). As the temperature is raised, however, photorespiration increases relative to photosynthesis, and so the quantum yield declines for C_3 plants, but is essentially unchanged for C_4 plants. On the other hand, lowering the ambient O_2 level would raise the quantum yield for C_3 (photorespiring) plants, since the oxygenase activity of ribulose-1,5-bisphosphate carboxylase/oxygenase (see Fig. 8.8) would then be suppressed, while such changes would have little effect on C_4 plants until the O_2 level fell below about 2%, where mitochondrial respiration would be affected.

Because of the avid binding of HCO_3^- (relatively low Michaelis constant of about 10 μM for HCO_3^-) and high activity of the enzyme responsible for the initial fixation of CO_2 in C_4 plants (phosphoenolpyruvate carboxylase, which occurs in the cytosol of the mesophyll cells), CO_2 uptake by C_4 plants is CO_2-saturated at a relatively low CO_2 level in the intercellular air spaces.[*] For instance, an $N_{CO_2}^{ias}$ equivalent to 100 to 150 ppm usually leads to over 90% of the maximum J_{CO_2} for C_4 plants (Fig. 8.11), so increasing the ambient CO_2 level usually has little effect on the quantum yield. But the quantum yield for CO_2 fixation by C_3 plants progressively increases as the ambient CO_2 level is raised, and at an $N_{CO_2}^{ias}$ corresponding to 700 ppm it approaches to within 10% of the value occurring when the ambient O_2 level is reduced 10-fold (about 0.081). Such raising of the CO_2 level is another way of favoring the carboxylase activity of ribulose-1,5-bisphosphate carboxylase/oxygenase (see footnote, p. 435). Also, the requirement for a high $N_{CO_2}^{ias}$ for C_3 plants is consistent with the $K_{M_{CO_2}}$ of 5 to 15 μM CO_2 for CO_2 fixation by the carboxylase/oxygenase (noted on p. 431).

For a series of both C_3 and C_4 plants, the stomata open to a degree that gives an approximately constant CO_2 level in the intercellular air spaces, the level differing between plants representing the two photosynthetic pathways (Fig. 8.15; see Wong et al.). A similar adjustment in stomatal conductance also occurs as PAR increases for a particular plant (see Fig. 8.14). We must therefore conclude that stomata regulate the entry of CO_2, J_{CO_2}, to match the photosynthetic capability of the mesophyll region. The slope of $g_{CO_2}^{st}$ versus J_{CO_2} gives the drop in CO_2 across the stomata; i.e., $J_{CO_2} = g_{CO_2}^{st} \Delta N_{CO_2}^{st}$, and so the slope is $J_{CO_2}/g_{CO_2}^{st} = \Delta N_{CO_2}^{st}$. For the C_3 plants in Figure 8.15, the slope is 74×10^{-6} (i.e., 74 ppm CO_2 by volume), and for the C_4 plants it is 189×10^{-6}. For an ambient CO_2 level of 340 ppm, and ignoring the CO_2 drop across the boundary layer (about 10 to 30 ppm),

---

[*] An HCO_3^- concentration of 10 μM at pH 7 and 20°C is in equilibrium with about 2 μM CO_2, which corresponds to 50 ppm CO_2 in the gas phase. Hence, the K_M for the binding of HCO_3^- by phosphoenolpyruvate carboxylase occurs at a low atmospheric CO_2 level. A K_M of 10 μM for CO_2, as occurs for the carboxylase/oxygenase, would correspond to about 250 ppm in the gas phase.

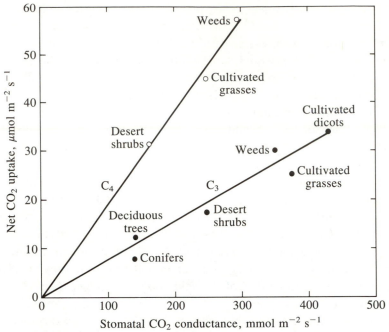

Figure 8.15

Relation between stomatal CO_2 conductance ($g^{st}_{CO_2}$) and net CO_2 uptake (J_{CO_2}) for various categories of C_3 and C_4 plants under optimal conditions and 340 ppm CO_2. (Data obtained from references cited in Tables 8.1 and 8.4.)

$N^{ias}_{CO_2}$ corresponds to 270 ppm for C_3 plants and 150 ppm for C_4 plants. Even though the level is lower for C_4 plants, it is still high enough to lead to CO_2 saturation of their CO_2 fixation pathway. A higher $N^{ias}_{CO_2}$ brought about by a higher $g^{st}_{CO_2}$ would thus not benefit photosynthesis for them, but the accompanying greater stomatal conductance would lead to more water loss. For a C_3 plant, photosynthesis does not approach saturation until $N^{ias}_{CO_2}$ exceeds 700 ppm (see Fig. 8.11). However, opening stomata further than that required to maintain an $N^{ias}_{CO_2}$ of about 270 ppm would not enhance photosynthesis very much, but it would considerably increase transpiration (see Fig. 8.14). Thus, the adjustment of stomatal opening to meet the conflicting demands of photosynthesis and transpiration, using feedback and feedforward control by microclimatic and leaf parameters, leads to a remarkable regulation that minimizes water loss while maximizing CO_2 uptake.

Problems

8.1. Consider a leaf 0.5 mm thick with 64 stomata per mm². Approximate the stomatal opening by a rectangle 6 μm by 20 μm; the depth of the stomatal pore is 25 μm. Assume that the leaf and air temperatures are both 20°C and the ambient air pressure is one atmosphere. (a) What are g_{wv}^{bl} (in mm s^{-1}) and r_{wv}^{bl} if the boundary layer is 0.8 mm thick? (b) What are na^{st} and the effective r^{st}? (c) What is the average flux density of water vapor within the stomatal pores compared with that across the boundary layer? (d) What is g_{wv}^{st} in mm s^{-1} and mmol m^{-2} s^{-1}? What are the values if the ambient air pressure is reduced to 0.9 atmosphere? (e) What is g_{wv}^{ias} in the two units if the effective path length in the intercellular air spaces equals the leaf thickness? (f) Suppose that each stoma is sunken in a cylindrical cavity 50 μm across and 100 μm deep. What additional resistance to water vapor diffusion does this provide?

8.2. Suppose that $g_{wv}^{bl_1}$ is 20 mm s^{-1}, $g_{wv}^{st_1}$ is 6 mm s^{-1}, g_{wv}^{c} is 0.1 mm s^{-1}, and g_{wv}^{ias} is 40 mm s^{-1}. (a) What is g_{wv}^{total} if water vapor diffuses out only across the lower epidermis of the leaf? (b) What are the three g_{wv}^{total}'s in (a) if the cuticular pathway is ignored, if the intercellular air spaces are ignored, and if both g_{wv}^{c} and g_{wv}^{ias} are ignored? (c) What is g_{wv}^{total} if the stomata in the upper epidermis are of the same size and frequency as in the lower one? (d) What is g_{wv}^{total} if 28% of J_{wv} is through the upper epidermis? (e) Suppose that the leaf temperature is 30°C, the air in the cell wall pores where the water evaporates is at 99% relative humidity, and c_{wv}^{ta} is 7.5 g m^{-3}. What is J_{wv} through the lower epidermis? (f) What are g_{wv}^{total} in mmol m^{-2} s^{-1} and J_{wv} (in mmol m^{-2} s^{-1}) under the conditions of (e)? Assume that the air pressure is one atmosphere and that N_{wv}^{ta} is 0.0100. (g) Under the conditions of (e) and (f), what is the drop in water vapor concentration and mole fraction along the stomatal pores (ignore cuticular transpiration)?

8.3. Suppose that a shade leaf has a layer of tightly packed palisade mesophyll cells with rectangular sides that are externally 30 μm by 100 μm and with square ends 30 μm by 30 μm (the long dimension is perpendicular to the leaf surface). Suppose that there are two spherical spongy mesophyll cells (30 μm in diameter) under each palisade cell. Let the cell wall thickness of mesophyll cells be 1.5 μm, the mean distance from the plasmalemma to the chloroplasts be 0.1 μm, and the average distance that CO_2 diffuses in the chloroplasts before reaching the photosynthetic enzymes be 0.5 μm. (a) What is A^{mes}/A if essentially the entire surface area of the mesophyll cells is exposed to the intercellular air spaces? (b) Assume that a sun leaf on the same plant has two layers of palisade cells and half as many spongy mesophyll cells. If the dimensions of the cells are the same as for the shade leaf, what is A^{mes}/A for the sun leaf? (c) If $D_{CO_2}^{cw}$ is 5.0×10^{-10} m^2 s^{-1}, what is the maximum value for $r_{CO_2}^{cw}$ at 20°C for the shaded leaf? (d) If P_{CO_2} is 2.0×10^{-4} m s^{-1} for the plasmalemma and the chloroplast limiting membranes, what are $r_{CO_2}^{pl}$ and $r_{CO_2}^{clm}$ (shade leaf)? (e) If $D_{CO_2}^{cyt}$ and $D_{CO_2}^{stroma}$ are 1.0×10^{-9} m^2 s^{-1}, what are $r_{CO_2}^{cyt}$ and $r_{CO_2}^{stroma}$ (shade leaf)? Assume that the relevant partition coefficients for the various forms of CO_2 are unity. (f) What is the resistance to CO_2 diffusion from the intercellular air spaces to the photosynthetic enzymes for the sun leaf? Assume that $J_{CO_2}^{r+pr}$ is negligible and that $r_{CO_2}^{cw}$ has its maximal, 20°C value.

8.4. Let us suppose that $c_{CO_2}^{ta}$ is 13 mmol m^{-3}, $K_{M_{CO_2}}$ is 5 μM, $r_{CO_2}^{bl}$ is 60 s m^{-1}, $r_{CO_2}^{leaf1}$ is 250 s m^{-1}, $r_{CO_2}^{mes}$ is 300 s m^{-1}, and $r_{CO_2}^{chl}$ is 250 s m^{-1} for a leaf that CO_2 enters only across the lower epidermis. (a) If the rate of gross photosynthesis is 1 mol CO_2 fixed m^{-3} s^{-1} when $c_{CO_2}^{chl}$ is 2 μM, what is $V_{max_{CO_2}}$? (b) What is $c_{CO_2}^{chl}$ when v_{CO_2} is 90% of $V_{max_{CO_2}}$? (c) If the rate of respiration plus photorespiration is 45% of that of gross photosynthesis, what are $r_{CO_2}^{total}$ and J_{CO_2}? Assume that $c_{CO_2}^{chl}$ is 4 μM. (d) Repeat (c) for a nonphotorespiring plant where the rate of respiration is 5% of $J_{CO_2}^{ps}$. Assume that $c_{CO_2}^{chl}$ is 2 μM. (e) Let us place a small transparent bag completely around a leaf of the nonphotorespiring plant. What would $c_{CO_2}^{chl}$ be if the CO_2 concentration in the bag in the steady state were 10 ppm? Assume that all resistances and the rate of respiration are unchanged. (f) What is the concentration of CO_2 in the mitochondria at night for the nonphotorespiring plant? Let $r_{CO_2}^{i}$ be 800 s m^{-1}, and assume that the rate of respiration as well as the resistances remain the same as the daytime values. What is the mitochondrial c_{CO_2} at night if stomatal closure causes $r_{CO_2}^{leaf1}$ to become 5 000 s m^{-1}?

8.5. Consider a sunlit leaf at 35°C with a g_{wv}^{bl} of 15 mm s^{-1}, stomata only in the lower epidermis, a g_{wv}^{ias} of 30 mm s^{-1}, and a $g_{CO_2}^{total}$ of 0.70 mm s^{-1}. Assume that the ambient air is at 30°C and 32% relative humidity, that cuticular transpiration is negligible and total transpiration is 5.0 mmol m^{-2} s^{-1}, and that the air in the intercellular air spaces reaches 100% relative humidity. (a) What are g_{wv}^{total} and g_{wv}^{st}? (b) What is the essentially immediate effect on g_{wv}^{total} and J_{wv} of decreasing the stomatal opening 4-fold, as can occur during wilting? (c) What is the qualitative effect of the action in (b) on T^{leaf}? (d) Neglecting effects caused by leaf temperature, what are the percentage changes of photosynthesis and water use efficiency caused by the action in (b)? Assume that $c_{CO_2}^{chl}$ is unchanged. (e) What is the essentially immediate effect on g_{wv}^{total} and J_{wv} of increasing the wind speed 4-fold? (f) What is the qualitative effect of the action in (e) on heat conduction across the boundary layer $[J_H^C = 2K^{air}(T^{leaf} - T^{air})/\delta^{bl}$, Eq. 7.10] and on T^{leaf}?

References

Aylor, D. E., J.-Y. Parlange, and A. D. Krikorian. 1973. Stomatal mechanics. *American Journal of Botany* 60:163–171.

Berry, J., and O. Björkman. 1980. Photosynthetic response and adaptation to temperature in higher plants. *Annual Review of Plant Physiology* 31:491–543.

Björkman, O. 1981. Responses to different quantum flux densities. In *Physiological Plant Ecology*, O. L. Lange, P. S. Nobel, C. B. Osmond, and H. Ziegler, eds. *Encyclopedia of Plant Physiology, New Series*, Vol. 12A. Springer-Verlag, Berlin. Pp. 57–107.

Clark, J. A., and G. Wigley. 1975. Heat and mass transfer from real and model leaves. In *Heat and Mass Transfer in the Biosphere, I, Transfer Processes in Plant Environment*, D. A. de Vries and N. H. Afgan, eds. Halsted Press, Wiley, New York. Pp. 413–422.

Cooke, J. R., J. G. De Baerdemaeker, R. H. Rand, and H. A. Mang. 1975. A finite element shell analysis of guard cell deformations. *Transactions of the American Society of Agricultural Engineers 19*:1107–1121.

Cooke, J. R., and R. H. Rand. 1980. Diffusion resistance models. In *Predicting Photosynthesis for Ecosystem Models*, Vol. 1, J. D. Hesketh and J. W. Jones, eds. CRC Press, Boca Raton, Florida. Pp. 93–121.

Cowan, I. R. 1972. Oscillations in stomatal conductance and plant functioning associated with stomatal conductance: observations and a model. *Planta 106*:185–219.

Cowan, I. R. 1977. Stomatal behavior and environment. *Advances in Botanical Research 4*:117–227.

Cowan, I. R. 1981. Coping with water stress. In *Biology of Australian Native Plants*, J. S. Pate and A. J. McComb, eds. University of Western Australia Press, Perth. Pp. 1–32.

Cowan, I. R., and F. L. Milthorpe. 1968. Plant factors influencing the water status of plant tissues. In *Water Deficits and Plant Structure*, Vol. I, T. T. Kozlowski, ed. Academic Press, New York. Pp. 137–193.

Cowan, I. R., and G. D. Farquhar. 1977. Stomatal function in relation to leaf metabolism and environment. In *Integration of Activity in the Higher Plant*, D. H. Jennings, ed. Society for Experimental Biology Symposium No. 31. Cambridge University Press, Cambridge.

Dacey, J. W. H. 1981. Pressurized ventilation in the yellow waterlily. *Ecology 62*: 1137–1147.

Edwards, G. E. and S. C. Huber. 1980. The C_4 pathway. In *The Biochemistry of Plants*, Vol. 8, M. D. Hatch and N. K. Boardman, eds. Academic Press, New York. Pp. 237–281.

Ehleringer, J., and O. Björkman. 1977. Quantum yields for CO_2 uptake in C_3 and C_4 plants: dependence on temperature, CO_2, and O_2 concentration. *Plant Physiology 59*: 86–90.

El-Sharkawy, M., and J. Hesketh. 1965. Photosynthesis among species in relation to characteristics of leaf anatomy and CO_2 diffusion resistances. *Crop Science 5*: 517–521.

Esau, K. 1965. *Plant Anatomy*, 2nd ed. Wiley, New York.

Evans, L. T. 1963. *Environmental Control of Plant Growth*. Academic Press, New York.

Farquhar, G. D., D. R. Dubbe, and K. Raschke. 1978. Gain of the feedback loop involving carbon dioxide and stomata. Theory and measurement. *Plant Physiology 62*:406–412.

Farquhar, G. D., E.-D. Schulze, and M. Küppers. 1980a. Responses to humidity by stomata of *Nicotiana glauca* L. and *Corylus avellana* L. are consistent with the optimization of carbon dioxide uptake with respect to water loss. *Australian Journal of Plant Physiology 7*:315–327.

Farquhar, G. D., S. von Caemmerer, and J. A. Berry. 1980b. A biochemical model of photosynthetic CO_2 assimilation in leaves of C_3 species. *Planta 149*:78–90.

Farquhar, G. D., and S. von Caemmerer. 1982. Modelling of photosynthetic response to environmental conditions. In *Physiological Plant Ecology*, O. L. Lange, P. S. Nobel, C. B. Osmond, and H. Ziegler, eds. *Encyclopedia of Plant Physiology, New Series*, Vol. 12B. Springer-Verlag, Berlin. Pp. 549–587.

Fischer, R. A. 1972. Aspects of potassium accumulation by stomata of *Vicia faba. Australian Journal of Biological Sciences 25*: 1107–1123.

Forster, R. E., J. T. Edsall, A. B. Otis, and F. J. W. Roughton, eds. 1969. *CO₂: Chemical, Biochemical, and Physiological Aspects*. National Aeronautics and Space Administration, Washington, D.C.

Gaastra, P. 1959. Photosynthesis of crop plants as influenced by light, carbon dioxide, temperature and stomatal diffusion resistance. *Mededelingen van de Landbouwhogeschool te Wageningen, Nederland 59*: 1–68.

Gale, J. 1972a. Availability of carbon dioxide for photosynthesis at high altitudes: theoretical considerations. *Ecology 53*: 494–497.

Gale, J. 1972b. Elevation and transpiration: some theoretical considerations with special reference to Mediterranean-type climate. *Journal of Applied Ecology 9*: 691–702.

Hall, A. E. 1982. Mathematical models of plant water loss and plant water relations. In *Physiological Plant Ecology*, O. L. Lange, P. S. Nobel, C. B. Osmond, and H. Ziegler, eds. *Encyclopedia of Plant Physiology, New Series*, Vol. 12B. Springer-Verlag, Berlin. Pp. 231–261.

Hall, A. E., and E.-D. Schulze. 1980. Stomatal response to environment and a possible interrelation between stomatal effects on transpiration and CO_2 assimilation. *Plant, Cell, and Environment 3*: 467–474.

Hatch, M. D., and C. R. Slack. 1970. Photosynthetic CO_2-fixation pathways. *Annual Review of Plant Physiology 21*: 141–162.

Holm, L. G., D. L. Plucknett, J. V. Pancho, and J. P. Herberger. 1977. *The World's Worst Weeds: Distribution and Biology*. University of Hawaii Press, Honolulu.

Jarvis, P. G., and T. A. Mansfield, eds. 1981. *Stomatal Physiology*. Cambridge University Press, Cambridge.

Körner, C. H., J. A. Scheel, and H. Bauer. 1979. Maximum leaf diffusive conductance in vascular plants. *Photosynthetica 13*: 45–82.

Kramer, P. J. 1969. *Plant and Soil Water Relationships*. McGraw-Hill, New York.

Lake, J. V. 1967. Respiration of leaves during photosynthesis. *Australian Journal of Biological Sciences 20*: 487–499.

Lommen, P. W., C. R. Schwintzer, C. S. Yocum, and D. M. Gates. 1971. A model describing photosynthesis in terms of gas diffusion and enzyme kinetics. *Planta 98*: 195–220.

Longstreth, D. J., T. L. Hartsock, and P. S. Nobel. 1980. Mesophyll cell properties for some C_3 and C_4 species with high photosynthetic rates. *Physiologia Plantarum 48*: 494–498.

Lorimer, G. H. 1981. The carboxylation and oxygenation of ribulose 1,5-bisphosphate: the primary events in photosynthesis and photorespiration. *Annual Review of Plant Physiology 32*: 349–383.

Maren, T. H. 1967. Carbonic anhydrase: chemistry, physiology, and inhibition. *Physiological Reviews 47*: 595–781.

Marrero, T. R., and E. A. Mason. 1972. Gaseous diffusion coefficients. *Journal of Physical and Chemical Reference Data 1*: 3–118.

Meidner, M., and T. A. Mansfield. 1968. *Physiology of Stomata.* McGraw-Hill, New York.

Nobel, P. S. 1976. Water relations and photosynthesis of a desert CAM plant, *Agave deserti. Plant Physiology 58*: 576–582.

Nobel, P. S. 1980. Leaf anatomy and water use efficiency. In *Adaptation of Plants to Water and High Temperature Stress*, N. C. Turner and P. J. Kramer, eds. Wiley, New York. Pp. 43–55.

Nobel, P. S., L. J. Zaragoza, and W. K. Smith. 1975. Relation between mesophyll surface area, photosynthetic rate, and illumination level during development for leaves of *Plectranthus parviflorus* Henckel. *Plant Physiology 55*: 1067–1070.

O'Leary, M. H. 1982. Phosphoenolpyruvate carboxylase: an enzymologist's view. *Annual Review of Plant Physiology 33*: 297–315.

Osmond, C. B., O. Björkman, and D. J. Anderson. 1980. *Physiological Processes in Plant Ecology. Toward a Synthesis with Atriplex.* Springer-Verlag, Berlin.

Parkhurst, D. 1977. A three-dimensional model for CO_2 uptake by continuously distributed mesophyll in leaves. *Journal of Theoretical Biology 67*: 471–488.

Parlange, J.-Y., and P. E. Waggoner. 1970. Stomatal dimensions and resistance to diffusion. *Plant Physiology 46*: 337–342.

Patterson, D. T. 1980. Light and temperature adaptation. In *Predicting Photosynthesis for Ecosystem Models*, Vol. 1, J. D. Hesketh and J. W. Jones, eds. CRC Press, Boca Raton, Florida. Pp. 205–235.

Penman, H. L., and R. K. Schofield. 1951. Some physical aspects of assimilation and transpiration. *Symposia of the Society for Experimental Biology 5*: 115–129.

Raschke, K. 1975. Stomatal action. *Annual Review of Plant Physiology 26*: 309–340.

Reid, R. C., J. M. Prausnitz, and T. K. Sherwood. 1977. *The Properties of Gases and Liquids*, 3rd ed. McGraw-Hill, New York.

Schulze, E.-D., O. L. Lange, U. Buschbom, L. Kappen, and M. Evenari. 1972. Stomatal responses to changes in humidity in plants growing in the desert. *Planta 108*: 259–270.

Šetlik, I., ed. 1970. *Prediction and Measurement of Photosynthetic Productivity.* Pudoc, Centre for Agricultural Publishing and Documentation, Wageningen, The Netherlands.

Smith, W. K., and G. N. Geller. 1979. Plant transpiration at high elevations: theory, field measurements, and comparisons with desert plants. *Oecologia 41*: 109–122.

Szarek, S. R., and I. P. Ting. 1975. Photosynthetic efficiency of CAM plants in relation to C_3 and C_4 plants. In *Environmental and Biological Control of Photosynthesis*, R. Marcelle, ed. W. Junk, The Hague. Pp. 289–297.

Tanton, T. W., and S. H. Crowdy. 1972. Water pathways in higher plants. III. The transpiration stream within leaves. *Journal of Experimental Botany 23*: 619–625.

Tenhunen, J. D., J. D. Hesketh, and D. M. Gates. 1980. Leaf photosynthesis models. In *Predicting Photosynthesis for Ecosystem Models*, Vol. 1, J. D. Hesketh and J. W. Jones, eds. CRC Press, Boca Raton, Florida. Pp. 123–181.

Tolbert, N. E. 1981. Metabolic pathways in peroxisomes and glyoxysomes. *Annual Review of Biochemistry 50*: 133–157.

Tyree, M. T., and P. Yianoulis. 1980. The site of water evaporation from sub-stomatal cavities, liquid path resistances and hydroactive stomatal closure. *Annals of Botany 46*:175–193.

Waggoner, P. E. 1969. Predicting the effect upon net photosynthesis of changes in leaf metabolism and physics. *Crop Science 9*:315–321.

Wong, S. C., I. R. Cowan, and G. D. Farquhar. 1979. Stomatal conductance correlates with photosynthetic capacity. *Nature 282*:424–426.

Zelitch, I. 1971. *Photosynthesis, Photorespiration, and Plant Productivity*. Academic Press, New York.

CHAPTER 9

Plants and Fluxes

In the previous chapter we analyzed gas fluxes for single leaves. We repeatedly used Fick's first law in the following form: flux density equals concentration difference divided by resistance, or, conductance times concentration difference. This approach can be extended to cover some of the physiological aspects of an entire plant community. We will first describe certain fluxes in the air above the plants. Although the fluxes of water vapor and CO_2 in the air above the vegetation resemble diffusion, in that the net migration of these gases is toward regions of lower concentration, we are here dealing not with the random thermal motion of molecules, but with the random motion of relatively large packets of air in the turbulent region above the plants.

Our next task will be to discuss concentrations and fluxes *within* a plant community. When we begin to analyze the water vapor and CO_2 fluxes from the soil up to the tops of plants, we are immediately confronted by the great structural diversity among different types of vegetation. Each plant community has its own unique spatial patterns for water vapor and CO_2 concentration. The possible presence of many layers of leaves and changing illumination also greatly complicates the analysis. In fact, even approximate descriptions of the gas fluxes within carefully selected plant communities involve complex calculations based on models incorporating numerous simplifying assumptions. To illustrate how CO_2 and water vapor concentrations and fluxes within a plant community can be handled in practical calculations, we will consider a cornfield as a specific example.

461

Three quarters of the water vaporized on land is transpired by plants. This water comes from the soil, and the soil also affects the CO_2 fluxes for vegetation. So, after we consider gas fluxes within a plant community, we will examine some of the hydraulic properties of soil. For instance, water in the soil is removed from larger pores before smaller ones. This removal decreases the soil conductivity for subsequent water movement, and a greater drop in water potential from the bulk soil up to a root is therefore necessary for a given water flux density.

Our final topic will be the flow of water in a continuous stream from the soil, to the root, into the root xylem, up to the leaves, and eventually out through the stomata into the atmosphere. As a useful first approximation, we can use the negative gradient of the water potential to represent the driving force for the flux across any segment where water moves as a liquid. We usually replace $-\partial\Psi/\partial x$ by $\Delta\Psi/\Delta x$. The greater the resistance—or, alternatively, the lower the conductance—the larger is the $\Delta\Psi$ required to maintain a given water flux across a particular component. However, $\Delta\Psi$ does not always represent the driving force on water. Furthermore, we must be prepared to recognize interactions between water movement in the xylem and in the other major transport system in plants, the phloem.

GAS FLUXES ABOVE THE LEAF CANOPY

When we considered the fluxes of H_2O and CO_2 for individual leaves (Ch. 8), we assumed that outside the boundary layers on each side of a leaf there is a turbulent region where both water vapor and CO_2 have uniform concentrations. Actually, as we will find, gradients in both CO_2 and H_2O exist within this turbulent region around plants. We will also find that the ambient wind speed in not constant, but varies with distance above the vegetation.

Wind Speed Profiles

Because of frictional interactions between moving air and a leaf, the air immediately adjacent to a leaf surface is stationary. As we move short distances away from the leaf surface, there is a transition from laminar flow parallel to the leaf in the lower part of the boundary layer to a turbulent flow with eddying motion (discussed in Ch. 7; see Fig. 7.5). The wind speed increases as we move still further away from the leaf (Fig. 9.1), often increasing

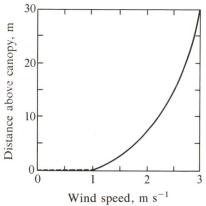

Figure 9.1

Change in wind speed with distance above a leaf at the top of a canopy. At the leaf surface v is 0, while at a distance δ^{bl} (of the order of mm) it is 1.0 m s^{-1}. At 0.5 m into the turbulent air, v increases to 1.1 m s^{-1}, and it can triple at 30 m above the canopy.

logarithmically for a few hundred meters above the leaf canopy.* Since the wind speed varies with distance in the turbulent region above vegetation, there is some ambiguity in deciding where to measure the ambient wind speed. Wind speed, however, generally does not increase very much until we are at least 1 m above the foliage (see Fig. 9.1). Thus, wind speed measured about 0.2 m above the vegetation may be used as the ambient value, which is needed to calculate the boundary layer thickness for an exposed leaf at the top of the canopy.

Plants exert a frictional drag on moving air masses and thereby modify the local wind patterns (Fig. 9.1). The fractional interaction between trees and wind is quite different from that of a flexible crop like wheat, which leads to different wind patterns in the overlying turbulent air (see Evans, and Oliver). Air flow within plant communities is also complicated by the three-dimensional architecture of the plants. For instance, wind speed does not necessarily decrease toward the ground—air in an open forest can "tunnel" under the branches and hence the wind speed can be greater there than further up in the canopy.

---

* The turbulent region generally extends 0.5 to 1.0 km above the earth's surface, at which point there is a region of more or less laminar flow in the direction of the prevailing wind.

Flux Densities

A transpiring and photosynthesizing plant community as a whole can have a net vertical flux density of CO_2 (J_{CO_2}) toward it and a net flux density of water vapor (J_{wv}) away from it into the turbulent air above the canopy. These flux densities are expressed per unit area of the ground or, equivalently, per unit area of the (horizontal) leaf canopy. Each of the flux densities is related to the appropriate gradient. The vertical flux density of water vapor, for example, depends on the rate of change of the water vapor concentration in the turbulent air, c_{wv}^{ta}, with respect to distance, z, above the vegetation:

$$J_{wv} = -K_{wv} \frac{\partial c_{wv}^{ta}}{\partial z} \tag{9.1}$$

In Equation 9.1 we are again employing the relation, flux density equals a proportionality coefficient times a force, where the force here is the negative gradient of water vapor concentration. Since J_{wv} from the plant community can be expressed in mol m^{-2} s^{-1} and $\partial c_{wv}^{ta}/\partial z$ in mol m^{-4}, the coefficient K_{wv} in Equation 9.1 can have units of m^2 s^{-1}, the same as for diffusion coefficients. In fact, K_{wv} in Equation 9.1 is analogous to D_j in Fick's first law ($J_j = -D_j \partial c_j/\partial x$, Eqs. 1.1 and 8.2), except for one very important distinction: it does not reflect the random thermal motion of water vapor molecules, but rather the irregular swirling motion of packets, or eddies, of air in the turbulent region (see Fig. 9.2). This makes the coefficients much larger than for the molecular case.

Before further considering the random eddying motion of air packets, let us write down a general expression for the CO_2 flux density in the turbulent air above the canopy. By analogy with Equation 9.1, we can represent the flux density of CO_2 as

$$J_{CO_2} = -K_{CO_2} \frac{\partial c_{CO_2}^{ta}}{\partial z} \tag{9.2}$$

where K_{CO_2} is the "air packet" or *eddy* diffusion coefficient for CO_2. (K_j is also referred to as a transfer coefficient, an exchange coefficient, or a diffusivity coefficient.) Similarly, the vertical flux density of O_2 in the turbulent air, J_{O_2}, could be equated to $-K_{O_2} \partial c_{O_2}^{ta}/\partial z$. We will let the positive direction for z be increasing altitude; consequently, the positive direction for a net flux density is from the leaf canopy upward into the turbulent air.

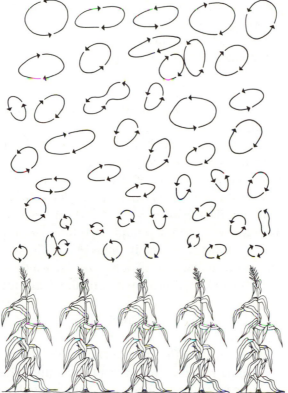

Figure 9.2
Schematic illustration of small packets or eddies of air swirling
about in the turbulent region above vegetation. The eddies,
which tend to increase in size with height, carry all molecules
they contain more or less as a unit. They are continuously
changing size—breaking up, or coalescing with other eddies—
making their actual size somewhat hypothetical.

Eddy Diffusion Coefficients

The eddy diffusion coefficients, K_{wv} and K_{CO_2}, unlike the ordinary diffusion
coefficients, D_{wv} and D_{CO_2}, have the *same* value in a given situation. A small
packet of air moves more or less as a unit, and thus carries with it the H_2O,
CO_2, and other molecules that it contains (see Fig. 9.2). Although we cannot
really assign actual volumes to these eddies—which are constantly changing
in size and shape, because of shearing effects or coalescence with neighboring
packets—they are indeed very large compared with intermolecular distances,

and contain enormous numbers of molecules. The random motion of an air packet is caused by random fluctuations in pressure in local regions of the turbulent air. The eddying motions of the air packets promote a mixing, formally like the mixing due to diffusion, and thus lead to relations such as Equations 9.1 and 9.2. Besides their eddying motion, air packets have an average drift velocity represented by the local wind velocity. Pressure gradients over large distances cause the winds and the resulting horizontal drift of the air packets.

The actual values for K_{wv} and K_{CO_2} describing the "diffusion" of air packets vary with the wind speed above the canopy. Also, eddy diffusion coefficients are affected by the rates of change of both the wind speed and air temperature with altitude. For instance, hot air tends to rise and become replaced by cooler air. Such buoyancy effects, which are encouraged when $\partial T/\partial z$ is steeply negative above the canopy, lead to more rapid mixing and higher values for K_j. Usually, the eddy diffusion coefficients are approximately proportional to the local wind speed. As the wind speed increases, there is more opportunity for turbulent mixing of the air, and thus K_j becomes larger. Since the wind speed varies with height (see Fig. 9.1), and since K_j also depends on the gradient in wind speed, we often employ an eddy diffusion coefficient averaged over an appropriate distance to describe the vertical fluxes in some region of the turbulent air above the canopy. Morever, the wind speed, its gradient, and the vertical temperature gradient all vary during the day. Consequently, our value for K_j should also be averaged over a suitable time interval, e.g., an hour.

For a moderate wind speed of 2 m s^{-1}, the eddy diffusion coefficient is generally 0.05 to 0.2 m^2 s^{-1} just above the leaf canopy. Under these conditions, K_j might be about 1.5 m^2 s^{-1} at 30 m above the canopy, and in excess of 3 m^2 s^{-1} at or above 300 m, where turbulent mixing is even greater (see van Wijk). By comparison, D_{wv} is 2.4×10^{-5} m^2 s^{-1} and D_{CO_2} is 1.5×10^{-5} m^2 s^{-1} in air at 20°C. Thus, K_j is 10^4 to 10^5 times larger in the turbulent air above the canopy than are these D_j's. The random motion of air packets is indeed much more effective than the random thermal motion of molecules in moving H_2O and CO_2.

Since K_j increases rapidly with altitude as we move into turbulent regions with higher wind speeds, the steady-state concentration gradients become less steep with increasing height above the vegetation. Specifically, J_{wv} equals $-K_{wv}\partial c_{wv}^{ta}/\partial z$ by Equation 9.1, and as K_{wv} increases with altitude, the absolute value of $\partial c_{wv}^{ta}/\partial z$ must become smaller at greater heights above the canopy. K_j may increase, for example, by a factor of ten in the first 30 m above the vegetation, in which case the gradient in water vapor concentration could decrease 10-fold in this interval.

The equality of the eddy diffusion coefficients for different gaseous species results from the air packet and the molecules within it moving as a unit. In fact, K_j is often assumed to be the same for the transfer of gases, heat, and momentum (expressed in the same units). K_j is therefore generally measured for the most convenient quantity in some situation, and is then assumed to be the same (or at least similar) for all others (see Evans; Oke, Ch. 7 reference; Rosenberg, Ch. 7 reference; C. B. Tanner in Kozlowski 1968, Vol. I; and van Wijk).

Resistance of Air Above the Canopy

As in our use of resistances for gaseous diffusion in the previous chapter, we will identify a resistance to the flow of water vapor in the turbulent air by r_{wv}^{ta}, and that for CO_2 by $r_{CO_2}^{ta}$. To derive such quantities, we will replace the negative gradient by the difference in concentration of species j, Δc_j^{ta}, across a given distance, Δz, in the turbulent air, i.e., $-\partial c_j^{ta}/\partial z \cong \Delta c_j^{ta}/\Delta z$. By analogy with our previous definition of conductance and resistance ($g_j^{bl} = J_j/\Delta c_j^{bl} = D_j/\delta^{bl} = 1/r_j^{bl}$, Eq. 8.3), we can identify resistances from the flux density expressions in Equations 9.1 and 9.2. Since K_{wv} has the same value as K_{CO_2}, we obtain the following equalities:

$$r_{wv}^{ta} = \frac{\Delta c_{wv}^{ta}}{J_{wv}} = \frac{\Delta z}{K_{wv}}$$

$$= \frac{\Delta z}{K_{CO_2}} = \frac{\Delta c_{CO_2}^{ta}}{J_{CO_2}} = r_{CO_2}^{ta} \tag{9.3}$$

As with the analogous relations in Chapter 8 (e.g., Eqs. 8.3 and 8.5), Equation 9.3 describes the steady state condition. Equation 9.3 indicates that r_{wv}^{ta} has the same value as $r_{CO_2}^{ta}$, as we would indeed expect based on the random motions of whole packets of air (also, $g_{wv}^{ta} = 1/r_{wv}^{ta} = 1/r_{CO_2}^{ta} = g_{CO_2}^{ta}$).

Let us now estimate the resistance of the turbulent air immediately above a leaf canopy. For simplicity, we will let K_j average 1.0 $m^2\ s^{-1}$ for the first 30 m above the plants, a typical value in a moderate wind during the daytime. By Equation 9.3, the resistances over this 30 m interval then are

$$r_{wv}^{ta} = r_{CO_2}^{ta} = \frac{(30\ m)}{(1\ m^2\ s^{-1})}$$

$$= 30\ s\ m^{-1}$$

Measured values for these resistances generally range from 20 to 40 s m^{-1} for moderate wind speeds, as do predicted values from computer analyses of r_j^{ta} using models incorporating the variation of K_j with altitude. Wind speeds, and therefore K_j, tend to be lower at night, so r_j^{ta} tends to be somewhat higher then than during the daytime (see J. L. Monteith in Evans).

Transpiration and Photosynthesis

As mentioned, we will express flux densities above plants per unit area of the ground or, equivalently, per unit area of the canopy. For many agricultural as well as ecological considerations, such a measure of the average transpiration or photosynthesis of the whole plant community is far more useful than the water vapor or CO_2 flux densities of an individual leaf. Environmental measurements in the turbulent air above vegetation can thus be a very important approach to measuring the overall rates of transpiration and photosynthesis, especially if the extent of similar plants is fairly large, as might occur for a cornfield or a grassland. Moreover, such measurements can generally be made without disturbing the plants or their leaves. On this large scale, however, we unavoidably lose sight of certain factors, such as the effect of stomatal opening or leaf size on the gas fluxes. Also, the turbulent air above the canopy is greatly influenced by the terrain as well as by the vegetation, so we must reckon with other factors not involved in our study of leaves. For example, both K_j and the gradients in water vapor or CO_2 depend on whether we are at the edge or the center of a field, whether and what types of trees are present, and whether the region is flat or hilly.

For simplicity, we are considering a one-dimensional situation where the net fluxes of water vapor and CO_2 occur only in the vertical direction above the canopy, as would be the case near the center of a large uniform plant community. Just as our assumption of a boundary layer of uniform thickness breaks down at the leading and trailing edges of a leaf, we must also consider air packets transferring H_2O and CO_2 horizontally in and out at the sides of certain areas of vegetation. Such net horizontal transfer of various gases is referred to as *advection*. Instead of using Equations 9.1 through 9.3 to analyze net gas flux densities, we may have to use much more cumbersome three dimensional equations to handle advection for small fields or individual plants.

As we considered in Chapter 8 (p. 444), J_{wv}/J_{CO_2} might be about 500 H_2O/CO_2 for a representative sunlit mesophytic leaf, and 100 to 200 for a photosynthetically efficient C_4 species such as corn. For an entire plant community, however, the water lost per CO_2 fixed is often considerably higher than for a single, well-illuminated leaf. In particular, J_{wv} measured above the canopy also includes water vapor coming from the soil and from leaves that are not well illuminated and which therefore contribute little to net photosynthesis. Some of the CO_2 taken up by the plant community is evolved by soil microorganisms, root cells, and leaves that do not receive much sunlight and so are below the light compensation point—all of which decrease the amount of CO_2 that needs to be supplied from above the plant canopy. These effects tend to raise J_{wv}/J_{CO_2} above the values for an exposed leaf. Although the absolute value of J_{wv}/J_{CO_2} above a leaf canopy depends on the ambient relative humidity and the physiological status of the plants, it is generally between 300 and 2 000 H_2O/CO_2 when averaged over a day in the growing season. Moreover, mainly because J_{CO_2} for C_4 plants is often about twice as large as for C_3 plants, the absolute value of J_{wv}/J_{CO_2} is lower and daily growth tends to be greater for C_4 plants (see Kozlowski 1968, Vol. II; Kramer; and Zelitch).

Values for Fluxes and Concentrations

We will use representative values of J_{CO_2} to calculate the concentration drops of CO_2 that might be expected over a certain vertical distance in the turbulent atmosphere. When there is net photosynthesis, the net flux density of CO_2 is directed from the turbulent air down into the canopy. J_{CO_2} above the vegetation is then negative by our sign convention, which means that $c_{CO_2}^{ta}$ increases as we go vertically upward ($J_{CO_2} = -K_{CO_2} \, \partial c_{CO_2}^{ta}/\partial z$, Eq. 9.2). But this is just as we would expect if CO_2 is to be transferred downward toward the plants by the random motion of the eddies in the turbulent air.

A commonly observed value for J_{CO_2} above the plant canopy is $-20 \, \mu$mol m^{-2} s^{-1} at midday. For comparison, the flux density of CO_2 into an exposed *leaf* of a mesophyte at a moderate light level might be 9 μmol m^{-2} s^{-1} (p. 440). Using Equation 9.3 and a resistance of 30 s m^{-1} for the lower 30 m of the turbulent air ($r_{CO_2}^{ta}$), we calculate that the CO_2 concentration drop across this region might be[*]

$$\Delta c_{CO_2}^{ta} = J_{CO_2} r_{CO_2}^{ta} = (-20 \times 10^{-6} \text{ mol m}^{-2} \text{ s}^{-1})(30 \text{ s m}^{-1})$$

$$= -0.6 \text{ mmol m}^{-3}$$

---

[*] By analogy with Ohm's law ($\Delta E = IR$), in this chapter we will use the form $\Delta c_j^{ta} = J_j r_j^{ta}$ for our calculations of concentration drops.

Employing a conversion factor from Table 8.2, this $\Delta c_{CO_2}^{ta}$ corresponds to (0.6)(24.4), or 15 ppm by volume (a CO_2 mole fraction of 15×10^{-6}; or a CO_2 partial pressure of 1.5 Pa, or 15 μbar) at 20°C and a pressure of 0.1 MPa (1 bar). Thus, CO_2, which generally has a concentration near 340 ppm well into the turbulent air, e.g., 30 m above vegetation, could be at $340 - 15$, or 325 ppm just above the canopy (see Loomis et al., Partridge, Šetlik, San Pietro et al., and Zelitch).

For a rapidly photosynthesizing corn crop at noon, J_{CO_2} can be -60 μmol $m^{-2} s^{-1}$. For the above $r_{CO_2}^{ta}$, $\Delta c_{CO_2}^{ta}$ would be -44 ppm by Equation 9.3. Thus, the CO_2 concentration at the top of the canopy might be 296 ppm, i.e., 44 ppm lower than the 340 ppm in the turbulent air tens of meters above the corn plants. In fact, measurement of $c_{CO_2}^{ta}$ at the canopy level can indicate the net rate of photosynthesis by the plants. At night respiration occurs, but not photosynthesis, and so vegetation then acts like a *source* of CO_2. Thus, the concentration of CO_2 just above the canopy at night is usually a few ppm greater than it is higher up in the turbulent air, as we will see below.

The flux density of water vapor just above the canopy, which represents transpiration from the leaves plus evaporation from the soil, is often referred to as *evapotranspiration*. For fairly dense vegetation this evapotranspiration is appreciable, generally amounting to 60% to 90% of the flux density of water vapor from an exposed water surface at the ambient air temperature. The daily evapotranspiration from a forest is often equivalent to a layer of water 3 to 5 mm thick, which averages out to 2 to 3 mmol $m^{-2} s^{-1}$ (see Kozlowski 1968, Vol. II; Kramer; and Slatyer). At noon on a sunny day with a moderate wind, J_{wv} above a leaf canopy can be 7 mmol water $m^{-2} s^{-1}$. Using Equation 9.3 ($\Delta c_{wv}^{ta} = J_{wv} r_{wv}^{ta}$), and employing our above value for r_{wv}^{ta} of 30 s m^{-1}, we then calculate that water vapor decreases in concentration by 0.21 mol m^{-3} over the first 30 m of the turbulent air. We indicated in Chapter 8 that the turbulent air immediately outside the boundary layer adjacent to a leaf contains 0.48 mol water m^{-3} when it is at 20°C and 50% relative humidity (Fig. 8.5). Our calculation indicates that c_{wv}^{ta} could drop by 0.21 mol m^{-3}, or from 0.48 to 0.27 mol m^{-3} (28% relative humidity at 20°C), as we move 30 m upward into the turbulent air above the canopy. Such a marked decrease in absolute and relative humidity is indeed often observed in the turbulent air above vegetation.

Condensation

What appears to be steam is often seen rising from leaves or other surfaces (Fig. 9.3) when the sun breaks through the clouds following a rainstorm or

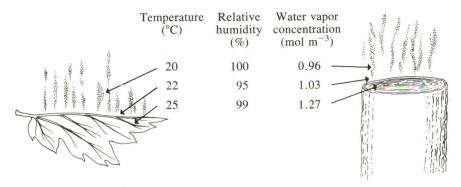

Temperature (°C)	Relative humidity (%)	Water vapor concentration (mol m^{-3})
20	100	0.96
22	95	1.03
25	99	1.27

Figure 9.3
"Steam" rising from a leaf and a fencepost that are rapidly heated by the sun after a rainstorm. Moisture-laden air next to the objects is swept in an eddying motion into a cooler region, where the water vapor condenses out.

at sunrise following a night with a heavy dew. To analyze this phenomenon, we will assume that the sun warms up the leaves at the top of the canopy to 25°C and that the concentration of water vapor in the cell wall pores of their mesophyll cells is then 1.27 mol m^{-3} (see p. 412). Suppose that the air just outside the boundary layer adjacent to a leaf is at 22°C and has a high relative humidity of 95% just after the rainstorm; c_{wv}^{ta} is then 1.03 mol m^{-3} ($c_{wv}^{*} = 1.08$ mol m^{-3} at 22°C). Hence, water vapor will diffuse from the leaf, across the boundary layer, and into the turbulent air (Fig. 9.3). Now let us suppose that the air at a greater distance from the leaf is somewhat cooler, e.g., 20°C at 10 mm from the leaf. At 20°C c_{wv}^{*} is 0.96 mol m^{-3}. Thus, as the air with 1.03 mol water m^{-3} moves away from the unstirred layer adjacent to the leaf in an eddy, or air parcel, it will be cooled and some of its water vapor will condense, since c_{wv}^{ta} cannot exceed c_{wv}^{*} for the local air temperature. This condensation leads to the fog, or "steam," seen moving away from the plants into the surrounding cooler turbulent air (Fig. 9.3).

GAS FLUXES WITHIN PLANT COMMUNITIES

The pattern of concentrations and fluxes obviously depends on the particular plant community being considered. We will not attempt to examine all types of vegetation, but instead we will focus on a cornfield representing a mono-specific stand of high productivity (see Lemon et al.). The same general principles apply to other fairly uniform plant communities, but isolated plants provide special difficulty for analysis, since the gas concentrations and fluxes then vary in three dimensions (also see App. VIII).

Eddy Diffusion Coefficient and Resistance

Let us begin by considering how the eddy diffusion coefficient might vary within a plant community. Near the ground there is a boundary layer that can be rather thick, since the air there is generally quite still. In fact, K_j often averages 5×10^{-5} m^2 s^{-1} in the first 10 mm above the ground, a value only 2 to 3 times larger than the diffusion coefficients of water vapor and CO_2 in air.* As we move up to the top of the canopy, the eddy diffusion coefficient increases, often more or less logarithmically with altitude in the upper part of many plant communities. It may reach a value of 0.2 m^2 s^{-1} at the top of

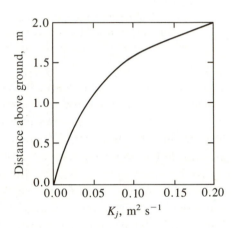

Figure 9.4
Idealized representation of the variation in the eddy diffusion coefficient within a uniform corn crop 2 m in height. The wind speed is 2 m s^{-1} at the top of the canopy.

a canopy in a moderate wind (Fig. 9.4). Since K_j is approximately proportional to wind speed, v^{wind} within the plant community varies in a manner similar to the variation described for the eddy diffusion coefficient (Fig. 9.4). For instance, the wind speed about 0.2 m above the ground might be 0.1 m s^{-1}, increasing to 2 m s^{-1} at the top of the canopy (see Millington and Peters).

Two aspects concerning K_j within the plant community deserve special emphasis. First, transfer of gaseous substances within the vegetation takes place by the random motion of relatively large parcels or eddies of air, just as in the turbulent region above the canopy. Second, because of frictional drag with the many leaves, branches, and other plant parts, the eddy diffusion

---

* When K_j is of the same order of magnitude as diffusion coefficients, differences in movement between individual gases may become apparent. Effects of individual D_j's on K_j are noticeable only in the first 10 mm or so above the ground and they are ignored in the present example.

Table 9.1
Summary of gas exchange parameters within a 2-m-tall corn crop at noon on a sunny day.
Also indicated are the flux densities at ground level and just above the canopy.

Height above ground (m)	av K_j (m² s⁻¹)	r_j^{ta} (s m⁻¹)	av J_{wv} (mmol m⁻² s⁻¹)	Δc_{wv} (mol m⁻³)	av J_{CO_2} (μmol m⁻² s⁻¹)	Δc_{CO_2} (ppm)
above canopy			7		−60	
1 to 2	1×10^{-1}	10	4	0.04	−30	−7
0.1 to 1	2×10^{-2}	50	1	0.05	3	4
0.01 to 0.1	1×10^{-3}	90	0.5	0.05	2	5
0.00 to 0.01	5×10^{-5}	200	0.5	0.10	2	10
ground level			0.5		2	

coefficient within the vegetation is considerably less than in the air above the canopy.

We will now estimate the resistance of the turbulent air from the ground to the top of a corn crop 2 m in height (Fig. 9.4). To illustrate the relative contributions of various air layers, we will let K_j average 5×10^{-5} m² s⁻¹ for the first 0.01 m, 1×10^{-3} m² s⁻¹ for the next 0.09 m, 2×10^{-2} m² s⁻¹ for the next 0.9 m, and 1×10^{-1} m² s⁻¹ for the last 1 m, consistent with the plot of K_j versus height in Figure 9.4 (also see Table 9.1). We can use Equation 9.3, $r_{wv}^{ta} = \Delta z / K_{wv} = r_{CO_2}^{ta}$, to estimate the resistance of each of the four air layers in series; e.g., r_j^{ta} for the lowermost layer is (0.01 m)/(5 × 10⁻⁵ m² s⁻¹), or 200 s m⁻¹ (see Table 9.1). The total resistance from the ground up to the top of the canopy equals $200 + 90 + 50 + 10$, or 350 s m⁻¹ (summarized in Table 9.1). (In a sense, we are performing a numerical integration to determine the resistance.) Computer analyses using models describing the turbulent air within such a crop also indicate that the resistance is generally 300 to 400 s m⁻¹. Most of the resistance within a plant community is generally due to the relatively still air next to the ground. For instance, just over half (200 s m⁻¹ out of 350 s m⁻¹) of the resistance for the 2 m pathway is provided by the lowest 0.01 m, while the entire upper half of the corn crop accounts for a resistance of only 10 s m⁻¹ (see Table 9.1).

Water Vapor

A considerable amount of water can evaporate from the soil and move by air packets up through the vegetation. For instance, J_{wv} from a moist,

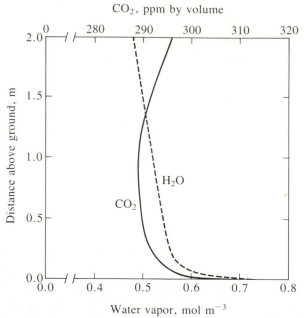

CO$_2$, ppm by volume

Water vapor, mol m^{-3}

Figure 9.5

Possible variation of water vapor and CO$_2$ concentrations within a 2-m-high corn crop at noon on a sunny day. At the top of the canopy the wind speed is 2 m s^{-1}. In the turbulent air 30 m above the vegetation, c_{wv}^{ta} is 0.3 mol m^{-3}, and $c_{CO_2}^{ta}$ corresponds to 340 ppm by volume. (See Lemon et al. for actual values under field conditions.)

intermittently illuminated soil, such as commonly occurs for a temperate forest, can be 0.2 to 1 mmol m^{-2} s^{-1}. (For comparison, 0.5 mmol m^{-2} s^{-1} corresponds to a depth of water of 0.8 mm/day or 280 mm/year.) If a flow of 0.5 mmol m^{-2} s^{-1} occurs across a resistance of 290 s m^{-1} to reach a distance 0.1 m above the ground, we can calculate using Equation 9.3 (by which $\Delta c_{wv}^{ta} = J_{wv} r_{wv}^{ta}$) that the drop in water vapor from the ground to this level is 0.15 mol m^{-3}. At 20°C the saturation concentration is 0.96 mol m^{-3} (App. II), so a water vapor drop of 0.15 mol m^{-3} then corresponds to a 16% decrease in relative humidity. Thus, an appreciable drop in water vapor concentration can occur across the relatively still air near a moist soil under a canopy (Fig. 9.5).

Because of the addition of water vapor from transpiration by the leaves, J_{wv} increases as we move from the ground up through a corn crop. On a sunny day the water vapor flux density might be 1 mmol m^{-2} s^{-1} at 0.5 m, 2 at 1.0 m, 4 at 1.5 m, and 7 mmol m^{-2} s^{-1} at 2.0 m, the top of the canopy.

(On a cloudy humid day, J_{wv} for a corn crop might be only 1 mmol m^{-2} s^{-1} at the top of the canopy.) If J_{wv} averages 1 mmol m^{-2} s^{-1} from 0.1 to 1.0 m above the ground, where the resistance is 50 s m^{-1} (Table 9.1), then by Equation 9.3 Δc_{wv}^{ta} for this part of the pathway is 0.05 mol m^{-3}. For the upper 1 m the resistance is 10 s m^{-1}, and so, for an average J_{wv} of 4 mmol m^{-2} s^{-1}, the decrease in water vapor concentration would be 0.04 mol m^{-3} (Table 9.1). Thus, Δc_{wv}^{ta} is 0.15 mol m^{-3} over the 0.1 m just above the ground, 0.05 from 0.1 to 1.0 m, and 0.04 from 1.0 to 2.0 m, or 0.24 mol m^{-3} overall (see Table 9.1 and Fig. 9.5). If the turbulent air at the top of the canopy is at 20°C and 50% relative humidity, it would contain 0.48 mol water m^{-3}. The air near the soil would then contain approximately 0.48 + 0.24, or 0.72 mol H$_2$O m^{-3}, which corresponds to 75% relative humidity at 20°C (c_{wv}^* = 0.96 mol m^{-3} at 20°C; App. II). In summary, we note that (1) air close to the soil under a (fairly dense) leaf canopy can have an appreciable relative humidity, (2) c_{wv}^{ta} continuously decreases as we move upward from the ground, (3) most of the overall drop in water vapor concentration occurs near the ground, and (4) most of the water vapor actually comes from the upper half of the corn crop in the present example.

PAR Attenuation

Before discussing J_{CO_2} within a plant community, we should consider how the amount of light varies down through the various layers of vegetation, since the PAR (photosynthetically active radiation, 400 to 700 nm) at each level helps determine the rate of photosynthesis there. The net rate of CO$_2$ fixation approaches light-saturation near a PAR of 600 μmol m^{-2} s^{-1} for leaves of many C$_3$ plants, and it decreases to zero at the light compensation point. A comprehensive formulation—including effects of leaf angle, sun elevation in the sky, the finite width of the sun's disc, changes in spectral distribution of PAR at various levels within the plant community, multiple reflections of PAR from leaves and other surfaces, as well as clumping versus uniform arrangement of leaves—would lead to a nearly hopeless complication of the algebra. Instead, we will assume that the decrease in PAR is due to absorption by the foliage in a manner analogous to Beer's law, $\ln(J_b/J_0) = -k_\lambda cb$ (Eq. 4.14). This approximation is particularly useful when there is a random distribution of leaves horizontally, as can occur in certain moderately dense plant communities.

As we move downward into the vegetation, PAR decreases more or less exponentially with the amount of absorbing material encountered. For some canopies the greatest leaf area occurs near the center (e.g., many grasses), and for others it occurs about three-quarters of the way up from the ground (e.g., many crops and trees). We will let F be the average cumulative total leaf area per unit ground area as we move down through the plant community. The dimensionless parameter F uses the area of only one side of a leaf, and thus it is expressed on the same basis as our flux densities. F is zero at the top of the canopy, and takes on its maximum value at ground level, a value generally referred to as the *leaf area index*. If the leaves in a particular plant community were horizontal, the leaf area index would equal the average number of leaves above any point on the ground.

We will represent the PAR incident on the leaf canopy by J_0. Primarily because of absorption by photosynthetic pigments, PAR is attenuated as it moves down through the plant community. At any level in the vegetation, J, the photon flux density from 400 to 700 nm, is related to J_0 and F as follows:

$$\ln \frac{J_0}{J} = kF \tag{9.4}$$

where k is a dimensionless parameter describing the absorption properties of a particular type of foliage and is referred to as the *foliar absorption coefficient*. Since we are ignoring changes in spectral distribution at different levels in the vegetation, J_0 and J in Equation 9.4 can actually be measured as the flux density of photons from 400 to 700 nm or as an energy flux density for these photons. Equation 9.4 was introduced into plant studies by Monsi and Saeki in 1953 (see Eastin et al., Monsi and Saeki, T. Saeki in Evans, Šetlik, and Zelitch).

Values of Foliar Absorption Coefficients

The foliar absorption coefficient k has values between 0.3 and 1.3 for most leaf canopies. Light penetrates the vertically oriented blades of grasses rather easily; in such cases, k often has a low value near 0.4. At what cumulative leaf area per ground area would the incident PAR be reduced by 95% for grasses with a foliar absorption coefficient of 0.4? By Equation 9.4, the accumulated leaf area per unit ground area in this case would be

$$F = \frac{\ln \left(\dfrac{J_0}{0.05\, J_0} \right)}{0.4} = 7.5$$

Thus, when the average leaf area index is 7.5 for such grasses, 5% of the PAR would reach the soil surface. For 95% of the PAR to be absorbed for a

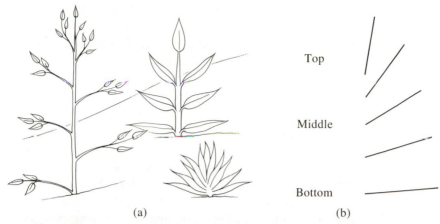

Figure 9.6
Variation in leaf angle and hence foliar absorption coefficient with distance
above the ground for (a) various idealized plants and (b) sugar beet measured at various canopy positions (see Hodáňová). The greater erectness
of the uppermost leaves leads to a lower k for them, and hence better
penetration of PAR down to the lower leaves.

leaf area index of 3, k must be 1.0 by Equation 9.4. Such a relatively high
foliar absorption coefficient applies to horizontal leaves with at least 0.5 g
chlorophyll m^{-2}, which can occur for crop plants like potato, soybean,
sunflower, and white clover.

When the sun is overhead, vertical leaves absorb less sunlight and reflect
more of it down into the vegetation per unit leaf area than do horizontal
leaves. This accounts for the low values of k for grasses, since their leaves are
generally rather erect. Moreover, leaves tend to be vertical near the top of
certain plants, e.g., sugar beet, becoming on the average more horizontal
toward the ground (Fig. 9.6). This orientation reduces the foliar absorption
coefficient of the upper leaves, and therefore more of the light incident on the
plants is available for the lower leaves. In fact, optimal light utilization for
photosynthesis generally occurs when the incident PAR is distributed as
uniformly as possible over the leaves, since the fraction of leaves exposed to
PAR levels above light saturation or below light compensation is then
usually minimized (see Monsi et al., Moss and Musgrave, and Yoshida).
Indeed, canopies with erect leaves or angles varying from erect near the top
of the canopy to horizontal near the bottom (Fig. 9.6; also see McMillen
and McClendon) tend to have higher productivities when the leaf area index
exceeds about 3 (see Duncan, Monsi et al., and Turitzin and Drake). Our
arguments about the effect of leaf orientation on k presuppose that essentially
all of the light is incident on the top of the canopy. When much PAR comes

in from the sides, as for an isolated tree, foliar absorption coefficients determined for vertically incident light should not be used in Equation 9.4—indeed, k can be determined for other sun angles. Also, a foliar absorption coefficient can be determined for shortwave irradiation, instead of just for PAR, as considered here (see Norman, and Ross).

Light Compensation Point

Let us now consider the light compensation point for CO_2 fixation by leaves. As we mentioned in the last chapter (p. 438), the light compensation point generally occurs at a PAR of about 8 μmol m^{-2} s^{-1} for a leaf temperature near 20°C and a CO_2 concentration of 340 ppm ($c_{CO_2}^{ta}$ is somewhat below this level within the plant community, as we will show below). Suppose that a moderate PAR of 400 μmol m^{-2} s^{-1} occurs on trees whose leaves have a foliar absorption coefficient of 0.8. At what cumulative area of leaves per unit ground area is a light compensation point of 8 μmol m^{-2} s^{-1} reached? By Equation 9.4, F is then 4.9. Thus, only the upper five "layers" of leaves in a dense forest might be above the light compensation point for that part of the day when the PAR on the canopy is 400 μmol m^{-2} s^{-1}. Of course, for a lower PAR on the leaf canopy, more leaves would be below the light compensation point. Occasional sunflecks of high PAR reach the lower parts of the vegetation, which can complicate our analysis of where light compensation occurs.

Leaves that are below the light compensation point for most of the day do not contribute to the net photosynthesis of the plant because of insufficient PAR penetrating down to them. Such leaves generally lose 30% to 50% of their dry weight before actually dying and abscising. Following this loss of leaves on the lower branches of trees, the branches themselves die and eventually fall off or are blown off by the wind. Thus, tall trees in a dense forest often have few or no branches over the lower part of their trunks.

CO_2 Concentrations and Fluxes

In contrast to the concentration of water vapor, which continuously decreases with distance above the ground, the CO_2 concentration generally achieves a minimum somewhere within the plant community on a sunny day (Fig. 9.5). This occurs because both the turbulent air above the canopy and the soil can serve as sources of CO_2. During the day CO_2 diffuses toward lower concentrations from the soil up into the vegetation and from the overlying turbulent air down into the leaf canopy (see Eastin et al., and Zelitch).

Respiration in root cells and in soil microorganisms can lead to a net upward CO_2 flux density from the ground of 1 to 3 μmol m^{-2} s^{-1} during the growing season. (An O_2 flux density of similar magnitude occurs in the opposite direction.) The actual value of J_{CO_2} from the soil varies diurnally in phase with the soil temperature, which is higher during the day (see p. 374). We have already estimated that $r^{ta}_{CO_2}$, which is the same as r^{ta}_{wv}, might be 200 s m^{-1} for the first 0.01 m and 90 s m^{-1} for the next 0.09 m above the ground for a corn crop 2 m tall (Table 9.1). Using Equation 9.3 ($\Delta c^{ta}_{CO_2} = J_{CO_2} r^{ta}_{CO_2}$), we calculate for a moderate CO_2 flux density of 2 μmol m^{-2} s^{-1} emanating from the soil that the decrease in CO_2 concentration across the first 0.01 m above the ground is 0.4 mmol m^{-3}, which by the conversion factor in Table 8.2 represents a drop of 10 ppm CO_2 by volume at 20°C and 0.1 MPa air pressure. In the next 0.09 m, the CO_2 concentration might decrease by [(90 s m^{-1})/(200 s m^{-1})](10 ppm), or about 5 ppm (Table 9.1). Thus, the CO_2 level might drop by 15 ppm from 308 ppm at the soil surface to 293 ppm at 0.1 m above the ground (Fig. 9.5).

As we move further upward from the ground, the flux density of CO_2 initially increases as we encounter leaves that are below the light compensation point and thus have a net evolution of CO_2. For instance, J_{CO_2} directed upward may increase from 2 μmol m^{-2} s^{-1} at 0.1 m to 5 μmol m^{-2} s^{-1} at 0.5 m. As we move even higher and encounter leaves with net photosynthesis, the net flux density of CO_2 in the turbulent air decreases, and may become zero at 1.0 m above the ground in a 2-m-tall corn crop with a high photosynthetic rate. Thus J_{CO_2} may average 3 μmol m^{-2} s^{-1} from 0.1 to 1.0 m above the ground, an interval that has a resistance of 50 s m^{-1} (Table 9.1). This would lead to a $\Delta c^{ta}_{CO_2}$ of 0.15 mmol m^{-3}, which corresponds to a CO_2 decrease of 4 ppm. Hence, the CO_2 concentration may reach its lowest value of 289 ppm at the midway point in the crop (see Fig. 9.5).

At noon on a sunny day, J_{CO_2} down into a cornfield might be 60 μmol m^{-2} s^{-1}. Essentially all of the net CO_2 flux from the turbulent air above the canopy is directed into the leaves in the upper half of the corn crop; e.g., J_{CO_2} may become -30 μmol m^{-2} s^{-1} at 1.5 m above the ground and zero at 1.0 m. Thus, the average CO_2 flux density in the upper half of the vegetation is about -30 μmol m^{-2} s^{-1}, and the resistance is 10 s m^{-1} (Table 9.1). Consequently, $\Delta c^{ta}_{CO_2}$ for this upper portion of an actively photosynthesizing cornfield might be -0.3 mmol m^{-3}, which corresponds to a 7 ppm decrease in CO_2 from the top of the canopy to 1.0 m below (Table 9.1, Fig. 9.5). We calculated on p. 470 that $c_{CO_2}^{ta}$ at the top of this canopy might be 296 ppm. The CO_2 concentration at 1.0 above the ground would then be 289 ppm, the same value we estimated by working our way up from the ground (see Fig. 9.5). Two-thirds or more of the net photosynthesis generally occurs in the upper one-third of most canopies, as it does here for corn.

CO_2 concentrations in the air can vary over quite a wide range for different plant communities. For a cornfield exposed to a low wind speed (below 0.3 m s^{-1} at the top of the canopy), for a rapidly growing plant community, or for other dense vegetation where the eddy diffusion coefficient may be relatively small, the CO_2 concentration in the turbulent air within the plant stand can drop below 200 ppm during a sunny day. On the other hand, for sparse desert vegetation, especially on windy or overcast days, $c_{CO_2}^{ta}$ generally does not decrease even 1 ppm from the value at the top of the canopy.

CO_2 at Night

The CO_2 concentration at night is highest near the ground and continuously decreases as we go upward through the plants into the turbulent air above. J_{CO_2} emanating from the soil might be 1 μmol m^{-2} s^{-1}, while the respiratory flux density of CO_2 from the above-ground parts of plants can be 3 μmol m^{-2} s^{-1} at night. Respiration averaged over a 24-hour period can be 20% of gross photosynthesis for a rapidly growing plant community, and can increase to over 50% as the community matures. For certain climax communities, respiration can become nearly 100% of gross photosynthesis. When considered over a growing season, respiration for an entire plant is generally 30% to 50% of gross photosynthesis for agronomic crops.

We will now estimate CO_2 concentration drops that might occur in a 2-m-high corn crop at night. For purposes of calculation, let us assume that J_{CO_2} vertically upward averages 1 μmol m^{-2} s^{-1} over the first 0.1 m, where r_j^{ta} is 400 s m^{-1}; 2 μmol m^{-2} s^{-1} over the next 0.9 m, where the resistance is 80 s m^{-1}; and 3 μmol m^{-2} s^{-1} over the final 1.0 m, where r_j^{ta} may be 20 s m^{-1} at night (see Fig. 9.7; lower wind speeds at night can lead to these higher resistances compared with daytime values—Table 9.1). Using Equation 9.4 ($\Delta c_{CO_2}^{ta} = J_{CO_2} r_{CO_2}^{ta}$), the drop in CO_2 concentration from the ground to the top of the canopy is 0.40 + 0.16 + 0.06, or 0.62 mmol m^{-3}, which corresponds to 15 ppm CO_2 by volume at 20°C (Fig. 9.7; conversion factor in Table 8.2). The resistance of the first 30 m of turbulent air above the plants might be 50 s m^{-1} at night (compare the value of 30 s m^{-1} during the daytime that we used earlier). If we let the total CO_2 flux density from the

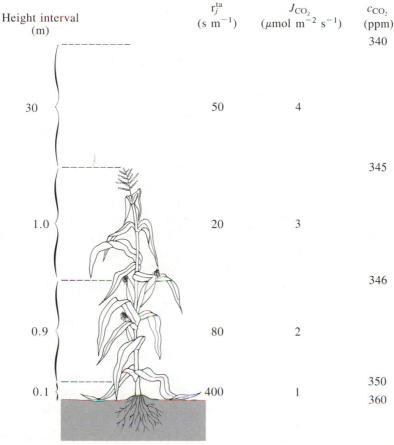

Height interval (m)	r_j^{ta} (s m^{-1})	J_{CO_2} (μmol m^{-2} s^{-1})	c_{CO_2} (ppm)
			340
30	50	4	
			345
1.0	20	3	
			346
0.9	80	2	
			350
0.1	400	1	360

Figure 9.7
CO_2 resistances, flux densities, and concentrations within and above a cornfield at night.

canopy be 4 μmol m^{-2} s^{-1}, then $\Delta c_{CO_2}^{ta}$ for the first 30 m above the canopy would be 0.2 mmol m^{-3}, which corresponds to 5 ppm CO_2. Therefore, assuming that $c_{CO_2}^{ta}$ is 340 ppm 30 m up in the turbulent air, the CO_2 concentration at night would be 345 ppm at the top of the canopy and 360 ppm at the soil surface (Fig. 9.7). These estimates are in agreement with measurements of CO_2 concentrations at night.

SOIL

Soils vary tremendously in their physical properties, such as the size of the individual particles. In a sandy soil many particles are over 1 mm in diameter, while in a clay most of the particles are less than 2 μm in diameter. Small soil particles have a much greater surface area per unit weight than do large particles. Sand, for example, can have under 1 m^2 of surface area/g, while most clays have 100 to 1 000 m^2 of surface area/g. Most soil minerals are aluminosilicates, with negatively charged surfaces that act like Donnan phases (see p. 134) with the mobile cations in the adjacent soil water. Because of their large surface areas per unit weight, clays dominate the ion exchange properties of many soils. The clay montmorillonite, for example, has 800 m^2 of surface area/g, and can hold nearly 1 mmol of monovalent cations/g (see Childs, and Hillel). Nutrient concentrations vary tremendously with soil type and water content; for moist agricultural soil, the phosphate level can be about 2 μM, with K$^+$ and NO$_3^-$ being about 1 000-fold more concentrated (Ca^{2+}, Mg^{2+}, Na$^+$, and Cl$^-$ can be even higher in concentration).

The irregularly shaped pores between the numerous particles in the soil contain both air and water (see Fig. 9.8). The soil pores, or voids, vary from somewhat under 40% to about 60% of the soil by volume. Thus, a soil whose pores are completely filled with water will contain 40% to 60% water by volume. In the vicinity of most roots, moist soil contains 8% to 30% water by volume, while the rest of the pore space is filled with air. The pores therefore provide many air-liquid interfaces where surface tension effects can lead to a negative hydrostatic pressure in the soil water. Such a negative P is generally the main contributor to the water potential in the soil, especially as the soil dries out. Since the thermal properties of soil have already been discussed in Chapter 7, we will focus on water relations aspects here.

Figure 9.8
Schematic indication of the gaseous, liquid, and solid phases in a soil. The radii of curvature at the air-water interfaces help determine the negative hydrostatic pressures or tensions in the liquid phase.

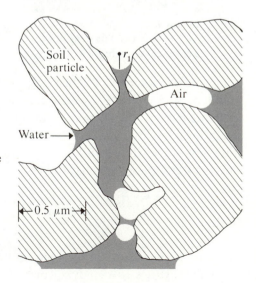

Soil Water Potential

The predominant influence on the soil water potential is usually the many air-liquid interfaces present in the soil (Fig. 9.8). Of course, the soil water contains dissolved solutes, which generally lead to an osmotic pressure (Π^{soil}) of between 0.01 and 0.2 MPa (0.1 and 2 bars); the magnitude of Π^{soil} depends on the relative wetness of the soil, which varies greatly. Because of the relatively low osmotic pressures, we will often refer to the solution in the soil simply as water. In contrast to the generally small values for Π^{soil}, the many interfaces present in the rather small soil pores can lead to hydrostatic pressures of -2 MPa or lower, i.e., more negative. A few days after being saturated, a wet clay soil might retain 45% water by volume ("field capacity") and have a soil water potential (Ψ^{soil}) of -0.01 MPa, whereas permanent wilting from which many agricultural crops will not recover occurs when Ψ^{soil} is about -1.5 MPa and the volumetric water content of the clay is 25% (field capacity and permanent wilting occur at about 35% and 15%, respectively, for loam, and 10% and 3% for sand).

The surfaces of the air-water interfaces in the pores between the soil particles are usually concave when viewed from the air side, just as for capillaries (see Fig. 2.3). But the surfaces generally are not spherical or otherwise regularly shaped (the same restriction applies to the air-liquid interfaces in the pores of a cell wall). Nevertheless, it is possible to define two principal radii of curvature for a surface.* Let us designate these radii, which occur in planes perpendicular to each other and to the liquid surface, by r_1 and r_2. The hydrostatic pressure in the soil water is then:

$$P = -\sigma \left(\frac{1}{r_1} + \frac{1}{r_2} \right) \tag{9.5}$$

where σ is the surface tension at an air-liquid interface.† The negative hydrostatic pressure or positive tension described by Equation 9.5 and

---

* If the arc formed by the intersection of a surface and a plane perpendicular to it is continued around to form a circle, the radius of the circle is the radius of curvature of the surface, r. A slightly curved surface has a large r, and r becomes infinite if the surface is flat in a particular direction. By convention, r is positive for a concave surface (see r_1 in Fig. 9.8) and negative for a convex surface.

† To connect Equation 9.5 with Equation 2.22 ($P = -2\sigma \cos \alpha / r$, where r is the radius of the *capillary*), note that for a cylindrical capillary the two principal radii of the *surface* are the same ($r_1 = r_2$), and hence the factor ($1/r_1 + 1/r_2$) is $2/r_1$, which equals $2 \cos \alpha / r$.

resulting from the presence of air-water interfaces is often called the *soil matric potential*.

Instead of being concave, the water surface extending between adjacent soil particles may assume a semi-cylindrical shape, i.e., like a trough or channel. This means that one of the radii of curvature becomes infinite—e.g., $r_2 = \infty$; in such a case, the pressure is $-\sigma/r_1$ by Equation 9.5. If the air-liquid surface is convex when viewed from the air side, the radii as defined are negative; we would then have a positive hydrostatic pressure in the water. In the intermediate case—one radius positive and one negative (a so-called "saddle-shaped" surface)—whether the pressure is positive or negative depends on the relative sizes of the two radii of curvature.

Let us now estimate the hydrostatic pressure in the soil water within a wedge-shaped crevice between two adjacent soil particles, as illustrated at the top of Figure 9.8. We will assume that the air-liquid surface is cylindrical, so r_2 equals ∞, while r_1 is 0.1 μm. If we let σ be 0.0728 Pa m, the value for water at 20°C (App. II), then by Equation 9.5 the hydrostatic pressure is

$$P = \frac{-(0.0728 \text{ Pa m})}{(10^{-7} \text{ m})} = -7 \times 10^5 \text{ Pa}$$

$$= -0.7 \text{ MPa}$$

As the amount of soil water decreases, the air-water surface retreats into the crevice between the particles, the radius of curvature becomes less, and the pressure accordingly becomes more negative. Since by Equation 2.13a Ψ^{soil} equals $P^{soil} - \Pi^{soil} + \rho_w gh$, the soil water potential also becomes more negative as water is lost from such crevices.

Darcy's Law

Darcy in 1856 recognized that the flow of water through the soil was driven by a gradient in hydrostatic pressure. We can represent this relation, known as Darcy's law, by the following expression:

$$J_V = -L^{soil} \frac{\partial P^{soil}}{\partial x} \tag{9.6}$$

where J_V is the volume of solution crossing unit area in unit time and L^{soil} is the soil hydraulic conductivity coefficient.

Although Equation 9.6 is in a familiar form (flux density equals a proportionality coefficient times a force), we have used $-\partial P^{soil}/\partial x$ instead of the possibly more general force, $-\partial \Psi^{soil}/\partial x$ ($\Psi^{soil} = P^{soil} - \Pi^{soil} + \rho_w gh$, Eq. 2.13a). In Chapter 3 we derived an expression for J_V that incorporated a reflection coefficient, $J_V = L_P(\Delta P - \sigma \Delta \Pi)$ (Eq. 3.38). When σ is zero, as occurs for a porous barrier like soil, $\Delta \Pi$ does not lead to any volume flux density. Hence, we do not expect $\partial \Pi^{soil}/\partial x$ to influence the movement of water in the soil, and it is not included in Darcy's law. Since $\partial(\rho_w gh)/\partial x$ is not incorporated into Equation 9.6, the indicated form of Darcy's law really applies only to horizontal flow in the soil ($\partial h/\partial x = 0$ when x is in a horizontal direction). Actually, $\rho_w g$ is only 0.01 MPa m^{-1} (App. II), so changes of $\rho_w gh$ in the vicinity of a root are relatively small. However, for percolation of water down appreciable distances into the soil, P^{soil} in Equation 9.6 should be replaced by $P^{soil} + \rho_w gh$. In fact, for drainage of very wet soil, the gravitational term can be the dominant factor in such a formulation of Darcy's law.

Soil Hydraulic Conductivity Coefficient

L^{soil} in Equation 9.6 does not have the same units as L_P in Equation 3.38 $[J_V = L_P(\Delta P - \sigma \Delta \Pi)]$ or L_w in Equation 2.23 $[J_{V_w} = L_w(\Psi^o - \Psi^i)]$. L_P and L_w have units of volume flux density/pressure—e.g., m s^{-1} Pa^{-1}. But L^{soil} in Equation 9.6 has units of volume flux density/pressure gradient—e.g., (m s^{-1})/(Pa m^{-1}), or m^2 s^{-1} Pa^{-1}. We could use other self-consistent sets of units for Darcy's law; indeed, the soil hydraulic conductivity coefficient is expressed in many different ways in the literature.

The soil hydraulic conductivity coefficient depends on the geometry of the pores in the soil. For geometrically similar pore shapes, and ignoring certain surface effects, the hydraulic conductivity coefficient is approximately proportional to the square of the pore size. However, the pores are so complex in shape that, in general, we cannot directly calculate L^{soil}. As soil dries, its water potential decreases and P^{soil} becomes less. When P^{soil} decreases below the minimum hydrostatic pressure that can occur in some fairly large pores, water flows out of them, but it will remain in pores which have smaller dimensions and which can therefore have even more negative pressures— see Equation 9.5, $P = -\sigma(1/r_1 + 1/r_2)$. Not only is the higher conductivity of the larger pores thus lost, but the remaining pathway for water flow becomes more tortuous, and so L^{soil} decreases as the soil dries out.

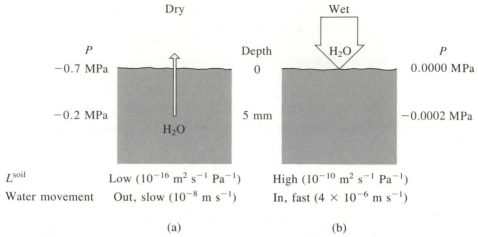

Figure 9.9
Summary of valve-like properties of upper layer of soil in (a) a fairly dry state and (b) while being wet by rain. Water fairly readily enters wet soil but is lost only gradually across a dry crust.

The soil hydraulic conductivity coefficient can be 10^{-17} m^2 s^{-1} Pa^{-1} (10^{-8} cm^2 s^{-1} bar^{-1}) or lower for a dry nonporous soil, and 10^{-13} m^2 s^{-1} Pa^{-1} or higher for a wet porous one. Specifically, for a porous clay whose pores are nearly filled with water, L^{soil} is usually 10^{-13} to 10^{-10} m^2 s^{-1} Pa^{-1}; and for a water saturated sandy soil, the soil hydraulic conductivity coefficient is generally 10^{-8} to 10^{-7} m^2 s^{-1} Pa^{-1} (see Childs, Hillel, and Rose). (We will reconsider L^{soil} in terms of Poiseuille flow on p. 495.)

The ground is often covered by a rather dry crust where the soil hydraulic conductivity coefficient is quite low. Specifically, L^{soil} may average 10^{-16} m^2 s^{-1} Pa^{-1} in the upper 5 mm of the soil. If P^{soil} is -0.7 MPa near the surface and -0.2 MPa 5 mm beneath it (Fig. 9.9), then by Darcy's law (Eq. 9.6), the volume flux density of water would be

$$J_V = -(10^{-16} \text{ m}^2 \text{ s}^{-1} \text{ Pa}^{-1})\left[\frac{(-0.7 \times 10^6 \text{ Pa}) - (-0.2 \times 10^6 \text{ Pa})}{(5 \times 10^{-3} \text{ m})}\right]$$

$$= 1 \times 10^{-8} \text{ m s}^{-1}$$

This water flux density directed vertically upward at the soil surface equals $(1 \times 10^{-8}$ m^3 m^{-2} s$^{-1})(1$ mol$/18 \times 10^{-6}$ m$^3)$, or 0.6×10^{-3} mol m^{-2} s^{-1}.

When discussing water vapor movement in the previous section, we indicated that J_{wv} emanating from a shaded soil is generally 0.2 to 1 mmol $m^{-2} s^{-1}$, and so our calculated flux density is consistent with the range of measured values. The calculation also indicates that a fairly large gradient in hydrostatic pressure can exist near the surface of a moist soil.

During a rainstorm the upper part of the soil can become nearly saturated with water. As a result, L^{soil} there might increase 10^6-fold from 10^{-16} m^2 $s^{-1} Pa^{-1}$ all the way up to $10^{-10} m^2 s^{-1} Pa^{-1}$ (Fig. 9.9). This facilitates the entry or infiltration of water into the soil, which often has a volume flux density of $4 \times 10^{-6} m s^{-1}$, or 0.014 m hour^{-1}. Such an infiltration of a 14-mm depth of water per hour can be maintained for a few hours by many soils (see Childs, Hillel, Slatyer, and Taylor and Ashcroft). Also, the uppermost layer or crust acts somewhat like a valve, retarding the outward movement of water when the soil is fairly dry (low L^{soil}), but promoting the infiltration of water upon moistening by rain (high L^{soil}), as is indicated in Figure 9.9.

Besides moving as a liquid, water can also move as a vapor in the soil. Since water is continually evaporating from and condensing onto the many air-liquid interfaces in the soil, such movement can be relatively important, especially for dry soils where the liquid phases are discontinuous. The saturation value for water vapor partial pressure or concentration increases nearly exponentially with temperature (Fig. 8.4). Since the air between the soil particles generally is nearly saturated with water vapor (except near the surface), the amount of water vapor in the soil air increases rapidly with the soil temperature. Consequently, the movement of water in the form of vapor tends to be greater at higher soil temperatures.

Fluxes for Cylindrical Symmetry

Although we have been mainly considering one-dimensional cases in Cartesian coordinates, the flow of soil water toward a root may be more appropriately described using cylindrical coordinates, since the length of a root is generally much greater than its diameter. We will restrict our attention to the steady state condition where the fluxes do not change with time. Also, we will handle the cylindrically symmetrical case where the fluxes and forces depend only on the radial distance r from the axis of the cylinder, and not on any angle around or location along the axis of the cylinder.

For a cylindrical surface of length l along the axis, the total volume of solution crossing per unit time is $J_V 2\pi rl$, where J_V is the volume flux density directed radially at a distance r from the axis of the cylinder. In the steady state, $J_V 2\pi r l$ is constant, and so the magnitude of the flux density depends inversely on the radial distance. When L^{soil} is constant, we can represent the flux density for the cylindrically symmetrical case by

$$J_V = \frac{1}{r} \frac{L^{soil}(P_a - P_b)}{\ln(r_a/r_b)} \tag{9.7}$$

where P_a is the hydrostatic pressure at a distance r_a from the axis of the cylinder, and P_b is the value at r_b. J_V is considered to be positive when the net flux density is directed into the root, as occurs when the hydrostatic pressure is higher $(P_a > P_b)$ the further we are from the root $(r_a > r_b)$. Equation 9.7 represents a quite general form for steady state cases with cylindrical symmetry; e.g., Fick's first law (p. 12) then is $J_j = (1/r)D_j(c_j^a - c_j^b)/ \ln(r_a/r_b)$, and Equation 7.11 gives the heat flux density for cylindrical symmetry.

The uptake of water by a young root 1 mm in diameter is usually 1×10^{-5} to 5×10^{-5} m^3 day^{-1} per m of root length (0.1 to 0.5 cm^3 day^{-1} per cm; see Kramer, and Slatyer). This uptake occurs over a root surface area of $2\pi r l$, and so the volume flux density of water at the root surface for a moderate uptake of 3×10^{-5} m^3 day^{-1} per m of root length would be

$$J_V = \frac{(3 \times 10^{-5} \text{ m}^3 \text{ day}^{-1})}{(8.64 \times 10^4 \text{ s day}^{-1})[(2\pi)(0.5 \times 10^{-3} \text{ m})(1 \text{ m})]}$$

$$= 1.1 \times 10^{-7} \text{ m s}^{-1}$$

The influx of water is enhanced by the root hairs, which protrude from the epidermal cells (Fig. 1.4). They are often about 12 μm in diameter, up to 1 mm long, and generally vary in frequency from 0.5 to 50 per mm^2. J_V may be 10- to 100-fold less for older roots, since their outer surfaces generally become extensively cutinized and suberized.

For representative root spacing in a soil, water may move toward a root over a radial distance of about 10 mm. Let us next estimate the drop in hydrostatic pressure that might occur over this interval. We will assume that Ψ^{soil} at an r_a of 10.5 mm is -0.3 MPa, made up of an osmotic pressure of 0.1 MPa and a P^{soil} of -0.2 MPa ($\Psi = P - \Pi + \rho_w gh$, Eq. 2.13a). We will let L^{soil} be 10^{-15} m^2 s^{-1} Pa^{-1}, as might apply to a loam of moderately low water content, and we will suppose that J_V at the surface of a root 1 mm in diameter (i.e., $r_b = 0.5$ mm) is 1.1×10^{-7} m s^{-1}. Using Equation 9.7, we can then calculate that the hydrostatic pressure near the root surface would be

$$P_b^{soil} = -\frac{rJ_V \ln (r_a/r_b)}{L^{soil}} + P_a^{soil}$$

$$= -\frac{(0.5 \times 10^{-3} \text{ m})(1.1 \times 10^{-7} \text{ m s}^{-1}) \ln [(10.5 \text{ mm})/(0.5 \text{ mm})]}{(10^{-15} \text{ m}^2 \text{ s}^{-1} \text{ Pa}^{-1})}$$

$$+ (-0.2 \text{ MPa})$$

$$= -0.4 \text{ MPa}$$

Thus, the hydrostatic pressure drops 0.2 MPa across a distance of 10 mm in the soil next to the root. Assuming that the solute content of the soil water does not change appreciably over this interval, Π_b^{soil} is 0.1 MPa, and thus Ψ_b^{soil} adjacent to the root is -0.4 MPa $- 0.1$ MPa, or -0.5 MPa (see Table 9.2).

When the soil dries out, L^{soil} decreases; therefore, the drops in hydrostatic pressure and water potential would have to be larger to maintain a given volume flux density toward a root. For example, L^{soil} is rather high for a wet soil with a Ψ^{soil} of -0.02 MPa; ΔP^{soil} over a radial distance of 10 mm from a root is generally 0.01 MPa or less in such a case. To maintain the same J_V for Ψ^{soil} equal to -1.5 MPa in the bulk soil, a drop in P^{soil} of about 1.0 MPa might have to occur over the 10 mm interval adjacent to the root.

Fluxes for Spherical Symmetry

Sometimes the fluxes of water in the soil toward plant parts can approximate spherical symmetry. For instance, water movement can occur radially over a 10-mm interval toward a seed or a recently initiated root. For roots most water and nutrient uptake apparently takes place over the region containing the root hairs, which is proximal to the elongation region (see Fig. 1.4) and distal to where the periderm begins. The external cell layers of the periderm have suberin in the cell walls, which greatly limits water uptake. In the suberized regions lenticels occur, which consist of a loose aggregation of cells facilitating gas exchange (lenticels also occur on stems). Lenticels and other interruptions of the suberized periderm can act as local sites toward which the flux of water converges. In such a case, the water movement from the surrounding soil toward the root can also be approximately spherically symmetrical (see Caldwell).

Using the same definitions as in the case of cylindrical symmetry (Eq. 9.7), we obtain the following steady-state relation describing the volume flux density J_V at distance r from the center of a sphere when J_V varies only in the radial direction and L^{soil} is constant:

$$J_V = \frac{1}{r^2} \left(\frac{r_a r_b}{r_a - r_b} \right) L^{soil}(P_a - P_b) \tag{9.8}$$

Equation 9.8 is similar to Equation 7.12 $[J_H^C = (r + \delta^{bl})K^{air}(T^{surf} - T^{ta})/(r\delta^{bl})]$ describing the heat flux density across an air boundary layer for spherical symmetry ($r_a = r + \delta^{bl}$ and $r_b = r$). We should note that a steady state is often not achieved in soils, and so a "steady rate" is often used, where the rate of volumetric water depletion is constant, which leads to relations considerably more complicated than Equations 9.7 and 9.8 (see Caldwell).*

During germination the volume flux density of water into a seed is often limited by a seed coat (testa) of thickness δ^{sc}. The seed coat, which generally contains an inner and an outer cuticle, is often two to four cell layers thick, and thus is relatively thin compared to the radius of the seed (see Bewley and Black). For the volume flux density at the seed surface ($r = r_s$, $r_a = r_s$, and $r_b = r_s - \delta^{sc}$), Equation 9.8 becomes

$$J_V = \frac{1}{r_s^2} \frac{(r_s)(r_s - \delta^{sc})}{[r_s - (r_s - \delta^{sc})]} L^{sc} \Delta P^{sc}$$

$$= \frac{r_s - \delta^{sc}}{r_s} \frac{L^{sc}}{\delta^{sc}} \Delta P^{sc} \tag{9.9}$$

$$\cong \frac{L^{sc}}{\delta^{sc}} \Delta P^{sc}$$

---

* In a steady state the water content of the soil would not change with time, whereas for a steady rate situation the rate of change is set equal to a constant. In the general time-dependent case, a relation similar to $\dfrac{\partial c_j}{\partial t} = -\dfrac{\partial}{\partial x}\left(-D_j \dfrac{\partial c_j}{\partial x}\right)$ (Eq. 1.4), but in the proper co-ordinate system, must be satisfied to describe the water flow in the soil for cylindrical or spherical symmetry; D_j is replaced by a quantity analogous to L^{soil} that varies with water content and hence location in the soil.

where L^{sc} is the hydraulic conductivity coefficient of a seed coat of thickness δ^{sc}, and the bottom line incorporates the supposition that $\delta^{sc} \ll r_s$. Equation 9.9 indicates that, when the region of interest is thin compared with the radial distance, the flow can be approximated by an equation appropriate for a one-dimensional movement across a seed coat of conductance L^{sc}/δ^{sc} (see Eq. 8.1b). Also, ΔP^{sc} should be replaced in Equation 9.9 by $\Delta \Psi^{sc}$, since osmotic pressures can also influence water uptake by seeds.

As we have indicated for L^{soil}, L^{sc} also depends on water content, increasing more than 10^6-fold upon water uptake by the seed and subsequent rupture of the seed coat. L^{sc} then approaches values found for the L^{soil} of a moist soil (see Shaykewich and Williams). When dry, seeds can have a very negative water potential, e.g., -100 to -200 MPa. As the seeds imbibe water, the internal water potential rises, germination being initiated only when Ψ^{seed} and hence Ψ^{soil} are above about -1.0 MPa. Water uptake is generally not uniform over the whole seed surface, initially being higher near the micropyle (a small opening in the integuments of an ovule through which the pollen tube enters and which remains as a minute pore in the testa). Another crucial factor in the germination of seeds is the area of contact between the seed coat and the soil particles, which becomes progressively more important as the seed imbibes water and L^{sc} increases, leading to a higher J_V (see Eq. 9.9; during water uptake, L^{sc} increases more than $\Delta \Psi^{sc}$ decreases).

WATER MOVEMENT IN THE XYLEM AND PHLOEM

Under usual conditions, essentially all of the water entering a land plant comes from the soil by way of the root. The water is then conducted to the other parts of the plant mainly by the xylem. To reach the xylem in the mature part of a root, water must cross the root epidermis, the cortex, and then the endodermis. We will briefly describe the various cells encountered by water as it moves from the soil to the root xylem (see Fig. 1.4) and then discuss the dependence of water flow in the xylem on the gradient in hydrostatic pressure. Flow in the phloem may also depend on $\partial P/\partial x$, although osmotic pressures are also important for interpreting water movement in this other "circulatory" system. The divisions and subdivisions of the vascular tissue in a leaf generally result in individual mesophyll cells being no more than 3 or 4 cells away from the xylem or the phloem. This proximity facilitates the movement of photosynthetic products into the phloem, such "loading" being necessary for the distribution of sugars like sucrose to "sinks" located in different parts of the plant.

Root Tissues

Longitudinal and cross-sectional views near the tip of a root are presented in Figure 1.4, indicating the various types of cells that can act as barriers to flow. Water may fairly easily traverse the single-cell layer of the root epidermis to reach the cortex. The root cortex often consists of 5 to 10 cell layers, the cytoplasm of adjacent cells being continuous because of the plasmodesmata. The collective protoplasm of the interconnected cells is referred to as the *symplasm* (see Ch. 1). In the symplasm, permeability barriers in the form of plasmalemmas and cell walls do not have to be surmounted for diffusion to occur from cell to cell, and this facilitates the movement of water and solutes across the cortex. Much of the water flow across the root cortex actually occurs in an alternative pathway, the cell walls, which form the *apoplast* (see Weatherley).

Roots of most woody and some herbaceous plants have mycorrhizal associations. The fungal hyphae invade the cortical region of the young roots and can extend over 10 mm away from the root surface. The relation appears to be mutualistic, since the fungus obtains organic matter from the plant and in return increases the effective area of contact of the plant with soil particles. The hyphae can remain active for older roots whose outer surface has become suberized and hence less conductive. The fungus thus increases the availability of certain nutrients like N and P to the host plant (see Kramer and Kozlowski).

Along that part of a root where cells have differentiated but where there is little root thickening due to secondary conducting tissue, an endodermis occurs inside the cortex (Fig. 1.4 and p. 8). Upon reaching the endodermis, the cell wall pathway for water movement is blocked by the casparian strip, as realized by Caspary in 1865. This is a continuous hydrophobic band around the radial walls of the endodermal cells that blocks the apoplastic pathway along which water and solutes could have crossed the endodermal layer (see p. 8). The casparian strip is impregnated with suberin (a polymer of fatty acids) and lignin. It is generally considered part of the primary cell wall, but suberin and lignin are also deposited in the secondary wall (which often becomes quite thick) and all the way across the middle lamella. Water and ions therefore cannot flow across the endodermal cells in their cell walls but rather must move through their cytoplasm. Thus, the endodermis, with its casparian strip, serves to regulate the passage of solutes and water from the cortex to the xylem in the affected parts of the root. Inside the endodermis is a layer of parenchyma cells known as the *pericycle*, which surrounds the vascular tissue containing the conducting vessels of the xylem and the phloem (see Fig. 1.4). During water movement through a plant accompanying transpiration, the hydrostatic pressure in the root xylem is reduced, and can become quite negative. This decreases the water potential ($\Psi = P - \Pi + \rho_w gh$, Eq. 2.13a) in the xylem and promotes water movement from the soil down a water potential gradient to the root xylem.

The Xylem

Before discussing the characteristics of flow in the xylem, we will briefly review some of the anatomical features of its cells. In general, the conducting xylem elements have thick, lignified secondary walls and contain no protoplasts; i.e., the xylem cells serve their special function of providing the plant with a low resistance conduit only when they are dead.

Two types of conducting cells are distinguished in the xylem: the vessel members (found in angiosperms) and the phylogenetically more primitive tracheids (in angiosperms, gymnosperms, and the lower vascular plants). Tracheids are typically tapered at the ends, while the generally shorter and broader vessel members often abut each other with rather blunt ends. The end cell walls of the vessel members are perforated, so the vessel members arranged end-to-end form a continuous tube called a *xylem vessel*. The end cell wall of a vessel member bearing the holes is referred to as a *perforation plate*; a simple perforation plate (Fig. 1.3) essentially eliminates the end walls between the individual members in a vessel. Although xylem elements actually vary considerably in their widths (from 8 μm all the way up to 500 μm), we will represent them by cylinders with radii of 20 μm for purposes of calculation. Conducting cells of the xylem generally range in length from a few hundred microns to a few mm or even more.

Besides vessel members and tracheids, other types of cells in the xylem are parenchyma cells and fibers (Fig. 1.3). Xylem fibers, which contribute to the structural support of a plant, are long thin cells having heavily lignified cell walls; they are generally devoid of protoplasts at maturity but are nonducting. The living parenchyma cells in the xylem are important for the storage of carbohydrates and for the lateral movement of water and solutes into and out of the conducting cells.

Poiseuille's Law

To describe fluid movement in the xylem quantitatively, we need to relate the flow to the particular driving force causing the motion. For cylindrical tubes, an appropriate relationship is the one determined experimentally by

Hagen in 1839, and independently by Poiseuille in 1840. They found that the volume of fluid moving in unit time along a cylinder is proportional to the fourth power of its radius, and that the movement depends linearly on the drop in hydrostatic pressure. Wiedemann in 1856 showed that this rate of volume movement per tube could be represented as follows:

$$\text{Volume flow rate per tube} = -\frac{\pi r^4}{8\eta}\frac{\partial P}{\partial x} \qquad (9.10a)$$

where r is the radius of the cylinder, η is the viscosity of the solution, and $-\partial P/\partial x$ is the negative gradient of the hydrostatic pressure. Throughout this text we have been concerned with the volume flowing per unit time and *area*, J_V. For flow in a cylinder of radius r, and hence area πr^2, J_V would be

$$J_V = -\frac{r^2}{8\eta}\frac{\partial P}{\partial x} \qquad (9.10b)$$

Since positive flow ($J_V > 0$) occurs in the direction of decreasing hydrostatic pressure ($\partial P/\partial x < 0$), the minus sign is necessary in Equation 9.10.

Equation 9.10 is generally referred to as *Poiseuille's law* and sometimes as the Hagen–Poiseuille law. It was derived assuming that the fluid in the cylinder moves in layers, or *laminas*, each layer gliding over the adjacent one (see Streeter). Such laminar movement occurs only if flow is slow enough to meet a criterion deduced by Reynolds in 1883. Specifically, the dimensionless quantity $\rho J_V d/\eta$ (another form of the Reynolds number; Eq. 7.15) must be less than 2 000 (ρ is the solution density and d is the cylinder diameter).[*] Otherwise, a transition to turbulent flow occurs, and Equation 9.10 is no longer valid (we will shortly consider the Reynolds number for flow in the xylem). Another important point is that the fluid in Poiseuille (laminar) flow is actually stationary at the wall of the cylinder; the speed of solution flow increases in a parabolic fashion to a maximum value in the center of the tube. Thus, the flows in Equation 9.10 are actually the mean flows averaged over the entire cross section of cylinders of radius r.[†] The dimensions for

---

[*] Equation 7.15 indicates that Re is vd/v, where the kinematic viscosity v equals η/ρ (η is properly known as the dynamic viscosity). The footnote on p. 97 indicates that the volume flux density J_V equals the mean velocity of fluid movement v. Thus, Re can be represented by $J_V d/(\eta/\rho)$, or $\rho J_V d/\eta$.

[†] The maximum speed for Poiseuille flow occurs along the cylinder axis and is twice the average speed, J_V (see Vogel 1981, Ch. 7 reference).

viscosity η are pressure $\times$ time, e.g., Pa s (the cgs unit, dyn s cm^{-2}, is termed a *poise*, which equals 0.1 Pa s); the viscosity of water at 20°C is 1.002×10^{-3} Pa s (values of η_w at other temperatures are given in App. II).

As we have already indicated, the volume flux densities described by both Poiseuille's law $[J_V = -(r^2/8\eta)\,\partial P/\partial x$, Eq. 9.10b] and Darcy's law ($J_V = -L^{\text{soil}}\,\partial P^{\text{soil}}/\partial x$, Eq. 9.6) depend on the negative gradient of the hydrostatic pressure. Can we establish any correlation between these two equations? If the soil pores were cylinders of radius r, all aligned in the same direction, we could have Poiseuille flow in the soil, in which case L^{soil} would equal $r^2/8\eta$—we are assuming for the moment that the pores occupy the entire soil volume. When r is 1 μm and η equals 1.002×10^{-3} Pa s, we obtain

$$\frac{r^2}{8\eta} = \frac{(1 \times 10^{-6}\ \text{m})^2}{(8)(1.002 \times 10^{-3}\ \text{Pa s})} = 1.2 \times 10^{-10}\ \text{m}^2\ \text{s}^{-1}\ \text{Pa}^{-1}$$

which corresponds to the L^{soil} for a very wet soil (Fig. 9.9). Actually, L^{soil} for water-filled porous clay with average pore radii of 1 μm is about 10^{-11} m^2 s^{-1} Pa^{-1}, which is about 10-fold smaller than $r^2/8\eta$ calculated using the average pore radius. As a result, J_V in the soil is considerably less than it is for Poiseuille flow through pores of the same average radius. This is mainly because the soil pores are not in the shape of cylinders aligned in the direction of flow; also, the pores between soil particles occupy somewhat less than half the soil volume.

Applications of Poiseuille's Law

Next, we will use Poiseuille's law (Eq. 9.10b) to estimate the pressure gradient necessary to cause a specified volume flux density in the conducting cells of the xylem. The speed of sap ascent in the xylem of a transpiring tree can be about 3.6 m hour^{-1}, which equals 1 mm s^{-1} (see Briggs, Slatyer, and Zimmermann and Brown). On p. 97 we indicated that v_w equals the volume flux density of water, J_{V_w}; the average speed of the *solution* equals J_V, the volume flux density of the solution.* Thus, J_V in the xylem of a transpiring tree might be 1 mm s^{-1}. For a viscosity of 1.0 mPa s and a xylem element with a 20 μm radius for the lumen, the pressure gradient required to satisfy

---

* In any case, for a dilute aqueous solution such as occurs in the xylem, the volume flux density of solution, J_V, is essentially the same as the volume flux density of water, J_{V_w} (see p. 165).

Equation 9.10b is

$$\frac{\partial P}{\partial x} = \frac{-8\eta J_V}{r^2} = \frac{-(8)(1.0 \times 10^{-3}\text{ Pa s})(1.0 \times 10^{-3}\text{ m s}^{-1})}{(20 \times 10^{-6}\text{ m})^2}$$

$$= -2 \times 10^4\text{ Pa m}^{-1} = -0.02\text{ MPa m}^{-1}$$

As expected, the pressure decreases along the direction of flow (Fig. 9.10).

The estimate of -0.02 MPa m^{-1} is consistent with experimental observations of the $\partial P/\partial x$ accompanying solution flow in *horizontal* xylem vessels. For the vertical vessels in a tree, however, a static hydrostatic pressure gradient of -0.01 MPa due to gravity exists even in the absence of flow.* For transpiring plants, the total $\partial P/\partial x$ in the vertical sections of the xylem is often about -0.03 MPa m^{-1} (Fig. 9.10). Based on the above calculations, this pressure gradient is sufficient to overcome gravity (-0.01 MPa m^{-1}) and to cause Poiseuille flow in the xylem vessels (-0.02 MPa m^{-1} needed in the above example where $r = 20\ \mu$m; see Briggs). The above gradients refer to vessel members joined by simple perforation plates that offer little obstruction to flow (Fig. 1.3). Gradients would be greater for tracheids, since

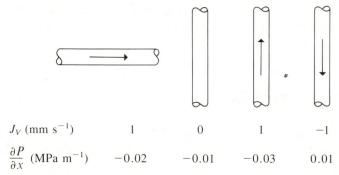

| J_V (mm s^{-1}) | 1 | 0 | 1 | -1 |
| $\frac{\partial P}{\partial x}$ (MPa m^{-1}) | -0.02 | -0.01 | -0.03 | 0.01 |

Figure 9.10
Flows and pressure gradients in cylindrical tubes. The tube radius is 20 μm and $\partial P/\partial x$ is described by Poiseuille's law (Eq. 9.10b) as modified by gravitational effects. For vertical tubes, x is considered positive upward. Arrows indicate the direction of flow.

---

* We can appreciate this by considering a vertical column of pure water ($\Pi = 0$) at equilibrium, where the water potential Ψ is $P + \rho_w gh$; $\rho_w gh$ increases vertically upward, so P must decrease by the same amount for Ψ to remain unchanged, as it does at equilibrium. Since $\rho_w g$ equals 0.0098 MPa m^{-1} (App. II), the additional pressure gradient caused by gravity amounts to -0.01 MPa m^{-1} (see Fig. 9.10).

they are connected by numerous small pits containing locally thin regions of the cell wall. These pits can greatly restrict the flow, since some cell wall material must be traversed to move from one tracheid to another. For instance, in conifers half of the pressure drop along the xylem may occur across the pits (see Jarvis), and hence the overall $\partial P/\partial x$ for the same lumen diameter and flow rate would be about twice as large as that calculated here.

We have been approximating xylem vessels by cylinders 20 μm in radius, as might be appropriate for diffuse-porous (small-porous) trees. For ring-porous (large-porous) trees, the mean radii of xylem vessels are often about 100 μm (see Zimmermann and Brown). For a given pressure gradient, Poiseuille's law (Eq. 9.10b) indicates that a much larger speed of xylem sap movement would occur in the ring-porous trees, as is indeed observed. Using our above values and a water density of 1 000 kg m^{-3}, the Reynolds number for the diffuse-porous case would be

$$\text{Re} = \frac{\rho J_v d}{\eta} = \frac{(1\ 000\ \text{kg m}^{-3})(1 \times 10^{-3}\ \text{m s}^{-1})(20 \times 10^{-6}\ \text{m})}{(1.0 \times 10^{-3}\ \text{Pa s})}$$

$$= 0.02\ \text{kg m}^{-1}\ \text{s}^{-2}\ \text{Pa}^{-1} = 0.02$$

Such a low value indicates that no turbulence is expected. Even if r were 5-fold larger and J_v were 10-fold higher, as can occur in a ring-porous tree, the Reynolds number would still be far less than the value 2 000 where turbulence generally sets in.

It is interesting to estimate the sort of pressure gradient necessary to promote a given flow along a cell wall. This use of Poiseuille's law will allow us to compare such a pathway with water movement in the lumens of xylem vessels. Since the interfibrillar spaces, or interstices, in a cell wall have diameters near 10 nm (100 Å), we will let r be 5 nm for purposes of calculation. (Complications due to the tortuosity of the aqueous channels through the interstices will be omitted here. We will also pretend that the pores occupy the entire cell wall.) For J_v equal to 1 mm s^{-1}, Equation 9.10b indicates that the pressure gradient required for solution flow through the cell walls is -3×10^5 MPa m^{-1}. In contrast, a $\partial P/\partial x$ of only -0.02 MPa m^{-1} is needed for the same J_v in the xylem element having a 20 μm radius. Thus, the $\partial P/\partial x$ for Poiseuille flow through the small interstices of a cell wall is over 10^7 times greater than for the same flux density through the lumen of the above xylem element. Because of the tremendous pressure gradients required to force water through the small interstices available for solution conduction in the cell wall, fluid could not flow rapidly enough up a tree in the cell walls—as has been suggested—to account for observed rates of water movement.

To help compare the relative effects on flow of a cell wall, a plasmalemma, and the lumen of a xylem vessel, we will calculate the hydrostatic pressure drops across each of them in a hypothetical case. Let us consider a xylem vessel 1 mm long and 20 μm in radius. We have just calculated that the pressure gradient necessary for a Poiseuille flow of 1 mm s^{-1} in the lumen of this xylem vessel would be -0.02 MPa m^{-1}, which amounts to (-0.02 MPa m^{-1}) (10^{-3} m), or -2×10^{-5} MPa for the 1 mm length of the cell. Let us suppose that a cell wall 3 μm thick is placed on both ends of the cell. In the previous paragraph we indicated that a pressure gradient of -3×10^5 MPa m^{-1} is necessary for a J_V of 1 mm s^{-1} through the interstices of a cell wall. To cross this cell wall (a total distance of 6 μm for both ends) would require a pressure change of (-3×10^5 MPa m^{-1})(6 $\times$ 10^{-6} m), or -2 MPa. Finally, let us imagine that a plasmalemma with a fairly high hydraulic conductivity coefficient of 10^{-12} m s^{-1} Pa^{-1} is also placed around the cell. Using Equation 3.38 $[J_V = L_P(\Delta P - \sigma \Delta \Pi)]$ with J_V equal to 1 mm s^{-1}, the difference in hydrostatic pressure required for such flow would be $(1 \times 10^{-3}$ m s$^{-1})/(10^{-12}$ m s^{-1} Pa$^{-1})$, or 1×10^9 Pa, which means a total pressure drop of 2×10^3 MPa to traverse the plasmalemma at each end of the cell. In summary, the pressure drops needed in this hypothetical case are 2×10^{-5} MPa to flow through the lumen, 2 MPa to cross the cell walls, and 2×10^3 MPa to cross the membranes.

In fact, membranes generally serve as the main barrier to water flow across plant cells. The interstices of the cell walls provide a much easier pathway for solution flow (such as across the root cortex or from leaf xylem terminals to mesophyll cells), while a hollow xylem vessel presents the least obstacle to flow (such as up the stem). Consequently, the evolution of xylem provides a plant with a tube, or conduit, well suited to moving water over long distances. The region of a plant made up of cell walls and the hollow xylem vessels is often called the apoplast, as noted in Chapter 1. Water and the solutes it contains can move fairly readily in the apoplast, but they must cross a membrane to enter the symplast (symplasm), the interconnected cytoplasm of the cells.

The Phloem

Water and solute movement in the phloem involves cooperative interactions among several types of cells. The conducting cells of the phloem generally have a high internal hydrostatic pressure, which often leads to artifacts during cytological investigations. In part because of the observational difficulties involved, there is a lack of agreement on the actual internal structure of the cells and the mechanism for solute movement in the phloem. Our brief treatment will be primarily concerned with certain anatomical features of the phloem cells and an overall description of flow (see Zimmermann and Milburn). Although a final decision on mechanism must await future research, certain implications of a pressure-driven flow will be presented.

The conducting cells of the phloem are the sieve cells in lower vascular plants and gymnosperms, and the more advanced sieve-tube members in angiosperms (Fig. 1.3). Both types of cells are collectively called *sieve elements*. Mature sieve elements almost invariably have lost their nuclei, and apparently the tonoplast has broken down, so there is no large central vacuole, although the plasmalemma remains intact. However, unlike the conducting elements of the xylem, the sieve elements contain cytoplasm and are living. The sieve elements in most plants vary from 0.1 to several mm in length, and tend to be somewhat longer in gymnosperms than in angiosperms. Typical radii of the lumens are 6 to 25 μm. Specialized cells are generally found adjacent to and in close association with the sieve elements, called *companion cells* in angiosperms (see Fig. 1.3) and *albuminous cells* in gymnosperms. These cells have nuclei and generally contain many mitochondria. Although their precise functions are still unclear, such cells are metabolically related to the conducting cells and may supply them with carbohydrates, ATP, and other materials. Moreover, companion and albuminous cells accumulate sugars and other solutes, which may either passively diffuse— e.g., through the plasmodesmata that are present—or be actively transported into the sieve elements (see Crafts and Crisp, Esau, and Zimmermann and Milburn).

Sieve-tube members usually abut end-to-end to form the *sieve tubes*. The pair of generally inclined end walls between two sequential sieve-tube members forms the so-called *sieve plate* (Fig. 1.3). Sieve plates have many pores, ranging in diameter from less than 1 μm up to about 5 μm, and ordinarily have strands of cytoplasmic material passing through them. The pores are lined, not crossed, by the plasmalemmas. Solution moving in the phloem therefore does not have to cross any membranes as it flows from cell to cell along a sieve tube. A distinguishing feature of sieve-tube members in dicots and some monocots is the presence of *P protein* (for *phloem protein*, formerly termed *slime*), which often occurs in tubular as well as fibrillar form (see Cronshaw). Although the exact functions of this material are unclear, it is known that P protein can plug the sieve plate pores upon injury of the sieve tube. Specifically, when the phloem is opened by cutting a stem, the positive hydrostatic pressure in the phloem forces the contents of the conducting cells toward the incision, carrying the P protein into the sieve plate pores. As a further wound response, callose (a glucose polysaccharide) can also be deposited in the sieve plate pores, thereby closing them within a matter of minutes following injury.

The products of photosynthesis—*photosynthate*—are distributed throughout the plant by the phloem. Changes in phloem flow can affect the distribution of photosynthate, which in turn can affect the metabolism of mesophyll cells (see De Michele et al.). For high transpiration rates or a moderate moisture stress leading to reduced leaf water potentials, translocation of photosynthate can be decreased more than photosynthesis in some plants, with the consequence that leaves and nearby parts of the stem accumulate more starch. In C_4 plants the profuse vascularization and close proximity of bundle sheath cells to the phloem (see Fig. 8.9) can lead to rapid removal of photosynthate from the leaves and little deposition of starch in the bundle sheath cells over the course of a day, whereas certain C_3 plants can have a large increase of starch in the mesophyll cells during the daytime. The practice of "ringing" or "girdling" a branch (cutting away a band of phloem-containing bark all the way around) can stop the export of photosynthate, with the consequence that fruit on such a branch can become considerably larger than it otherwise would. The number of sieve elements in a petiole can reflect the productivity of the leaf; e.g., petioles of sun leaves tend to have more sieve elements per unit leaf area than petioles of shade leaves.

Phloem Contents and Speed of Movement

An elegant way of studying the contents of certain sieve elements is by means of aphid stylets (see Zimmermann and Milburn). An aphid feeds on the phloem by inserting its stylet into an individual sieve element. After the aphid has been anesthetized and its body removed, the remaining mouth part forms a tube that leads the phloem solution from the sieve element to the outside. Solutes in the phloem solution extracted using this technique are over 90% carbohydrates, mainly sucrose and some other oligosaccharides. The concentration of sucrose is generally 0.2 to 0.7 M; values near 0.3 M are typical for small plants, whereas some tall trees have sucrose concentrations near the upper limit. The types and concentrations of the solutes exhibit diurnal and seasonal variations, and also depend on the tissues that the phloem solution is flowing toward or away from. For example, the solution in the phloem moving out of senescent leaves is low in sucrose but contains an appreciable concentration of amino acids and amides, sometimes as high as 0.5 M—an amino acid concentration near 0.05 M is representative of phloem solution in general. The initial movement of ions from the root to the rest of the plant occurs mainly in the xylem; subsequent recirculation can take place in the phloem, such as the movement of ions out of leaves just before their abscission. (See Crafts and Crisp.)

Solutes can move rapidly in the phloem, flow being toward regions of lower osmotic pressure. Thus, photosynthetic products move from the leaves to storage tissues in the stem and root, where they are generally converted to starch. At other times, sugars produced from the hydrolysis of such starch may move in the opposite direction, from the storage tissue to meristematic areas at the top of the plant. The speed of solute movement is ordinarily 0.2 to 1 m hour^{-1}, the rate varying, among other things, with the plant species and vigor of growth (see Canny, and Crafts and Crisp). Although a particular sugar generally moves in the phloem in the direction of a decrease in its concentration, diffusion is not the mechanism. First, the rate of solute movement far exceeds that of diffusion (see Ch. 1). Second, when a radio-active solute, a dye, or a heat pulse is introduced into conducting sieve elements, the "front" moves with a fairly constant speed (distance moved is proportional to time), whereas in a one-dimensional diffusional process, distance moved is proportional to the square root of time in such a case ($x_e^2 = 4D_j t_e$, Eq. 1.6).

Rather than the solute speed in the phloem, we are often more interested in how much dry matter is actually translocated. For example, if the sieve elements contain 0.5 M (500 mol m^{-3}) sucrose moving at an average speed of 0.6 m hour^{-1}, what would be the transfer rate of sucrose in kg m^{-2} hour^{-1} (1 kg m^{-2} hour^{-1} = 0.1 g cm^{-2} hour^{-1})? By Equation 3.6 ($J_j = \bar{v}_j c_j$), the flux density of sucrose is

$$J_{sucrose} = (0.6 \text{ m hour}^{-1})(500 \text{ mol m}^{-3})$$

$$= 300 \text{ mol m}^{-2} \text{ hour}^{-1}$$

Since sucrose has a molar mass of 0.342 kg mol^{-1}, this flux density corresponds to 100 kg m^{-2} hour^{-1}. In the present example, the flow is per m^2 of sieve-tube *lumens*; the rate of flow per unit area of *phloem tissue* would be less by the ratio of the lumen cross-sectional area to the total phloem cross-sectional area, generally 0.2 to 0.5.

Mechanism of Phloem Flow

What causes the movement of solutes in the phloem? Primarily because of observational difficulties, this question has proved hard to answer. A secondary complication is that water may readily enter and leave the various types of cells in the phloem and surrounding tissue. The phloem therefore cannot be viewed as an isolated independent system. For example, when the water potential in the xylem decreases, as during rapid transpiration, the speed of fluid movement in the phloem generally becomes less. Some water may move upward in the xylem and, later, downward in the phloem; but this is not the whole story, since movement in the phloem can be in either direction.

Around 1930 Münch proposed that fluid movement in the phloem was caused by a gradient in hydrostatic pressure. This would lead to flow analogous to that in the xylem as described by Poiseuille's law (Eq. 9.10b), $J_V = -(r^2/8\eta)(\partial P/\partial x)$. To examine this hypothesis, we will assume that the average speed of flow in the lumen of the conducting cells of the phloem, J_V, is 0.6 m hour^{-1} (0.17 mm s^{-1}). Our sieve-tube members will be 12 μm in radius and 1 mm long, and the sieve plates 5 μm thick, with pores 1.2 μm in radius covering one-third of their surface area. Based on the relative areas available for conduction, J_V is three times as high in the sieve plates as in the lumen of the sieve tube—namely, 0.51 mm s^{-1} across the sieve plates. The viscosity of the solution in a sieve tube is greater than that of water; e.g., η^{phloem} may be about 1.7 mPa s at 20°C, the value for 0.5 M sucrose. The pressure change per cell expected from Poiseuille's law is then

$$\Delta P = \frac{-8\eta J_V}{r^2} \Delta x = \frac{-(8)(1.7 \times 10^{-3} \text{ Pa s})(0.17 \times 10^{-3} \text{ m s}^{-1})(1 \times 10^{-3} \text{ m})}{(12 \times 10^{-6} \text{ m})^2}$$

$$= -16 \text{ Pa}$$

in the lumen and -24 Pa across the sieve plate. For cells 1 mm long, the pressure gradient needed to cause Poiseuille flow is then -40×10^3 Pa m^{-1}, or -0.040 MPa m^{-1}.

In the above case, somewhat over half the hydrostatic pressure gradient in the phloem is necessary to overcome the resistance of the sieve plate pores. When the end walls are steeply inclined to the axis of the sieve element, the pores of the sieve plate can actually occupy an area greater than that of the cross section of the sieve tube. This causes J_V in the pores to be less than in the lumen, which tends to reduce the resistance to flow in the phloem.

Values for Components of the Phloem Water Potential

We will now examine some of the consequences of the plausible, but un-proved, hypothesis of a pressure-driven flow in the phloem. We will let the water potential in the xylem be -0.6 MPa at ground level and -0.8 MPa 10 m above the ground (see Fig. 9.11 and Table 9.2). Since there is no evidence

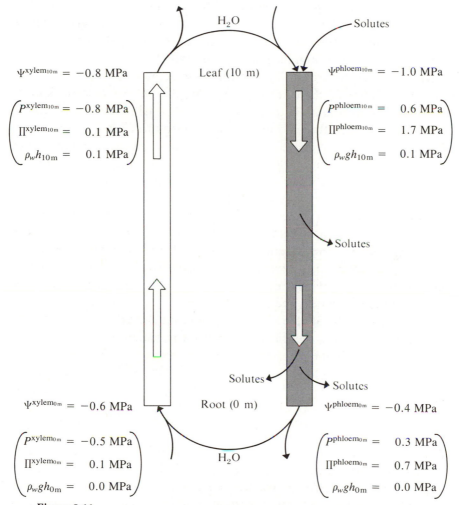

$\Psi^{xylem}_{10m} = -0.8$ MPa

$\begin{pmatrix} p^{xylem}_{10m} = -0.8 \text{ MPa} \\ \Pi^{xylem}_{10m} = 0.1 \text{ MPa} \\ \rho_w h_{10m} = 0.1 \text{ MPa} \end{pmatrix}$

H$_2$O

Leaf (10 m)

Solutes

$\Psi^{phloem}_{10m} = -1.0$ MPa

$\begin{pmatrix} p^{phloem}_{10m} = 0.6 \text{ MPa} \\ \Pi^{phloem}_{10m} = 1.7 \text{ MPa} \\ \rho_w g h_{10m} = 0.1 \text{ MPa} \end{pmatrix}$

Solutes

Solutes

Solutes

$\Psi^{xylem}_{0m} = -0.6$ MPa

Root (0 m)

H$_2$O

$\Psi^{phloem}_{0m} = -0.4$ MPa

$\begin{pmatrix} p^{xylem}_{0m} = -0.5 \text{ MPa} \\ \Pi^{xylem}_{0m} = 0.1 \text{ MPa} \\ \rho_w g h_{0m} = 0.0 \text{ MPa} \end{pmatrix}$

$\begin{pmatrix} p^{phloem}_{0m} = 0.3 \text{ MPa} \\ \Pi^{phloem}_{0m} = 0.7 \text{ MPa} \\ \rho_w g h_{0m} = 0.0 \text{ MPa} \end{pmatrix}$

Figure 9.11
Idealized representation of xylem and phloem flow driven by gradients in hydrostatic pressure. Water enters and leaves the phloem by passively moving toward regions of lower water potential. Solutes either diffuse or are actively transported into and out of the sieve elements, leading to a decrease in the phloem osmotic pressure from 1.7 MPa in the leaf to 0.7 MPa in the root.

for active transport of water across membranes into the phloem, the water potential in the phloem 10 m above the ground, $\Psi^{phloem\,10\,m}$, must be lower than -0.8 MPa if water is to enter passively—water entering the phloem in the leaves can come from the xylem. The water potential there might be -1.0 MPa, for example. By analogous reasoning, Ψ^{phloem} in the root must be higher than Ψ^{xylem} if water is to spontaneously leave the phloem in the region where the solutes, such as sucrose, are removed; an appropriate value for the water potential in the phloem at ground level, $\Psi^{phloem\,0\,m}$, might be -0.4 MPa. Water may then flow down the phloem from a water potential of -1.0 MPa to one of -0.4 MPa. This may seem at first like an energetically uphill movement. However, as in the xylem, the term representing the force on solution moving within the phloem is ΔP, not $\Delta \Psi$.

The fact that the drop in hydrostatic pressure, not the change in water potential, represents the driving force along the phloem deserves special emphasis. By Equation 3.38, J_V equals $L_P(\Delta P - \sigma \Delta \Pi)$, where the reflection coefficient σ within the phloem is zero because there are no membranes intervening between sequential members of a sieve tube. Differences in the osmotic pressure at various locations in the phloem, which help determine the overall change in water potential, therefore do not directly affect the movement of solution *along* the phloem. On the other hand, substances entering or leaving the phloem generally pass across a plasmalemma, which most likely has a mean reflection coefficient close to unity for solutes such as sucrose. J_V given by Equation 3.38 is then $L_P(\Delta P - \Delta \Pi)$, which equals $L_P \Delta \Psi$ by Equation 2.13a ($\Psi = P - \Pi + \rho_w g h$). Consequently, the change in water potential probably does represent the driving force on water moving into or out of the phloem (see Fig. 9.11).

Next we will estimate the various components of the water potential at the upper end of the phloem tissue under consideration (Fig. 9.11). The osmotic pressure in the phloem of a leaf 10 m above the ground $\Pi^{phloem\,10\,m}$, might be due to the following solutes: 0.5 M sucrose, 0.1 M other sugars, 0.05 M amino acids, and 0.05 M inorganic ions, i.e., $\sum c_j = 0.5$ M $+ 0.1$ M $+ 0.05$ M $+ 0.05$ M $= 0.7$ M, which is 700 mol m^{-3}. Using the Van't Hoff relation—$\Pi_s = RT \sum_j c_j$ (Eq. 2.10), where RT is 2.437×10^{-3} m^3 MPa mol^{-1} at 20°C (App. II)—we estimate that the osmotic pressure in the phloem 10 m above the ground is $(2.437 \times 10^{-3}$ m^3 MPa mol$^{-1})(700$ mol m$^{-3})$, or 1.7 MPa. Such a large osmotic pressure, caused by the high concentrations of sucrose and other solutes, suggests that active transport is necessary at some stage to move certain photosynthetic products from leaf mesophyll cells to the sieve elements of the phloem. From the definition of water potential, $\Psi = P - \Pi + \rho_w gh$ (Eq. 2.13a), we conclude that the hydrostatic pressure in the phloem of a leaf 10 m above the ground must be

$$P^{\text{phloem}_{10\text{ m}}} = \Psi^{\text{phloem}_{10\text{ m}}} + \Pi^{\text{phloem}_{10\text{ m}}} - \rho_w gh_{10\text{ m}}$$

$$= -1.0 \text{ MPa} + 1.7 \text{ MPa} - 0.1 \text{ MPa}$$

$$= 0.6 \text{ MPa}$$

(see Fig. 9.11). This appreciable hydrostatic pressure would cause sieve-tube members to exude when cut, as indeed they do.

As we indicated on p. 502, the hydrostatic pressure gradient might be -0.04 MPa m^{-1} for Poiseuille flow in representative (horizontal) conducting cells of the phloem. P would therefore drop 0.4 MPa for a distance of 10 m. Gravity also leads to a pressure gradient in a column of water (0.1 MPa increase in P when descending 10 m), so in the present case the ΔP needed to cause Poiseuille flow vertically *downward* is -0.3 MPa. Thus, the hydrostatic pressure in the phloem at ground level is 0.3 MPa. The osmotic pressure there equals $P^{\text{phloem}_{0\text{ m}}} + \rho_w gh_{0\text{ m}} - \Psi^{\text{phloem}_{0\text{ m}}}$ (Eq. 2.13a), which is 0.7 MPa. The values of Ψ^{phloem} and its components at $h = 0$ m and $h = 10$ m are summarized in Figure 9.11.

In our present example, the osmotic pressure of the phloem solution decreases from 1.7 MPa in the leaf to 0.7 MPa in the root (Fig. 9.11). Such a decrease in Π is consistent with the phloem's function of delivering photosynthetic products to different parts of the plant. Moreover, our calculations indicate that flow is in the direction of decreasing concentration, but that diffusion is not the mechanism. (Although the total concentration decreases in the direction of flow, this does not mean that the c_j of every solute necessarily does so.) Finally, we note the importance of removing solutes, either by active transport or by diffusion into the cells near the conducting cells of the phloem.

The known involvement of metabolism in translocation in the phloem could be due to active transport of solutes into the phloem of the leaf or other *source*, which is often referred to as *loading*, and/or to their removal, or *unloading*, in the root or other *sink*, such as the fruit. Indeed, loading appears to involve proton-sucrose cotransport via a carrier located in the plasmalemma of sieve elements, which moves this sugar from the apoplast into the conducting cells (movement of photosynthate among mesophyll cells and possibly to companion cells is mainly in the symplast). Low concentrations of translocated solutes in the cells outside the phloem in sink regions, which would favor the diffusion of these solutes out of the sieve elements, could also be maintained by metabolic conversions of phloem solutes, such as by starch formation in the cells adjacent to the conducting cells of the phloem. In any case, active loading or unloading of the phloem sets up a gradient in osmotic pressure, which in turn generates the hydrostatic pressure gradient that could lead to Poiseuille flow along the phloem. However, much remains to be learned about the actual details of fluid movement in the phloem (see Gifford and Evans).

THE SOIL-PLANT-ATMOSPHERE CONTINUUM

As our final topic, we will examine certain aspects of the movement of water from the soil, through a plant, out into the surrounding atmosphere. During a growing season, about 100 times more water is transpired by a plant than remains in it for growth. Thus, the amount of transpiration is actually a fairly accurate estimate of the water uptake by the roots. Although the rate that water crosses each section of the pathway is essentially the same in the steady state, the resistances and the areas across which water flows differ markedly for the various components.

Values of Water Potential Components

Possible values for the water potential and its components in various parts of the soil-plant-atmosphere system are given in Table 9.2. The values do not apply to all plants, nor even to the same plant at all times. Rather, they serve to indicate representative contributions of P, Π, $\rho_w gh$, and relative humidity to Ψ in various parts of the soil-plant-atmosphere continuum.

First, we will consider the soil water potential, Ψ^{soil}. As we indicated earlier in this chapter, Ψ^{soil} is usually dominated by P^{soil}, which is negative because of the surface tension effects at the numerous air-liquid interfaces in the soil. The actual magnitude of the water potential varies greatly with environmental conditions and the type of soil. In freshly irrigated soil it may be about -0.01 MPa, while permanent wilting of plants often occurs when the soil water potential decreases below about -1.5 MPa.

The value of the soil water potential at which wilting of a plant occurs depends on the osmotic pressure in the vacuoles of its leaf cells, as the following argument will show.* Let us consider the case in Table 9.2 where

---

* $\Pi^{vacuole}$ is generally the same as $\Pi^{cytosol}$; see p. 81.

Table 9.2

Representative values for the various components of the water potential in the soil-plant-atmosphere system. Ψ equals $P - \Pi + \rho_w gh$ in the liquid phases (Eq. 2.13a), and $(RT/\overline{V}_w)\ln(\%$ relative humidity$/100) + \rho_w gh$ in the gas phases (Eq. 2.21), all at 25°C.

Location	Ψ (MPa)	P (MPa)	$-\Pi$ (MPa)	$\rho_w gh$ (MPa)	$\frac{RT}{\overline{V}_w}\ln\left(\frac{\% \text{ RH}}{100}\right)$ (MPa)
Soil 0.1 m below ground and 10 mm from root	-0.3	-0.2	-0.1	0.0	
Soil adjacent to root	-0.5	-0.4	-0.1	0.0	
Xylem of root near ground surface	-0.6	-0.5	-0.1	0.0	
Xylem in leaf at 10 m above ground	-0.8	-0.8	-0.1	0.1	
Vacuole of leaf mesophyll cell at 10 m	-0.8	0.2	-1.1	0.1	
Cell wall of leaf mesophyll cell at 10 m	-0.8	-0.4	-0.5	0.1	
Air in cell wall pores at 10 m (water vapor assumed to be in equilibrium with water in cell wall)	-0.8			0.1	-0.9
Air just inside stomata at 95% relative humidity	-6.9			0.1	-7.0
Air just outside stomata at 60% relative humidity	-70.0			0.1	-70.1
Air just across boundary layer at 50% relative humidity	-95.1			0.1	-95.2

the water potential in the vacuole of a leaf cell 10 m above the ground is initially -0.8 MPa. As the soil dries out, Ψ^{soil} decreases and eventually becomes -1.0 MPa. When the soil water potential becomes -1.0 MPa, Ψ^{leaf} generally must be less than this for water movement to continue from the soil to the leaf. Ψ^{leaf} could be -1.0 MPa when $P^{\text{vacuole}\,10\,m}$ is 0.0 MPa, $\rho_w gh^{\text{vacuole}\,10\,m}$ remains 0.1 MPa, and $\Pi^{\text{vacuole}\,10\,m}$ remains 1.1 MPa (the latter two are values in Table 9.2; actually, as the hydrostatic pressure in a leaf cell decreases, the cell will shrink somewhat because of the elastic properties of the cell wall, and so Π^{vacuole} will increase slightly, see p. 88). Zero hydrostatic pressure in the vacuole means that the cell has lost its turgor, and the leaf thus wilts in response to this low Ψ^{soil}. For xerophytes in arid areas, the osmotic pressure in the leaves is often 2.5 to 5.0 MPa under normal physiological conditions. The value of the soil water potential at which wilting occurs for such plants would be considerably lower (i.e., more negative) than for the plant indicated in Table 9.2. A high osmotic pressure in the vacuoles of the leaf cells can therefore be viewed as an adaptation to the low soil water potentials in an arid climate (see Slatyer).

Let us suppose that the soil dries out even further from the level causing wilting. Since at least some transpiration still occurs because of the very low Ψ^{air}, Ψ^{leaf} will be less than Ψ^{soil}, and some cellular water will be lost from the leaf. The vacuolar contents then become more concentrated, and $\Pi^{vacuole}$ increases. For instance, if Ψ^{leaf} became -2.0 MPa for a Ψ^{soil} of -1.8 MPa, the osmotic pressure in the vacuole of a leaf cell 10 m above the ground would be 2.1 MPa, which represents a loss of nearly half of the cellular water ($\Pi^{vacuole}$ 10 m was originally 1.1 MPa for this leaf cell; Table 9.2).

As we have already indicated, the driving force for water movement in the xylem is the negative gradient in hydrostatic pressure, which can lead to a flow describable by Poiseuille's law (Eq. 9.10). In Table 9.2, P^{xylem} decreases by 0.3 MPa from the root to a leaf 10 m above the ground. The xylary sap, which contains chiefly water plus some minerals absorbed from the soil, usually does not have an osmotic pressure in excess of 0.2 MPa. The hydrostatic pressure, on the other hand, can have much larger absolute values, and generally changes markedly during the day. When there is extremely rapid transpiration, large tensions (negative hydrostatic pressures) can develop in the xylem. These tensions are maintained by the cohesion of water molecules resulting from the intermolecular hydrogen bonding (Ch. 2). When transpiration essentially ceases, as it can at night or under conditions of very high relative humidity in the air surrounding the plant, the tension in the xylem becomes very small—in fact, the hydrostatic pressure can even become positive (reflecting water movement from the surrounding cells into the root xylem in response to Π^{xylem}). Such a positive hydrostatic pressure (termed *root pressure*) can cause guttation as xylem fluid is then exuded through specialized structures called hydathodes located near the ends of veins on the margins of leaves.

Water is conducted to and across the leaves in the xylem vessels. It then moves to the individual leaf cells, presumably by flowing mainly in the cell walls (only short distances are involved, since the xylem ramifies extensively in a leaf). The water potential is usually about the same in the vacuole, the cytosol, and the cell wall of a particular leaf mesophyll cell (see values in Table 9.2). If this were not the case, water would redistribute itself by flowing energetically downhill toward lower water potentials. The water in the cell wall pores is in contact with air, where evaporation can take place, and so a flow occurs along the interstices to replace lost water. This flow can be approximately described by Poiseuille's law (Eq. 9.10), which indicates that a (small) hydrostatic pressure gradient exists across such cell walls.

As water continually evaporates from the cell walls of mesophyll cells, the accompanying solutes originally in the xylary sap are left behind and can accumulate in the cell wall water. Some solutes are of course needed for cell growth. For halophytes and xerophytes growing in soils of high salinity, excess inorganic ions can be actively excreted from a leaf by salt glands on its surface (see Lüttge, and Rains). The periodic abscission of leaves is another way of "preventing" an excessive buildup of solutes in the cell wall water, as well as returning mineral nutrients to the soil (see Epstein).

Resistances and Areas

We will now consider the resistances to water flow in those parts of the soil-plant-atmosphere continuum where water moves as a liquid. (We have already considered the gaseous parts of the pathway in Ch. 8.) If we let the flux of water equal the drop in water potential across some component divided by its resistance, we would have only part of the story, since we should also consider the relative areas of each component as well as whether $\Delta\Psi$ does indeed represent the driving force. Also, $\Delta\Psi$ represents the relative energies of water only at constant T, and thus ideally we should compare water potentials only between locations at the same temperature. The root and adjacent soil are usually at the same temperature, as are the leaf mesophyll cells and air in the intercellular air spaces. But roots and leaves generally are *not* at the same temperature. Nevertheless, using the pressure gradient (Eq. 9.10) turns out to be sufficiently accurate for describing the flow in the xylem or the phloem, even when a temperature difference exists along the pathway (see Spanner).

Let us designate the average volume flux density of water across area A^j of component j by $J_{V_w}^j$. A^j could be the root surface area, the effective cross-sectional area of the xylem, or the area of one side of the leaves. In the steady state, the product $J_{V_w}^j A^j$ is essentially constant, since nearly all of the water taken up by the root is lost by transpiration, and thus the same volume of water moves across each component along the pathway per unit time. We will represent the drop in water potential across component j by $\Delta\Psi^j$, and define the resistance of component j (R^j) as follows:

$$J_{V_w}^j A^j = \frac{\Delta\Psi^j}{R^j} \cong \text{constant} \tag{9.11}$$

where $J_{V_w}^j A^j$ is the volume of water crossing component j in unit time. A relation similar to Equation 9.11 was proposed by van den Honert in 1948.

Equation 9.11 can be used to describe certain overall characteristics of water flow in the soil-plant-atmosphere system. However, $\Delta\Psi^j$ does not always represent the driving force on water—e.g., a change in the osmotic pressure component of Ψ has no direct effect on the flow in the xylem or the phloem. Also, such a relation is not useful for a gas phase, since the value of the resistance R^j depends on the concentration of water vapor (see p. 402). When $J_{V_w}^j$ is in m s^{-1}, A^j in m^2, and $\Delta\Psi^j$ in MPa, Equation 9.11 indicates that the units of R^j are MPa s m^{-3}. For young sunflower and tomato plants, the resistance from the root surface to the leaf mesophyll cells, R^{plant}, is about 1.0×10^8 MPa s m^{-3} when the plants are approximately 0.3 m tall and $\Delta\Psi^{\text{plant}}$ is close to 0.2 MPa (see Kramer). Using Equation 9.11, we obtain

$$J_{V_w}^j A_j = \frac{(0.2 \text{ MPa})}{(1.0 \times 10^8 \text{ MPa s m}^{-3})} = 2 \times 10^{-9} \text{ m}^3 \text{ s}^{-1}$$

which gives the volume of water flowing across each component per unit time. For sunflower, bean, and tomato, R^{root}: R^{stem}: R^{leaves} is about 2:1:1.5, while R^{root} is relatively higher for soybean, and R^{leaves} is relatively higher for safflower (see Jarvis). Considering the above value for R^{plant}, R^{stem} would be about 2×10^7 MPa s m^{-3} for young sunflower or tomato plants. Resistances can be considerably higher for other plants—e.g., R^{plant} can be 10^{10} MPa s m^{-3} for wheat and even higher for barley (see Jones).

The resistance for water movement along the stem can be separated into (1) a quantity expressing some inherent flow properties of the xylem and (2) the geometrical aspects. By analogy with Ohm's law, where R equals $\rho\,\Delta x/A$ (pp. 117 and 390), we thus obtain

$$\rho^j = \frac{R^j A^j}{\Delta x^j} = \frac{1}{J_{V_w}^j} \frac{\Delta\Psi^j}{\Delta x^j} \tag{9.12}$$

where ρ^j is here the hydraulic resistivity of the xylem tissue of length Δx^j and cross-sectional area A^j, and the second equality incorporates Equation 9.11. For many plants containing xylem vessels, ρ^{stem} equals 100 to 500 MPa s m^{-2}; hydraulic resistivities can be somewhat higher for plants with tracheids, such as conifers; e.g., it can be 1 600 MPa s m^{-2} for certain ferns (see Milburn, Woodhouse and Nobel, and Zimmermann and Brown).* By Equation 9.12, the R^{stem} of 2×10^7 MPa s m^{-3} for the young sunflower and tomato could

---

* Hydraulic conductivities $(1/\rho^j)$ can also be used for the xylem as well as the phloem, e.g., the phloem hydraulic conductivity can be about 1×10^{-4} m^2 MPa^{-1} s^{-1} for herbaceous species and 4×10^{-3} m^2 MPa^{-1} s^{-1} for trees (see J. A. Milburn in Zimmermann and Milburn).

occur for a ρ^{stem} of 300 MPa s m^{-2}, a length of 0.3 m, and a total xylem cross-sectional area of 5 mm^2. A tree 10 m tall might have 600-times more xylem area, which for the same hydraulic resistivity would lead to an R^{stem} of 10^6 MPa s m^{-3}. The smaller resistance results from the many more xylem vessels in the tree stem, even though the overall xylem length would be much longer than for the tomato or sunflower plant.

We can relate the hydraulic resistivity of the xylem to flow characteristics predicted by Poiseuille's law [Eq. 9.10b, $J_V = -(r^2/8\eta)\, \partial P/\partial x$]. Specifically, by comparing Poiseuille's law with Equation 9.12 and identifying $-\partial P/\partial x$ with $\Delta \Psi^j/\Delta x^j$, we note that ρ^j equals $8\eta/r^2$. Using a representative value for ρ^{stem} and the viscosity of water at 20°C (App. II), we then have

$$ r = \sqrt{\frac{8\eta}{\rho^{stem}}} = \sqrt{\frac{(8)(1.0 \times 10^{-3}\ \text{Pa s})}{(300 \times 10^6\ \text{Pa s m}^{-2})}} $$

$$ = 5 \times 10^{-6}\ \text{m} $$

which is a possible effective radius of xylem elements. Owing to the presence of cell walls and other cells, the lumen of the xylem vessels might correspond to only about $\frac{1}{4}$ of the cross-sectional area of the xylem tissue in the stem, A^{stem}. A lumen radius of 10 μm for $\frac{1}{4}$ the area would have the same J_V and $\partial P/\partial x$ as pores 5 μm in radius occupying the entire area (see the r^2 factor in Eq. 9.10b). Other complications, such as the resistance of the perforation plates, cause the actual radius to be even greater than the effective radius. We should also note that, because of the inverse relationship between ρ^{stem} and r^2, the larger xylem vessels tend to conduct proportionally more than the smaller ones in a given stem. Finally, recall that we used Poiseuille's law to equate the soil hydraulic conductivity coefficient L^{soil} to $r^2/8\eta$ (p. 495), which is analogous to our present consideration of a reciprocally related quantity, the xylem hydraulic resistivity. In fact, the xylem hydraulic conductivity could be equated to $r^2/8\eta$, where r is the effective radius of the xylem vessels averaged over the entire cross-sectional area of the xylem.

For the example in Table 9.2, the drop in water potential is 0.2 MPa across the soil part of the pathway, 0.1 MPa from the root surface to the root xylem, and 0.2 MPa along the xylem—values that suggest the relative magnitudes of the three resistances involved ($\Delta\Psi^j \cong$ constant $\times$ R^j, Eq. 9.11). As the soil dries out, the relative size of the water potential drop in the soil usually becomes larger (see p. 489). Also, the rapid uptake of soil water by the root during those parts of the day when transpiration is particularly high can lead to a large hydrostatic pressure gradient (consider Darcy's law, Eq. 9.6), and $\Delta\Psi^{soil}$ from the bulk soil to the root then increases. In fact R^{soil}

is often the largest resistance for that part of the soil-plant-atmosphere system where water moves predominantly as a liquid (water can also move as a vapor in the soil; see p. 487). This is particularly true as the soil dries and its hydraulic conductivity consequently decreases, although for soybean in pots R^{plant} was greater than R^{soil} down to a Ψ^{soil} of -1.1 MPa (see Blizzard and Boyer). R^{plant} can also increase as Ψ^{leaf} decreases, perhaps because the entry of air breaks the water continuity (cavitation) in some of the xylem vessels, which thus become nonconducting. The resistance of the root epidermis, cortex, and endodermis is generally somewhat less than R^{soil}. We also note that, as the soil dries, a hydraulic resistance at the root-soil interface appears to develop, possibly representing a shrinking of the root away from the soil particles and hence less contact with the soil water (see Passioura, and Weatherley). The resistance of the xylem depends on its length, since the drop in water potential along the xylem is often about 0.02 MPa m^{-1} (see p. 496). In conifers and other plants, the conducting area of the stem xylem is often proportional to the leaf area (see Jarvis); if the transpiration rate per unit leaf area and the xylem element dimensions are the same, then the pressure gradient for Poiseuille flow would be the same; i.e., a higher J_V^{leaves} A^{leaves} is compensated for by a higher A^{stem} for the xylem tissue (see Eqs. 9.11 and 9.12), so $\Delta\Psi/\Delta x$ along the stem would be approximately the same.

In the last chapter we indicated that transpiration by an exposed leaf of a C_3 mesophyte might be 4.3 mmol m^{-2} s^{-1} (p. 415), which corresponds to a volume flux density of water of 7.7×10^{-8} m s^{-1} (1 mol H$_2$O = 0.018 kg = 18×10^{-6} m^3). In this chapter we calculated that J_{V_w} of a young root can be 1.1×10^{-7} m s^{-1} (p. 488). For a plant having leaves and roots with such volume flux densities, equality of water flow across each component indicates that the area of one side of the leaves must be $(1.1 \times 10^{-7}$ m s$^{-1})/(7.7 \times 10^{-8}$ m s$^{-1})$, or 1.4 times larger than the surface area of the young roots (see Eq. 9.11). $J_{V_w}^{root}$ depends markedly on root age and hence varies along the length of a root; e.g., it is usually considerably lower for older roots (see Epstein, and Slatyer). Since the total root system of many perennials can be quite extensive, A^{root} including the relatively nonconducting regions can actually be 20 to 100 times larger than the surface area of mature leaves. J_V can be 1 mm s^{-1} in the xylem (p. 495). Again using Equation 9.11, we conclude that the surface area of the young roots would then be $(1 \times 10^{-3}$ m s$^{-1})/(1.1 \times 10^{-7}$ m s$^{-1})$, or 10^4 times larger than the cross-sectional area of the conducting parts of the xylem. Relatively few measurements of A^j and $J_{V_w}^j$ have been made for the sequential components involved in water flow in a plant.

Capacitance and Time Constants

The marked diurnal changes in hydrostatic pressure in the xylem can cause plants to have observable fluctuations in their stem diameters. When the transpiration rate is high, the large tension within the xylem vessels is transmitted to the water in the cell walls of the xylem vessels, then to communicating water in adjacent cells, and eventually all the way across the stem. The decrease in hydrostatic pressure in its trunk can therefore cause a whole tree to contract during the day. At night, the hydrostatic pressure in the xylem increases and may even become positive, and the trunk diameter then increases, generally by about 1%. Such changes in diameter, and therefore volume, represent net release of water during the day and storage at night. In fact, the diurnal change in water content of a plant can be equal to the amount of water transpired in a few minutes to a few hours during the daytime. These changes in water content correspond to a capacitance effect superimposed on the resistance network for water flow (see Jarvis; Jarvis et al.; Jordan and Ritchie; Kozlowski, Vol. III; Running; and Slatyer). Since we have already discussed electrical capacitance in Chapter 3 (pp. 108–110), we will here focus on the capacity of various parts of a plant to store water and the time course for its release.

Let us begin by comparing the average daily evapotranspiration per unit ground area to the diurnal fluctuations in water content of a tree trunk. We will suppose that the trunks average 0.2 m in diameter, 10 m in height, 4 m apart in a rectangular grid, and transpire a depth of water of 4 mm day^{-1} (see p. 470), nearly all of which occurs during the daytime. If a 1% diurnal change in diameter reflects volumetric changes in water content, each trunk would daily vary 0.006 m^3 in volume; each tree transpires 0.06 m^3 H$_2$O daily (4 m $\times$ 4 m $\times$ 0.004 m day^{-1}). Thus, the change in water content of the trunk could supply about 1 hour's worth of transpired water during the daytime. Absorption of water by the roots can thus lag behind the loss of water by leaf transpiration, which reflects water coming from the storage capacity or capacitance of the trunk.

As transpiration increases following stomatal opening at dawn, the leaf water potential decreases. This lowers the hydrostatic pressure in the leaves, and eventually the water content of the leaves decreases, raising the osmotic pressure and thereby further decreasing the water potential ($\Psi = P - \Pi + \rho_w g h$, Eq. 2.13a). During this period, water uptake from the soil does not balance transpiration by the leaves, and so Equation 9.11 is not obeyed. The relative water content of the leaves may decrease 10% during the daytime.

For a leaf area index of 5 and leaves 400 μm thick consisting of half water by volume (p. 341), the change in water content corresponds to a depth of water of $(5)(400 \times 10^{-6}$ m$)(0.5)(0.1)$, or 1×10^{-4} m (0.1 mm). This corresponds to 3% of the daily transpiration of 4 mm, and thus could represent about 20 minutes of water loss during the daytime. Roots can also exhibit diurnal changes in water content. Since for some species they can have more biomass than the leaves and trunk together, roots may store the equivalent of a few hours' worth of transpiration (see Jarvis).

Capacitance effects can also be seen on a longer time scale. For instance, the sapwood between the vascular cambium on the outside and the heartwood* on the inside can represent 20% to 40% of the radial dimension of a mature tree and even more for a young tree or sapling. The sapwood can store about 1 week's worth of water at moderate transpiration rates (even longer in drought periods with lowered rates of transpiration). A cactus stem (see Fig. 7.9) can store many months' worth of transpired water during drought periods when there is limited stomatal opening.

We will define the water capacitance C^j of plant part j as follows:

$$C^j = \frac{\text{change in water content of component } j}{\text{change in average water potential along component } j} \quad (9.13)$$

$$= \frac{\Delta V^j_{\text{water}}}{\Delta \bar{\Psi}^j}$$

where V^j_{water} is the volume of water in component j. We estimated above that a tree trunk might diurnally change its water content by 0.006 m^3. This could be accompanied by a change in average xylem water potential from -0.3 MPa at dawn (the value 10 mm from root; Table 9.2) to -0.7 MPa (average of -0.6 MPa and -0.8 MPa, the values at the two ends of the xylem indicated in Table 9.2), or -0.4 MPa overall. By Equation 9.12, C^{xylem} would then be

$$C^{\text{xylem}} = \frac{(-0.006 \text{ m}^3)}{(-0.4 \text{ MPa})} = 1.5 \times 10^{-2} \text{ m}^3 \text{ (MPa)}^{-1}$$

The much smaller stem of a young tomato or sunflower would have a much lower water capacitance, e.g., its C^{stem} might be 1.5×10^{-5} m^3 (MPa)$^{-1}$.

---

   * The heartwood represents the central xylem, which is generally rather darkly pigmented. It has no living cells, conduction capacity, or food storage.

Upon comparing Equation 9.13 with Equation 3.1 ($Q = C\,\Delta E$, where Q is the net charge accumulated that leads to an electrical potential change ΔE across a region of capacitance C), we note that $\Delta\Psi^j$ in Equation 9.13 takes the place of ΔE in electrical circuits. In fact, we can again borrow from electrical circuit analysis to indicate how the initial average water potential along some component Ψ_0^j will approach a final average water potential Ψ_∞^j:

$$\Psi^j = (\Psi_0^j - \Psi_\infty^j)e^{-t/\tau^j} + \Psi_\infty^j \qquad (9.14)$$

which is identical to Equation 7.22 except that Ψ^j replaces T^{surf}. Similar to the time constant for thermal changes (Eq. 7.23), we can identify a time constant τ^j for changes in the average water potential of component j:

$$\tau^j = R_s^j C^j \qquad (9.15)$$

where C^j is defined by Equation 9.13 and R_s^j represents the effective resistance from the water storage region to the main transpiration pathway (Fig. 9.12). Since water can be stored all along the stem, C^{stem} equals the sum of the capacitances for individual parts of the pathway arranged in parallel ($C^j = \sum C_i^j$ for capacitances in parallel).[*] The accompanying resistances of individ-

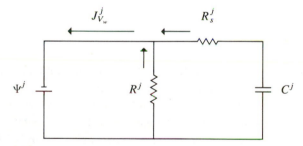

Figure 9.12
Electrical circuit portraying a plant component that can store water. The battery represents the drop in water potential along component j. The current represents the flux density of water, which can come across R^j as well as from storage in capacitor C^j and then across R_s^j.

---

[*] If $\Delta\Psi^j$ is the same all along the pathway, then we can simply add the component capacitances. Otherwise, we can let $C^j = \sum \Delta\Psi_i^j C_i^j / \Delta\Psi^j$, where $\Delta\Psi_i^j$ is the average water potential change along component i that has capacitance C_i^j.

ual parts of the pathway would be greater than the overall resistance ($1/R_s^j = \sum 1/R_{s_i}^j$ for resistances in parallel). The overall resistance for water movement from the water storage region to the xylem in the stem may be about 20% of the stem xylem resistance defined by Equation 9.11.

The time constant $R_s^j C^j$ indicates how rapidly the average water potential along component j will change following some perturbation that leads to changes in the water storage in C^j. Using the values we introduced above, we obtain a time constant for the stem part of the pathway of

$$\tau^{\text{stem}} = (0.2)(2 \times 10^7 \text{ MPa s m}^{-3})[1.5 \times 10^{-5} \text{ m}^3 (\text{MPa})^{-1}] = 60 \text{ s}$$

for the young tomato or sunflower, indicating that water potentials in the stem would respond fairly rapidly to environmentally induced changes in xylary water flow. For the 10-m-tall tree and again assuming that R_s^{stem} equals $0.2 \, R^{\text{stem}}$, the time constant would be 3×10^3 s (50 minutes). Thus, an appreciable amount of time is required to move the water from the sapwood and the vascular cambium into the trunk xylem of a tree. Also, the trunk has a relatively high water capacitance, and so water released from the trunk can affect the stem water potential for an extended period. Because of this, the peak xylary sap flow at the base of the trunk can lag several hours behind the peak transpiration by a tree.

Diurnal Changes

We have already indicated certain diurnal changes that take place in the soil-plant-atmosphere system. For instance, the soil temperature changes diurnally, which in turn affects the CO_2 evolution by respiration in soil microorganisms as well as in root cells. Also, the hydrostatic pressure in the xylem is more negative when the rate of transpiration is high, causing plants to decrease slightly in diameter during the day. All such diurnal changes are direct or indirect consequences of the daily variation in sunlight. Of course, the rate of photosynthesis also changes during the day. Photosynthesis is affected not only by the PAR level but also by the leaf temperature, which depends on the diurnally varying air temperature and net radiation balance for a leaf (Ch. 7).

Marked diurnal changes also occur in the water potentials in the soil-plant-atmosphere system. Let us consider a plant in a well-watered soil (Fig. 9.13). At night transpiration essentially ceases, since the stomata then close; the water potentials in the soil, root, and leaf may then all become

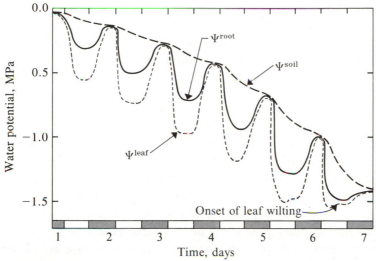

Figure 9.13
Schematic representation of daily changes in the water potentials in the soil, root, and leaf of a plant in an initially wet soil that dries out over a one week period. Ψ^{soil} is the water potential in the bulk soil, Ψ^{root} is that in the root xylem, and Ψ^{leaf} is the value in a leaf mesophyll cell. (Adapted by permission from R. O. Slatyer, *Plant-Water Relationships*, Academic Press, New York, 1967, p. 276.)

nearly equal (and close to zero). At dawn, the stomata open. Transpiration then removes water from the leaf and Ψ^{leaf} decreases (Fig. 9.13). After a short lag, the length of which depends on the capacitances involved and the transpiration rate, Ψ^{root} begins to decrease, but only to about -0.3 MPa (Fig. 9.13), since plenty of water is initially available in the wet soil. At dusk, these changes in Ψ^{leaf} and Ψ^{root} are reversed. As soil moisture is lost, Ψ^{soil} becomes more negative day by day, so that Ψ^{root} and Ψ^{leaf} also become more negative. (We are assuming that Ψ^{air} here remains essentially unchanged and is much lower than the other water potentials. Moreover, we are also assuming isothermal conditions, which though unrealistic will fortunately give us a useful approximation.) As the soil becomes drier, a steeper gradient in water potential is necessary to sustain the water flow up to the root, and therefore the difference between Ψ^{soil} and Ψ^{root} becomes larger day by day (Fig. 9.13). As Ψ^{leaf} becomes more negative on ensuing days, the leaf turgor pressure decreases, and eventually the hydrostatic pressure in the mesophyll cells of the leaf becomes zero some time during the day, e.g., when Ψ^{leaf} becomes -1.5 MPa in response to the decreasing

Ψ^{soil}. The leaf thus wilts, but recovers at night. Permanent wilting and damage to the leaf may result when Ψ^{soil} becomes lower on subsequent days. This portrayal of successive diurnal changes of Ψ^{soil}, Ψ^{root}, and Ψ^{leaf} illustrates that even such nonequilibrium processes in the soil-plant-air system can be analyzed in terms of the water potential.

What new applications of physics and chemistry might we expect in plant physiology and ecology in the future? Our quantitative approach should be expanded to include the interdependence of forces and fluxes as presented by irreversible thermodynamics. Cross-coefficients relating fluxes of charge to those of neutral species such as water need further consideration. Non-isothermal conditions must also be adequately handled. Of even greater potential impact is the application of a mathematical approach to the very complex field of plant growth and development. Hormone action, differentiation, photomorphogensis, reproduction, and senescence are all areas of plant physiology for which new insights can be provided by physics and chemistry. Many aspects of plant ecology are also ripe for physiochemical explanations. Mastery of the basic principles emphasized in this text as well as the adoption of a problem-solving approach will help us deal quantitatively with unresolved questions in the above areas and also allow us to predict plant responses to new environmental situations. Progress in any field requires some good fortune. But, in the words of Pasteur "Chance favors only the prepared mind."

Problems

9.1. Suppose that J_{wv} above some canopy reaches a peak value equivalent to 1.0 mm of water hour^{-1} during the daytime when the air temperature is 30°C and is 0.10 mm hour^{-1} at night when T^{ta} corresponds to 20°C. Assume that during the daytime the relative humidity decreases by 20% across the first 30 m of turbulent air and that the eddy diffusion coefficient halves at night because of an ambient wind speed lower than during the daytime. (a) What are J_{wv} in mmol m^{-2} s^{-1} and Δc_{wv}^{ta} in mol m^{-3} during peak transpiration? (b) What is r_{wv}^{ta} (over the first 30 m) during peak transpiration and at night? (c) What is Δc_{wv}^{ta} at night in mol m^{-3}? To what drop in relative humidity does this correspond? (d) If J_{wv}/J_{CO_2} is -700 H$_2$O/CO$_2$ during peak transpiration, what are J_{CO_2} and $\Delta c_{CO_2}^{ta}$ then?

(e) What is K_{CO_2} for the first 30 m of turbulent air at night? (f) How long would it take for water vapor to diffuse 1 m at night by eddy diffusion and by ordinary diffusion? Assume that Equation 1.6 ($x_e^2 = 4D_j t_e$) applies, where D_{wv} is 2.4×10^{-5} m^2 s^{-1} at 20°C, and K_{wv} has the value averaged over the first 30 m of turbulent air.

9.2. Suppose that the foliar absorption coefficient is 0.7 for trees with an average leaf area index of 8.0. (a) If the light compensation point for CO_2 fixation is at a PAR of 8 μmol m^{-2} s^{-1}, what are the cumulative leaf areas for light compensation when 2 000, 200, 20, and 0 μmol m^{-2} s^{-1} are incident on the canopy? (b) Suppose that J_{wv}/J_{CO_2} for the soil is 200 H_2O/CO_2, that 90% of the water vapor passing out of the canopy comes from the leaves, and that the net photosynthetic rate for the forest (using CO_2 from above the canopy as well as from the soil) corresponds to 20 kg of carbohydrate hectare^{-1} hour^{-1}. If J_{wv} from the soil is 0.6 mmol m^{-2} s^{-1}, what are the J_{CO_2}'s up from the soil and down into the canopy? (c) What is the absolute value of J_{wv}/J_{CO_2} above the canopy under the conditions of (b)? (d) When the leaves are randomly distributed with respect to distance above the ground and the trees are 16 m tall, at what level does the maximum upward flux of CO_2 occur when 200 μmol m^{-2} s^{-1} is incident on the canopy? (e) If the net rate of photosynthesis is proportional to PAR, about where does $c_{CO_2}^{ta}$ achieve a minimum? Assume that the conditions are as in (b) and (d) and that the maximum upward J_{CO_2} is 5 μmol m^{-2} s^{-1}. Note that in the present case the PAR halves every 2 m downward in the vegetation.

9.3. A horizontal xylem element has a cross-sectional area of 0.004 mm^2 and conducts water at 20°C at a rate of 20 mm^3 hour^{-1}. (a) What is the mean speed of the fluid in the xylem element? (b) What pressure gradient is necessary to cause such a flow? (c) If the pressure gradient remained the same, but cell walls with interstices 20 nm across filled the xylem element, what would be the mean speed of fluid movement? Assume that the entire area of the cell walls is available for conduction.

9.4. A horizontal phloem vessel has an effective radius of 10 μm. Assume that $D_{sucrose}$ in the phloem solution, which has a viscosity of 1.5 mPa s, is 0.3×10^{-9} m^2 s^{-1} at 20°C. (a) If there is no flow in the vessel and a thin layer of ^{14}C-sucrose is inserted, how long would it take by diffusion for the radioactive label at 10 mm and at 1 m to be 37% of the value at the plane of insertion? (b) If there is a pressure gradient of -0.02 MPa m^{-1} that leads to Poiseuille flow in the phloem, what is J_V there? (c) For the conditions of (b), how long does it take on the average for sucrose to move 10 mm and 1 m in the phloem vessel (ignore concomitant diffusion)? Compare your values with answers to (a). (d) For the pressure gradient in (b), what would J_V be in a vertical phloem vessel?

9.5. Suppose that L^{soil} is 1×10^{-11} and 2×10^{-16} m^2 s^{-1} Pa^{-1} when Ψ^{soil} is -0.01 and -1.4 MPa, respectively. For a plant in the wet soil, J_V^{xylem} is 2 mm s^{-1} and Ψ^{leaf} is -0.2 MPa for a leaf 3 m above the ground. Assume that the roots are 3 mm in diameter and that their surface area is 5 times that of one side of the leaves and 10^5 times larger than the conducting area of the xylem. Assume that all temperatures are 20°C, and ignore all osmotic pressures. (a) What are the radii of curvature at hemispherical air-liquid interfaces in the wet and the dry soils? (b) If water moves a distance of 8 mm to reach the root, what is the drop in hydrostatic pressure over that interval in the wet soil? (c) Suppose that the root xylem is arranged concentrically in a ring 500 μm below the root surface. If the average conductivity of the epidermis, cortex, endodermis, and xylem cell walls in the root is like that of the dry soil, what is the drop in hydrostatic pressure across them? (d) If the water in the leaf xylem is in equilibrium with that in the vacuoles of mesophyll cells, what is the average hydrostatic pressure gradient in the xylem for the wet soil case (assume that A^{xylem} is constant throughout the plant)? (e) If A^{mes}/A is 20, and the cell wall pores are 10 nm in diameter and 1 μm long, what ΔP along them will account for the rate of transpiration for the plant in the wet soil? (f) If the relative humidity above the dry soil is increased to 99%, what is J_V^{xylem} then?

9.6. Consider a tree with a leaf area index of 6 and a crown diameter of 6 m. The trunk is 3 m tall, has a mean cross-sectional area of 0.10 m^2, of which 5% is xylem tissue, and varies from an average water potential along its length of -0.1 MPa at dawn to -0.5 MPa in the steady state during the daytime. (a) If the average water vapor flux density of the leaves is 1 mmol m^{-2} s^{-1}, what is the transpiration rate of the tree in m^3 s^{-1}? (b) How long could such a transpiration rate be supported by water from the leaves, if the volume of water in the leaves per unit ground area changes from an equivalent depth of 1.0 mm to 0.8 mm? (c) What is R^{trunk}? (d) What is the hydraulic resistivity and conductivity of the trunk? (e) What is the capacitance and the time constant for water release from the trunk, if water equivalent to 0.8% of the trunk volume enters the transpiration stream? Assume that the resistance involved is the same as R^{trunk}.

References

Bewley, J. D., and M. Black. 1978. *Physiology and Biochemistry of Seeds in Relation to Germination*, Vol. 1. Springer-Verlag, Berlin.

Blizzard, W. E., and J. S. Boyer. 1980. Comparative resistance of the soil and the plant to water transport. *Plant Physiology 66*:809–814.

Briggs, G. E. 1967. *Movement of Water in Plants*. Blackwell Scientific Publications, Oxford.

Caldwell, M. M. 1976. Root extension and water absorption. In *Water and Plant Life: Problems and Modern Approaches*, O. L. Lange, L. Kappen, and E.-D. Schulze, eds. Ecological Studies 19. Springer-Verlag, Berlin. Pp. 63–85.

Canny, M. J. 1971. Translocation: mechanisms and kinetics. *Annual Review of Plant Physiology 22*:237–260.

Childs, E. C. 1969. *The Physical Basis of Soil Water Phenomena*. Wiley, New York.

Crafts, A. S., and C. E. Crisp. 1971. *Phloem Transport in Plants*. W. H. Freeman, San Francisco.

Cronshaw, J. 1981. Phloem structure and function. *Annual Review of Plant Physiology 32*:465–484.

De Michele, D. W., P. J. H. Sharpe, and J. D. Goeschl. 1978. Toward the engineering of photosynthetic productivity. *CRC Critical Reviews in Bioengineering 3*, Issue 1: 23–91.

Duncan, W. G. 1971. Leaf angles, leaf area, and canopy photosynthesis. *Crop Science* *11*:482–485.

Eastin, J. D., F. A. Haskins, C. Y. Sullivan, and C. H. M. van Bavel, eds. 1969. *Physiological Aspects of Crop Yield*. American Society of Agronomy, Madison.

Epstein, E. 1972. *Mineral Nutrition of Plants: Principles and Perspectives*. Wiley, New York.

Esau, K. 1969. *The Phloem. Encyclopedia of Plant Anatomy*, 2nd ed., Vol. 5, Part 2. Gebrüder Borntraeger, Berlin.

Evans, L. T. ed. 1963. *Environmental Control of Plant Growth*. Academic Press, New York.

Gifford, R. M., and L. T. Evans. 1981. Photosynthesis, carbon partitioning, and yield. *Annual Review of Plant Physiology 32*:485–509.

Hillel, D. 1971. *Soil and Water—Physical Principles and Processes*. Academic Press, New York.

Hodáňová, D. 1979. Sugar beet canopy photosynthesis as limited by leaf age and irradiance. Estimation by models. *Photosynthetica 13*:376–385.

Jarvis, P. G. 1975. Water transfer in plants. In *Heat and Mass Transfer in the Biosphere, I, Transfer Processes in Plant Environment*, D. A. de Vries and N. H. Afgan, eds. Halsted Press, Wiley, New York. Pp. 369–394.

Jarvis, P. G., W. R. N. Edwards, and H. Talbot. 1981. Models of plant and crop water use. In *Mathematics and Plant Physiology*, D. A. Rose and D. A. Charles-Edwards, eds. Academic Press, London. Pp. 151–194.

Jones, H. G. 1978. Modelling diurnal trends in leaf water potential in transpiring wheat. *Journal of Applied Ecology 15*:613–626.

Jordan, W. R., and J. T. Ritchie. 1971. Influence of soil water stress on evaporation, root absorption, and internal water status of cotton. *Plant Physiology 48*:783–788.

Kozlowski, T. T., ed. 1968. *Water Deficits and Plant Growth*, Vols. I and II. Academic Press, New York.

Kozlowski, T. T., ed. 1972. *Water Deficits and Plant Growth*, Vol. III. Academic Press, New York.

Kramer, P. J. 1969. *Plant and Soil Water Relationships: A Modern Synthesis*. McGraw-Hill, New York.

Kramer, P. J., and T. T. Kozlowski. 1979. *Physiology of Woody Plants*. Academic Press, New York.

Lemon, E., D. W. Stewart, and R. W. Shawcroft. 1971. The sun's work in a cornfield. *Science 174*:371–378.

Loomis, R. S., W. A. Williams, and A. E. Hall. 1971. Agricultural productivity. *Annual Review of Plant Physiology 22*:431–468.

Lüttge, U. 1975. Salt glands. In *Ion Transport in Plant Cells and Tissues*, D. A. Baker and J. L. Hall, eds. North Holland, Amsterdam. Pp. 335–376.

McMillen, C. G., and J. H. McClendon. 1979. Leaf angle: an adaptive feature of sun and shade leaves. *Botanical Gazette 140*:437–442.

Milburn, J. A. 1979. *Water Flow in Plants*. Longman, London.

Millington, R. J., and D. B. Peters. 1969. Exchange (mass transfer) coefficients in crop canopies. *Agronomy Journal 61*:815–819.

Monsi, M., and T. Saeki. 1953. Über den Lichtfaktor in den Pflanzengesellschaften und seine Bedeutung für die Stoffproduktion. *Japanese Journal of Botany 14*: 22–52.

Monsi, M., Z. Uchijima, and T. Oikawa. 1973. Structure of foliage canopies and photosynthesis. *Annual Review of Ecology and Systematics 4*: 301–327.

Moss, D. N., and R. B. Musgrave. 1971. Photosynthesis and crop production. *Advances in Agronomy 23*: 317–336.

Norman, J. M. 1980. Interfacing leaf and canopy light interception models. In *Predicting Photosynthesis for Ecosystem Models*, Vol. II, J. D. Hesketh and J. W. Jones, eds. CRC Press, Boca Raton, Florida. Pp. 49–67.

Oliver, H. R. 1971. Wind profiles in and above a forest canopy. *Quarterly Journal of the Royal Meteorological Society 97*: 548–553.

Paltridge, G. W. 1972. Experiments on a mathematical model of a pasture. *Agricultural Meteorology 10*: 39–54.

Passioura, J. B. 1982. Water in the soil-plant-atmosphere continuum. In *Physiological Plant Ecology*, O. L. Lange, P. S. Nobel, C. B. Osmond, and H. Ziegler, eds. *Encyclopedia of Plant Physiology, New Series*, Vol. 12B. Springer-Verlag, Berlin. Pp. 5–33.

Rains, D. W. 1972. Salt transport by plants in relation to salinity. *Annual Review of Plant Physiology 23*: 367–388.

Rose, C. W. 1966. *Agricultural Physics*. Pergamon, Oxford.

Ross, J. 1975. Radiative transfer in plant communities. In *Vegetation and the Atmosphere*, Vol. 1, J. L. Monteith, ed. Academic Press, London. Pp. 13–55.

Running, S. W. 1980. Relating plant capacitance to the water relations of *Pinus contorta*. *Forest Ecology and Management 2*: 237–252.

San Pietro, A., F. A. Greer, and T. J. Army, eds. 1967. *Harvesting the Sun: Photosynthesis in Plant Life*. Academic Press, New York.

Šesták, Z., J. Čatský, and P. G. Jarvis, eds. 1971. *Plant Photosynthetic Production: Manual of Methods*. Junk, The Hague.

Šetlik, I., ed. 1970. *Prediction and Measurement of Photosynthetic Productivity*. Pudoc, Centre for Agricultural Publishing and Documentation, Wageningen, The Netherlands.

Shaykewich, C. F., and J. Williams. 1971. Resistance to water absorption in germinating rapeseed (*Brassica napus* L.). *Journal of Experimental Botany 22*: 19–24.

Slatyer, R. O. 1967. *Plant-Water Relationships*. Academic Press, New York.

Spanner, D. C. 1972. Plants, water, and some other topics. In *Psychrometry in Water Relations Research*, R. W. Brown and B. P. Van Haveren, eds. Utah Agricultural Experiment Station, Utah State University, Logan. Pp. 29–39.

Streeter, V. L. 1966. *Fluid Mechanics*, 4th ed. McGraw-Hill, New York.

Taylor, S. A., and G. L. Ashcroft. 1972. *Physical Edaphology*. W. H. Freeman, San Francisco.

Thom, A. S. 1975. Momentum, mass and heat exchange of plant communities. In *Vegetation and the Atmosphere*, Vol. 1, *Principles*, J. L. Monteith, ed. Academic Press, New York. Pp. 57–109.

Turitzin, S. N., and B. G. Drake. 1981. The effect of a seasonal change in canopy structure on the photosynthetic efficiency of a salt marsh. *Oecologia 48*: 79–84.

van Wijk, W. R., ed. 1966. *Physics of Plant Environment*, 2nd ed. North-Holland, Amsterdam.

Weatherley, P. E. 1982. Water uptake and flow in roots. In *Physiological Plant Ecology*, O. L. Lange, P. S. Nobel, C. B. Osmond, and H. Ziegler, eds. *Encyclopedia of Plant Physiology, New Series*, Vol. 12B. Springer-Verlag, Berlin. Pp. 79–109.

Woodhouse, R. M., and P. S. Nobel. 1982. Stipe anatomy, water potentials, and xylem conductances in seven species of ferns (Filicopsida). *American Journal of Botany* 69:135–140.

Yoshida, S. 1972. Physiological aspects of grain yield. *Annual Review of Plant Physiology* 23:437–464.

Zelitch, I. 1971. *Photosynthesis, Photorespiration, and Plant Productivity.* Academic Press, New York.

Zimmermann, M. H., and C. L. Brown. 1971. *Trees—Structure and Function.* Springer-Verlag, New York.

Zimmermann, M. H., and J. A. Milburn, eds. 1975. *Transport in Plants I, Phloem Transport. Encyclopedia of Plant Physiology, New Series*, Vol. 1. Springer-Verlag, Berlin.

APPENDIXES

Symbols and Abbreviations

Where appropriate, magnitudes or typical units are indicated in parentheses.

Quantity	Description
a	absorptance or absorptivity (dimensionless)
a^{st}	mean area of stomata (m^{-2})
a_{IR}	absorptance or absorptivity in infrared region (dimensionless)
a_j	activity of species j (same as concentration)*
a.t.	subscript indicating active transport
Å	ångström (10^{-10} m)
A	electron acceptor
A	area (m^2)
A	absorbance (also called "optical density") (dimensionless)
ADP	adenosine diphosphate
ATP	adenosine triphosphate
b	nonosmotic volume (m^3)
b	optical path length (m)
bl	superscript for boundary layer
c	centi (as a prefix), 10^{-2}
c	superscript for cuticle
c_j	concentration of species j (mol m^{-3})†
$\bar{c}_s$	a mean concentration of solute s
cal	calorie
chl	superscript for chloroplast
clm	superscript for chloroplast limiting membranes
cw	superscript for cell wall
cyt	superscript for cytosol
C	superscript for conduction
C	capacitance, electrical (F)
C^j	capacitance for water storage in component j (m^3 MPa^{-1})
C'	capacitance/unit area (F m^{-2})

* The activity, a_j, is often considered to be dimensionless, in which case the activity coefficient, γ_j, has the units of reciprocal concentration ($a_j = \gamma_j c_j$, Eq. 2.5).

† We note that mol litre^{-1}, or molarity (M), is a concentration unit of widespread use, although it is not an SI unit.

527

Quantity	Description
Chl	chlorophyll
Cl	subscript for chloride ion
C_P	volumetric heat capacity (J m^{-3} °C^{-1})
Cyt	cytochrome
d	deci (as a prefix), 10^{-1}
d	depth or distance (m)
d	diameter (m)
dyn	dyne
D	electron donor
D	dielectric constant (dimensionless)
D_j	diffusion coefficient of species j (m^2 s^{-1})
e	electron
e	superscript for water evaporation site
e_{IR}	emissivity or emittance in infrared region (dimensionless)
eV	electron volt
E	light energy (J)
E	kinetic energy (J)
E	electrical potential (mV)
E_j	redox potential of species j (mV)
$E_j^{*,\mathrm{H}}$	midpoint redox potential of species j referred to standard hydrogen electrode (mV)
E_M	electrical potential difference across a membrane (mV)
E_{N_j}	Nernst potential of species j (mV)
f	femto (as a prefix), 10^{-15}
F	farad
F	average cumulative leaf area/ground area (dimensionless)
FAD	flavin adenine dinucleotide (oxidized form)
FADH$_2$	reduced form of flavin adenine dinucleotide
g	gram
g_j	conductance of species j (mm s^{-1} with Δc_j and mmol m^{-2} s^{-1} with ΔN_j)
G	giga (as a prefix), 10^9
G	Gibbs free energy (J)
Gr	Grashof number (dimensionless)
G/n_j	Gibbs free energy/mole of some product or reactant (J mol^{-1})
h	hour
h_c	heat convection coefficient (W m^{-2} °C^{-1})
h	height (m)
$h\nu$	a quantum of light energy
H	subscript for heat
i	superscript for inside
i	electrical current (ampere)
ias	superscript for intercellular air spaces
in	superscript for inward

Quantity	Description
in vitro	in a test tube, beaker, flask (literally, in glass)
in vivo	in a living organism (literally, in the living)
I	electrical current (ampere)
IR	infrared
j	subscript for species j
J	joule
J_j	flux density of species j (mol m^{-2} s^{-1})
J_j^{in}	inward flux density (influx) of species j (mol m^{-2} s^{-1})
J_j^{out}	outward flux density (efflux) of species j (mol m^{-2} s^{-1})
J_{V_j}	volume flux density of species j (m^3 m^{-2} s^{-1}, i.e., m s^{-1})
J_V	total volume flux density (m s^{-1})
k	kilo (as a prefix), 10^3
k	foliar absorption coefficient (dimensionless)
k_j	first-order rate constant for the jth process (s^{-1})
K	temperature on kelvin scale
K	subscript for potassium ion
K^j	thermal conductivity coefficient of region j (W m^{-1} °C^{-1})
K_j	partition coefficient of species j (dimensionless)
K_j	concentration for half-maximal uptake of species j (Michaelis constant) (mol m^{-3}, or M)
K_j	eddy diffusion coefficient of gaseous species j (m^2 s^{-1})
K	equilibrium constant (concentration raised to some power)
$K_{pH\,7}$	equilibrium constant at pH 7
l	litre
l	lower
l	length (m), e.g., mean distance across leaf in wind direction
ln	natural or Napierian logarithm (to the base e, where e is 2.71828 . . .)
log	common or Briggsian logarithm (to the base 10)
L^{soil}	hydraulic conductivity coefficient of the soil (m^2 Pa^{-1} s^{-1})
L_{jk}	Onsager or phenomenological coefficient (flux density per unit force)
L_P	hydraulic conductivity coefficient (in irreversible thermodynamics) (m Pa^{-1} s^{-1})
L_w	water conductivity coefficient (m Pa^{-1} s^{-1})
m	milli (as a prefix), 10^{-3}
m	metre
m_j	mass/mole of species j (molar mass)(kg mol^{-1})
max	subscript for maximum
memb	superscript for membrane
mes	superscript for mesophyll
min	subscript for minimum
min	minute
mol	mole, a mass equal to the molecular weight of the species in grams; contains Avogadro's number of molecules

Quantity	Description
M	mega (as a prefix), 10^6
M	molar (mol litre^{-1})
M_j	amount of species j/unit area (mol m^{-2})
n	nano (as a prefix), 10^{-9}
n	number of stomata/unit area
$n(E)$	number of moles with energy of E or greater
n_j	amount of species j (mol)
N	newton
N	normal (equivalents litre^{-1})
Na	subscript for sodium ion
NAD	nicotinamide adenine dinucleotide (oxidized form; also symbolized by NAD$^+$)
NADH	reduced form of nicotinamide adenine dinucleotide
NADH$_2$	NADH + H$^+$ (used in diagrams and for the redox couple)
NADP	nicotinamide adenine dinucleotide phosphate (oxidized form; also symbolized by NADP$^+$)
NADPH	reduced form of nicotinamide adenine dinucleotide phosphate
NADPH$_2$	NADPH + H$^+$ (used in diagrams and for the redox couple)
N_j	mole fraction of species j (dimensionless)
Nu	Nusselt number (dimensionless)
o	superscript for outside
0	subscript for initial value (at $t = 0$)
out	superscript for outward
p	pico (as a prefix), 10^{-12}
p	period (s)
pH	$-\log(a_{H^+})$
pl	superscript for plasmalemma
ps	superscript for photosynthesis
P	pigment
Pa	pascal
PAR	photosynthetically active radiation (photon flux density for 400 to 700 nm)
P	hydrostatic pressure (MPa)
P_j	permeability coefficient of species j (m s^{-1})
P_j	partial pressure of gaseous species j (kPa)
q	number of electrons transferred/molecule (dimensionless)
Q	charge (coulomb)
Q_{10}	temperature coefficient (dimensionless)
r	radius (m)
r	reflectivity (dimensionless)
r + pr	superscript for respiration plus photorespiration
r_j	resistance for gaseous species j (s m^{-1})
R	electrical resistance (ohm)
R^j	resistance of component j across which water moves as a liquid (MPa s m^{-3})

Quantity	Description
Re	Reynolds number (dimensionless)
RH	relative humidity (%)
s	subscript for solute
s	second
s_j	amount of species j (mol)
st	superscript for stoma(ta)
surf	superscript for surface
surr	superscript for surroundings
S	singlet
$S_{(\pi,\pi)}$	singlet ground state
$S_{(\pi,\pi^*)}$	singlet excited state where a π electron has been promoted to a π^* orbital
S	magnitude of net spin (dimensionless)
S	total flux density of solar irradiation, i.e., global irradiation (W m^{-2})
t	time (s)
ta	superscript for turbulent air
T	superscript for transpiration
T	triplet
$T_{(\pi,\pi^*)}$	excited triplet state
T	temperature (K, °C)
u	upper
u_j	mobility of species j (velocity per unit force)
u_+	mobility of monovalent cation
u_-	mobility of monovalent anion
U	kinetic energy (J mol^{-1})
U_B	minimum kinetic energy to cross barrier (J mol^{-1})
UV	ultraviolet
v	magnitude of velocity (m s^{-1})
v	wind speed (m s^{-1})
v^{wind}	wind speed (m s^{-1})
v_j	magnitude of velocity of species j (m s^{-1})
v_{CO_2}	rate of photosynthesis/unit volume (mol m^{-3} s^{-1})
V	volt
V	subscript for volume
V	volume (m^3)
$\bar{V}_j$	partial molal volume of species j (m^3 mol^{-1})
w	subscript for water
wv	subscript for water vapor
W	watt
x	distance (m)
z	altitude (m)
z_j	charge number of ionic species j (dimensionless)
α	contact angle (°)
γ_j	activity coefficient of species j (dimensionless, but see a_j)
$\gamma_\pm$	mean activity coefficient of cation-anion pair (dimensionless)

Quantity	Description
δ	delta, a small quantity of something, e.g., δ^- refers to a small fraction of an electronic charge
δ	distance (m)
δ^{bl}	thickness of air boundary layer (mm)
Δ	delta, the difference or change in the quantity which follows it
ε	volumetric elastic modulus (MPa)
ε_λ	absorption coefficient at wavelength λ ($m^2\ mol^{-1}$)
η	viscosity ($N\ s\ m^{-2}$, Pa s)
λ	wavelength of light (nm)
λ_{max}	wavelength position for the maximum absorption coefficient in an absorption band or for the maximum photon (or energy) emission in an emission spectrum
μ	micro (as a prefix), 10^{-6}
μ_j	chemical potential of species j ($J\ mol^{-1}$)
ν	frequency of electromagnetic radiation (s^{-1}, hertz)
ν	kinematic viscosity ($m^2\ s^{-1}$)
π	an electron orbital in a molecule, or an electron in such an orbital
π^*	an excited or antibonding electron orbital in a molecule, or an electron in such an orbital
Π	total osmotic pressure (MPa)
Π_j	osmotic pressure of species j (MPa)
Π_s	osmotic pressure due to solutes (MPa)
ρ	density ($kg\ m^{-3}$)
ρ	resistivity, electrical (ohm m)
ρ^j	hydraulic resistivity of component j ($MPa\ s\ m^{-2}$)
σ	surface tension ($N\ m^{-1}$)
σ	reflection coefficient (dimensionless)
σ_j	reflection coefficient of species j (dimensionless)
σ_L	longitudinal stress (MPa)
σ_T	tangential stress (MPa)
τ	matric potential (MPa)
τ	lifetime (s)
τ_j	lifetime for the jth de-excitation process (s)
φ_j	osmotic coefficient of species j (dimensionless)
φ_i	quantum yield or efficiency for ith de-excitation pathway (dimensionless)
Ψ	water potential (MPa)
Ψ_Π	osmotic potential (MPa)
$^\circ C$	degree Celsius (see p. 16)
$^\circ$	angular degree
$*$	superscript for a standard or reference state
$*$	superscript for a molecule in an excited electronic state
$*$	superscript for saturation of air with water vapor
∞	infinity

Numerical Values of Constants and Coefficients

Symbol	Description	Magnitude
c	speed of light in vacuum	2.998×10^8 m s^{-1}
c_{wv}^*	saturation concentration of water vapor (i.e., at 100% relative humidity)	2.358 g m^{-3} (0.1309 mol m^{-3}) at $-10°$C 3.407 ,, 0.1891 ,, $-5°$C 4.847 ,, 0.2690 ,, 0°C 6.797 ,, 0.3773 ,, 5°C 9.399 ,, 0.5217 ,, 10°C 12.83 ,, 0.7121 ,, 15°C 17.30 ,, 0.9603 ,, 20°C 23.05 ,, 1.279 ,, 25°C 30.38 ,, 1.686 ,, 30°C 39.63 ,, 2.200 ,, 35°C 51.19 ,, 2.841 ,, 40°C 65.50 ,, 3.636 ,, 45°C 83.06 ,, 4.610 ,, 50°C
C_P^{water}	volumetric heat capacity of water at constant pressure (1 atmosphere, 0.1013 MPa)	4.217 MJ m^{-3} °C^{-1} at 0°C 4.175 ,, 20°C 4.146 ,, 40°C
C_P^{air}	volumetric heat capacity of dry air at constant pressure (1 atmosphere)	1.29 kJ m^{-3} °C^{-1} at 0°C 1.21 ,, 20°C 1.13 ,, 40°C
D_{CO_2}	diffusion coefficient of CO_2 in air (1 atmosphere, 0.1013 MPa)	1.33×10^{-5} m^2 s^{-1} at 0°C 1.42×10^{-5} ,, 10°C 1.51×10^{-5} ,, 20°C 1.60×10^{-5} ,, 30°C 1.70×10^{-5} ,, 40°C
D_{O_2}	diffusion coefficient of O_2 in air (1 atmosphere, 0.1013 MPa)	1.95×10^{-5} m^2 s^{-1} at 20°C
D_{wv}	diffusion coefficient of water vapor in air (1 atmosphere, 0.1013 MPa)	2.13×10^{-5} m^2 s^{-1} at 0°C 2.27×10^{-5} ,, 10°C 2.42×10^{-5} ,, 20°C 2.57×10^{-5} ,, 30°C 2.72×10^{-5} ,, 40°C

Symbol	Description	Magnitude
e	base for natural logarithm	2.71828 ($1/e = 0.368$)
F	faraday	9.649×10^4 coulomb mol^{-1} 9.649×10^4 J mol^{-1} V^{-1} 2.306×10^4 cal mol^{-1} V^{-1} 23.06 kcal mol^{-1} V^{-1}
g	gravitational acceleration	9.780 m s^{-2} (sea level*, 0° latitude) 9.807 " " 45° " 9.832 " " 90° " 978.0 cm s^{-2} " 0° " 980.7 " " 45° " 983.2 " " 90° "
h	Planck's constant	6.626×10^{-34} J s 6.626×10^{-27} erg s 0.4136×10^{-14} eV s 1.584×10^{-37} kcal s
hc		1.986×10^{-25} J m 1 240 eV nm
H_{sub}	heat of sublimation of water	51.37 kJ mol^{-1} (2.847 MJ kg^{-1}) at $-10°$C 51.17 " 2.835 " $-5°$C 51.00 " 2.826 " $0°$C 12.27 kcal mol^{-1} (680 cal g^{-1}) at $-10°$C 12.22 " 677 " $-5°$C 12.18 " 675 " $0°$C
H_{vap}	heat of vaporization of water	45.06 kJ mol^{-1} (2.501 MJ kg^{-1}) at $0°$C 44.63 " 2.477 " $10°$C 44.21 " 2.454 " $20°$C 44.00 " 2.442 " $25°$C 43.78 " 2.430 " $30°$C 43.35 " 2.406 " $40°$C 42.91 " 2.382 " $50°$C 40.68 " 2.258 " $100°$C 10.76 kcal mol^{-1} (597 cal g^{-1}) at $0°$C 10.67 " 592 " $10°$C 10.56 " 586 " $20°$C 10.50 " 583 " $25°$C 10.45 " 580 " $30°$C 10.36 " 575 " $40°$C 10.25 " 569 " $50°$C 9.71 " 539 " $100°$C

* The correction for height above sea level is -3.09×10^{-6} m s^{-2} per m of altitude.

Symbol	Description	Magnitude
k	Boltzmann's constant	1.381×10^{-23} J molecule^{-1} K^{-1} 1.381×10^{-16} erg molecule^{-1} K^{-1} 8.617×10^{-5} eV molecule^{-1} K^{-1}
kT		0.0235 eV molecule^{-1} at 0°C 0.0253 „ 20°C 0.0257 „ 25°C
K^{air}	thermal conductivity coefficient of dry air (1 atmosphere)*	0.0237 W m^{-1} °C^{-1} at -10°C 0.0243 „ 0°C 0.0250 „ 10°C 0.0257 „ 20°C 0.0264 „ 30°C 0.0270 „ 40°C 0.0277 „ 50°C
	thermal conductivity coefficient of moist air (100% relative humidity, 1 atmosphere)	0.0242 W m^{-1} °C^{-1} at 0°C 0.0255 „ 20°C 0.0264 „ 40°C
K^{water}	thermal conductivity coefficient of water	0.565 W m^{-1} °C^{-1} at 0°C 0.599 „ 20°C 0.627 „ 40°C
$\ln 2$		0.693
N	Avogadro's number	6.0220×10^{23} molecules mol^{-1}
Nhc		0.1196 J mol^{-1} m 119 600 kJ mol^{-1} nm 28.60 kcal mol^{-1} μm 28 600 kcal mol^{-1} nm
Nk		R
N_{wv}^*	saturation mole fraction of water vapor (i.c., at 100% relative humidity) at 1 atmosphere (0.1013 MPa)	0.00256 at -10°C 0.00396 „ -5°C 0.00603 „ 0°C 0.00861 „ 5°C 0.01211 „ 10°C 0.01682 „ 15°C 0.02307 „ 20°C 0.03126 „ 25°C 0.04187 „ 30°C 0.05550 „ 35°C 0.07281 „ 40°C 0.09459 „ 45°C 0.1218 „ 50°C

* The pressure sensitivity is very slight, K^{air} increasing only about 0.0001 W m^{-1} °C^{-1} per atmosphere (0.1013 MPa) increase in pressure.

Symbol	Description	Magnitude
P_{wv}^*	saturation vapor pressure of water	0.260 kPa (2.60 mbar) at $-10°C$
		0.401 " 4.01 " $-5°C$
		0.611 " 6.11 " $0°C$
		0.872 " 8.72 " $5°C$
		1.227 " 12.27 " $10°C$
		1.704 " 17.04 " $15°C$
		2.337 " 23.37 " $20°C$
		3.167 " 31.67 " $25°C$
		4.243 " 42.43 " $30°C$
		5.623 " 56.23 " $35°C$
		7.378 " 73.78 " $40°C$
		9.585 " 95.85 " $45°C$
		12.34 " 123.4 " $50°C$
R	gas constant	8.3143 J mol^{-1} K^{-1}
		1.987 cal mol^{-1} K^{-1}
		8.3143 m^3 Pa mol^{-1} K^{-1}
		8.3143 $\times$ 10^{-6} m^3 MPa mol^{-1} K^{-1}
		0.082054 litre atmosphere mol^{-1} K^{-1}
		0.083143 litre bar mol^{-1} K^{-1}
		83.143 cm^3 bar mol^{-1} K^{-1}
RT		2.271 $\times$ 10^3 J mol^{-1} (m^3 Pa mol^{-1}) at $0°C$
		2.437 $\times$ 10^3 " $20°C$
		2.479 $\times$ 10^3 " $25°C$
		2.271 $\times$ 10^{-3} m^3 MPa mol^{-1} at $0°C$
		2.437 $\times$ 10^{-3} " $20°C$
		2.479 $\times$ 10^{-3} " $25°C$
		543 cal mol^{-1} at $0°C$
		582 " $20°C$
		2.271 litre MPa mol^{-1} at $0°C$
		2.437 " $20°C$
		22.71 litre bar mol^{-1} at $0°C$
		24.37 " $20°C$
		22 710 cm^3 bar mol^{-1} at $0°C$
		24 370 " $20°C$
		22.41 litre atmosphere mol^{-1} at $0°C$
		24.05 " $20°C$
2.303 RT		5.62 kJ mol^{-1} at 20°C
		5.71 " 25°C
		1.342 kcal mol^{-1} at 20°C
		1.364 " 25°C
		56 120 cm^3 bar mol^{-1} at 20°C
RT/F		25.3 mV at 20°C
		25.7 " 25°C
2.303 RT/F		58.2 mV at 20°C
		59.2 " 25°C
		60.2 " 30°C

Symbol	Description	Magnitude
$RT/\bar{V}_w$		135.0 MPa at 20°C 137.3 " 25°C 32.4 cal cm^{-3} at 20°C 135.0 J cm^{-3} at 20°C 1 350 bars at 20°C 1 330 atmospheres at 20°C
$2.303\ RT/\bar{V}_w$		310.9 MPa at 20°C 316.2 " 25°C 3 063 atmospheres at 20°C 3 109 bars at 20°C
	solar constant (see p. 195)	1 360 W m^{-2} 1.95 cal cm^{-2} min^{-1} 1.36×10^5 erg cm^{-2} s^{-1} 0.136 W cm^{-2}
	specific heat of water (mass basis)	4 187 J kg^{-1} °C^{-1} 1.00 cal g^{-1} °C^{-1}
	specific heat of water (mole basis)	75.4 J mol^{-1} °C^{-1} 18.0 cal mol^{-1} °C^{-1}
$\bar{V}_w$	partial molal volume of water	1.805×10^{-5} m^3 mol^{-1} at 20°C 18.05 cm^3 mol^{-1} at 20°C
ε_0	permittivity of a vacuum	8.854×10^{-12} coulomb^2 m^{-2} N^{-1} 8.854×10^{-12} coulomb m^{-1} V^{-1}
η_w	viscosity of water	1.781×10^{-3} Pa s at 0°C 1.306 " 10°C 1.002 " 20°C 0.798 " 30°C 0.653 " 40°C 0.01002 dyn s cm^{-2} at 20°C
π		3.14159
ρ_{air}	density of dry air (1 atmosphere, 0.1013 MPa)	1.293 kg m^{-3} at 0°C 1.205 " 20°C 1.128 " 40°C
	density of saturated air (1 atmosphere)*	1.290 kg m^{-3} at 0°C 1.194 " 20°C 1.097 " 40°C

* Moist air is less dense than dry air at the same temperature and pressure, since the molecular weight of water (18.0) is less than the average for air (29.0).

Symbol	Description	Magnitude
ρ_w	density of water	999.8 kg m^{-3} (0.9998 g cm^{-3}) at 0°C
		1 000.0 " 1.0000 " 4°C
		999.7 " 0.9997 " 10°C
		998.2 " 0.9982 " 20°C
		995.6 " 0.9956 " 30°C
		992.2 " 0.9922 " 40°C
$\rho_w g$		0.00979 MPa m^{-1} (20°C, sea level, 45° latitude)
		0.0979 bar m^{-1} (20°C, sea level, 45° latitude)
		979 dyn cm^{-3} (20°C, sea level, 45° latitude)
		0.0966 atmosphere m^{-1} (20°C, sea level, 45° latitude)
σ	Stefan–Boltzmann constant	5.670 × 10^{-8} W m^{-2} K^{-4}
		5.670 × 10^{-12} W cm^{-2} K^{-4}
		8.130 × 10^{-11} cal cm^{-2} min^{-1} K^{-4}
		5.670 × 10^{-5} erg cm^{-2} s^{-1} K^{-4}
σ_w	surface tension of water	0.0756 N m^{-1} (Pa m) at 0°C
		0.0742 " 10°C
		0.0728 " 20°C
		0.0712 " 30°C
		0.0696 " 40°C
		7.28 × 10^{-8} MPa m at 20°C
		72.8 dyn cm^{-1} at 20°C
		7.18 × 10^{-5} atmosphere cm at 20°C
		7.28 × 10^{-5} bar cm at 20°C

Conversion Factors

Quantity	Equals
acre	$4\,047 \text{ m}^2$ 0.4047 hectare
ampere	1 coulomb s^{-1} V ohm^{-1}
Å	10^{-10} m 0.1 nm 10^{-8} cm $10^{-4} \text{ } \mu\text{m}$
atmosphere	0.1013 MPa $1.013 \times 10^5 \text{ N m}^{-2}$ $1.013 \times 10^5 \text{ Pa}$ $1.013 \times 10^5 \text{ J m}^{-3}$ 1.013 bar $1.033 \times 10^4 \text{ kg m}^{-2}$ (at sea level, 45° latitude)* $1.013 \times 10^6 \text{ dyn cm}^2$ 1.033 kg cm^{-2} (at sea level, 45° latitude)* 76.0 cm or 760 mm Hg (at sea level, 45° latitude)*
bar	0.1 MPa 100 kPa 10^5 Pa 10^5 N m^{-2} 10^5 J m^{-3} 10^6 dyn cm^{-2} 0.987 atmosphere 1.020 kg cm^{-2} (at sea level, 45° latitude)* 75.0 cm or 750 mm Hg (at sea level, 45° latitude)*
cal	4.187 J $4.187 \times 10^7 \text{ ergs}$

* Sometimes it proves convenient to express forces in units of mass. To see why this is possible, consider the force F exerted by gravity on a body of mass m. This force is equal to mg, g being the gravitational acceleration (see App. II). Thus F/g can be used to represent a force, but the units are those of mass. One atmosphere is quite often defined as 760 mm Hg (or 1.033 kg cm^{-2}), but the elevational and latitudinal effect on g should also be considered (see p. 534)

Quantity	Equals
cal cm^{-2}	1 langley $4.187 \times 10^4 \text{ J m}^{-2}$
$\text{cal cm}^{-2} \text{ min}^{-1}$	697 W m^{-2} $6.97 \times 10^5 \text{ ergs cm}^{-2} \text{ s}^{-1}$ 1 langley min^{-1}
$\text{cal cm}^{-1} \,{}^\circ\text{C}^{-1} \text{min}^{-1}$	$6.97 \text{ W m}^{-1} \,{}^\circ\text{C}^{-1}$
$\text{cal cm}^{-1} \,{}^\circ\text{C}^{-1} \text{s}^{-1}$	$418 \text{ W m}^{-1} \,{}^\circ\text{C}^{-1}$
cal cm^{-3}	41.87 bars
cal g^{-1}	$4\,187 \text{ J kg}^{-1}$
cal min^{-1}	0.0697 J s^{-1} 0.0697 W
cal s^{-1}	4.187 J s^{-1} 4.187 W
$\text{cm}^2 \text{ bar}^{-1} \text{ s}^{-1}$	$10^{-9} \text{ m}^2 \text{ Pa}^{-1} \text{ s}^{-1}$ $10^{-3} \text{ m}^2 \text{ MPa}^{-1} \text{ s}^{-1}$
cm s^{-1}	0.01 m s^{-1} 10 mm s^{-1} $0.0360 \text{ km hour}^{-1}$ $0.0224 \text{ mile hour}^{-1}$
cm^3	10^{-6} m^3
coulomb	1 J V^{-1}
coulomb V	1 J
dalton	$1.660 \times 10^{-24} \text{ g}$ (1/12 mass of carbon atom)
day	86 400 s
degree (angle)	0.01745 radian
dyn	1 erg cm^{-1} 1 g cm s^{-2} 10^{-5} N
dyn cm	1 erg 10^{-7} J

Quantity	Equals
dyn cm^{-1}	10^{-3} N m^{-1}
dyn cm^{-2}	0.1 N m^{-2} 10^{-6} bar
einstein	1 mol 6.023×10^{23} photons
erg	1 dyn cm 1 g cm^2 s^{-2} 10^{-7} J 6.242×10^{11} eV 2.390×10^{-8} cal 2.390×10^{-11} kcal
erg cm^{-2} s^{-1}	10^{-3} J m^{-2} s^{-1} 10^{-3} W m^{-2} 10^{-7} W cm^{-2} 1.434×10^{-6} cal cm^{-2} min^{-1}
eV	1.602×10^{-19} J 1.602×10^{-12} erg
eV molecule^{-1}	96.5 kJ mol^{-1} 23.06 kcal mol^{-1}
F	1 coulomb V^{-1}
footcandle	10.76 lux
g	1 dyn s^2 cm^{-1}
g cm^{-2}	10 kg m^{-2}
g cm^{-3}	1 000 kg m^{-3}
g dm^{-2} hour^{-1}	27.8 mg m^{-2} s^{-1}
hectare	10^4 m^2 2.47 acres
hertz	1 cycle s^{-1}
horsepower	745.7 W
hour	3 600 s
inch	25.4 mm

Quantity	Equals
J	1 N m 1 m^3 Pa 1 kg m^2 s^{-2} 1 coulomb V 10^7 ergs 0.2388 cal 10 cm^3 bar
J kg^{-1}	2.388×10^{-4} cal g^{-1}
J mol^{-1}	0.2388 cal mol^{-1}
J s^{-1}	W
kcal	4.187 kJ 4.187×10^{10} ergs
kdalton	1.660×10^{-21} g (corresponds to molar mass in kg mol^{-1})
kcal mol^{-1}	4.187 kJ mol^{-1} 0.0434 eV molecule^{-1}
kg m^{-2}	0.1 g cm^{-2} 9.81 Pa (at sea level, 45° latitude; see g) 9.81 N m^{-2} (at sea level, 45° latitude; see g)
kg m^{-3}	10^{-3} g cm^{-3}
kJ mol^{-1}	0.01036 eV molecule^{-1}
km	0.6214 mile
km hour^{-1}	0.278 m s^{-1} 27.8 cm s^{-1} 0.6214 mile hour^{-1}
knot	1.8532 km hour^{-1} 1.1516 mile hour^{-1}
kW hour	3.60×10^6 J 8.60×10^5 cal
langley	1 cal cm^{-2} 4.187×10^4 J m^{-2}
litre	0.001 m^3

Quantity	Equals
litre atmosphere	24.2 cal
litre bar	100 J 23.9 cal
ln	2.303 log
log	0.434 ln
lux	1 lumen m^{-2}
mbar	0.1 kPa 10^{-3} bar 10^3 dyn cm^{-2} 0.987×10^{-3} atmosphere
mg cm^{-3}	1 kg m^{-3}
m s^{-1}	100 cm s^{-1} 3.60 km hour^{-1} 2.24 miles hour^{-1}
micron	1 μm 10^{-6} m 10^4 ångström 10^{-3} mm 10^{-4} cm
mile	1 609 m
mile hour^{-1}	0.447 m s^{-1} 44.70 cm s^{-1} 1.609 km h^{-1}
MPa	10^6 N m^{-2} 10^6 J m^{-3} 10 bars 1.020×10^5 kg m^{-2} (at sea level, 45° latitude; see g)
mol m^{-3}	1 mM 1 μmol cm^{-3}
ng cm^{-2} s^{-1}	10 μg m^{-2} s^{-1}
N	1 kg m s^{-2} 10^5 dyn

Quantity	Equals
N m	1 J
N m^{-1}	10^3 dyn cm^{-1}
N m^{-2}	1 Pa 10^{-6} MPa 10 dyn cm^{-2} 10^{-2} mbar
Pa	1 N m^{-2} 1 J m^{-3} 1 kg m^{-1} s^{-2} 10^{-5} bar
poise	1 dyn s cm^{-2} 0.1 N s m^{-2} 0.1 Pa s
pound	0.4536 kg
radian	57.30°
s cm^{-1}	100 s m^{-1}
s m^{-1}	10^{-2} s cm^{-1}
torr	1 mm Hg 133 Pa (at sea level, 45° latitude; see g) 0.00133 bar (at sea level, 45° latitude; see g) 1 330 dyn cm^{-2} (at sea level, 45° latitude; see g)
V	1 ampere ohm 1 J coulomb^{-1}
W	1 J s^{-1} 1 kg m^2 s^{-3} 10^7 ergs s^{-1} 14.34 cal min^{-1}
W cm^{-2}	10^4 W m^{-2}
W m^{-2}	1 J m^{-2} s^{-1} 10^3 ergs cm^{-2} s^{-1} 1.433 × 10^{-3} cal cm^{-2} min^{-1} 1.433 × 10^{-3} langley min^{-1} 2.388 × 10^5 cal cm^{-2} s^{-1}
W m^{-1} °C^{-1}	2.388 × 10^{-3} cal cm^{-1} °C^{-1} s^{-1} 3.981 × 10^{-5} cal cm^{-1} °C^{-1} min^{-1}

Quantity	Equals
W s	1 J
year (calendar)	3.15×10^7 s 5.26×10^5 min 8 760 h 365 days
μg cm^{-2} s^{-1}	10 mg m^{-2} s^{-1}

Logarithms and Trigonometric Functions

The following relations are presented to facilitate the use of natural and common logarithms, their antilogarithms, and trigonometric functions. For those readers who are completely unfamiliar with such quantities, a text or handbook should be consulted first.

$$\ln (xy) = \ln x + \ln y \qquad\qquad \ln x^a = a \ln x$$

$$\ln (x/y) = \ln x - \ln y \qquad\qquad \ln (1/x^a) = -a \ln x$$

$$\ln (1/x) = -\ln x \qquad\qquad\quad \ln x = 2.303 \log x$$

$$\ln e = 1, \quad \ln 10 = 2.303 \qquad \log 10 = 1, \quad \log e = \frac{1}{2.303}$$

$$\ln e^y = y \qquad\qquad\qquad\qquad \log 10^y = y$$

$$e^{\ln y} = y \qquad\qquad\qquad\qquad 10^{\log y} = y$$

$$\ln (1 + x) = x - \frac{x^2}{2} + \frac{x^3}{3} - \frac{x^4}{4} + \cdots \qquad -1 < x \leq 1$$

$$e^x = 1 + x + \frac{x^2}{2!} + \frac{x^3}{3!} + \cdots$$

547

Consider a right triangle of hypotenuse r:

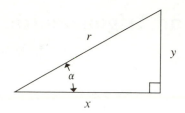

α in radians or °
$(2\pi$ radians $= 360°)$

$$\sin \alpha = \frac{y}{r}$$

$$-1 \leq \sin \alpha \leq 1$$

$$\cos \alpha = \frac{x}{r}$$

$$-1 \leq \cos \alpha \leq 1$$

$$\tan \alpha = \frac{y}{x}$$

$$-\infty \leq \tan \alpha \leq \infty$$

$$\sin \alpha = \alpha - \frac{\alpha^3}{3!} + \frac{\alpha^5}{5!} - \frac{\alpha^7}{7!} + \cdots \qquad \alpha \text{ in radians}$$

$$\cos \alpha = 1 - \frac{\alpha^2}{2!} + \frac{\alpha^4}{4!} - \frac{\alpha^6}{6!} + \cdots \qquad \alpha \text{ in radians}$$

Calculus

Functions. Derivatives of functions. Partial derivatives. Total differentials and derivatives. Well-defined and continuous functions, higher order derivatives. Examples of derivatives. Integration. Definite integrals, indefinite integrals, and the integration constant. Examples of integrals. Differential equations.

The purpose of this appendix is to bring together some of the more useful relations from calculus pertinent to the quantitative aspects of physical chemistry in particular and biology in general. A few important definitions are provided to help remove some of the mystery that often surrounds the use of calculus in biology as well as to provide a convenient refresher for those readers whose formal coursework in this area was either rather short or long ago.

Functions

A *function* is a mathematical way of expressing the relationship between a number of quantities referred to as *variables*. One of the parameters is referred to as a *dependent* variable, which depends on one or more *independent* variables. For instance, we might have $y = mx + b$, which says that the dependent variable y equals a constant m times the value of the independent variable x plus an additive constant b. In other words, y is a function of x, and the analytical expression of the functional form is $y = mx + b$. We often treat such functional relationships graphically. For example, we can plot x on the x-axis, or abscissa, and y on the y-axis, or ordinate. We then obtain a straight line of slope m—an aspect to which we will return later—with an intercept on the y-axis of b; i.e., when x is zero (as it is for the y-axis), y then equals $m \times 0 + b$, which is simply b.

The chemical potential of species j, μ_j, is a dependent variable used

549

throughout this text and depends on the independent variables referring to position, e.g., x for the one-dimensional case considered here, and time, t:

$$\mu_j = \mu_j(x, t) \qquad \qquad \text{(V.1)}$$

where the notation $\mu_j(x, t)$ means that this function depends on x and t as independent variables. In many cases we find it convenient to express the functional relationship of the chemical potential in an analytical form with independent variables other than x and t. In particular, from Equation 2.4 we see that we can also represent the function μ_j as follows:

$$\mu_j = \mu_j(n_i, a_j, T, P, E, h) = \mu_j^* + RT \ln a_j + \bar{V}_j P + z_j FE + m_j gh \quad \text{(V.2)}$$

where n_i is necessary since the so-called constant μ_j^* actually depends on the number of moles of any species present other than species j. On the other hand, R, $\bar{V}_j$, z_j, F, m_j, and g are here assumed to be constants, and do not therefore appear in the functional form for μ_j given in Equation V.2, $\mu_j(n_i, a_j, T, P, E, h)$.

Each of the independent variables in Equation V.2 in turn depends on position and time. For instance, if we knew the value of a_j at every location in our system for all the times of interest, we could in principle determine $a_j(x, t)$, where a_j is now a dependent variable which can be expressed in terms of the independent variables x and t. Likewise, we could find the temperature at every position and time and thereby determine $T(x, t)$. By suitably expressing the position (x) and time (t) dependence of all the independent variables in Equation V.2 (namely, n_i, a_j, T, P, E, and h), we could convert this equation into the form given by Equation V.1, $\mu_j = \mu_j(x, t)$. Thus, we can express the chemical potential either in terms of space and time as independent variables, or in terms of another set of physical parameters. In the latter case we obtain a useful, if approximate, analytical form for the chemical potential, while the functional form $\mu_j(x, t)$ is of great utility in representing the forces on various species and the resultant fluxes.

Derivatives of Functions

We are often interested in how much a function changes in value when one of the independent variables changes by a certain amount. For example, when y equals $mx + b$, by how much does y change when x increases by 1 unit? The change in y, which we will represent by Δy, where delta is used to

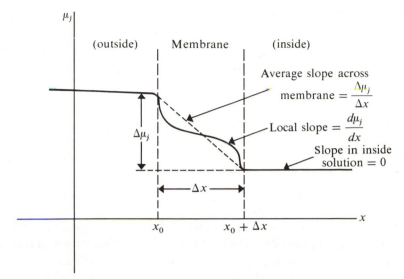

Figure V.1
Dependence of the chemical potential of species j on position.
The value of μ_j is indicated in two solutions and in a membrane
(Δx thick) intervening between them.

indicate in finite change, is equal here to $m \times 1$ or m. In the more general case, when x changes by Δx, y changes by $m\,\Delta x$. We might also want to know the change in chemical potential of species j, μ_j, when we move across a membrane of thickness Δx; i.e., what is $\Delta \mu_j$ for a change along the x-axis of Δx? On the other hand, we might wish to know not the gross change in μ_j across the membrane, but how the chemical potential varies over an extremely short, in fact, infinitesimal distance. In calculus this short distance is represented by dx (the differential of x), and the dependent variable μ_j changes by $d\mu_j$ in the infinitesimal distance dx.

Next, let us consider the relationship of deltas and differentials to the slopes of lines on a graph. Let us begin by considering the functional form $y = mx + b$. What is the rate of change of y for a finite change in x? This rate, represented by $\Delta y/\Delta x$, is $m\,\Delta x/\Delta x$, or simply m in this case. We have already indicated that m is the slope of the straight line represented by $y = mx + b$, so we should suspect that rates of change and slopes of curves are related. We also recognize that the total change in chemical potential across a membrane ($\Delta \mu_j$) divided by the distance over which this change occurs (Δx) gives us the average rate of change of chemical potential in the x-direction in the membrane, $\Delta \mu_j/\Delta x$. This average rate of change is illustrated by a dashed line in Figure V.1. The dashed line has a slope $\Delta \mu_j/\Delta x$, since the dependent variable changes by $\Delta\mu_j$ when the independent variable changes by Δx. We also see that the actual values of μ_j in moving across the membrane do not fall on the dashed line of slope $\Delta\mu_j/\Delta x$ in Figure V.1. Rather, we notice that μ_j varies in a more complicated way with a change in the x-direction. This then leads us to a consideration of how μ_j varies over distances much shorter than Δx.

What is the rate of change of μ_j with position over an infinitesimally short distance dx? This is simply $d\mu_j/dx$, which is called the *derivative* of μ_j with respect to x. When we divide one infinitesimal or differential, e.g., $d\mu_j$, by another, e.g., dx, the quotient is not an infinitesimal—rather, it is the *slope of the local tangent* to the curve representing μ_j versus x at the particular location to which the $d\mu_j$ refers. For instance, to the right of the membrane in Figure V.1 μ_j is constant, so that $d\mu_j$ is zero for any dx in this region. Hence, the slope of the line $(d\mu_j/dx)$ is $0/dx$, which is zero. Within the membrane, on the other hand, the slope of μ_j as a function of x is sometimes steeper and sometimes shallower than the average rate of change of μ_j with position, $\Delta\mu_j/\Delta x$.

If we choose successively smaller values for Δx, we obtain correspondingly smaller values for the change in the dependent variable, $\Delta\mu_j$. In fact, we approach a limiting value for the ratio, $\Delta\mu_j/\Delta x$, as the value of Δx becomes vanishingly small. This then leads us to the fundamental definition of a derivative:

$$\lim_{\Delta x \to 0} \frac{\Delta\mu_j}{\Delta x} = \frac{d\mu_j}{dx} = \text{derivative of } \mu_j \text{ with respect to } x \qquad (V.3)$$

We can geometrically interpret the derivative as the slope of the tangent to the curve relating how the dependent variable (here μ_j) changes for a given change in the independent variable (here x).

Partial Derivatives

In the example referred to above, namely, the straight line represented by $y = mx + b$, the derivative dy/dx equals m, the slope of the line. In this case, the dependent variable (y) depends on only one independent variable (x), and there is no ambiguity as to the value of the slope. On the other hand, Equation V.1, $\mu_j = \mu_j(x, t)$, indicates that the chemical potential can depend on time t in addition to x (in principle, μ_j can also depend on y and z, but for simplicity we will concern ourselves only with the one-dimensional case). We expect then that the derivative $d\mu_j/dx$ might actually have different values at different times, since μ_j can vary with time. Thus, when we were using the derivative $d\mu_j/dx$ to give the slope of the curve in Figure V.1 (where the dependence of μ_j only on position x was shown), we must have been implicitly referring to a specific value for t or to the special case where μ_j does not depend on time.

Whenever a dependent variable depends on more than one independent variable, there are a number of different ways in which a derivative can be defined. It is convenient, however, to determine the change in the dependent variable when only one of the independent variables is allowed to vary, which leads us to a quantity known as the *partial derivative*. This is a derivative in the sense that it equals the infinitesimal change in the dependent variable divided by the infinitesimal change in some independent variable which brought about the change. It is not a total derivative, however, only a partial one, since only one of the independent variables is allowed to change. Thus, a partial derivative indicates some function's rate of change due to a change in one particular variable, all other independent variables being held constant.

Although we have indicated that the slope of μ_j versus x in Figure V.1 is given by the derivative $d\mu_j/dx$, the slope should really be represented by the partial derivative of μ_j with respect to x, $\partial\mu_j/\partial x$, at some fixed value of time. To indicate which independent variables are held constant when we obtain a partial derivative, the symbols for a derivative are usually modified to the following form:

$$\left(\frac{\partial\mu_j}{\partial x}\right)_t = \text{partial derivative of } \mu_j \text{ with respect to } x \text{ with } t \text{ held constant}$$

$$(\text{V.4})$$

We often find it convenient to specify the actual value of t at which the differentiation is made; e.g., $(\partial\mu_j/\partial x)_{t_0}$ indicates that the partial derivative is for a specific time, t_0.

The force causing species j to move in the x-direction is represented by the negative gradient of the chemical potential, $-\partial\mu_j/\partial x$ (see p. 118). The force is positive in the direction of decreasing μ_j, so the force in the example in Figure V.1 is causing species j to move from the outside across the membrane to the inside. In Figure V.1 we note that the value of $\partial\mu_j/\partial x$ changes as we cross the membrane. In particular, $-\partial\mu_j/\partial x$ is large on each side just within the membrane, indicating that the force on species j is great there. In the center of the membrane, however, the slope of the curve and consequently $-\partial\mu_j/\partial x$ are considerably less. Thus, the actual force, $-\partial\mu_j/\partial x$, differs from the average force, $-\Delta\mu_j/\Delta x$, although the latter is essentially all that we can determine in the case of biological membranes. On the other hand, the theoretical development of the relationship between forces and fluxes (pp. 116–118) requires that we deal with the actual forces as represented by the partial derivatives of the chemical potentials of the species involved. This is one example among many where we represent forces by gradients in the appropriate potentials, a gradient being simply a partial derivative of some function with respect to position.

Partial derivatives are critically important for an adequate treatment of physical chemistry, particularly in the field of thermodynamics. For example, in Appendix VI we must evaluate the partial derivative of the chemical potential with respect to pressure, $(\partial\mu_j/\partial P)_{T,E,h,n_i,n_j}$, which we find equals $\bar{V}_j$ (Eq. VI.13). We then use this partial derivative to obtain the pressure dependence of the chemical potential in the liquid phase (Eq. VI.14), and the quite different dependence of μ_j on P that applies to a gas phase (Eq. VI.16).

Total Differentials and Derivatives

Initially, we presented the slope of μ_j versus x as $d\mu_j/dx$ and then refined the derivative to $(\partial\mu_j/\partial x)_t$. We will now return to the idea of an unrestricted, or total derivative; for convenience, however, we will first consider a total differential. Is it possible to represent $d\mu_j$ without any restrictions as to which variables must be held constant? The answer is yes, and $d\mu_j$ assumes the following form based on Equation V.1, $\mu_j = \mu_j(x, t)$:

$$d\mu_j = \left(\frac{\partial\mu_j}{\partial x}\right)_t dx + \left(\frac{\partial\mu_j}{\partial t}\right)_x dt \qquad \text{(V.5)}$$

where $d\mu_j$ is referred to as the total differential of the function μ_j, and x and t are the independent variables on which the function depends. In other words, to obtain the *total differential* of some quantity (the dependent variable) we must add up contributions from each of the ways in which that quantity might change (all the independent variables). The form for the total differential is the partial derivative of the function with respect to some independent variable times the differential of that variable summed over all independent variables, as illustrated in Equation V.5.

Having established the notion of a total differential, we will next consider a couple of examples of total derivatives. If, in a particular case, μ_j does not depend on t, $(\partial \mu_j / \partial t)_x$ equals zero. Equation V.5 then becomes $d\mu_j = [(\partial \mu_j / \partial x)_t] \, dx$. Upon dividing both sides of this relation by dx, we obtain $d\mu_j / dx = (\partial \mu_j / \partial x)_t$. In fact, the total derivative equals the partial derivative only when the function whose derivative is being taken depends on but one independent variable. For instance, the slope of the curve in Figure V.1 would indeed be $d\mu_j / dx$, if μ_j depended only on x.

Let us now examine the total derivative, $d\mu_j / dt$, when μ_j depends on both x and t. If we divide both sides of Equation V.5 by dt, we obtain $d\mu_j / dt = (\partial \mu_j / \partial x)_t (dx/dt) + (\partial \mu_j / \partial t)_x$. The time rate of change of chemical potential is thus the sum of the partial derivative of μ_j with respect to t at constant x, $(\partial \mu_j / \partial t)_x$, plus a contribution depending on how much our position changes in time dt and how such a change affects the chemical potential. Thus, the total derivative, $d\mu_j / dt$, is the sum of the intrinsic dependence of μ_j on t—the partial derivative, $(\partial \mu_j / \partial t)_x$—plus our velocity of movement (dx/dt) times how rapidly μ_j changes with position $[(\partial \mu_j / \partial x)_t]$.

Well-Defined and Continuous Functions, Higher Order Derivatives

To obtain the derivatives indicated in Equations V.3, V.4, and V.5, certain basic criteria must be met by the functions with which we are dealing. In the first place, μ_j must have a unique value at the particular point of interest; i.e., the chemical potential cannot have a number of different magnitudes at a given x and t, but must be a *well-defined* function. Also, if we move an infinitesimal distance along the x-axis, μ_j must change by an infinitesimal, not finite amount, a property referred to as *continuity*. If, for instance, we advance a distance dx (at a fixed time t), μ_j must change by an infinitesimal amount, $d\mu_j$. On the other hand, if μ_j changed by a finite amount, say $\Delta \mu_j$, when we moved a distance dx, then the rate of change of the chemical potential with position would be $\Delta \mu_j / dx$, which is infinite (a finite quantity divided by an infinitesimal one). This would in turn imply that we had an infinite force— the force on species j is $-\partial \mu_j / \partial x$—which is certainly not the case for a situation of biological interest. We have assumed in this text that we are dealing with well-defined and continuous functions.

Often, not only must μ_j itself be well defined and continuous, but also its partial derivatives. In other words, both $(\partial \mu_j/\partial x)_t$ and $(\partial \mu_j/\partial t)_x$ must be single-valued functions and must change by an infinitesimal amount when x changes by dx, or t changes by dt. The quantity $(\partial \mu_j/\partial x)_t$ is referred to as a *first-order* partial derivative. If such a derivative is well defined and continuous, we can take its partial derivative. For instance, $(\partial \mu_j/\partial x)_t$ has two partial derivatives. $[\partial(\partial \mu_j/\partial x)_t/\partial x]_t$ and $[\partial(\partial \mu_j/\partial x)_t/\partial t]_x$, which are known as *second-order* partial derivatives. We generally represent the first of these by $(\partial^2 \mu_j/\partial x^2)_t$, which is called a *homogeneous* second-order partial derivative, while the other is referred to as a *mixed* second-order partial derivative, since the derivative is taken with respect both to x and to t. When we have well-defined and continuous first-order partial derivatives, we can switch the order of differentiation in a mixed second-order partial derivative. For instance, $[\partial(\partial \mu_j/\partial x)_t/\partial t]_x$ is then equal to $[\partial(\partial \mu_j/\partial t)_x/\partial x]_t$. Such a switch in the order of differentiation is crucial for indicating how the chemical potential depends on pressure (consider Eqs. VI.11 and VI.13).

Since the concept of higher-order derivatives has been introduced, we will briefly comment on the maxima and minima of functions. Let us suppose that a function increases up to a maximum value and then decreases again as we move in the positive direction along the x-axis (note the dependence of concentration on position in Fig. 1.6). When the function reaches a maximum value, the tangent to the curve is then parallel to the x-axis and the slope of the tangent is therefore zero. We thus conclude that the first-order derivative, which is equal to the slope, is zero whenever a function attains a local maximum, e.g., at the origin in Figure 1.6. Analogously, the slope and the first-order derivative are also both zero at the point where a function has a local minimum. This useful correlation between *extrema* (maxima or minima) and the first-order derivative of a function finds many uses in physical chemistry. For example, the statement that entropy is at a maximum at equilibrium means that we can describe an equilibrium situation as one where the derivative of entropy is zero (see App. VI). Another criterion for equilibrium (at constant T and P) is that it occurs for a minimum in the Gibbs free energy (see p. 292 and App. VI), where again the derivative—in this case, of G—is zero, e.g., $dG/dt = 0$ at equilibrium.

Whether an extremum represents a minimum or maximum can be determined from the second-order derivative. A homogeneous second-order partial derivative has a positive value at a local minimum and a negative value at a local maximum, e.g., $\partial^2 c/\partial x^2 < 0$ at the local maximum at the origin in Figure 1.6. Extrema are therefore conveniently described using the values for first-order and second-order partial derivatives.

Examples of Derivatives

The following equations show various derivatives used in biological applications of calculus. In a sense, the equations are simply rules that must be followed in order to differentiate the functions indicated. Each relation can be rigorously proved. For example, we have already indicated that the derivative of a constant is zero (p. 552), which we will reconsider here for the case where the dependent variable y equals a constant b ($y = b$, which represents the equation of a straight line parallel to the x-axis at a distance b). When we move a distance dx in the positive x-direction along this line, there is no change in the value of y ($dy = 0$), hence dy/dx, which is the same as db/dx in this case, is zero. This fundamental property, that the derivative of a constant is zero, will be used when we discuss indefinite integrals and constants of integration in the next section. In the following equations, a and n are constants, x, x_j, y, and t are independent variables, and u, v, w, and f are functions that can depend on x as well as on other independent variables.

$$\frac{\partial(a)}{\partial x} = \frac{d(a)}{dx} = 0$$

$$\frac{\partial(x^n)}{\partial x} = \frac{d(x^n)}{dx} = nx^{n-1}$$

$$\text{e.g.,} \quad \frac{\partial(x^1)}{\partial x} = 1x^0 = 1$$

$$\frac{\partial(x^{-1/2})}{\partial x} = -\frac{1}{2x^{3/2}}$$

$$\frac{\partial \ln x}{\partial x} = \frac{d \ln x}{dx} = \frac{1}{x}, \quad \frac{\partial \log x}{\partial x} = \frac{d \log x}{dx} = \frac{1}{2.303}\frac{d \ln x}{dx} = \frac{1}{2.303\,x}$$

$$\frac{\partial(e^x)}{\partial x} = \frac{d(e^x)}{dx} = e^x$$

$$\frac{\partial u}{\partial x} = 1 \left/ \frac{\partial x}{\partial u} \right.$$

$$\frac{\partial(au)}{\partial x} = a\frac{\partial u}{\partial x}$$

$$\frac{\partial(u^n)}{\partial x} = nu^{n-1}\frac{\partial u}{\partial x}$$

$$\frac{\partial \ln u}{\partial x} = \frac{1}{u}\frac{\partial u}{\partial x}, \qquad \frac{\partial \log u}{\partial x} = \frac{1}{2.303u}\frac{\partial u}{\partial x}$$

e.g., $\dfrac{\partial \ln ax}{\partial x} = \dfrac{1}{ax}\dfrac{\partial(ax)}{\partial x} = \dfrac{1}{x}\dfrac{\partial x}{\partial x} = \dfrac{1}{x}$

$$\frac{\partial(e^u)}{\partial x} = e^u\frac{\partial u}{\partial x}$$

e.g., $\dfrac{\partial(e^{ax^n})}{\partial x} = e^{ax^n}\dfrac{\partial(ax^n)}{\partial x} = ae^{ax^n}\dfrac{\partial(x^n)}{\partial x} = anx^{n-1}e^{ax^n}$

$$\frac{\partial(u \pm v)}{\partial x} = \frac{\partial u}{\partial x} \pm \frac{\partial v}{\partial x}$$

e.g., $\dfrac{\partial(\ln a + \ln x)}{\partial x} = \dfrac{\partial \ln a}{\partial x} + \dfrac{\partial \ln x}{\partial x} = 0 + \dfrac{1}{x} = \dfrac{1}{x}$

(note that $\ln a + \ln x = \ln ax$)

$$\frac{\partial(uv)}{\partial x} = u\frac{\partial v}{\partial x} + v\frac{\partial u}{\partial x}$$

e.g., $\dfrac{\partial(x^3 e^{ax})}{\partial x} = x^3\dfrac{\partial(e^{ax})}{\partial x} + e^{ax}\dfrac{\partial(x^3)}{\partial x} = ax^3 e^{ax} + 3x^2 e^{ax}$

$$\frac{\partial(uvw)}{\partial x} = uv\frac{\partial w}{\partial x} + uw\frac{\partial v}{\partial x} + vw\frac{\partial u}{\partial x}$$

$$\frac{\partial(u/v)}{\partial x} = \frac{1}{v}\frac{\partial u}{\partial x} - \frac{u}{v^2}\frac{\partial v}{\partial x} = \frac{v\dfrac{\partial u}{\partial x} - u\dfrac{\partial v}{\partial x}}{v^2}$$

$$\frac{\partial f(u)}{\partial x} = \frac{\partial f(u)}{\partial u}\frac{\partial u}{\partial x}$$

$$\frac{\partial}{\partial x}\left(\frac{\partial u}{\partial x}\right) = \frac{\partial^2 u}{\partial x^2}$$

$$\frac{\partial^2 f(u)}{\partial x^2} = \frac{\partial f(u)}{\partial u}\frac{\partial^2 u}{\partial x^2} + \frac{\partial^2 f(u)}{\partial u^2}\left(\frac{\partial u}{\partial x}\right)^2$$

$$\frac{\partial(a^u)}{\partial x} = a^u \ln a\frac{\partial u}{\partial x}$$

$$\frac{\partial}{\partial x}\left(\frac{\partial u}{\partial t}\right) = \frac{\partial}{\partial t}\left(\frac{\partial u}{\partial x}\right)$$

$$\text{e.g.,} \quad \frac{\partial}{\partial x}\left(\frac{\partial(x \ln at)}{\partial t}\right) = \frac{\partial}{\partial x}\left(x\,\frac{\partial \ln at}{\partial t}\right)$$

$$= \frac{\partial}{\partial x}\left(\frac{x}{t}\right) = \frac{1}{t} = \frac{\partial}{\partial t}\left(\frac{\partial(x \ln at)}{\partial x}\right)$$

$$= \frac{\partial}{\partial t}\left(\ln at\,\frac{\partial x}{\partial x}\right) = \frac{\partial}{\partial t}(\ln at) = \frac{1}{t}$$

$$du = \sum_j \frac{\partial u}{\partial x_j}\,dx_j$$

$$\text{e.g.,} \quad d(x^2 y^3) = 2xy^3\,dx + 3x^2 y^2\,dy$$

Integration

In addition to derivatives, in this text we also make use of a number of rather common integrals, which we will next briefly consider. In a sense, integration—symbolized by $\int$, a distorted S standing for summation—is the opposite of differentiation. For example, we can differentiate μ_j to obtain the total differential, $d\mu_j$. If we now *integrate* this quantity, we obtain $\int d\mu_j$, which is μ_j, our original function. In integration, however, we must reckon with another aspect, the integration constant. To be specific, let us return to Figure V.1, which will also help show that integration is in a sense a sophisticated adding procedure.

The chemical potential of species j is μ_j^o in the solution just outside the membrane, i.e., at x_0 in Figure V.1. As we move an infinitesimal distance dx from x_0 in the positive x direction, μ_j changes by $d\mu_j$; i.e., at $x_0 + dx$, μ_j has a value of $\mu_j^o + d\mu_j$ ($d\mu_j$ is negative in the present case, since μ_j decreases as we move across the membrane in Fig. V.1). As we go another infinitesimal distance in the positive x-direction, e.g., dx', μ_j changes by an additional infinitesimal amount, e.g., $d\mu_j'$. Thus, in going from x_0 to $x_0 + dx + dx'$, μ_j becomes $\mu_j^o + d\mu_j + d\mu_j'$. The process of summing up these and subsequent changes in μ_j is known as integration. We are dividing up the pathway into (infinitesimally) small pieces along the x-axis and adding up all (an infinite number) of the (infinitesimal) changes occurring in μ_j to obtain the overall change in the chemical potential. When we get across the membrane, we expect to achieve the value that μ_j has in the inside solution, μ_j^i. We can express this process of summation, or integration, as follows:

$$\int_{x_0}^{x_0 + \Delta x} d\mu_j = \mu_j^i - \mu_j^o = \Delta\mu_j$$

$$= \int_{x_0}^{x_0 + \Delta x} \left(\frac{\partial\mu_j}{\partial x}\right)_t dx \qquad (V.6)$$

where the lower limit on the integral, x_0, means that we begin the integration at the position where x equals x_0, and the upper limit, $x_0 + \Delta x$, means that we terminate the integration after we cross the membrane, where our x-position is $x_0 + \Delta x$ (see Fig. V.1).

We have performed the integration indicated in the upper line of Equation V.6 by considering limits imposed on the x-position, and it might seem reasonable to represent the *integrand* $d\mu_j$ for such an integration as a function of x. To do this, we will use Equation V.5 for the special case where μ_j does not depend on t, namely, $d\mu_j = (\partial\mu_j/\partial x)_t dx$. Thus, we can replace the integral in the upper line by $\int_{x_0}^{x_0 + \Delta x} (\partial\mu_j/\partial x)_t dx$, where $(\partial\mu_j/\partial x)_t$ is the slope of the curve representing μ_j versus x (illustrated in Fig. V.1). If we add up the products of the local values of $(\partial\mu_j/\partial x)_t$ (which gives the rate of change of chemical potential with position) times the (infinite) number of infinitesimal distances dx over which these various changes occur, then we also obtain the overall change in the function, $\Delta\mu_j$ (see the bottom line of Eq. V.6).

Instead of integrating to obtain a value for $\Delta\mu_j$, we could use an addition procedure. For example, we could subdivide the section between x_0 and $x_0 + \Delta x$ into a finite number, n, of pieces, each extending a distance $\Delta x/n$ along the x-axis. The local slope of the curve could be approximated by $\Delta\mu_j^k/\Delta x^k$ for the kth such subdivision, where Δx^k equals $\Delta x/n$, and where $\Delta\mu_j^k$ is the change in chemical potential of species j across the kth section. To obtain a value for $\Delta\mu_j$, we add up the products of such an average slope, $\Delta\mu_j^k/\Delta x^k$, times the finite width of each subdivision, $\Delta x/n$, for all n sections between x_0 and $x_0 + \Delta x$, which leads to $\sum_{k=1}^{n} (\Delta\mu_j^k/\Delta x^k)(\Delta x/n)$ (compare the integral in the bottom line of Eq. V.6). Since Δx^k equals $\Delta x/n$, the summation becomes $\sum_{k=1}^{n} \Delta\mu_j^k$, which is $\Delta\mu_j$, the overall change in chemical potential (see Eq. V.6). This addition process—which can become quite laborious for a rapidly changing function, since the number of subdivisions (n) must then be quite large to match the curve closely—is known as *numerical integration.* The alternate procedure of integration using integrals is simply an elegant way of summing up an infinite number of infinitesimal changes in some integrand along a particular pathway of interest.

Definite Integrals, Indefinite Integrals, and the Integration Constant

When an integral has both upper and lower limits as in Equation V.6, it is referred to as a *definite* integral. When such limits are not given we have an *indefinite* integral, which can only be evaluated to within an additive constant. For example, the indefinite integral $\int d\mu_j$ equals $\mu_j + C$, where C is known as the *constant of integration.* If we impose limits on the ends of the integration pathway and hence convert the indefinite integral into a definite one, the integral would have a definite value and we would not have an additive constant of integration.

To help understand how the integration constant arises, let us return to a consideration of the integrand. We will let $d[f(x)]$ be the integrand that we wish to integrate between the limits of $x = a$ and $x = b$. But the integral of the derivative of a function is simply that function; hence we obtain the following relations:

$$\int_a^b d[f(x)] = f(x) \Big]_a^b = f(b) - f(a) \tag{V.7}$$

where the expression $f(x) \Big]_a^b$ means that we must evaluate the function at the upper limit b, i.e., $f(b)$, and then subtract its value at the lower limit, i.e., $f(a)$. (We have implicitly performed this operation in Eq. V.6.) Thus, $\int_a^b d[f(x)]$ has a *definite* value that depends on the x-positions (a and b) taken for the limits of integration.

We might wish to consider the *indefinite* integral, $\int d[f(x)]$, which does not have defined upper and lower limits. We recognize that $f(x)$ is a solution to $\int d[f(x)]$, but so is $f(x) + C$, where C is an arbitrary constant. To see why this is so, let us use the rules for differentiation on p. 557: $d[f(x) + C]$ equals $d[f(x)] + dC$, which is $d[f(x)]$, since the derivative of a constant is zero ($dC = 0$). Therefore, $\int d[f(x)]$ apparently has the same value as $\int d[f(x) + C]$, which means we know the value of an indefinite integral only to within an additive constant, the so-called constant of integration. When we know the specific pathway for integration, i.e., when we have a definite integral, the constant of integration drops out, as in Equations V.6 and V.7. As a final example, let us consider an integration pathway beginning at $x = a$ (the lower limit) and going to some unspecified (variable) value of x, i.e., a definite integral with a variable upper limit. We then have $\int_a^x d[f(x)]$, which is $f(x)\Big]_a^x$, or $f(x) - f(a)$. Such an integral adds up all the changes which occur in the function $f(x)$ beginning at a specific position ($x = a$) and going to some arbitrary value of x.

Examples of Integrals

Equation V.7 not only suggests the relationship between differentiation and integration, it also embodies the basic method for evaluating integrals— namely, to find a function whose derivative is equal to the integrand. A couple of examples should help clarify this important point. For instance, what is $\int 3x^2\,dx$? We know—or can determine from p. 557 with a little practice—that $d(x^3)$ equals $3x^2\,dx$. Consequently, $\int 3x^2\,dx$ equals $\int d(x^3)$, which is $x^3 + C$, where C is the constant of integration.

In Chapter 1 (Eq. 1.11) we had the following integral to evaluate: $\int_{c^i(0)}^{c^i(t)} dc^i/(c^o - c^i)$. Since c^o is a constant, $d(c^o - c^i)$ is equal to $-dc^i$, or dc^i is the same as $-d(c^o - c^i)$. Thus, we can re-express the integral as $\int_{c^i(0)}^{c^i(t)} -d(c^o - c^i)/(c^o - c^i)$. From the examples of derivatives given on p. 557, we note that $d(\ln x)$ is equal to dx/x. Hence, the integral becomes $\int_{c^i(0)}^{c^i(t)} -d[\ln(c^o - c^i)]$, which is $-\ln(c^o - c^i)\Big]_{c^i(0)}^{c^i(t)}$, or $-\ln[c^o - c^i(t)] + \ln[c^o - c^i(0)]$, which in turn equals $\ln[c^o - c^i(0)]/[c^o - c^i(t)]$, as we indicated on p. 32.

In the following examples of integrals, a is a constant, x is an independent variable, and u and v are functions that can depend on x as well as on other independent variables. For simplicity, the additive integration constant has been omitted from the indefinite integrals presented.

$$\int d[f(x)] = f(x)$$

$$\int a\,dx = a\int dx = ax$$

$$\int x^n \, dx = \frac{x^{n+1}}{n+1} \qquad n \neq -1$$

$$\int \frac{dx}{x} = \ln x, \qquad \int \frac{dx}{ax + b} = \frac{1}{a} \ln (ax + b)$$

$$\int e^x \, dx = e^x, \qquad \int e^{ax} \, dx = \frac{1}{a} e^{ax}$$

$$\int x e^{ax} \, dx = e^{ax} \left(\frac{x}{a} - \frac{1}{a^2} \right)$$

$$\int a^x \, dx = \frac{a^x}{\ln a}$$

$$\int \ln x \, dx = x \ln x - x$$

$$\int (u \pm v) \, dx = \int u \, dx \pm \int v \, dx$$

$$\int u \, dv = uv - \int v \, du$$

$$\int_a^b f(x) \, dx = - \int_b^a f(x) \, dx$$

Differential Equations

As a final topic in this appendix, we will consider the application of the integral calculus to the solution of differential equations. A differential equation expresses the relationship between derivatives (first-order as well as higher-order) and various variables or functions. The procedure in solving differential equations is first to put the relation into a form which can be integrated and then to carry out suitable integrations such that the derivatives are eliminated. Finally, to complete the solution of a differential equation, we must incorporate the known values of the functions at particular values of the variables, the so-called boundary conditions. We will illustrate the handling of differential equations by a simple, but extremely useful, example.

One of the most important differential equations in biology has the following general form:

$$\frac{dy}{dt} = -ky \qquad\qquad (V.8)$$

where k is a positive constant and t represents time. We encountered this equation in Chapter 4 (Eq. 4.8) in discussing the various competing pathways for the de-excitation of an excited singlet state. Equation V.8 also describes the process of radioactive decay, where y is the amount of radioisotope present at any time t. The equation indicates that the rate of change in time of the amount of radioactive substance (dy/dt) is linearly proportional to the amount present at that time (y). Since the radioisotope decays away in time, dy/dt is negative, and hence there is a minus sign in Equation V.8. Any process that can be described by Equation V.8, such as a chemical reaction, is called a first-order rate process, and k is known as the first-order rate constant.

To put Equation V.8 into a form suitable for integration, we must separate the variables (y and t) so that each one appears on only one side of the equation:

$$\frac{dy}{y} = -k \, dt \tag{V.9}$$

which follows from Equation V.8 upon multiplying each side by dt/y. (Note that the same initial process of separation of variables applies to the integration of a more complicated example in Ch. 3, namely, Eq. 3.11b.) When the variables are separated so that a possible integrand, e.g., $-k \, dt$, is expressed in terms of only one variable, we can integrate that integrand. On the other hand, the integrand $-ky \, dt$ cannot be integrated as it stands—i.e., we cannot perform $\int -ky \, dt$—since we do not know how y depends on t. In fact, the very purpose of solving Equation V.8 is to determine the functional relationship between y and t.

Next, we will insert integral signs into Equation V.9 and perform the integration making use of the examples of integrals listed above (p. 563):

$$\int \frac{dy}{y} = \ln y = -\int k \, dt = -kt + C \tag{V.10}$$

When we take exponentials of both sides of Equation V.10, we obtain the following expression for y:

$$e^{\ln y} = y(t) = e^{-kt+C} = e^{C} e^{-kt} \tag{V.11a}$$

where e^{C} is a constant. In Equation V.11a we have replaced y with $y(t)$ to emphasize that y depends on the independent variable t. When t equals zero,

Equation V.11a indicates that $y(0) = e^C e^{-k \times 0}$, which is simply e^C. Thus, the constant e^C is the value taken on by the dependent variable y when t is 0. This latter relationship, $e^C = y(0)$, is referred to as a boundary condition (or initial condition, since we are here dealing with time). Incorporating this boundary condition into Equation V.11a, we obtain the following solution to the differential equation represented by Equation V.8:

$$y(t) = y(0)e^{-kt} \qquad\qquad (\text{V.11b})$$

Because of the factor e^{-kt}, Equation V.11b indicates that y decays away exponentially with time for a first-order rate process. Moreover, $y(t)$ decreases to $1/e$ of its initial value $[y(0)]$ when t satisfies the following relation:

$$y(\tau) = \frac{1}{e} y(0) = y(0)e^{-k\tau} \qquad\qquad (\text{V.12})$$

where the value of time, τ, that satisfies Equation V.12 is known as the *lifetime* of the process whose decay or disappearance is being considered. Equation V.12 indicates that e^{-1} equals $e^{-k\tau}$, so the first-order rate constant is equal to the reciprocal of the lifetime τ (see Eq. 4.11). Therefore, the solution (Eq. V.11b) of the partial differential equation (Eq. V.8) describing a first-order rate process becomes

$$y(t) = y(0)e^{-t/\tau} \qquad\qquad (\text{V.13})$$

Gibbs Free Energy
and Chemical Potential

Entropy and equilibrium. Gibbs free energy. Chemical potential.
Pressure dependence of μ_j. Concentration dependence of μ_j. References.

The concept of chemical potential is introduced in Chapter 2 and used throughout the rest of the book. In order not to overburden the text with mathematical details, certain points are stated without proof. Here we will justify the form of the pressure term in the chemical potential and also provide insight into how the expression for the Gibbs free energy arises.

Entropy and Equilibrium

A suitable point of departure is to reconsider the condition for equilibrium. The most general statement we can make concerning the attainment of equilibrium by a system is that it occurs when the entropy of the system plus its surroundings is at a maximum. Unfortunately, entropy has proved to be a rather elusive concept to master and a difficult quantity to measure. Moreover, reference to the surroundings—the "rest of the universe" in the somewhat grandiloquent language of physics—is a nuisance. Consequently, thermodynamicists sought a function that would help describe equilibrium, but would depend only on readily measurable parameters of the system under consideration. As we will see, the Gibbs free energy is such a function for most applications in biology.

The concept of entropy is really part of our day-to-day observations. We know that an isolated system will spontaneously change in certain ways—a system proceeds toward a state that is more random or less ordered than the initial one. For instance, neutral solutes will diffuse toward regions where

567

they are less concentrated. In so doing, the system lowers its capacity for further spontaneous change. For all such processes ΔS is positive, whereas ΔS becomes zero and S achieves a maximum at equilibrium. Equilibrium means that no more spontaneous changes will take place; entropy is therefore an *index for the capacity for spontaneous change*. It would be more convenient in some ways if entropy had been originally defined with the opposite sign. In fact, many authors introduce the quantity *negentropy*, which equals $-S$ and reaches a *minimum* at equilibrium. In any case, we must ultimately use a precise mathematical definition for entropy, such as $dS = dQ/T$, where dQ refers to the heat gain or loss in some reversible reaction taking place at temperature T.

We can represent the total entropy of the universe, S_u, as the entropy of the system under consideration, S_s, plus the entropy of the rest of the universe, S_r. We can express this in symbols as follows:

$$S_u = S_s + S_r \quad \text{or} \quad dS_u = dS_s + dS_r \tag{VI.1}$$

An increase in S_u accompanies all real processes—this is, incidentally, the most succinct way of stating the second law of thermodynamics. S_u is maximum at equilibrium.

The heat absorbed by a system during some process is equal to the heat given up by the rest of the universe. Let us represent the infinitesimal heat exchange of the system by dQ_s. For an isothermal reaction or change, dQ_s is simply $-dQ_r$, since the heat must come from the rest of the universe. From the definition of entropy,* $dS = dQ/T$, we can obtain the following relationship:

$$dS_r = \frac{dQ_r}{T} = -\frac{dQ_s}{T} = -\frac{dU_s + P\,dV_s}{T} \tag{VI.2}$$

The last step in Equation VI.2 derives from the principle of the conservation of energy for the case when the only form of work involved is mechanical—a common assumption in stating the first law of thermodynamics. It is thus possible to express dQ_s as the sum of the change in internal energy (dU_s) plus a work term ($P\,dV_s$). The internal energy (U_s) is a function of the state of a system; i.e., its magnitude depends on the characteristics of the system but is independent of how the system got to that state. PV_s is also a well-defined variable. However, heat (Q_s) is not a function of the state of a system.

---

* This definition really applies only to reversible reactions, which we can in principle use to approximate a given change; otherwise, dQ is not uniquely related to dS.

As we indicated above, equilibrium occurs when the entropy of the universe is maximum. This means that dS_u then equals zero (see p. 556 for a comment on derivatives and extrema of curves). By substituting Equation VI.2 into the differential form of Equation VI.1, we can express this equilibrium condition solely in terms of system parameters:

$$0 = dS_s + \left(-\frac{dU_s + P\,dV_s}{T}\right) \quad \text{or} \quad -T\,dS_s + dU_s + P\,dV_s = 0 \quad \text{(VI.3)}$$

Equation VI.3 suggests that there is some function of the *system* which has an extremum at equilibrium. In other words, we might be able to find some expression determined by the parameters describing the system and whose derivative is zero at equilibrium. If so, the rather abstract statement that the entropy of the universe is a maximum at equilibrium could then be replaced by a statement referring only to measurable attributes of the system—easily measurable ones, we hope.

In the 1870's J. W. Gibbs—perhaps the most brilliant thermodynamicist to date—chose a rather simple set of terms which turned out to have the very properties for which we are searching. This function is now referred to as the *Gibbs free energy,* and has the symbol G:

$$G = U + PV - TS \tag{VI.4a}$$

which, upon differentiating, yields

$$dG = dU + P\,dV + V\,dP - T\,dS - S\,dT \tag{VI.4b}$$

Equation VI.4b implies that, at constant temperature $(dT = 0)$ and pressure $(dP = 0)$, dG is

$$dG = dU + P\,dV - T\,dS \quad \text{at constant } T \text{ and } P \tag{VI.5}$$

By comparing Equation VI.5 with the equilibrium condition expressed by Equation VI.3, we see that dG for a system equals zero at equilibrium at constant temperature and pressure. Moreover, G depends only on U, P, V, T, and S of the *system*. It turns out that the extremum condition, $dG = 0$, actually occurs when G reaches a minimum at equilibrium. This useful attribute of the Gibbs free energy is strictly valid only when the overall system is at *constant temperature and pressure,* conditions that closely approximate those encountered in many biological situations. Thus, our criterion for equilibrium shifts from a maximum of the entropy of the universe to a minimum in the Gibbs free energy of the system.

Gibbs Free Energy

We will now consider how the internal energy, U, changes when material enters or leaves a system. This will help us derive an expression for the Gibbs free energy that is quite useful for biological applications.

The internal energy of a system changes when substances enter or leave it. For convenience, we will consider a system of fixed volume and at the same temperature as the surroundings, so that there are no heat exchanges. If dn_j moles of species j enter such a system, U increases by $\mu_j dn_j$, where μ_j is an intensive variable representing the free energy contribution to the system per mole of species j entering or leaving. Work is often expressed as the product of an intensive quantity (such as μ_j, P, T, E, h) times an extensive one (dn_j, dV, dS, dQ, dm, respectively); i.e., the amount of any kind of work depends on both some thermodynamic parameter characterizing the internal state of the system and the extent or amount of change for the system. In our present example the extensive variable describing the amount of change is dn_j, and μ_j represents the contribution to the internal energy of the system per mole of species j. When more than one species crosses the boundary of our system, which is at constant volume and the same temperature as the surroundings, the term $\sum_j \mu_j dn_j$ is added to dU, where dn_j is positive if the species enters the system and negative if it leaves. In the general case, when we consider all the ways that the internal energy of a system can change, we can represent dU as follows:

$$dU = dQ - P\,dV + \sum_j \mu_j dn_j \qquad \text{(VI.6)}$$

Let us now return to the development of a useful relation for the Gibbs free energy of a system. When dU as expressed by Equation VI.6 is substituted into dG as given by Equation VI.5, we obtain

$$dG = T\,dS - P\,dV + \sum_j \mu_j dn_j + P\,dV - T\,dS$$

$$= \sum_j \mu_j dn_j \qquad \text{(VI.7)}$$

where dQ has been replaced by $T\,dS$. Equation VI.7 indicates that the particular form chosen for the Gibbs free energy leads to a very simple expression for dG at constant T and P—namely, dG then depends only on μ_j and dn_j.

To obtain an expression for G, we must integrate Equation VI.7. To facilitate the integration we will define a new variable, α, such that dn_j equals $n_j\,d\alpha$, where n_j is the total number of moles of species j present in the final system; i.e., n_j is a constant describing the final system. The subsequent integration from $\alpha = 0$ to $\alpha = 1$ corresponds to building up the system by a simultaneous addition of all the components in the same proportions present in the final system. (The intensive variable μ_j is also held constant for this integration pathway; i.e., the chemical potential of species j does not depend on the size of the system.) Using Equation VI.7 and this rather easy integration pathway, we obtain the following expression for the Gibbs free energy:

$$G = \int dG = \int \sum_j \mu_j\,dn_j = \int_0^1 \sum_j \mu_j n_j\,d\alpha = \sum_j \mu_j n_j \int_0^1 d\alpha = \sum_j \mu_j n_j \quad \text{(VI.8)}$$

The well-known relation between G and the μ_j's in Equation VI.8 can also be obtained by a method that is more elegant mathematically, but somewhat involved.

In Chapter 6 we presented without proof an expression for the Gibbs free energy (Eq. VI.8 is similar to Eq. 6.1), and also noted some of the properties of G. For instance, at constant temperature and pressure, the direction for a spontaneous change is toward a lower Gibbs free energy; minimum G is achieved at equilibrium. Hence, ΔG is negative for such spontaneous processes. Spontaneous processes can in principle be harnessed to do useful work, where the maximum amount of work possible at constant temperature and pressure is equal to the absolute value of ΔG (some of the energy is dissipated by inevitable inefficiencies such as frictional losses, so that $-\Delta G$ represents the *maximum* work possible). To drive a reaction in the direction opposite to that in which it proceeds spontaneously requires a free energy input of at least ΔG.

Chemical Potential

Let us now examine the properties of the intensive variable μ_j. Equation VI.8 $\left(G = \sum_j \mu_j n_j\right)$ suggests a very useful way of defining μ_j. In particular, if we keep μ_i and n_i constant, we obtain the following expression:

$$\mu_j = \left(\frac{\partial G}{\partial n_j}\right)_{\mu_i,n_i} = \left(\frac{\partial G}{\partial n_j}\right)_{T,P,E,h,n_i} \tag{VI.9}$$

where n_i and μ_i refer to all species other than species j. Since μ_i can depend on T, P, E (the electrical potential), h (the height in a gravitational field), and n_i, the act of keeping μ_i constant during partial differentiation is the same as keeping T, P, E, h, and n_i constant, as indicated in Equation VI.9. Equation VI.9 indicates that the chemical potential of species j is the partial molal Gibbs free energy of a system with respect to that species, and that it is obtained when T, P, E, h, and the amount of all other species are held constant. Thus, μ_j corresponds to the intensive contribution of species j to the extensive quantity G, the Gibbs free energy of the system.

In Chapter 2 (pp. 61–68, 90) and Chapter 3 (p. 106) we argued that μ_j depends on T, a_j ($a_j = \gamma_j c_j$, Eq. 2.5), P, E, and h in a solution, and that the partial pressure of species j, P_j, is also involved for the chemical potential in a vapor phase. We can summarize the two relations as follows:

$$\mu_j^{\text{liquid}} = \mu_j^* + RT\ln a_j + \bar{V}_j P + z_j FE + m_j gh$$

$$\mu_j^{\text{vapor}} = \mu_j^* + RT\ln\frac{P_j}{P_j^*} + m_j gh \tag{VI.10}$$

The forms for the gravitational contribution ($m_j gh$) and the electrical one ($z_j FE$) can easily be understood. We showed in Chapter 3 (p. 121) that $RT\ln a_j$ is the correct form for the concentration term in μ_j. The reasons for the forms of the pressure terms in a liquid ($\bar{V}_j P$) and in a gas $[RT\ln(P_j/P_j^*)]$ are not so obvious. Therefore, we will examine the pressure dependence of the chemical potential of species j in some detail.

Pressure Dependence of μ_j

To derive the pressure terms in the chemical potentials of solvents, solutes, and gases, we must rely on certain properties of partial derivatives as well as on commonly observed effects of pressure. To begin with, we will differentiate the chemical potential in Equation VI.9 with respect to P:

$$\left(\frac{\partial \mu_j}{\partial P}\right)_{T,E,h,n_i,n_j} = \left[\frac{\partial}{\partial P}\left(\frac{\partial G}{\partial n_j}\right)_{T,P,E,h,n_i}\right]_{T,E,h,n_i,n_j} = \left[\frac{\partial}{\partial n_j}\left(\frac{\partial G}{\partial P}\right)_{T,E,h,n_i,n_j}\right]_{T,P,E,h,n_i} \tag{VI.11}$$

where we have reversed the order for partial differentiation with respect to P and n_j (this is permissible for functions such as G, which have well-defined and continuous first-order partial derivatives; see p. 555). The differential form of Equation VI.4 ($dG = dU + P\,dV + V\,dP - T\,dS - S\,dT$) gives us a suitable form for dG. If we substitute dU given by Equation VI.6 $\left(dU = T\,dS - P\,dV + \sum_j \mu_j\,dn_j, \text{ where } dQ \text{ is replaced by } T\,dS; \text{ see p. 568}\right)$ into this expression for the derivative of the Gibbs free energy, we can express dG in the following useful form:

$$dG = V\,dP - S\,dT + \sum_j \mu_j\,dn_j \qquad \text{(VI.12)}$$

Using Equation VI.12 we can readily determine the pressure dependence of the Gibbs free energy as needed in the last bracket of Equation VI.11—namely, $(\partial G/\partial P)_{T,E,h,n_i,n_j}$ equals V by Equation VI.12. Next, we have to consider the partial derivative of this V with respect to n_j (see last equality of Eq. VI.11). But Equation 2.6 indicates that $(\partial V/\partial n_j)_{T,P,E,h,n_i}$ is $\bar{V}_j$, the partial molal volume. Substituting these partial derivatives into Equation VI.11 leads us to the following useful expression:

$$\left(\frac{\partial \mu_j}{\partial P}\right)_{T,E,h,n_i,n_j} = \bar{V}_j \qquad \text{(VI.13)}$$

Equation VI.13 is of pivotal importance in deriving the form of the pressure term in the chemical potentials of both liquid and vapor phases.

Let us first consider an integration of Equation VI.13 appropriate for a liquid. We will make use of the observation that the partial molal volume of a species in a solution does not depend on the pressure to any significant extent. For a solvent this means that the liquid generally is essentially incompressible. If we integrate Equation VI.13 at constant T, E, h, n_i and n_j with $\bar{V}_j$ independent of P, we obtain the following relations:

$$\int \frac{\partial \mu_j}{\partial P}\,dP = \int_{\mu_j^*}^{\mu_j^{\text{liquid}}} d\mu_j = \mu_j^{\text{liquid}} - \mu_j^*$$

$$\qquad \text{(VI.14)}$$

$$= \int \bar{V}_j\,dP = \bar{V}_j \int dP = \bar{V}_j P + \text{"constant"}$$

where the definite integral in the top line is taken from the chemical potential of species j in a standard state as the lower limit up to the general μ_j in a liquid as the upper limit. The bottom line contains an indefinite integral—

see p. 561 for definitions of definite and indefinite integrals. The integration of $\int \bar{V}_j\, dP$ leads to our pressure term $\bar{V}_j P$ plus a constant. Since the integration was performed while holding T, E, h, n_i, and n_j fixed, the "constant" can depend on all of these variables, but not on P.

Equation VII.14 indicates that the chemical potential of a liquid contains a pressure term of the form $\bar{V}_j P$. The other terms (in particular, μ_j^*, $RT \ln a_j$, $z_j FE$, and $m_j gh$) do not depend on pressure, a condition used throughout this text. The experimental observation which gives us this very useful form for μ_j is that $\bar{V}_j$ is generally not influenced very much by pressure; e.g., a liquid is often essentially incompressible. If this should prove invalid under certain situations, $\bar{V}_j P$ would then not be a suitable term in the chemical potential of species j for expressing the pressure dependence in a solution.

Next, we turn to the form of the pressure term in the chemical potential of a gas, where the assumption of incompressibility that we used for a liquid is obviously not valid. Our point of departure is the perfect or ideal gas law:

$$P_j V = n_j RT \tag{VI.15}$$

where P_j is the partial pressure of species j and n_j is the number of moles of species j in volume V. Thus, we will assume that real gases behave like ideal gases, which appears to be justified for biological applications. Using Equations VI.15 and 2.6 $\left[\bar{V}_j = \left(\dfrac{\partial V}{\partial n_j} \right)_{n_i, T, P, E, h} \right]$, the partial molal volume $\bar{V}_j$ for gaseous species j is RT/P_j. We also note that the total pressure P equals $\sum\limits_j P_j$, where the summation is over all gases present (Dalton's law of partial pressures); hence, dP equals dP_j when n_i, T, and V are constant. When we integrate Equation VI.13, we thus find that the chemical potential of gaseous species j depends on the *logarithm* of its partial pressure:

$$\int \frac{\partial \mu_j}{\partial P}\, dP = \int_{\mu_j^*}^{\mu_j^{\text{vapor}}} d\mu_j = \mu_j^{\text{vapor}} - \mu_j^* = \int \bar{V}_j\, dP = \int \frac{RT}{P_j}\, dP_j$$

$$= RT \ln P_j + \text{"constant"} \tag{VI.16}$$

where the "constant" can depend on T, h, and n_i. In particular, the "constant" equals $-RT \ln P_j^* + m_j gh$, where P_j^* is the saturation partial pressure for species j at atmospheric pressure and some particular temperature. Hence, the chemical potential for species j in the vapor phase (μ_j^{vapor}) is

$$\mu_j^{\text{vapor}} = \mu_j^* + RT \ln P_j - RT \ln P_j^* + m_j gh$$

$$= \mu_j + RT \ln \frac{P_j}{P_j^*} + m_j gh \qquad (VI.17)$$

which is essentially the same as Equations 2.20 and VI.10.

We have defined the standard state for gaseous species j, μ_j^*, as the chemical potential when the gas phase has a partial pressure for species j (P_j) equal to the saturation partial pressure at atmospheric pressure (P_j^*), when we are at the zero level for the gravitational term ($h = 0$), and for some specified temperature. Many physical chemistry texts ignore the gravitational term (we calculated that it has only a small effect for water vapor on p. 92) and define the standard state for the condition when P_j equals 1 atmosphere. The chemical potential of such a standard state equals $\mu_j^* - RT \ln P_j^*$ in our symbols.

The partial pressure of some species in a vapor phase in equilibrium with a liquid depends slightly on the total pressure in the system—loosely speaking, when the pressure on the liquid is increased, more molecules are squeezed out of it into the vapor phase. The exact relationship between the pressures involved, which is known as the Gibbs equation, is as follows for water:

$$\frac{\partial (\ln P_{wv})}{\partial P} = \frac{\bar{V}_w}{RT} \quad \text{or} \quad \frac{\partial P_{wv}}{\partial P} = \frac{\bar{V}_w}{\bar{V}_{wv}} \qquad (VI.18)$$

where the second equality follows from the derivative of a logarithm $[\partial \ln u / \partial x = (1/u)(\partial u / \partial x)$, p. 558] and the perfect gas law ($P_{wv} V = n_{wv} RT$, Eq. VI.15, and so $\partial V / \partial n_{wv} = \bar{V}_{wv} = RT/P_{wv}$). Since $\bar{V}_w$ is much less than $\bar{V}_{wv}$, the effect is quite small (e.g., at 20°C and atmospheric pressure, $\bar{V}_w = 1.8 \times 10^{-5} \text{ m}^3 \text{ mol}^{-1}$ and $\bar{V}_{wv} = 2.4 \times 10^{-2} \text{ m}^3 \text{ mol}^{-1}$). From the first equality in Equation VI.18, we see that $\bar{V}_w dP = RT d \ln P_{wv}$. Hence, if the chemical potential of the liquid phase (μ_w) increases by $\bar{V}_w dP$ as an infinitesimal pressure is applied, then an equal increase, $RT d \ln P_{wv}$, occurs in μ_{wv} (see Eq. VI.13 for a definition of μ_{wv}), and hence we can still be in equilibrium ($\mu_w = \mu_{wv}$).

Concentration Dependence of μ_j

We will complete our discussion of chemical potential by using Equation VI. 17 to obtain the logarithmic term in concentration which is found for

μ_j in a liquid phase. First, it should be pointed out that Equation VI. 17 has no concentration term per se for the chemical potential of species j in a gas phase. However, the partial pressure of a species in a gas phase is really analogous to the concentration of a species in a liquid; e.g., $P_j V = n_j RT$ for gaseous species j (Eq. VI.15), and concentration means number/volume = $n_j/V = P_j/RT$.

Raoult's law states that at equilibrium the partial pressure of a particular gas above its volatile liquid is proportional to the mole fraction of that solvent in the liquid phase. A similar relation more appropriate for solutes is Henry's law, which states that P_j in the vapor phase is proportional to the N_j of that solute in the liquid phase. Although the proportionality coefficients in the two relations are different, they both indicate that P_j^{vapor} depends linearly on N_j^{solution}. For dilute solutions the concentration of species j, c_j, is proportional to its mole fraction, N_j (this is true for both solute and solvent). Thus, when P_j^{vapor} changes from one equilibrium condition to another, we expect a similar change in c_j^{solution}, since μ_j^{liquid} equals μ_j^{vapor} at equilibrium. In particular, Equation VI.17 indicates that μ_j^{vapor} depends on $RT \ln (P_j/P_j^*)$, and hence the chemical potential of a solvent or solute should contain a term of the form $RT \ln c_j$, as in fact it does (see Eqs. 2.4 and VI.10). As we discussed on p. 60, we should be concerned about the concentration that is thermodynamically active, a_j ($a_j = \gamma_j c_j$, Eq. 2.5), so that the actual term in the chemical potential for a solute or solvent is $RT \ln a_j$, not $RT \ln c_j$. In Chapter 3 (p. 121), instead of the present argument based on Raoult's and Henry's laws, we used a comparison with Fick's first law to justify the $RT \ln a_j$ term. Moreover, the Boyle–Van't Hoff relation, which was derived *assuming* the $RT \ln a_j$ term, has been amply demonstrated experimentally. Consequently, the $RT \ln a_j$ term in the chemical potential for a solute or solvent can be justified or derived in a number of different ways, all of which depend on agreement with experimental observations.

References

Castellan, G. W. 1971. *Physical Chemistry*, 2nd ed. Addison-Wesley, Menlo Park, California.

Denbigh, K. G. 1981. *The Principles of Chemical Equilibrium*, 4th ed. Cambridge University Press, Cambridge.

Katchalsky, A., and P. F. Curran. 1967. *Nonequilibrium Thermodynamics in Biophysics.* Harvard University Press, Cambridge.

Klotz, J. 1964. *Chemical Thermodynamics.* Benjamin, New York.

Morowitz, H. J. 1970. *Entropy for Biologists—An Introduction to Thermodynamics.* Academic Press, New York.

Further Comments on
Irreversible Thermodynamics

Conjugate forces and fluxes. Interdependence of forces and fluxes, reflection coefficients. Irreversible thermodynamic description of solute flux density. References.

The introduction to irreversible thermodynamics in the text (pp. 160 – 179) is primarily concerned with stating the overall principles, and includes a detailed discussion of reflection coefficients. Here we will consider a choice of forces and fluxes that is more common. In particular, we will replace the "forces" $\Delta\mu_s$ and $\Delta\mu_w$ by ΔP and $\Delta\Pi$, which leads to a replacement of the conjugate flux densities J_s and J_w by J_V and J_D, respectively. We will also derive a parameter analogous to the classical permeability coefficient of species j, P_j.

Conjugate Forces and Fluxes

We are not at liberty to pick any set of forces and fluxes in irreversible thermo-dynamics. Rather, we must make a correct choice of pairs in order for the Onsager reciprocity relation (p. 162) to be valid; this involves the so-called *conjugate* force and flux. In Chapter 3 we used $\Delta\mu_s$ as a force and J_s as its paired, or conjugate flux density, together with $\Delta\mu_w$ and J_w. Although the equations developed for L_P and σ (Eqs. 3.36 and 3.37) using such forces and fluxes are relatively straightforward, the expressions are rather unwieldy. The conventional approach, introduced by Kedem and Katchalsky in 1958 and adopted by essentially all subsequent treatments of irreversible thermo-dynamics, is to change to a more convenient set of conjugate forces and fluxes. A discussion of the criteria for deciding whether a particular force-flux pair is conjugate would take us into a detailed consideration of the dissipation function and the rate of entropy production, topics outside the scope of this text. It turns out that the total volume flux density, J_V, is conjugate to ΔP, and the diffusional flux density, J_D, which we will discuss below, is conjugate to $\Delta\Pi$. To help make this last statement at least a little more plausible, we will transform the dissipation function from our previous set of forces and fluxes to a more convenient one involving J_V, J_D, ΔP, and $\Delta\Pi$.

We will begin by simply stating how the dissipation function, Φ, can be expressed. This function, which is a measure of the rate at which energy is irreversibly dissipated in a system, equals the sum of the various fluxes, J_k, times their conjugate forces, X_k; i.e., $\Phi = \sum_k J_k X_k$. Let us consider the dissipation function for the conjugate forces and fluxes used in Chapter 3:

$$\Phi = J_w \Delta\mu_w + J_s \Delta\mu_s \tag{VII.1}$$

Using Equations 3.31 ($\Delta\mu_w = -\bar{V}_w \Delta\Pi + \bar{V}_w \Delta P$) and 3.32 $\left(\Delta\mu_s = \dfrac{1}{\bar{c}_s} \Delta\Pi + \bar{V}_s \Delta P\right)$, we can rewrite Equation VII.1 as follows:

$$\varphi = J_w(-\bar{V}_w \Delta\Pi + \bar{V}_w \Delta P) + J_s\left(\frac{1}{\bar{c}_s}\Delta\Pi + \bar{V}_s\Delta P\right)$$

$$= (J_w\bar{V}_w + J_s\bar{V}_s)\Delta P + \left(\frac{J_s}{\bar{c}_s} - J_w\bar{V}_w\right)\Delta\Pi$$

$$= J_V \Delta P + J_D \Delta\Pi \tag{VII.2}$$

where J_V is the total volume flux density as defined in Chapter 3 (Eq. 3.33b, $J_V = \bar{V}_w J_w + \bar{V}_s J_s$). Since the dissipation function is the sum of the various flux densities times their conjugate "forces," from the last line of Equation VII.2 we can identify J_V as being conjugate to ΔP, and J_D to $\Delta\Pi$. Thus, we have transformed Φ into our new set of conjugate forces and fluxes.

We can readily show that the newly introduced diffusional flux density, J_D, is the difference between the average velocities of the solute, v_s, and water, v_w. We begin by noting that $\bar{V}_w c_w$ is essentially unity for a dilute aqueous solution ($\bar{V}_w c_w + \bar{V}_s c_s = 1$ for a solution of a single solute, where $\bar{V}_s c_s \ll 1$ holds for a dilute solution); hence $J_w \bar{V}_w$ is approximately equal to J_w/c_w. But the average speed of species j, v_j, is simply J_j/c_j (see p. 118). Hence, J_D as defined by Equation VII.2 can be re-expressed as follows:

$$J_D = J_s/\bar{c}_s - J_w\bar{V}_w$$

$$\cong J_s/\bar{c}_s - J_w/c_w$$

$$= v_s - v_w \tag{VII.3}$$

The diffusional flux density given in Equation VII.3 is the difference between the mean velocities of solute and water. In mass flow, e.g., that described by Poiseuille's law (Eq. 9.10), v_s equals v_w, so J_D is zero; such flow is independent of $\Delta\Pi$ and depends only on the gradient in hydrostatic pressure. On the other hand, let us consider $\Delta\Pi$ across a membrane that greatly restricts the passage of some solute relative to the movement of water, i.e., a barrier that acts as a differential filter; v_s would then be considerably less than v_w, so J_D would have a nonzero value in response to its conjugate "force," $\Delta\Pi$. Thus, J_D expresses the tendency of the solute relative to water to diffuse in response to a gradient in osmotic pressure.

Interdependence of Forces and Fluxes, Reflection Coefficients

Next, we will use our new set of parameters to express the linear interdependence of conjugate forces and fluxes in irreversible thermodynamics $\left(\text{Eq. 3.28,} \ J_j = \sum_k L_{jk} X_k\right)$:

$$J_V = L_P \, \Delta P + L_{PD} \, \Delta\Pi \tag{VII.4}$$

$$J_D = L_{DP} \, \Delta P + L_D \, \Delta\Pi \tag{VII.5}$$

where the subscripts in both equations are those generally used in the literature. (Although these subscripts—P referring to pressure and D to diffusion—are not consistent with the L_{jk} convention, four coefficients are still needed to describe the dependence of two fluxes on their conjugate forces.) By the Onsager reciprocity relation, which is applicable in the present case of conjugate forces and fluxes, L_{PD} equals L_{DP}, and again only three different coefficients are needed to describe the movement of water and a single solute across a membrane (see Eqs. 3.29 and 3.30).

Equation VII.4 is quite similar to Equation 3.38 $[J_V = L_P(\Delta P - \sigma \Delta\Pi)]$. In particular, L_P is the same hydraulic conductivity coefficient in both cases. It describes the mechanical filtration capacity of a membrane or other barrier; i.e., when $\Delta\Pi$ is zero, L_P relates the total volume flux density, J_V, to the hydrostatic pressure difference, ΔP. When ΔP is zero, Equation VII.5 indicates that a difference in osmotic pressure leads to a diffusional flow characterized by the coefficient L_D. Membranes also generally exhibit a property called *ultrafiltration*, whereby they offer different resistances to the passage of the solute and water. For instance, in the absence of an osmotic pressure drop ($\Delta\Pi = 0$), Equation VII.5 indicates that there will be a diffusional flux density equal to $L_{DP}\Delta P$. Based on Equation VII.3 ($J_D = v_s - v_w$), v_s is then different from v_w, which would result if a membrane restricted the passage of solute more than that of water. Thus, the phenomenological coefficient L_{DP} helps describe the relative ease with which solute crosses a membrane, compared with water. But such a property of relative selectivity by a barrier is also embodied in the reflection coefficient, σ (pp. 166–172). In fact, Staverman in 1951 defined the reflection coefficient in terms of L_{DP} as follows:

$$\sigma = -L_{DP}/L_P = -L_{PD}/L_P \qquad \text{(VII.6)}$$

Equation VII.6 is a considerably simpler definition of the reflection coefficient than the equivalent one given by Equation 3.37.

When a membrane is nonselective, v_s equals v_w and $\Delta\Pi$ is zero (see p. 169). If v_s equals v_w, then J_D must be zero by Equation VII.3. For this to be true for any ΔP and a zero $\Delta\Pi$, Equation VII.5 indicates that L_{DP} must be zero; hence the reflection coefficient is also zero in this case ($\sigma = -L_{DP}/L_P$, Eq. VII.6). Thus, σ is zero when the membrane does not select between solute and solvent. At the opposite extreme, the solute does not cross the membrane ($v_s = 0$); hence J_D is $-v_w$ ($J_D = v_s - v_w$, Eq. VII.3). Moreover, J_V, which is the average velocity of the solution (see p. 495), is then simply v_w. J_V therefore equals $-J_D$ when the solute does not cross the membrane, and $L_P\Delta P + L_{PD}\Delta\Pi$ equals $-L_{DP}\Delta P - L_D\Delta\Pi$ by Equations VII.4 and VII.5. Since this is true for any ΔP, L_P must then equal $-L_{DP}$; hence $-L_{DP}/L_P$ equals unity in this case. The reflection coefficient is therefore unity ($\sigma = -L_{DP}/L_P$, Eq. VII.6) when the solute does not cross the membrane. Consequently, both the limits of nonselectivity ($\sigma = 0$) and impermeability ($\sigma = 1$) can easily be presented using the forces and fluxes adopted in this appendix.

Using the definition of σ (Eq. VII.6), we can rewrite Equation VII.4 to obtain the following form for the total volume flux density:

$$J_V = L_P(\Delta P - \sigma\,\Delta\Pi) \qquad \text{(VII.7)}$$

This equation is identical to Equation 3.38, further indicating that L_P and σ have the same meaning in the two different approaches. Equation VII.7 can likewise be generalized to include the case of more than one solute—see Equation 3.39, $J_V = L_P\left(\Delta P - \sum_j \sigma_j\Delta\Pi_j\right)$.

Irreversible Thermodynamic Description of Solute Flux Density

Our final objective is to obtain an expression for the solute flux density, J_s, which takes into consideration the coupling of forces and fluxes of irreversible thermodynamics. We will begin by showing that $J_D + J_V$ is approximately equal to J_s/c_s. By Equation VII.3, J_D equals $J_s/\bar{c}_s - J_w\bar{V}_w$, while J_V is $J_w\bar{V}_w + J_s\bar{V}_s$ (Eq. 3.33), and so

$$J_D + J_V = J_s\bar{c}_s - J_w\bar{V}_w + J_w\bar{V}_w + J_s\bar{V}_s$$

$$= J_s\left(\frac{1}{\bar{c}_s} + \bar{V}_s\right)$$

$$\cong J_s/\bar{c}_s \qquad \text{(VII.8)}$$

where the last step applies to a dilute solution ($\bar{V}_s\bar{c}_s \ll 1$, or $\bar{V}_s \ll 1/\bar{c}_s$). After multiplying both sides of Equation VII.8 by $\bar{c}_s$, and using Equation VII.5 for J_D, we obtain:

$$J_s = \bar{c}_s J_V + \bar{c}_s J_D$$

$$= \bar{c}_s J_V + \bar{c}_s L_{DP}\Delta P + \bar{c}_s L_D\,\Delta\Pi$$

$$= \bar{c}_s J_V + \bar{c}_s L_{DP}\frac{(J_V - L_{PD}\,\Delta\Pi)}{L_P} + \bar{c}_s L_D\,\Delta\Pi$$

$$= \bar{c}_s J_V\left(1 + \frac{L_{DP}}{L_P}\right) + \bar{c}_s\Delta\Pi\left(L_D - \frac{L_{DP}L_{DP}}{L_P}\right) \qquad \text{(VII.9)}$$

where Equation VII.4 is used in the second equality to express ΔP. Finally, we combine the factors multiplying $\Delta\Pi$ into a new coefficient for solute permeability, ω_s, and use the previous definition of σ (Eq. VII.6), so that we can rewrite the last line of Equation VII.9 as:

$$J_s = \bar{c}_s(1 - \sigma)J_V + \omega_s\,\Delta\Pi \qquad \text{(VII.10)}$$

Equation VII.10 indicates that J_s not only depends on $\Delta\Pi$, as would be expected on the basis of classical thermodynamics, but also that the solute flux density can be affected by the overall volume flux density, J_V. In particular, the classical expression for J_s for a neutral solute is $P_j\Delta c_j$ (Eq. 1.8), which equals $(P_j/RT)\Delta\Pi_j$ using the Van't Hoff relation $\Big($Eq. 2.10, $\Pi_s =$ $RT\sum_j c_j\Big)$. Thus, ω_s is analogous to P_j/RT of the classical thermodynamic description. The classical treatment indicates that J_s is zero if $\Delta\Pi$ is zero. On the other hand, when $\Delta\Pi$ is zero, Equation VII.10 indicates that J_s equals $\bar{c}_s(1 - \sigma)J_V$, and therefore P_j may not always be an adequate parameter by which to describe the permeability of species j. Finally, in the case of a single solute, movement across a membrane is fully characterized by the three experimentally convenient parameters introduced by irreversible thermodynamics: L_P, σ, and ω_s.

References

Katchalsky, A., and P. F. Curran. 1967. *Nonequilibrium Thermodynamics in Biophysics.* Harvard University Press, Cambridge.

Kedem, O., and A. Katchalsky. 1958. Thermodynamic analysis of the permeability of biological membranes to nonelectrolytes. *Biochimica et Biophysica Acta 27*: 229–246.

Stein, W. D. 1967. *The Movement of Molecules Across Cell Membranes.* Academic Press, New York.

Description of Plant Growth

Plant growth can be described using models based on leaf photosynthesis. For this, systems of equations with their own special abbreviations have been developed to relate leaf properties such as area and dry weight to the rate of dry weight gain by a whole plant, paying due regard to the allocation of photosynthate to photosynthetic and nonphotosynthetic tissues. The various parameters are related as follows:

$$\text{RGR} = \text{NAR} \times \text{LAR}$$

$$= \text{NAR} \times \text{LWR} \times \text{SLA} \qquad \text{(VIII.1)}$$

where

$$\text{RGR} = \text{relative growth rate} = \frac{1}{W_P} \frac{dW_P}{dt}$$

$$\cong \frac{\ln (W_{P_2} - W_{P_1})}{t_2 - t_1} \text{ in kg dry weight gained}$$

$$(\text{kg plant dry weight})^{-1} \text{ day}^{-1}$$

W_P = total plant dry weight in kg

$$\text{NAR} = \text{net assimilation rate} = \frac{1}{\text{LA}} \frac{dW_P}{dt} = \frac{1}{\text{LA}} \frac{W_{P_2} - W_{P_1}}{t_2 - t_1}$$

$$\text{in kg dry weight gained } (\text{m}^2 \text{ leaf area})^{-1} \text{ day}^{-1}$$

LA = total leaf area in m^2

$$\text{LAR} = \text{leaf area ratio} = \frac{\text{LA}}{\text{W}_P} \text{ in m}^2 \text{ (kg plant dry weight)}^{-1}$$

$$= \text{LWR} \times \text{SLA}$$

$$\text{LWR} = \text{leaf weight ratio} = \frac{\text{W}_L}{\text{W}_P} \text{ in kg leaf dry weight}$$

$$\text{(kg plant dry weight)}^{-1}$$

$$\text{W}_L = \text{total leaf dry weight in kg}$$

$$\text{SLA} = \text{specific leaf area} = \frac{\text{LA}}{\text{W}_L} \text{ in m}^2 \text{ (kg leaf dry weight)}^{-1}$$

$$\text{SLW} = \text{specific leaf weight} = \frac{1}{\text{SLA}} = \frac{\text{W}_L}{\text{LA}} \text{ in kg leaf dry weight m}^{-2}$$

Plant adaptations to the environment can be interpreted by variations in the relationships among these parameters. For instance, in a low PAR environment, deployment of large thin leaves would be expected (high LA/W_L, i.e., high SLA); also, plants might have a large fraction of their dry weight as leaves (a high W_L/W_P, i.e., a high LWR; see Fig. VIII.1). These two factors lead to a large LAR for plants developing in the shade (LAR = LWR $\times$ SLA). For plants developing under full sunlight, growth is not so much limited by the PAR as it is by the processing of the photons. Thus, SLW increases (SLA decreases, Fig. VIII.1) with the total daily PAR received during leaf development, which leads to the thicker sun leaves with a higher A^{mes}/A that we discussed in Chapter 8. The lower LWR for sun plants means that a greater fraction of their photosynthate is diverted to the roots or, perhaps, to the reproductive structures.

For the shade tolerant *Impatiens parviflora*, the compensating changes in LAR and NAR can cause the RGR of the plant to increase only 20% as the total daily PAR is increased 10-fold (leading to a 3-fold increase in NAR; see Björkman, Ch. 8 reference). In certain obligate shade plants, increasing total daily PAR has little beneficial effect on NAR but causes a reduction in LAR, with the consequence that RGR reaches an optimum at a relatively low total daily PAR and then decreases as the PAR is raised further. In an obligate sun plant, increasing the total daily PAR increases NAR proportionally more than it reduces LAR, with a result that RGR is greatest for high total daily PAR (see Fig. VIII.1). Also, RGR can increase with total daily PAR, since not all leaves are optimally oriented and those in the lower part of the canopy may be shaded by overlying leaves (see Ch. 9).

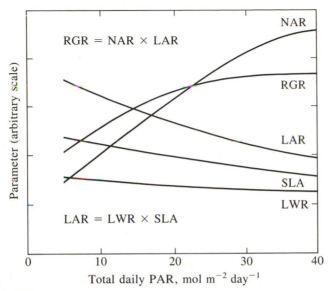

Figure VIII.1
Growth parameters for plants grown under different daily amounts of PAR.

Use of the above growth parameters can be useful for describing the influence of some environmental condition on partitioning. For instance, LWR is highest at some intermediate temperature and lower at temperature extremes for many species. In other species, the fraction of the dry weight in roots continuously decreases as the temperature during development is raised. For many growing plants, RGR is 0.05 to 0.3 kg dry weight gained (kg plant dry weight)$^{-1}$ day^{-1}, and NAR is 10 to 30 g dry weight (m^2 leaf area)$^{-1}$ day^{-1} (see Osmond et al., Ch. 8 reference).

References

Hesketh, J. D., and J. W. Jones. 1980. Integrating traditional growth analysis techniques with recent modeling of carbon and nitrogen metabolism. In *Predicting Photosynthesis for Ecosystem Models*, Vol. I, J. D. Hesketh and J. W. Jones, eds. CRC Press, Boca Raton, Florida. Pp. 51–92.

Thornley, J. H. M. 1976. *Mathematical Models in Plant Physiology*. Academic Press, London.

Answers to Problems

1.1. (a) 0.63×10^{-9} m^2 s^{-1}

(b) 3.2×10^6 s (37 days)

(c) 0.53 mol m^{-2}

(d) 9 mm from plane of insertion

1.2. (a) 2.0×10^8

(b) 5×10^{-13} m^2 s^{-1}

(c) 0.02 s, which is the approximate time necessary for the Brownian motion of a mitochondrion to be resolvable with a light microscope; 1 250 s (21 minutes).

(d) 2.1 s; for the ATP produced by respiration in the mitochondrion to be used 50 μm away, it is much more expedient for the ATP to diffuse than for the mitochondrion to so move.

1.3. (a) $t^{\text{air}} = 0.25\, t^{\text{cell wall}}$

(b) $D_{\text{CO}_2}^{\text{plasmalemma}} = 1.6 \times 10^{-5} D_{\text{CO}_2}^{\text{cell wall}}$

(c) 2.5×10^4 times larger in the cell wall than in the plasmalemma.

1.4. (a) 13×10^{-5} m s^{-1} for D$_2$O; 4.0×10^{-5} m s^{-1} for methanol; 1.0×10^{-5} m s^{-1} for L-leucine.

(b) 4.3×10^{-4} m s^{-1} for D$_2$O; 4.0×10^{-5} m s^{-1} for methanol; 3.0×10^{-8} m s^{-1} for L-leucine.

(c) The unstirred layer for water; both barriers provide the same resistance to methanol diffusion; the membrane for L-leucine.

(d) Same as (a), since the highest possible P_j^{total} occurs when only the unstirred layer (and not the membrane) is a significant barrier.

1.5. (a) 570 s (9.6 minutes)

(b) 1.2×10^5 s (32 hours)

(c) Double

(d) Increase 100-fold

1.6. (a) *Nitella*, 4×10^3 m^{-1}; *Valonia*, 6×10^3 m^{-1}; *Chlorella*, 1.5×10^6 m^{-1}

(b) *Chlorella*

(c) 370 s (6.3 minutes) for *Nitella* and 2 500 s (42 minutes) for *Valonia*

(d) *Valonia* (for a spherical cell, the cell wall stress is $rP/2t$).

2.1. (a) 15 mm
  (b) 12.5 mm
  (c) $60°$
  (d) 7.5 mm
  (e) 15 m
  (f) (a) and (b)

2.2. (a) 2.2 MPa, higher on 1 molal side
  (b) 140 MPa, higher on 0.1 molal side
  (c) 98 MPa, higher on 0.1 molal side
  (d) $\Delta P = 0$ MPa, $\Delta c_{solute} = 0$ molal
  (e) Indefinite, since μ_j^* is arbitrary

2.3. (a) Both 0.992
  (b) 1.07 MPa
  (c) 0.74%
  (d) 410 mol m^{-3}
  (e) 6×10^{-4} MPa (the polymer must be influencing the osmotic pressure mainly by decreasing the activity coefficient of water)
  (f) -1.9 MPa

2.4. (a) Both -0.600 MPa (at equilibrium)
  (b) All 0.600 MPa
  (c) P and $\rho_w gh$ are both 0.000 MPa at the upper surface; 0.1 m below the surface, $\rho_w gh$ is -0.001 MPa and P is 0.001 MPa; at the bottom of the tank, $\rho_w gh$ is -0.098 MPa and P is 0.098 MPa.
  (d) 99.6%

2.5. (a) 32 μm^3
  (b) 12 μm^3
  (c) 0.63
  (d) 3.3×10^{-15} mole/chloroplast (2.0×10^9 molecules/chloroplast)

2.6. (a) 4×10^{-7} m s^{-1} (flow is inward)
  (b) 28 s
  (c) None
  (d) -3.7×10^{-6} m s^{-1} (flow is outward)
  (e) $\cos \alpha = 0.027$ and so $\alpha = 88°$

3.1. (a) 3.3 mol m^{-3} (3.3 mM)
  (b) -129 mV
  (c) 3.7×10^{-13} J

3.2. (a) -128 mV (inside negative)
  (b) -121 mV
  (c) $\gamma_{K-ATP} = 0.36$; $a_{ATP} = 0.36$ mol m^{-3} (0.36 mM)

3.3. (a) K^+ is, but Mg^{2+} is not.
  (b) 10 mM (10 mol m^{-3}) Na^+ inside; 0.1 μM Ca^{2+} outside.
  (c) 59 mV; 13 mM (13 mol m^{-3})
  (d) 0.0010
  (e) μ_{Cl} is 17 kJ mol^{-1} higher inside the cell and μ_{Mg} is 11 kJ mol^{-1} lower there.

3.4. (a) 1 mV, negative toward the cell.
  (b) A Donnan potential of 77 mV, negative at the surface of the membrane.
  (c) -53 mV, i.e., negative inside the cell.
  (d) -55 mV; -59 mV

3.5. (a) 5 nmol m^{-2} s^{-1}
  (b) 4.0 kJ mol^{-1}, higher inside.
  (c) Efflux passive; influx active and requires 11.4 kJ mol^{-1}.
  (d) 15 μmol s^{-1} m^{-3}

3.6. (a) 1.36
  (b) 4

3.7. (a) 50.5
  (b) 1.01
  (c) 9.1 nmol m^{-2} s^{-1} (facilitated diffusion requires no metabolic energy source).

3.8. (a) 0.78
  (b) 0.65
  (c) Yes; -5×10^{-7} m s^{-1} (flow is outward).
  (d) 3×10^{-7} m s^{-1} (flow is inward).

3.9. (a) 0.75
  (b) 23 μm^3
  (c) 0.00, since the chloroplast volume did not change upon addition of glycine.
  (d) 0.5 molal; as the glycerol enters in time, the chloroplasts swell and eventually burst.

4.1. (a) 50 J
  (b) 16°C
  (c) 5.0 μmol m^{-2} s^{-1}
  (d) Units cannot be unambiguously interconverted.

4.2. (a) 333 nm in a vacuum, about 0.1 nm less in air, and 222 nm in the dense flint glass.
  (b) Yes, but transition is improbable.
  (c) Yes, it has enough energy.
  (d) 3.0×10^6 m^{-1}

4.3. (a) 139 μmol m^{-2} s^{-1}
  (b) 24 W m^{-2}
  (c) 60%
  (d) 3.0 μm

4.4. (a) 2.5×10^{-9} s
  (b) 0.75, when possible phosphorescence from the excited triplet is included.
  (c) Essentially 10^{-12} s

4.5. (a) 23 mmol m^{-3} (23 μM)
  (b) 0.46; absorbance unchanged.
  (c) 3.3×10^{18} molecules m^{-3} s^{-1} (5.5 μmol m^{-3} s^{-1}); -1.1×10^{-4} s^{-1}
  (d) 0.50

4.6. (a) 472 nm and 532 nm (distances in nm from main band are not equal).

(b) 67%

(c) 12 mmol m^{-3} (12 μM)

4.7. (a) 12 kJ mol^{-1}

(b) A transition from the first excited vibrational sublevel of the ground state to the same vibrational sublevel in the excited state as for the 450 nm transition; the Boltzmann factor, $e^{-E/RT}$, is 0.007 for an E of 12 kJ mol^{-1} at 20°C, so the transition is only 0.7% as probable as the 450 nm transition; compare the indicated ε_λ's. (Also, the population of the first excited vibrational sublevel of the ground state decreases as the temperature decreases.)

(c) From the lowest vibrational sublevel of the ground state to the third vibrational sublevel (second excited vibrational sublevel) of the excited state.

(d) 288 nm; 0.91 × 10^4 m^2 mol^{-1} (two conjugated systems of 5 bonds each).

5.1. (a) 0.05

(b) 33 hours

(c) 897.0

(d) Maximum absorption occurs along diameter (40 μm)—A is 0.54 in the blue region and 0.40 in the red region.

5.2. (a) 679 nm with 8 double bonds in conjugation; 878 nm with 10.

(b) Only the one with 8 double bonds can.

(c) Blue light excites the Soret band of Chl a, which rapidly becomes de-excited to the lower excited singlet (see Fig. 4.5); therefore the energy available is given by the red fluorescence of Chl a (see Fig. 5.3). The λ_{max} in the absorption spectrum of the pigment with 10 double bonds in conjugation is at 720 nm; thus it can become excited by resonance transfer from Chl a while the one with the λ_{max} at 580 nm essentially cannot.

5.3. (a) 0.24; 0.42

(b) 96 μmol m^{-2} s^{-1}; 40 μmol m^{-2}

(c) 2.4 times s^{-1}; 0.17

(d) 2.0 μmol m^{-2} s^{-1}; 25 × 10^{-17} mol s^{-1} per chloroplast

(e) 0.46 times s^{-1}; 0.88

5.4. (a) 10 $h\nu/O_2$

(b) 200 chlorophylls/photosynthetic unit

(c) 0.05 s

(d) The uncoupling of ATP formation from electron flow is analogous to disengaging a clutch, which allows a motor to run faster; the O_2 evolution therefore initially speeds up. After a while, the lack of ATP formation causes no CO_2 to be fixed; the electron acceptors therefore stay reduced and electron flow to them is curtailed. Moreover, electron flow may eventually switch over to the pseudocyclic type, which involves no net O_2 evolution.

5.5. (a) The 710 nm light absorbed by Photosystem I leads to cyclic electron flow and accompanying ATP formation. In the idealized example given, 550 nm light by itself leads to excitation of Photosystem II only, and no photophosphorylation.

(b) Accessory (or auxiliary) pigments, presumably carotenoids, which are isoprenoids.

(c) 1.00 at 710 nm and 0.23 at 550 nm

(d) 1.3 nmol m^{-2}

6.1. (a) $\Delta G = -17.1$ kJ mol^{-1} of reactant or product, and so the reaction proceeds in the forward direction.

(b) $\Delta G = 0$ kJ mol^{-1} of reactant or product, and so the reaction is at equilibrium and does not proceed in either direction.

(c) $\Delta G = 17.1$ kJ mol^{-1} of reactant or product, and so the reaction proceeds in the backward direction.

(d) 10^3 molal^{-1}

6.2. (a) 0.010

(b) 0.174 V

(c) 0.056 V positive for the $A-A^+$ half-cell; electrons flow toward the $A-A^+$ half-cell.

(d) $[A^+] = [B] = 0.54$ molal
$[A] = [B^+] = 1.46$ molal

(e) Answer unchanged

6.3. (a) 5×10^{-11} M

(b) 1 mM ATP; 44 kJ mol^{-1} ATP

(c) 305 mV

(d) 59 kJ/2 moles of electrons; yes.

6.4. (a) 0.081 V

(b) 4.5 μM

(c) 0.413 V

(d) 0.247 V

(e) 0.296 V

7.1. (a) 17°C for the cloudy sky, 42°C for the clear one.

(b) 10.2 μm for the leaf; 10.0 μm for the surroundings under the cloudy sky and 9.2 μm under the clear one.

(c) 970 W m^{-2}

(d) 72%

(e) 130 W m^{-2}

7.2. (a) 0.094 m (leaf area divided by diameter)

(b) 1.4 mm; 1.5 mm

(c) 190 W m^{-2}

(d) 2.5 mmol m^{-2} s^{-1}

7.3. (a) 112 W m^{-2}

(b) 1.1 mW; 4.2 mW m^{-2}

(c) -6 W m^{-2}

(d) 105 W m^{-2} ($\Delta J_{wv} H_{vap} = 13$ W m^{-2})

(e) 175°C m^{-1}

7.4. (a) 1.5 mm; 88 W m^{-2}

(b) 18 W m^{-2} K^{-1}

(c) 0.20 mm

(d) 20.2°C

(e) -209 W m^{-2} without spines; -118 W m^{-2} with spines

(f) -131 W m^{-2} (shortwave averaged over the stem surface is 75 W m^{-2}, since projected area $=$ one-fourth of total area for a sphere); -4.2°C (per hour)

8.1. (a) 30 mm s^{-1}; 33 s m^{-1}

   (b) 0.0077; 6 μm

   (c) 130 times higher in the pores ($A/A^{st} = 130$)

   (d) 6.0 mm s^{-1} and 250 mmol m^{-2} s^{-1}; 6.7 mm s^{-1} and 250 mmol m^{-2} s^{-1} (the latter conductance is independent of pressure)

   (e) 48 mm s^{-1}; 2.0 mol m^{-2} s^{-1}

   (f) 41 s m^{-1}

8.2. (a) 4.2 mm s^{-1}

   (b) 4.1 mm s^{-1}; 4.7 mm s^{-1}; 4.6 mm s^{-1}

   (c) 8.4 mm s^{-1}

   (d) 5.8 mm s^{-1}

   (e) 95 mg m^{-2} s^{-1}

   (f) 169 mmol m^{-2} s^{-1} (at 30°C, 1 mm s^{-1} corresponds to 40.2 mmol m^{-2} s^{-1}); 5.3 mmol m^{-2} s^{-1}

   (g) 15.8 g m^{-3} and 0.0220

8.3. (a) 22

   (b) 34

   (c) 160 s m^{-1}

   (d) 230 s m^{-1} for each

   (e) 5 s m^{-1}; 20 s m^{-1}

   (f) 420 s m^{-1}

8.4. (a) 3.5 mol m^{-3} s^{-1}

   (b) 45 μM

   (c) 1 060 s m^{-1}; 8.5 μmol m^{-2} s^{-1}

   (d) 870 s m^{-1}; 12.6 μmol m^{-2} s^{-1}

   (e) 0.24 μM ($J_{CO_2}^{r+pr} = 0.66$ μmol m^{-2} s^{-1}; 10 ppm corresponds to about 0.41 mmol CO$_2$ m^{-3})

   (f) 14 μM; 17 μM

8.5. (a) 3.0 mm s^{-1}; 4.3 mm s^{-1}

   (b) g_w^{total} decreases to 1.0 mm s^{-1}; J_{wv} decreases 68% to 1.6 mmol m^{-2} s^{-1}

   (c) because of lower J_{wv}, T^{leaf} increases.

   (d) photosynthesis decreased by 44%; WUE *increased* by 75%

   (e) g_{wv}^{total} increases to 3.5 mm s^{-1}; J_{wv} increases 17%

   (f) Because δ^{bl} is halved, J_H^C initially doubles; T^{leaf} becomes closer to T^{ta}. If T^{leaf} is above T^{ta}, as generally occurs during the day, the decrease in T^{leaf} in response to a higher wind will reduce c_{wv}^e (which is nearly at the saturation value). If the accompanying fractional reduction in $c_{wv}^e - c_{wv}^{ta}$ is more than the fractional increase in g_{wv}^{total}, the increase in wind speed will lead to a *decrease* in transpiration, since $J_{wv} = g_w^{total} (c_{wv}^e - c_{wv}^{ta})$ by Equations 8.15 and 8.16.

9.1. (a) 15.4 mmol m^{-2} s^{-1}; 0.34 mol m^{-3}

   (b) 22 s m^{-1} during peak transpiration, 44 s m^{-1} at night

   (c) 0.07 mol m^{-3}, which corresponds to a drop in relative humidity of 7% at 20°C

   (d) -22 μmol m^{-2} s^{-1}; -0.48 mmol m^{-3}

   (e) 0.68 m^2 s^{-1}

   (f) 0.4 s by eddy diffusion, 10^4 s (3 hour) by ordinary diffusion

9.2. (a) 7.9 at $2\,000\ \mu$mol m^{-2} s^{-1} (essentially none of the leaves are then below light compensation), 4.6 at 200 μmol m^{-2} s^{-1}, 1.3 at 20 μmol m^{-2} s^{-1}, and 0.0 at 0 μmol m^{-2} s^{-1} (all of the leaves are then below light compensation).

    (b) 3.0 μmol m^{-2} s^{-1} upward at the soil, -15.5 μmol m^{-2} s^{-1} downward into the canopy

    (c) 390 H_2O/CO_2

    (d) 6.8 m above the ground (only leaves beneath this level are below the light compensation point and thus have a net evolution of CO_2).

    (e) About 12 m above the ground.

9.3. (a) 1.4 mm s^{-1} (5 m hour^{-1})

    (b) -0.0087 MPa m^{-1}

    (c) 1.1×10^{-10} m s^{-1} (3.9×10^{-7} m hour^{-1})

9.4. (a) 8.3×10^4 s (23 hours) for 10 mm; 8.3×10^8 s (26 years) for 1 m

    (b) 0.17 mm s^{-1} (0.60 m hour^{-1})

    (c) 60 s for 10 mm; 1.7 hour for 1 m. These are considerably shorter times than are needed for diffusion—see (a). Also, the distance traversed is proportional to time.

    (d) 0.90 m hour^{-1} for flow downward or 0.30 m hour^{-1} for movement upward, depending on the direction of the pressure gradient

9.5. (a) 15 μm for the wet soil, 0.1 μm for the dry one

    (b) 6 Pa

    (c) 0.06 MPa

    (d) -0.05 MPa m^{-1}

    (e) -2 Pa

    (f) There is no net flow, since Ψ^{soil} then equals Ψ^{ta}.

9.6. (a) 3.1×10^{-6} m^3 s^{-1}

    (b) 1.9×10^3 s (31 minutes)

    (c) 1.3×10^5 MPa s m^{-3}

    (d) 220 MPa s m^{-2}; 4.6×10^{-3} m^2 s^{-1} (MPa)$^{-1}$

    (e) 6.0×10^{-3} m^3 (MPa)$^{-1}$; 790 s (13 minutes)

Index

Page numbers in **boldface** type refer to figures or to structural formulas.